AF593968

# GTAW handbook

by

**William H. Minnick**

Associate Professor
Palomar College
San Marcos, California

**South Holland, Illinois**

**THE GOODHEART-WILLCOX COMPANY, INC.**

**Publishers**

Library of Congress Catalog Card Number 84-27907
International Standard Book Number 0-87006-514-9

23456789-85-09876

**Library of Congress Cataloging in Publication Data**

**Minnick, William H.**
GTAW Handbook.

1. Gas tungsten arc welding. 1. Title.
TK4660.M53 1985 671.5′212 84-27907
ISBM 0-87006-514-9

# INTRODUCTION

The GTAW HANDBOOK provides complete and thorough coverage of the Gas Tungsten Arc Welding field. Theory, fundamentals, equipment, and safety information are provided to teach you how to make successful welds. Basic skills and proper procedures are presented in easy-to-understand language and combined with numerous illustrations to guide you in learning about GTAW.

The GTAW HANDBOOK is designed to teach the understanding and skills needed to enter the welding field. The sequence of chapters will lead you through GTAW principles and practices in logical order. You will be introduced to:

- the types and safe operation of GTAW equipment.
- setup and maintenance of equipment.
- types of welds, joint design, and tooling.
- welding techniques and procedures for various types of metals.
- semiautomatic and automatic welding systems.
- inspection practices and quality control.
- weld repair
- qualification and certification.
- estimating cost.

Each and every weld that you will make bears your personal trademark and represents your skill as a welder. Take pride in your work and make each weld as if your career depends upon it. The GTAW HANDBOOK will give you the information you need to make that weld.

**William H. Minnick**

# CONTENTS

# Chapter 1

# GAS TUNGSTEN ARC WELDING PROCESS

The GAS TUNGSTEN ARC WELDING (GTAW) process fuses metals by heating them between a non-consumable (does not melt) tungsten electrode and workpiece. The heat necessary for fusion (mixing or combining of molten metals) is provided by an arcing electric current between the tungsten electrode and the base metal.

This type of welding is usually done with a single electrode. However, it may be done with several electrodes. The tungsten electrode and the weld zone (area being welded) are shielded from the atmosphere (air around it) by an inert gas, such as argon or helium. Filler metal may or may not be used.

The discovery of the GTAW process is credited to two Americans, Russell Meredith and V.H. Pauleka who developed a method for welding magnesium. They used a direct current arc for heat and helium gas to protect the molten metal from the atmosphere. From this combination, the name "Heliarc" was born. Meredith later sold it to the Linde Company which now use the word Heliarc as a trademark on their equipment.

## TIG OR GTAW?

At one time, the American Welding Society (AWS) called the process "tungsten inert gas welding." The letters "TIG" were used to designate the process. Later, the definition was changed to "gas tungsten arc welding" and the letters "GTAW" came into popular use. Today, both of the names and letters are used. However, the American Welding Society recommends the use of Gas Tungsten Arc Welding or GTAW. This book will follow current practice and refer only to GTAW.

A typical manual GTA weld is shown in Fig. 1-1. The entire weld area and the tungsten electrode is shielded by a flow of an inert gas through the special designed welding torch and gas nozzle. The gas protects the tungsten, molten metal and heat affected area from contamination by oxygen, nitrogen, and hydrogen in the atmosphere. If the tungsten and the molten metal is not shielded properly, the atmosphere is absorbed into the weld and it becomes brittle and porous. Welds that have been contaminated do not usually have sufficient strength to carry the intended load or stress.

## WELDING MODES

GTA welding may be done in any position and in manual, semiautomatic, or automatic modes. The method used depends on the equipment and the application. In the MANUAL MODE, the welder controls every part of the GTAW process that can vary. In addition, the welder manipulates (handles) the welding torch and then adds filler wire as needed.

In the SEMIAUTOMATIC MODE, the welding torch may be mounted on a special holder or on a tractor. In another variation, the part may move under a stationary (not moving) torch. The welder controls the process wherever it varies.

In the AUTOMATIC MODE, all of the welding variables, as well as the sequence of steps, is programmed into the welding machine. The operator monitors (oversees) the overall operation.

## COMPONENTS AND ACCESSORIES

Basic equipment components (parts) and accessories (things that make the welding process work better) vary widely. The equipment selected depends upon:

1. The mode in which the process is used.
2. Type of material to be welded.
3. Number of welds needed.
4. Quality of the completed welds.

Each welding system must have a minimum of three major components. These three items include:

1. POWER SUPPLY for the type of desired current.

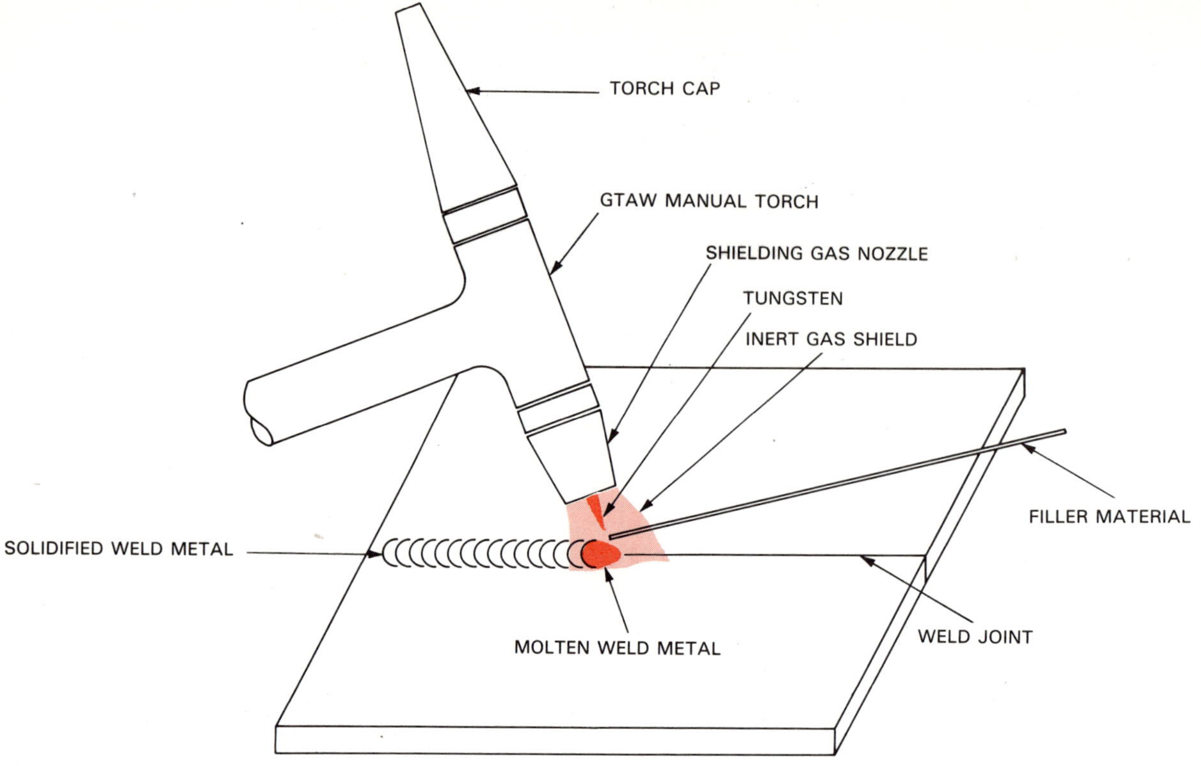


Fig. 1-1. The GTA welding torch is moved along the weld joint melting the base material. Filler material is added as needed by the operator.

2. INERT GAS SUPPLY and a flow control to regulate the flow of gas to the weld area.

3. GTAW TORCH, tungsten electrode, collect assembly, and gas nozzle.

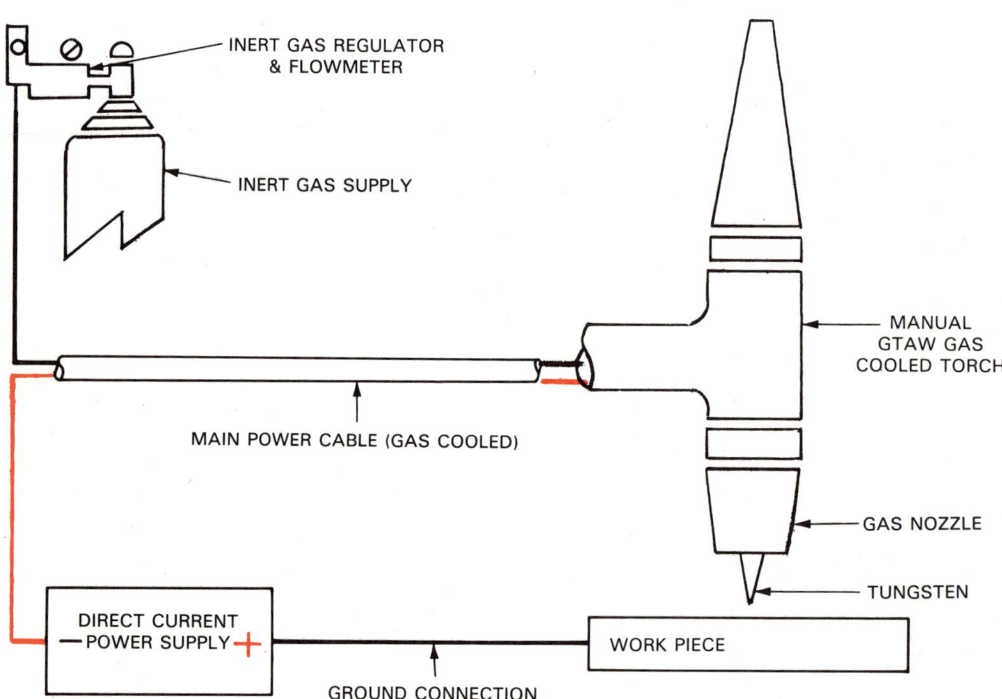


Fig. 1-2. Basic components for a simple GTAW system to weld steel, stainless steel, inconel, etc.

## Basic Welding System

A basic welding system is diagramed in Fig. 1-2. It is for the manual welding of thin gauge steel, stainless steel, inconel, titanium, etc., using direct current from the power supply. This type of system is limited to thin gauge materials because of the amperage (heat) capacity of the welding torch.

## Basic Welding System For Aluminum And Magnesium

A welding system for the manual welding of thin gauge aluminum and magnesium is illustrated in Fig. 1-3, using alternating current from the power supply.

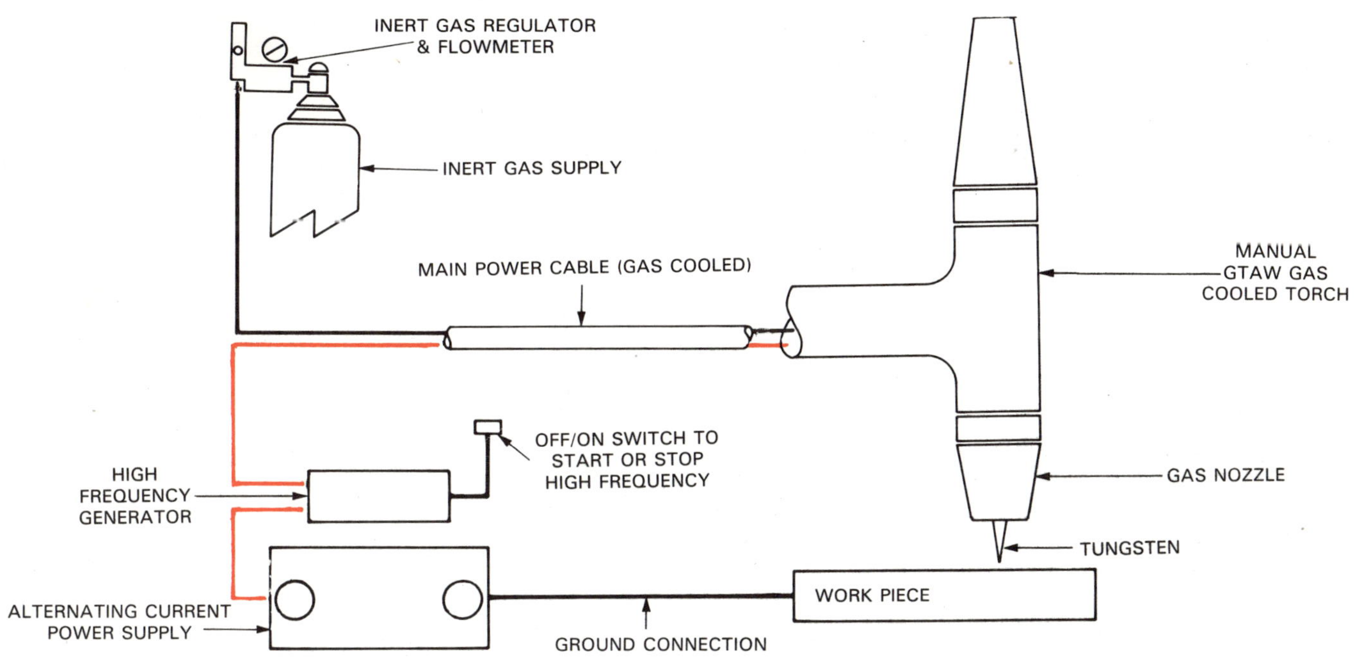


Fig. 1-3. An alternating current power supply and a high frequency voltage generator are required to weld aluminum and magnesium.

A high frequency voltage generator is required to maintain the arc to and from the workpiece, with this type of current. This type of system is limited to thin gauge materials because of the amperage (heat) capacity of the torch.

## Basic Welding System For All Types of Metals

The system shown in Fig. 1-4 uses a power supply which will supply either alternating or direct current. A high frequency voltage generator is also required when using alternating current for welding. This type of system is limited to thin gauge materials because of the amperage capacity of the torch.

## Basic Welding System For All Types of Metals and Thicknesses

The manual system shown in Fig. 1-5 can be used for all thicknesses of materials because the torch is water cooled. Heat which is generated in the tungsten electrode is removed by circulating water through the torch body and head.

## Semiautomatic Longseam Welding System

The system diagramed in Fig. 1-6 can be used on all types and thicknesses of metal. A tractor for longitudinal travel is added to move the machine welding torch along the weld joint. The arc gap is maintained along the joint by raising or lowering the torch with a gear and rack adjustment wheel.

A wire supply and feeder assembly may be used in this system. The addition of material is for filling groove weld joints and making a crown on the top of the weld. Longitudinal welds in cones, cylinders, channels, and tubes are welded using this type of system.

## Semiautomatic Rotational Welding System

Fig. 1-7 shows a system used to make welds between round parts, such as rings and cylinders. The machine welding torch is mounted above the part. Manual adjustments maintain the proper gap between the torch and the part.

A positioner rotates the part at a set speed under the torch during the weld. Tilt table positioners are used where the weld joint must be angled to the welding torch. A wire supply and feeder assembly may also be added in this system.

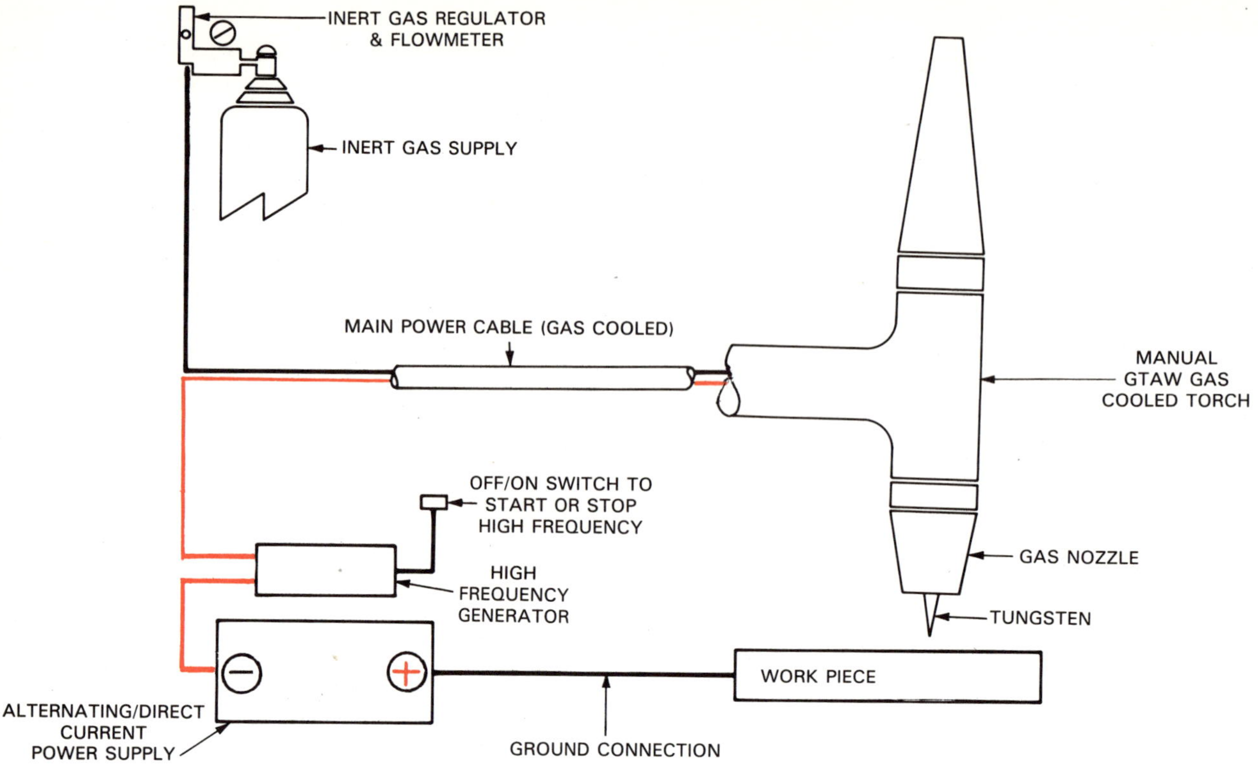


Fig. 1-4. Both aluminum and steel can be welded with a combination alternating and direct current welding power supply.

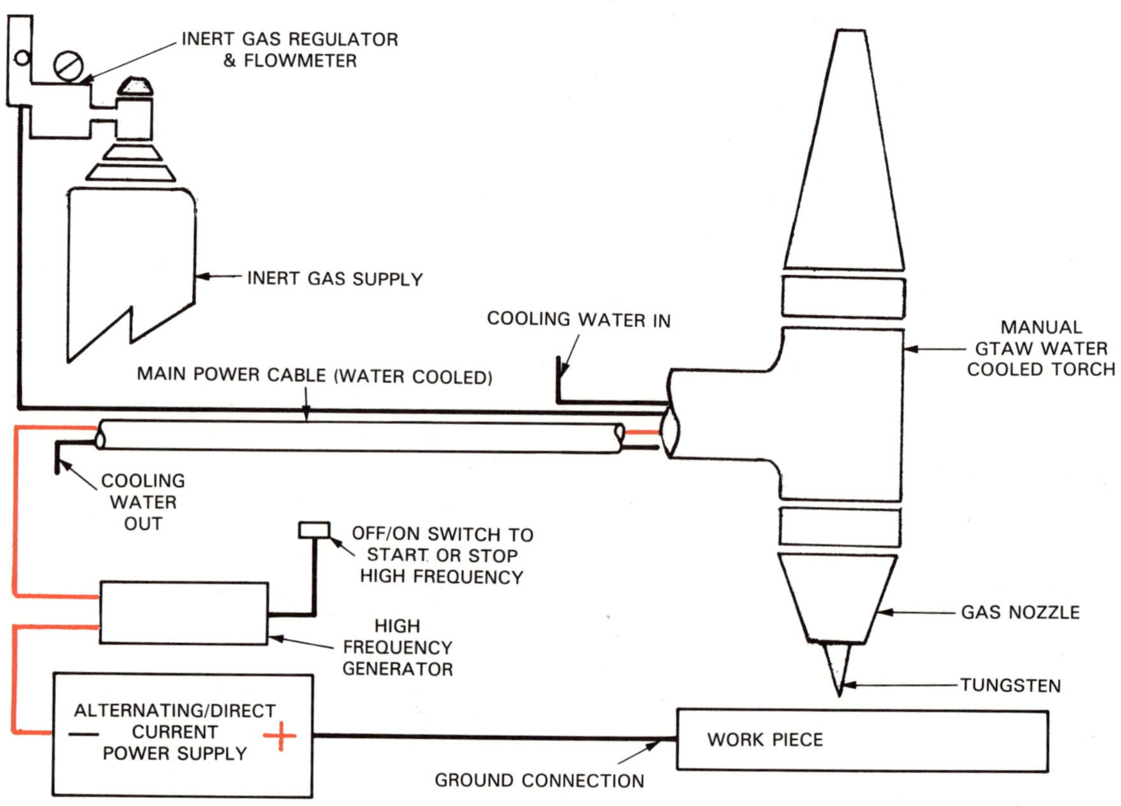


Fig. 1-5. Higher welding currents can be used in this GTAW system because the torch is water cooled.

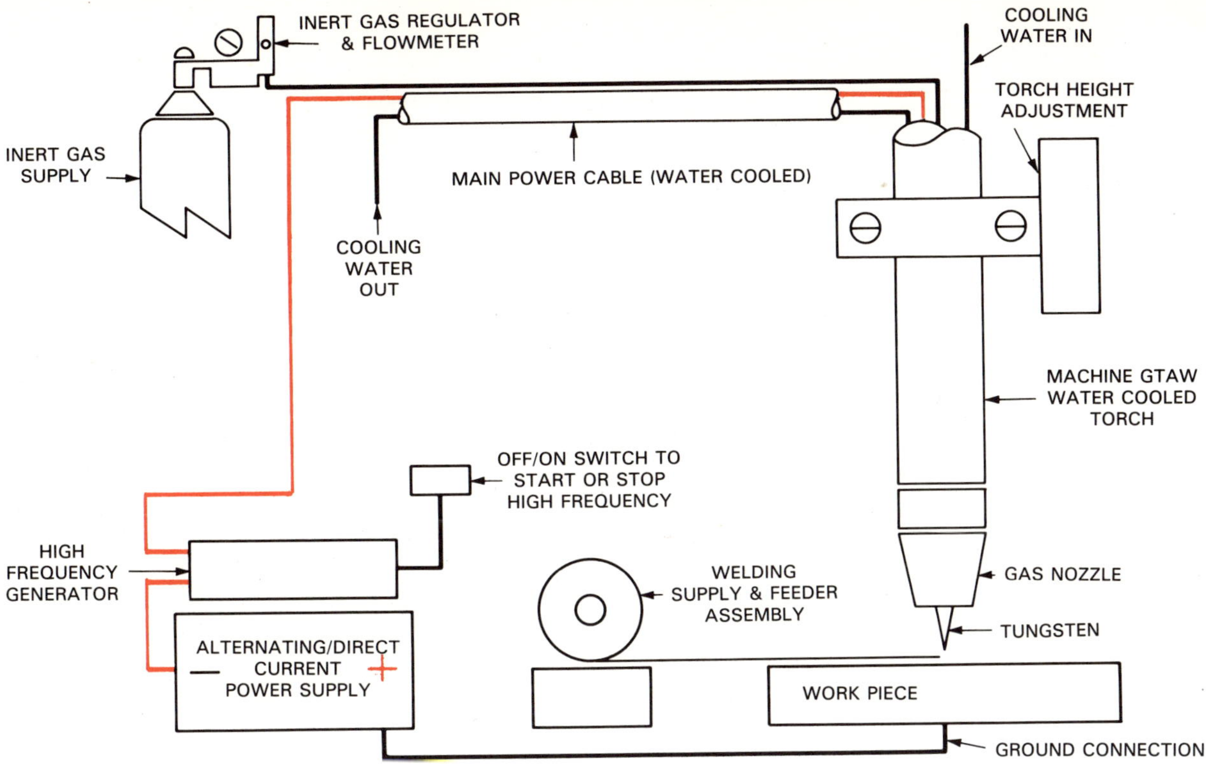


Fig. 1-6. Welds made with this semiautomatic type of equipment can be made much faster with less defects than manual welds. A wire supply and feeder assembly provide filler material.

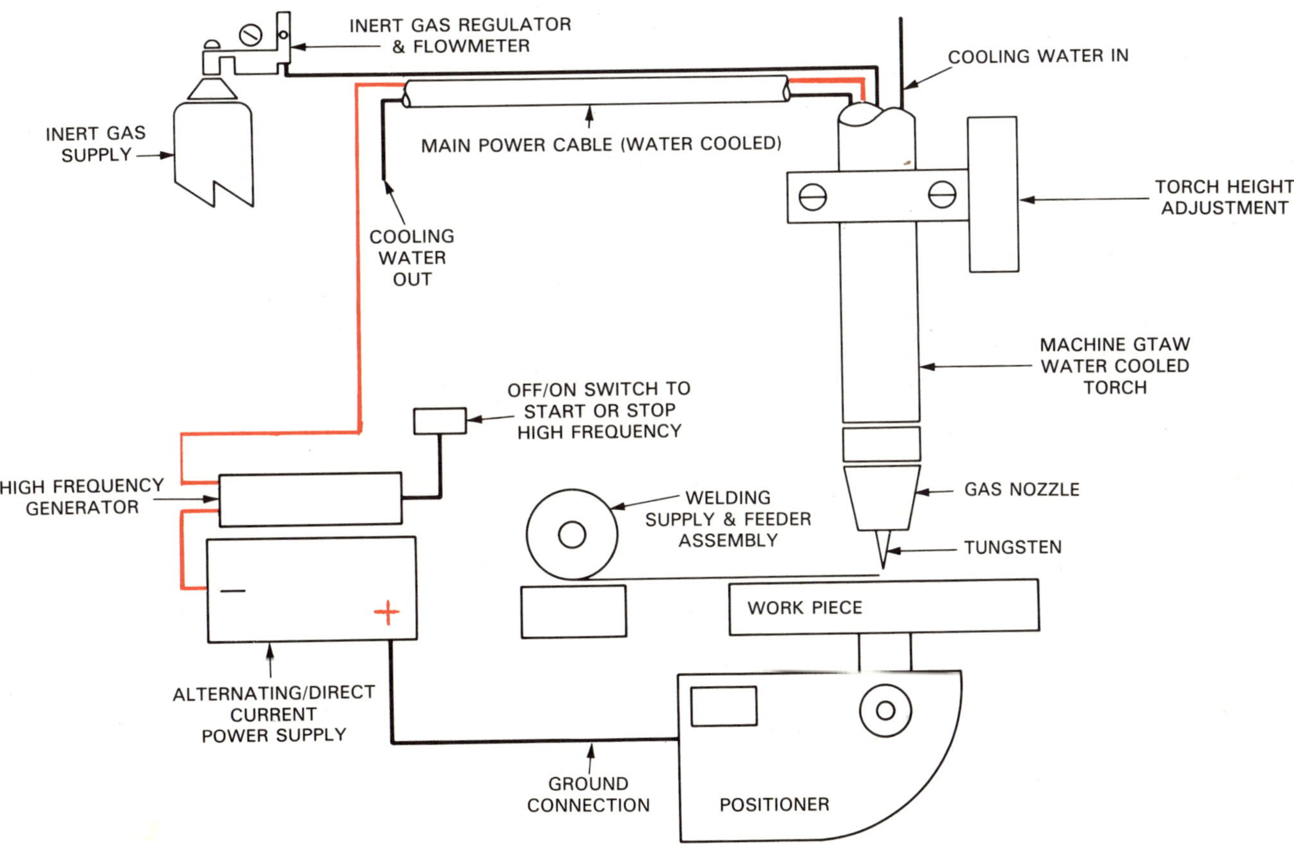


Fig. 1-7. Positioners are used to locate and rotate circular parts at an accurate speed to allow for semiautomatic rotational welding.

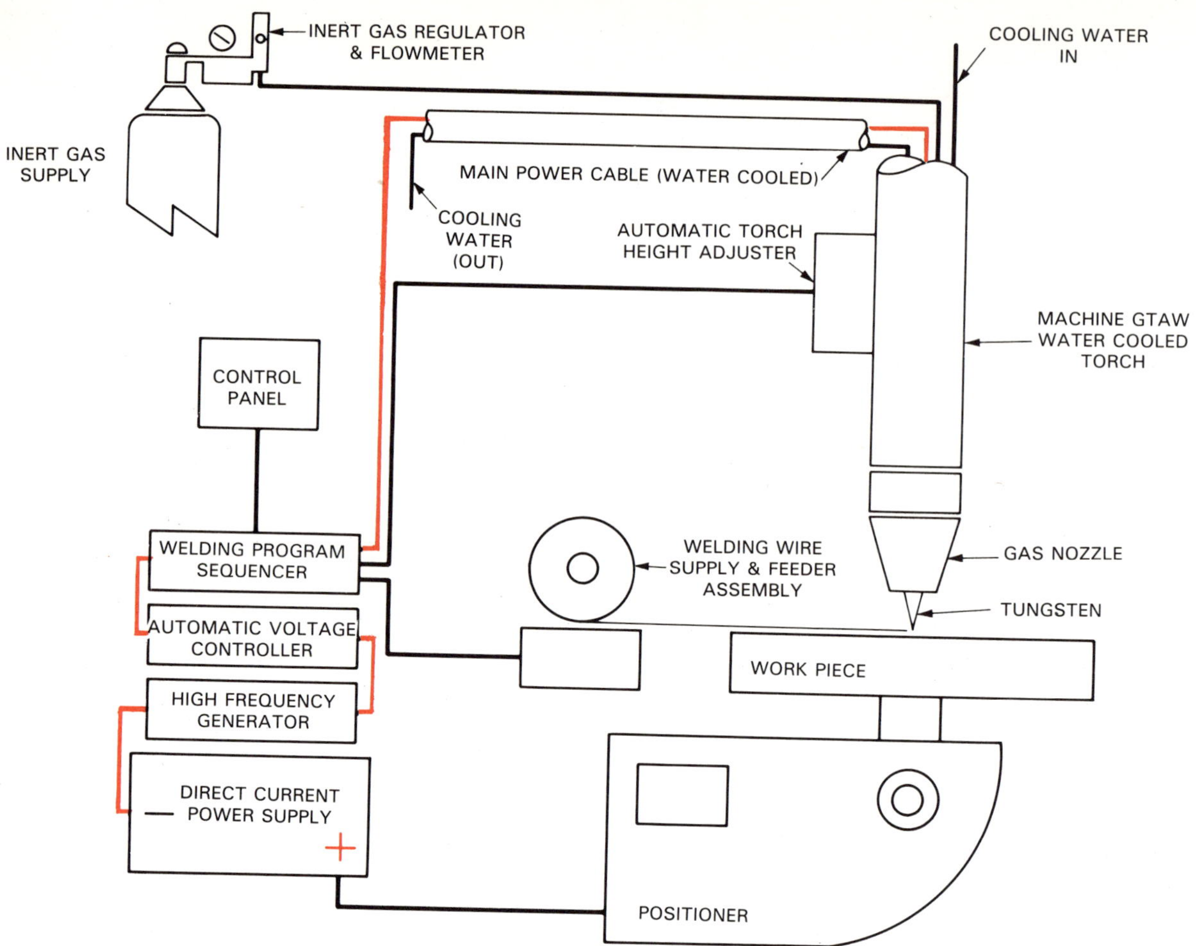


Fig. 1-8. Automatic welding systems are designed for high production rates and where high quality is required.

**Automatic Welding System**

The system shown in Fig. 1-8 may be used on longseam and/or rotational type welds. The components of the system are interconnected and programmed to operate as required during the operation. The welder oversees the operation and makes adjustments as required.

This type of equipment is typically very expensive and is used where high quality or high production welds are required.

**Semiautomatic Spot Welding System**

Spot welds may be used as an assembly tool to spot tack weld mating parts together. Spot welds may also be used as the main welding operation. Fig. 1-9 shows a basic spot welding system. In some cases, additional equipment may be required to start the arc or to add filler material to the weld puddle.

## PROCESS APPLICATIONS

Applications of the GTAW process are virtually unlimited in modern fabrication industry. The nuclear power, aircraft, missile, food processing, chemical and oil refining industries, today employ thousands of welders producing components using the GTAW process. The major advantages of the process include:

1. The welds are very clean and have good quality.
2. Arc heat is intense and highly concentrated.
3. There is no smoke, fumes, spatter or slag.
4. Weld zone is highly visible.
5. Weld chemistry is easily controlled.
6. GTAW can be used to weld most industrial metals.

The disadvantages of the process include:

1. Slow welding speed.
2. Equipment is expensive.
3. Inert gases are costly.
4. Requires a highly skilled welder for manual welding.

## REVIEW QUESTIONS

1. The GTAW process is a __GAS__ welding process.
2. __DIRECT__ current provides the heat for melting the base materials.

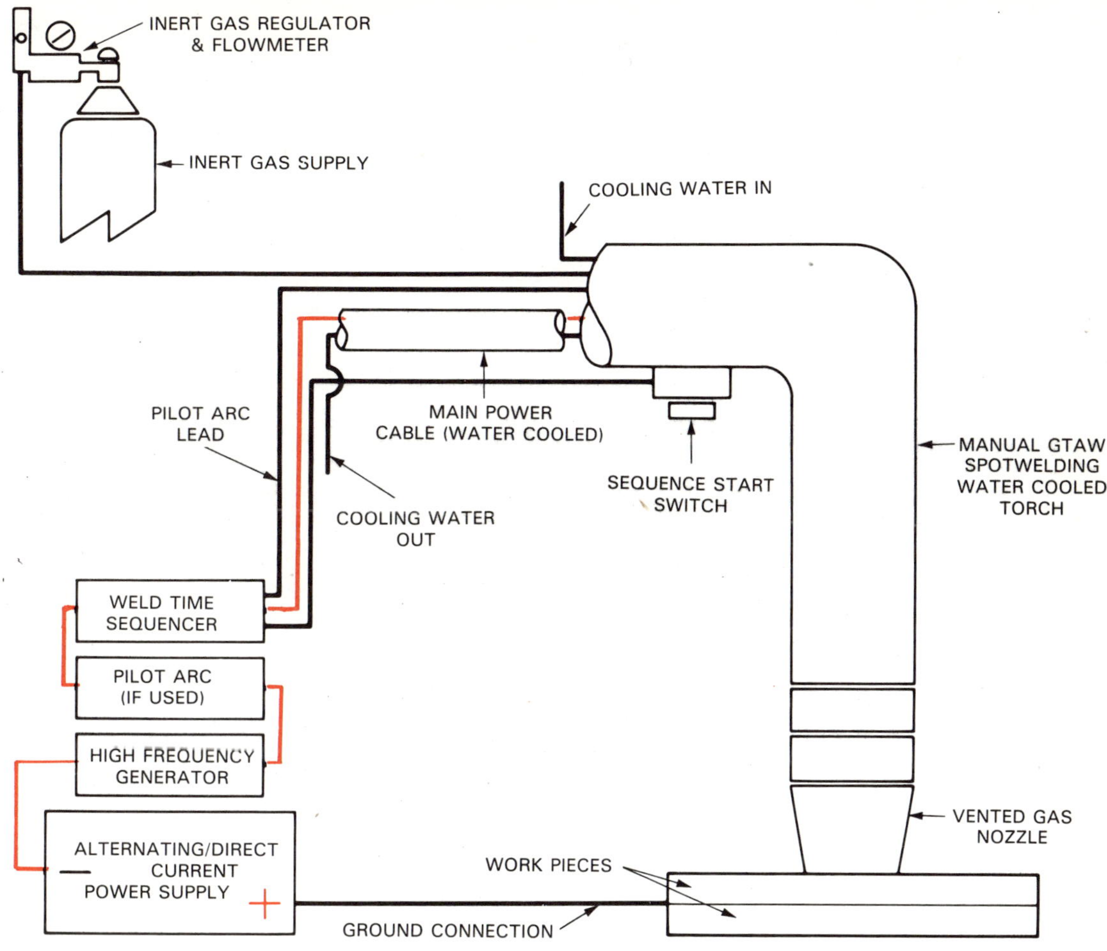


Fig. 1-9. Spot welding of overlapping materials is accomplished by heating the top plate and melting into the bottom plate.

3. ________ and ________ are typical gases used to shield the weld zone from the atmosphere.
4. The welding electrode is nonconsumable and is made from ________ material.
5. The original name for the GTAW was ________.
6. The letters AWS stand for ________ ________.
7. The letters TIG mean ________ ________ ________.
8. Weld metal that absorbs atmosphere become ________ and ________.
9. Three contaminates in the atmosphere are ________, ________ and ________.
10. In the ________ welding mode, the welder controls every part of the process that can vary.
11. In the automatic mode, the ________ oversee or monitors the entire operation.
12. Each welding system has a minimum of three major components. They are:
    1. ________
    2. ________
    3. ________
13. Another name for welding alternating/direct current is ________.
14. Steel, stainless steel, and inconel are welded using ________ current from the power supply.
15. Aluminum and magnesium is welded using ________ current.
16. A ________ ________ ________ generator must be used with alternating current to maintain the flow of welding current.
17. Heavier thicknesses of material are welded with ________ cooled torches.
18. To fill groove weld joints and make crowns on welds using the semiautomatic system, a ________ ________ and a ________ ________ must be added to the system.
19. Welding torches mounted on tractors are used to make ________ type welds.
20. GTAW spot welding of overlapping materials is accomplished by heating the ________ plate and melting into the ________ plate.

# Chapter 2

# GTAW PROCESS OPERATION AND SAFETY

In gas tungsten arc welding, the welder applies the arc to two workpieces that are butted together. The heat of the arc melts the metal along the butted edges of the two workpieces. (Melted filler metal may be added.) As the molten (melted) metal cools it becomes solid and permanently joins the workpieces. This GTAW process is called FUSION WELDING. Fig. 2-1 shows a manual GTAW operation. The movement of the torch is controlled by the welder.

Note in Fig. 2-1 that the welder is wearing an approved molded helmet with an extra large lens for maximum visibility. A lens shade number of 10 to 14 is recommended for GTAW. Dark clothes are worn to cut down reflections and glare. Trousers worn by welders should not have cuffs. Protective gloves are worn to prevent "sunburn" of the hands and wrist. Good ventilation is extremely important to all welding operations.

## PROCESS OPERATION

In the GTAW process, several components (parts) are combined to make welding systems. As you studied in Chapter 1, three items are required to produce GTAW results; a power supply for the type of desired current, an inert gas supply and a flow control to

Fig. 2-1. A 300 ampere AC/DC power supply is being used for this manual GTAW operation. Note the helmet, gloves, and dark clothing. (Miller Electric Manufacturing Co.)

regulate the flow of gas to the weld area, and the GTAW torch made up of the tungsten electrode, collet assembly, and gas nozzle. These components will be studied in more detail in this chapter.

When the systems are properly controlled, they will produce a carefully controlled arc. At the same time, a measured amount of inert gas will flow around the weld area.

Additional components may be used in semi-automatic or automatic systems. The added components are designed to:

1. Oscillate the torch. This means "move it back and forth."
2. Add filler wire.
3. Maintain the arc voltage.
4. Control other functions or aspects of the welding process.

Prior to welding, all the equipment must be set up to operate in a definite way. Each component in the system must be set to enter the operation and to work at the proper time. To properly set up the time and order of these events, a study must be made of the required weld. Definite characteristics are then determined before welding. These characteristics are called PARAMETERS and VARIABLES in the welding trade. These will vary depending on such items as types of metals, thickness of the weld joint, weld joint design and weld quality required.

Some of the major variables are:

1. Type of weld current.
2. Amount of weld current.
3. Arc voltage.
4. Power source characteristics.
5. Type of shielding gas.
6. Amount of gas flow.
7. Type of welding torch.
8. Electrode type.
9. Electrode diameter.
10. Electrode tip shape.
11. Gas nozzle size.

All of these variables will have to be carefully considered when setting up the components of any GTAW system.

## POWER SUPPLIES

Power supplies are specially designed and manufactured for GTAW. They must produce a variable low voltage and a constant electrical current to the welding arc. The various types of power supplies used in GTAW are shown in the chart in Fig. 2-2.

The welding machines are rated for the amount of current and length of time they can operate without damage to the power supply. This rating is called the MACHINE DUTY CYCLE. They should never be operated beyond these limits.

| WELDING CURRENT OUTPUT | | | |
|---|---|---|---|
| Machine Type | A/C ONLY | D/C ONLY | A/C or D/C |
| Transformer | X | | |
| Transformer-Rectifier | | X | X |
| Motor Generator | X | X | X |
| Motor Alternator | X | X | X |

Fig. 2-2. Four types of power supplies are used in Gas Tungsten Arc Welding.

Controls for a power supply may range from very simple to very complex. The simplest is a tap connection for changing the amount of welding current. One of the most complex is a series of controls that fully sequence the welding current throughout the entire operation.

Manual type industrial rated power supplies, like the one shown in Fig. 2-3, generally have two methods of controlling the arc welding current:

1. Machine mounted rheostat which is preset to the desired amperage prior to starting to weld.
2. Remote hand or foot rheostat which is adjusted by the welder during the operation to the desired current level. This control is very useful to the welder as the current may be started and ended at a very low level. This reduces the possibility of burn-through and craters at the start and end of the weld.

The remote amperage hand control used with a power supply is shown in Fig. 2-4. The remote amperage foot control used with a power supply is shown in Fig. 2-5.

The power supplies produce OPEN CIRCUIT VOLTAGE when started. They are ready to produce welding current as soon as the arc is struck. A CONTACTOR switch prevents accidental shortage of the arc welding current to the workpiece prior to the start of the weld. The contactor allows welding current to flow from the power supply to the torch.

The contactor is located in the power supply where it opens and closes the power circuit from the power supply to the torch. Switches located on the main power supply select desired modes (foot or machine) for amperage control and contactor operation. Refer to Fig. 2-3. Operation of the hand or foot control energizes the contactor and allows the arc to start.

Four major points to remember about the power

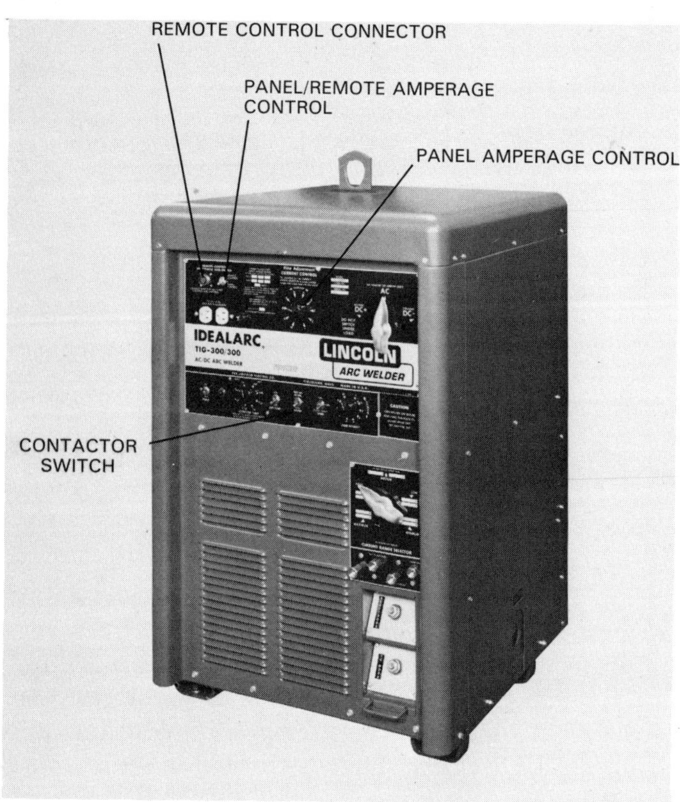


Fig. 2-3. This industrial type power supply has provisions for panel control of amperage or remote (hand or foot) control. (Lincoln Electric Co.)

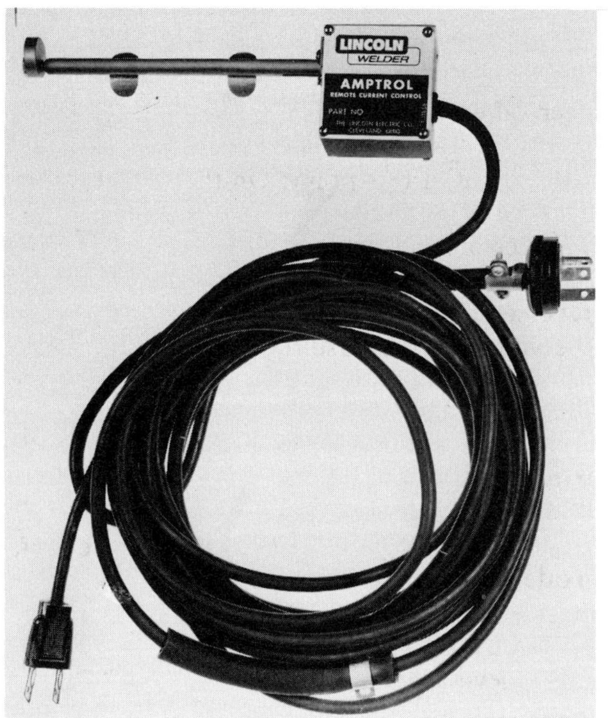


Fig. 2-4. The welder controls amperage by turning the shaft with a fingertip during welding. The assembly is mounted directly on the GTAW torch. (Lincoln Electric Co.)

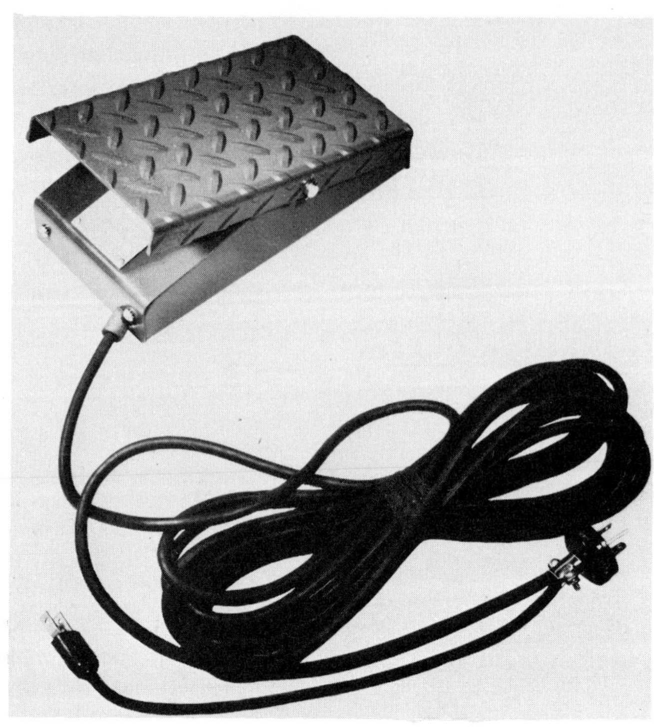

Fig. 2-5. The welder can raise or lower amperage by movement of the foot pedal. (Lincoln Electric Co.)

supply include:

1. NEVER use the power supply above the rated load.
2. NEVER change an output lead during welding.
3. NEVER change the current type while welding.
4. NEVER change the current range switch during welding.

## TYPES OF WELDING CURRENT

The GTAW process uses three types of current. Each is needed to weld different types of metals. ALTERNATING CURRENT is generally used for manual and semiautomatic welding of aluminum and magnesium. DIRECT CURRENT STRAIGHT POLARITY (DCSP), also referred to as Direct Current Electrode Negative (DCEN), is used to weld the steels, stainless steels, nickel, titanium, and many other materials. In some automatic welding of aluminum and magnesium, direct current straight polarity (DCSP or DCEN) is used with helium shielding gas. DIRECT CURRENT REVERSE POLARITY (DCRP) or Direct Current Electrode Positive (DCEP), type of current is usually used only on materials needing shallow penetration.

The American Welding Society has defined current flow as moving from the negative (−) terminal to the positive (+) terminal. When the current flows from the welding machine negative terminal to the electrode holder, across the arc gap, to the workpiece and back to the welding machine, the circuit is known as

DIRECT CURRENT ELECTRODE NEGATIVE (DCEN). This is the same as DIRECT CURRENT STRAIGHT POLARITY (DCSP). You will find many of the welders in the field use the DCSP markings, while newer equipment will be labeled as DCEN.

When the current flows from the welding machine to the workpiece, then across arc gap to the electrode, then back through the lead to the welding machine, the circuit is known as DIRECT CURRENT ELECTRODE POSITIVE (DCEP) or DIRECT CURRENT REVERSE POLARITY (DCRP).

These terms Electrode Negative and Straight Polarity will be used interchangeably in this book, as well as the terms Electrode Positive and Reverse Polarity.

## ACHF Using An Unbalanced Wave Transformer

The letters "AC" indicate the welding current is Alternating Current. The letters "HF" indicate that a HIGH FREQUENCY voltage is used to maintain the alternating current arc. These letters are used often in specifications, blue-prints, and fabrication orders throughout the industry.

Power companies supply high voltage and low amperage current. The transformer of an alternating current welding machine changes the current to lower voltage and higher current or amperage for welding. The frequency remains the same at 60 cycles per second.

Each complete cycle of current contains a half cycle of straight polarity and a half cycle of reverse polarity. In the STRAIGHT POLARITY portion of the cycle, the welding current flows from the tungsten electrode to the workpiece. In the REVERSE POLARITY portion of the cycle, the welding current flows from the workpiece to the tungsten. Fig. 2-6 shows a sine wave form of one complete cycle of alternating current.

Normally, the current is unbalanced in GTAW. This happens because hot tungsten can emit (send out) electrons better than molten metal. The current flows more readily in one direction than the other. As a result of the unbalance, current flow during the reverse polarity part of the cycle will not equal the current flow during the straight polarity part of the cycle.

During each part of the current cycle, electrons will carry heat across the arc gap. Gas ions are formed from the inert gas. Ions are atoms positively or negatively charged as a result of having gained or lost one or more electrons. An ionized path develops through which the welding current flows. Fig. 2-7 shows the electron and gas ion flow for one complete cycle of alternating current.

In the straight polarity (SP) part of the cycle, the electrons leave the tungsten tip and impinge (strike) on the metal. Enough heat is created to melt the workpiece. The reverse polarity (RP) part of the cycle removes oxides from the surface of the material as the electrons flow from the surface of the metal to the electrode. This part of the cycle will also cause the tip of the tungsten to melt. The tip will then become rounded because most of the heat is directed to the tungsten.

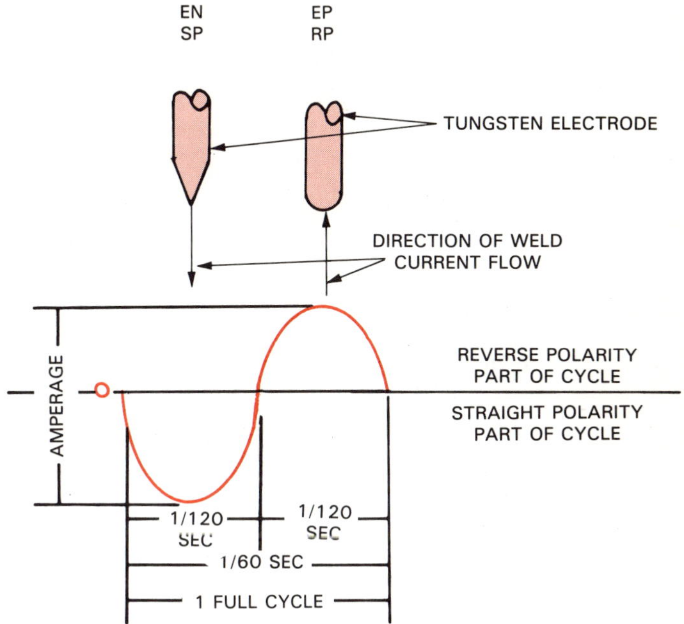


Fig. 2-6. The reverse polarity part of the cycle has less amperage output than the straight polarity part of the cycle in GTAW using AC. Alternating current will form a ball on the electrode on the reverse polarity part of the cycle.

## Rectification

When alternating current flows across the zero amperage point, as in Fig. 2-6, it changes polarity. The arc will go out at this point. On the straight polarity portion of the cycle, electrode emission (sending out) of the electrons is sufficient to ignite the arc. All of the electrons are concentrated at the electrode tip. On the reverse polarity portion of the cycle, the electron flow is very poor because the electrons are scattered all over the workpiece surface. As a result, the arc may not ignite on that half cycle. When the arc does not restart because of the resistance of electron flow from a surface to a point, the condition is called RECTIFICATION. This condition may last for one or more cycles. Without the reverse polarity from the surface, the oxide will be absorbed into the weld. Welds containing oxides are usually porous and weak.

The HIGH FREQUENCY VOLTAGE GENERATOR was developed to provide the spark necessary to ignite the reverse polarity part of the cycle and to stabilize the arc. The high voltage causes the arc gap to become ionized, thus becoming a path for the electrons to flow.

### ACHF Using a Balanced Wave Transformer

The balanced wave designation with this type of current means that the power supply has been changed. The sine wave form can now be changed from a normal unbalance of current to an even balance of current or any area inbetween. Fig. 2-8 shows a balanced wave form. A balanced flow of current is usually supplied by the use of series-connected condensors in the welding circuit. An alternative method is to place batteries in the circuit in such a way that the battery voltage will be additive to the reverse polarity half cycle. Balanced wave current flow will produce the following:

1. Best oxide cleaning action.
2. A more stable arc and uniform heating of the metal.
3. Minimum heating within the welding transformer.

Although sometimes desirable, a balanced wave is not essential for most manual welding applications. It is, however, desirable for many mechanized welding applications.

The electron flow is always from the tungsten to the workpiece. The gas ion flow is always from the workpiece to the electrode. In the DCSP or DCEN mode, the tungsten stays sharp or pointed and most of the heat is generated in the workpiece. See Fig. 2-7.

### DCRP Direct Current Reverse Polarity

DCRP (DCEP) is the common abbreviation for this type of current. It is produced the same way as DCSP. The electron flow is always from the workpiece to the electrode. The gas ion flow is always from the electrode to the workpiece.

In the DCRP or DCEP mode, the tungsten electrode will receive most of the heat. A large tungsten electrode will be required. A ball will usually form on the tip of the tungsten. Fig. 2-7 shows the electron and gas ion flow for DCRP.

DCRP has limited use in GTAW, as most of the heat in the arc is directed to the tungsten instead of the workpiece. The penetration is shallow.

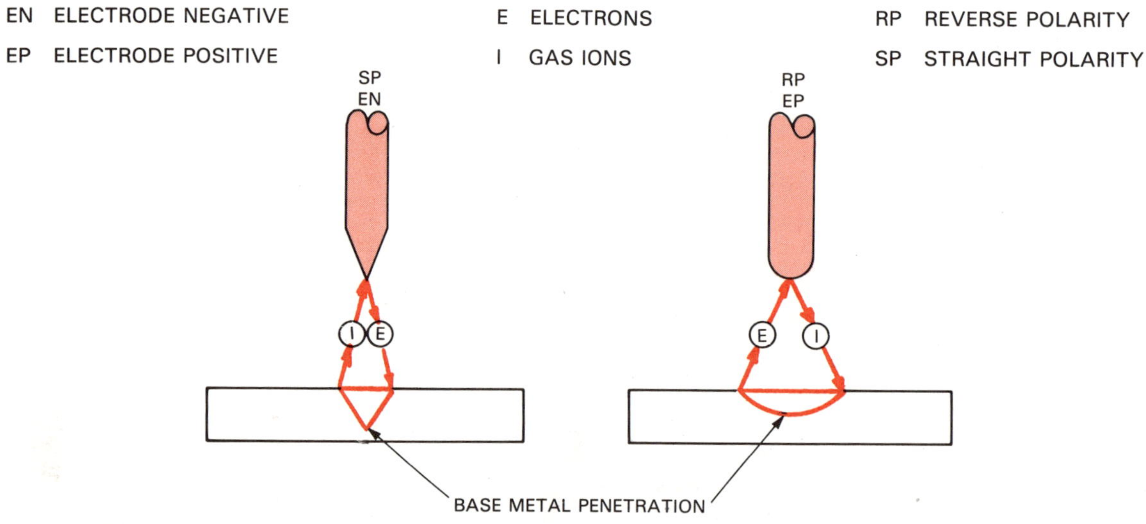


Fig. 2-7. Penetration into the base material is greater during the straight polarity part of the cycle. Electron flow is in the opposite direction of gas ion flow.

### DCSP Direct Current Straight Polarity

DCSP (DCEN) is the common abbreviation for this type of welding current. Welding current produced by generators is direct current, and may be either straight or reverse polarity. Polarity changes are made by switches, or by changing the output leads of the machine.

Welding current produced by transformer-rectifiers starts out as alternating current. It is changed to direct current by rectifiers which allow only one polarity or one-half of the alternating current cycle to pass through as welding current.

## STARTING THE ARC

How the arc will be started in GTAW depends on the system used and the type of power supply. Three methods are commonly used:

1. SCRATCH OR TAP START. This method is commonly used with motor generators or direct current rectifiers. The welder scratches or taps the workpiece creating a short circuit and an arc, allowing the current to flow. This type of arc start may contaminate the electrode or weld area if the

electrode sticks. (CONTAMINATE means it is unfit for use because undesirable material has entered it.) Therefore, extreme care must be used if this type of arc starting is used.

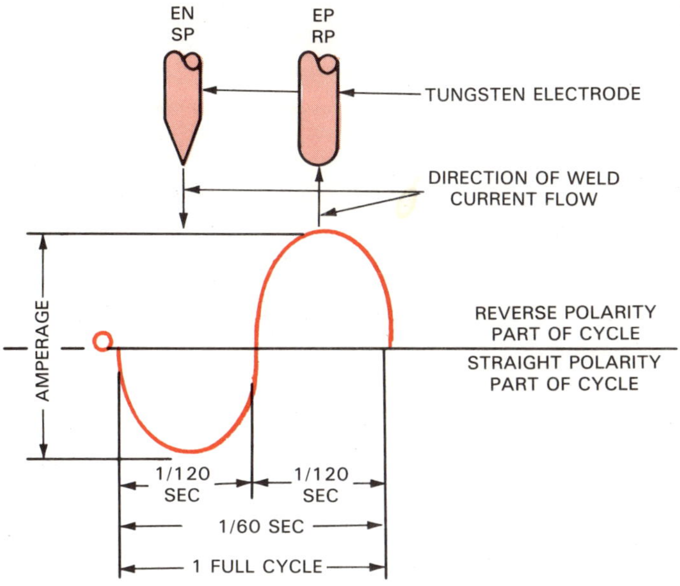


Fig. 2-8. Power supplies with wave balancing capabilities can produce equal amounts of current on both the straight and reverse part of the cycle.

2. HIGH FREQUENCY START. A high frequency voltage of 3,000 to 5,000 volts with a very low current flow is used to generate a spark when the tungsten nears the workpiece. The high frequency generator may be integral (a part of) with the welding machine, or it may be an accessory wired into the system. With this system, the high frequency spark will jump a gap of about one-half inch. Since the tungsten does not touch the workpiece, contamination of the workpiece and the tungsten is avoided. Fig. 2-9 shows a high frequency circuit. Note that the main power cable (in color) carries the high frequency voltage.
3. PILOT ARC START. A special pilot arc power supply is used to create a very small arc between the electrode and an insulated gas nozzle. The small arc heats the tungsten just before the starting of the main arc. It assures positive starting when the main arc is desired. This type of arc start is generally used in GTAW spot welding. Fig. 2-10 shows a pilot arc circuit.

## THE WELDING ARC

The welding arc produced in the GTAW operation consists of three main parts:

1. ANODE. The positive end of the arc.
2. PLASMA. The middle of the arc.

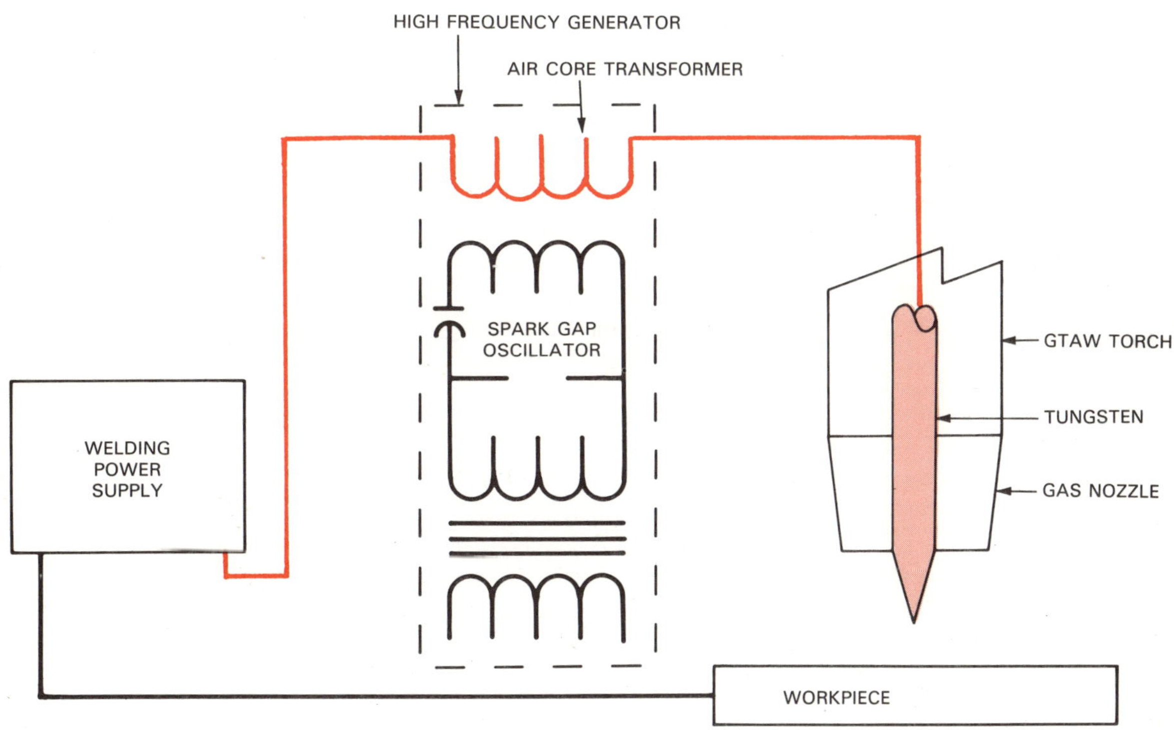


Fig. 2-9. The main power cable of the GTAW torch carries the high frequency voltage from the generator to the tungsten.

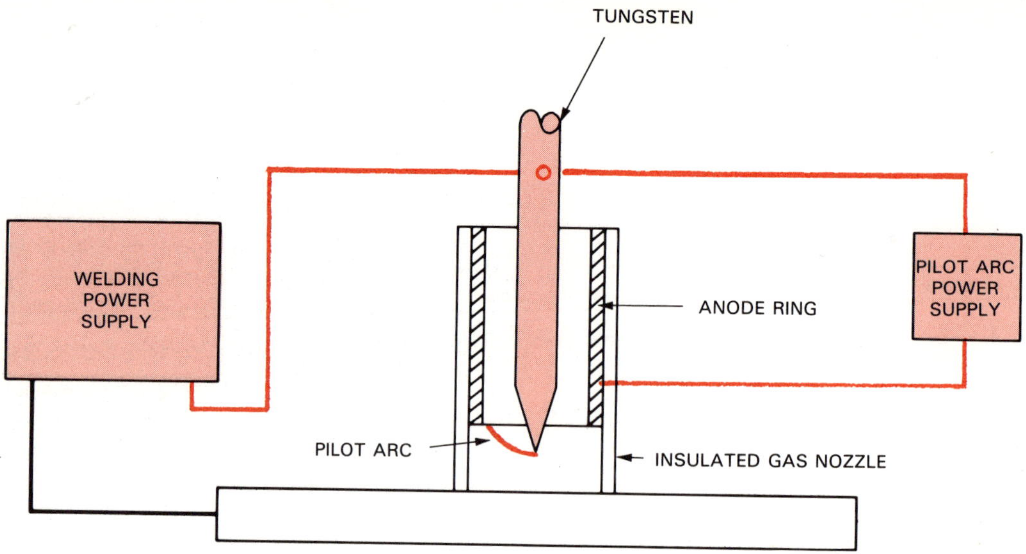


Fig. 2-10. The pilot arc generator produces voltage for maintaining a pilot arc to keep the tungsten electrode hot. The main arc is very easy to start when the tungsten is hot.

3. CATHODE. The negative end of the arc.

The arc is produced by an electrical circuit (spark) between the electrode and the workpiece, creating a small amount of gaseous particle ionization. The positively charged gaseous atoms are attracted to the electrode (cathode) where they give up their kinetic energy (energy of motion) in the form of heat. This heat is enough to bring about electron emission (flow) from the tip of the electrode.

The emitted electrons are attracted to the positively charged workpiece (anode). They raise the temperature of the shielding gas atoms through collisions with them. Repeated collisions of these extremely hot particles causes the atoms to change their makeup. They pick up or lose an electron and they become ions. This happens to more and more atoms and they form a plasma

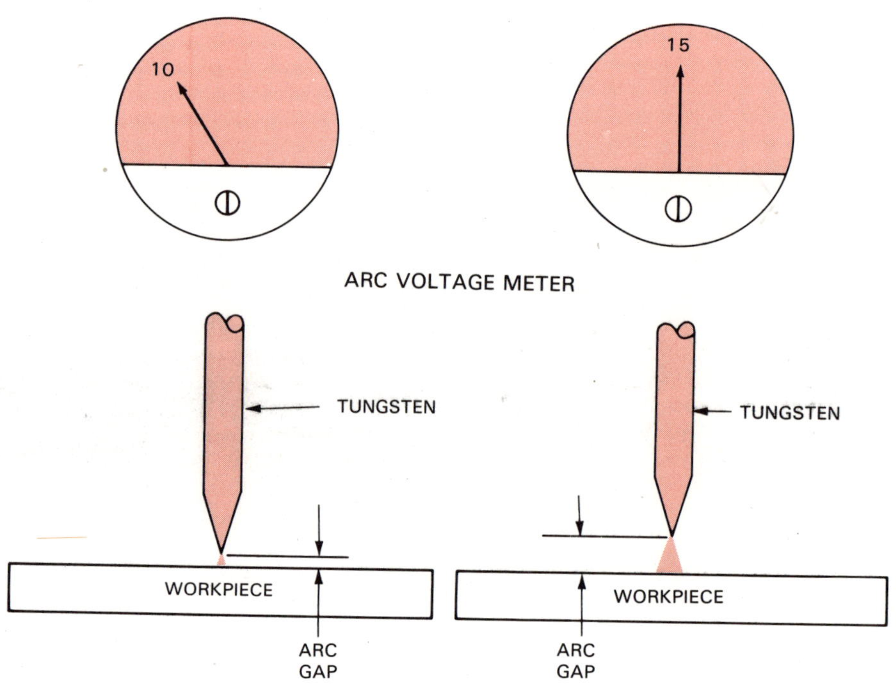


Fig. 2-11. Arc voltage meters are used to measure the gap between the tungsten electrode tip and the weld on the workpiece.

zone. These gaseous atoms or ions are also attracted to the electrode. This movement creates the required heat to sustain electron emission and thus the flow of welding current.

Electrical current produced by the power supply is a flow of negative charges (electrons). The amount of current flowing in the circuit is measured in amperes.

In GTAW, welding arc voltage is expressed as the distance (arc gap) between the end of the electrode and the workpiece. Fig. 2-11 shows an arc voltage meter reading along with the tungsten-to-work gap setting.

## GAS SUPPLY AND REGULATION

An inert gas must be used with the process to avoid contamination of the electrode and the weld area. Argon and helium gas are the most often used. They are supplied to the user as either a liquid or a gas. Liquid cylinders have additional equipment for converting the liquid to a pressurized gas prior to use. Argon gas is used more often because it is inexpensive, ionizes readily in the arc stream, and has a high purity. Being heavier than air, it tends to flow over the weld zone and the molten metal. Helium gas is lighter than air and requires a higher flow rate to cover the weld area. It is used in some weld operations where a hotter arc is desired. This has two advantages: 1. Increased penetration into the base material. 2. Increased welding speeds.

The use of helium gas alone causes many problems in the arc: 1. Hard starting. 2. Arc instability. 3. Black soot on the weld.

These problems can be reduced by using a mixture of gases. The cost is generally less with a reduced flow rate and some increased welding speeds are possible.

Since the cylinders contain gases under a high pressure, a way must be found to reduce the pressure to a usuable flow over the weld area. The solution is to use a:

1. Regulator to reduce the pressure.
2. Flowmeter to meter the amount of flow.

These two functions are usually built into one piece of equipment as shown in Fig. 2-12. The amount of gas delivered to the weld area is measured in either liters per minute (LPM) or cubic feet per hour (CFH).

For normal operation, enough gas must be delivered to the weld area for proper coverage. This will prevent contamination of the electrode and weld metal. Too little gas flow allows the atmosphere (air) to enter the weld zone and contaminate the weld. Too much gas flow causes turbulence in the gas envelope (shield), allowing the atmosphere to enter and mix with the gas shield. (TURBULENCE means great changes in speed and direction of flow.)

## THE WELDING TORCH

Welding torches for GTAW are a special design.

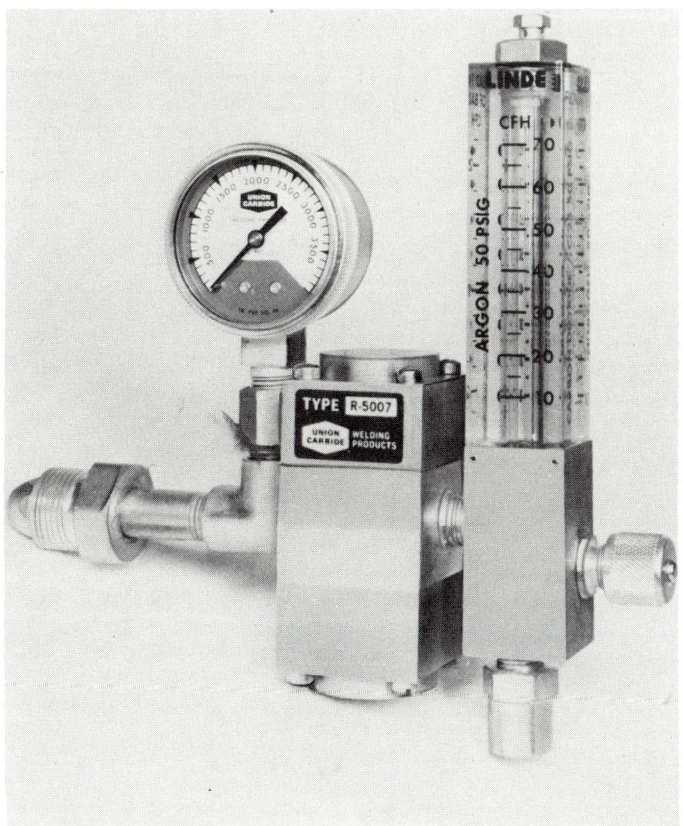


Fig. 2-12. Regulator/Flowmeters of this type are used on high pressure inert gas cylinders. The dial gauge shows cylinder pressure. The amount of inert gas flow is read on the exterior of the glass tube. (Linde Co.)

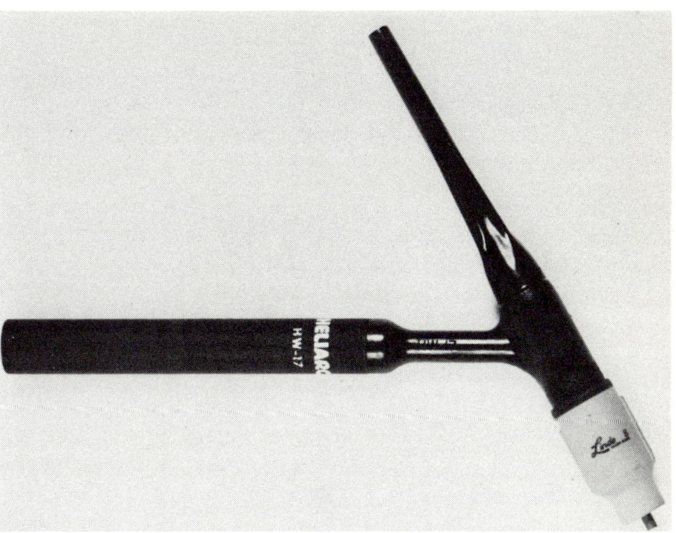


Fig. 2-13. GTAW torches are made with various head angles for the many different types of welds. This model torch is gas cooled and is used for the welding of thin gauge materials. (Linde Co.)

They have:

1. Clamps for holding the tungsten.
2. Gas nozzles for dispensing shielding gas at the weld area.
3. Conductors (wires) for bringing electrical current to the electrode
4. Tubing and chambers to carry gas or liquid for cooling of the welding torch.

Torches may be designed for low amperage, high amperage, manual or automatic welding, or for a special operation such as spot welding.

Torches are rated for the maximum (highest) amperage they may carry regardless of whether they are gas or water cooled. The welder is insulated from electrical current that torches carry.

Gas nozzles attach to the end of the torch either by a press fit or threads within the nozzle. Manufactured from many heat resistant materials, they are available in various types, styles, and sizes for all types of welds. A typical manual GTAW torch is shown in Fig. 2-13.

## GTAW SAFETY

GTAW is a skill which may be performed safely with a minimum of risk if the welder used good common sense and safety rules. It is recommended that you establish good safety habits as you work in this industrial area. Check your equipment regularly and be sure that your environment is safe. Safety in GTAW covers four major areas and includes:

1. ELECTRICAL CURRENT. Primary current to the electrical powered welding machine is usually 208 volts A/C or more, and this amount of voltage can cause extreme shock to the body and possible death. For this reason;
   - A. Never install fuses higher than specified.
   - B. Always ground the welding machine.
   - C. Install electrical components in compliance with all codes.
   - D. Be certain electrical connections are tight.
   - E. Never open a welding machine when it is operating.
   - F. Lock primary voltage switches open and remove fuses when working on electrical components inside the machine.
   - G. Welding current supplied by the power supply has a maximum of 80 open-circuit volts. At this low voltage, the possibility of lethal shock is very small. However, it will still produce a good shock. To reduce the possibility of this occurrence:
     - a. Keep the welding power supply dry.
     - b. Keep the power cable, ground cable, and torch dry.
     - c. Do not weld in damp area. If you must, wear rubber boots and gloves.
     - d. Make sure the ground clamp is securely attached to the power supply and the workpiece.
   - H. High frequency components in some GTAW machines produce a spark for starting the initial arc or maintenance of the arc during the alternating current operation. The high frequency voltage is very high, however the amperage is very low. Since the amperage is so low, the high frequency voltage will not usually travel through the body and is therefore not as dangerous as other currents.
2. INERT GASES. Inert gases used in GTAW are produced and distributed to the user in two forms: high pressure and liquid. All storage vessels used for inert gases are approved by the Department of Transportation and are so stamped on the vessel nameplate or the cylinder wall, as shown in Fig. 2-14

Most of the gases used in GTAW are inert, colorless, and tasteless. Therefore, special precautions must be taken when using them. Nitrogen, argon, and helium are non-toxic. However, they can cause asphyxiation (suffocation) in a confined or closed area that does not have adequate ventilation. Any atmosphere that does not contain at least 18 percent oxygen can cause dizziness, unconciousness, or even death. The gases cannot be detected by the human senses and will be inhaled like air. Never enter any tank or pit where gases may be present

Fig. 2-14. This cylinder has been made and tested to a Department of Transportation (DOT) specification. The letter "T" indicates the amount of gas which it will hold. The name imprinted is that of the owner.

until purged (cleaned) with air and checked for oxygen content.

High pressure gas cylinders contain gases under very high pressure (approximately 2,000 to 4,000 PSI) and must be handled with extreme care. Each of the following rules should be followed:

A. Store all cylinders in the vertical position.
B. Secure all cylinders with safety chains or cables. Fig. 2-15.
C. Do not use as rollers.
D. Know the contents of the cylinder before use. Fig. 2-16 shows an argon cylinder label.
E. Keep protective cap in place until ready to use.
F. Do not move cylinders without the protective cap. Use a cylinder cart, a cylinder cap, and safety chains to move any cylinder. See Fig. 2-15.
G. Check the outlet threads before attaching regulator and clean the valve opening by "cracking" (slightly opening) the cylinder valves as shown in Fig. 2-17.
H. Use only equipment designed for high pressure use.
I. Attach regulator securely and mount flowmeter in the vertical position as shown in Fig. 2-18.
J. Open cylinder outlet valve slowly.
K. Always stand aside when opening a cylinder tank valve.
L. When cylinder is empty, close the valve, and replace cap. Mark in chalk "MT."
M. Never tamper with a leaky valve, return to the supplier.
N. When using any mixture which includes hydrogen gas, remember that hydrogen gas will explode and burn. The gas is lighter than air and the flame is almost invisible.

Liquified gas cylinders are commonly called Dewars and are basically vacuum bottles. The gas has been reduced to a liquid at the supplier's plant for ease in handling, shipping, and storage. Conversion from a liquid to a gas for the user is handled by heat exchangers within or on the Dewars system. The following safety rules apply to a Dewars system:

A. Cylinders must be kept in an upright position.
B. Cylinders should be moved on special dollies as shown in Fig. 2-15.
C. Always use equipment designed for inert gases.
D. Do not interchange equipment components.
E. Liquid gases are extremely cold and can cause severe frostbite when exposed to the eyes or the skin. Do not touch frosted pipes or valves.

3. WELDING ENVIRONMENT SAFETY RULES:
A. Keep the welding area clean.
B. Keep combustibles out of the weld area.

Fig. 2-15. Secure all cylinders with safety chains.

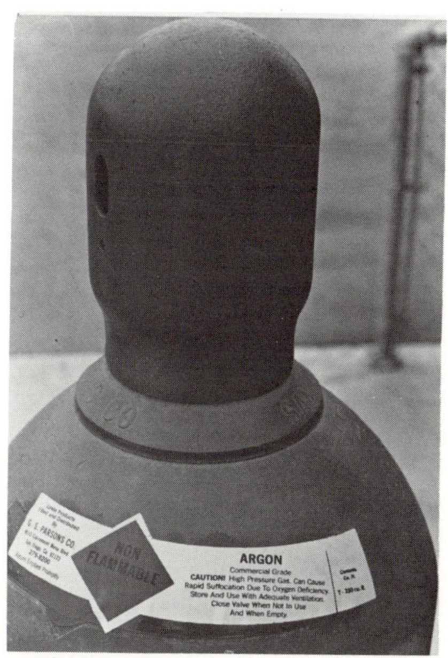


Fig. 2-16. The label indicates the type of gas stored in this cylinder. Notice the caution label and cylinder contents in cubic feet is also indicated on the label. The safety cap is in place until ready to use.

C. Maintain good ventilation in the weld area.
D. Repair or replace worn or frayed ground or power cables.
E. Make sure the part to be welded is securely grounded.
F. Welding helmets should have no light leaks.

Fig. 2-17. "Cracking" a cylinder before attaching a regulator blows out any dirt in the valve opening.

Fig. 2-18. This regulator and flowmeter have been attached to the cylinder properly. Note the pressure relief vent on the regulator and cylinder. These devices operate to release excess pressure in the cylinder. Never tamper with these devices.

G. Use the proper colored lens in the helmet.
H. Wear safety glasses when grinding or power brushing.
I. Wear tinted safety glasses when others are tack welding or welding near you.
J. Use safety screens or shields to protect your area.
K. Wear proper clothing. Your entire body should be covered to protect you from arc radiation.
L. When welding on cadmium coated steels, copper, or beryllium copper, use special ventilation to remove fumes from the weld area.
M. Do not weld near trichlorethylene vapor degreasers. The arc changes the vapor to phosgene gas. A sweet taste in the mouth indicates that this gas is being formed.

4. SPECIAL PRECAUTION AREAS.
   A. Fires may be started by the welder in a number of ways, such as combustible materials, fuel gases, short circuits, improper ground connections, etc. Make sure that you do not start a fire. If you accidently start a fire, make sure it is out before you leave the area.
   B. Never weld on a container that has held a fuel until you are sure that it has been purged with an inert gas and tested for fume content.
   C. Never enter a vessel or confined space that has been purged with an inert gas until the unit or space is checked, with an oxygen analyzer.
   D. Never use oxygen in place of compressed air. Oxygen supports combustion and will make a fire burn violently.
   E. Power wire brushes are very dangerous as they expel broken pieces of wire. Always wear safety glasses or safety shields.
   F. When working with mechanical, hydraulic or air clamps on tools, jigs and fixtures, be alert to the clamp operation. Serious injury may result if parts of the body are exposed to the clamp action.

### REVIEW QUESTIONS

1. GTAW is called ______ welding.
2. Welding characteristics are called parameters and ______.
3. Why are welding characteristics important to the welder?
4. What is the machine duty cycle?
5. Name two types of amperage control.
6. What do welding machines produce after they are started and before the arc is made?
7. What does the welding machine contactor do?
8. List four points to remember about the power supply.
9. ACHF is used to weld ______ and ______.

10. The steels, stainless steels, titanium, etc. are welded using __________ current.
11. Shallow penetration welds are made with __________ current.
12. Using an unbalanced wave transformer, does the reverse polarity cycle produce as much amperage as the straight polarity cycle?
13. Do gas ions carry amperage?
14. What removes the oxide film from the surface when welding with alternating current?
15. What does the high frequency generator do? Why is it required when using alternating current?
16. The welding arc produced in GTAW consists of three main parts. Name them.
17. Which inert gas has a hotter arc, argon or helium?
18. How is the inert gas flow measured?
19. GTAW torches are cooled by __________ and/or __________.
20. Welding power supplies produce a maximum of __________ open circuit volts.
21. Does the high frequency voltage have a high or low amperage?
22. The inert gases __________(can/cannot) be detected by the human senses.
23. Why should oxygen never be used in place of compressed air?
24. What operation should be done on the cylinder before attaching the regulator?

# Chapter 3

# EQUIPMENT

In the welding equipment industry, the term "POWER SUPPLIES" describes a machine for producing welding current. The machine may be a:

1. Power supply only.
2. Power supply with controls for manual welding.
3. Power supply with additional equipment such as slopers, pulsers, timers, etc.
4. Power supply with additional equipment for pipe welding, tube welding, spot welding, etc.

This chapter will cover the power supply requirements for GTAW.

## POWER SUPPLIES

The GTAW power supply must produce sufficient amperage (current) to provide heat needed for melting metal at a low voltage. The equipment must be able to control the power supply operation in the areas of:

1. Input voltage and open circuit voltage.
2. Duty cycle.
3. Output ratings and performance (minimum and maximum current).

The National Electrical Manufacturer's Association (NEMA) has established equipment guidelines called the Specification EW-1 Electric Arc Welding Apparatus. A power supply using utility supplied electricity is covered by this NEMA requirement.

The electrical power received from the utility line is high voltage alternating current and is called primary power or current. A power supply changes this current to low voltage alternating current or direct current. The major components of a typical system are shown in Fig. 3-1.

### Open Circuit Voltage (OCV)

Manual GTAW power supplies produce a maximum of 80 open circuit volts, according to NEMA specifications. Semiautomatic and automatic power supplies may have an OCV rating of 100 volts maximum. With these voltage ratings, the welder or welding operator may work on the low secondary (output) side of the welding machine, isolated from the high primary (input) voltage.

The OPEN CIRCUIT VOLTAGE readings are measured across the power supply terminals (output electrode and ground) while the machine is operating

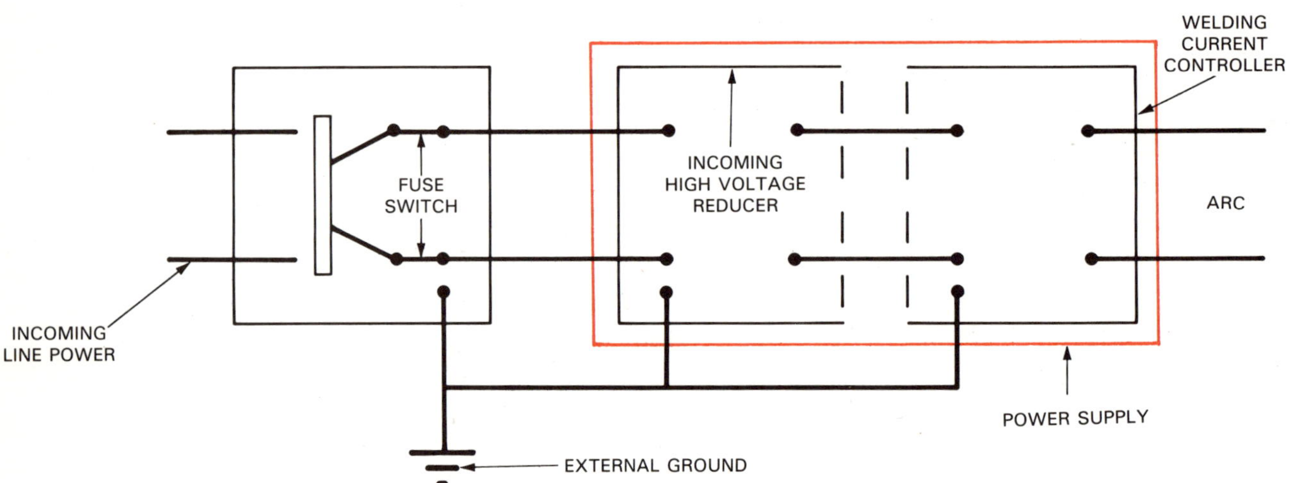


Fig. 3-1. Components required to change utility electrical power to welding current.

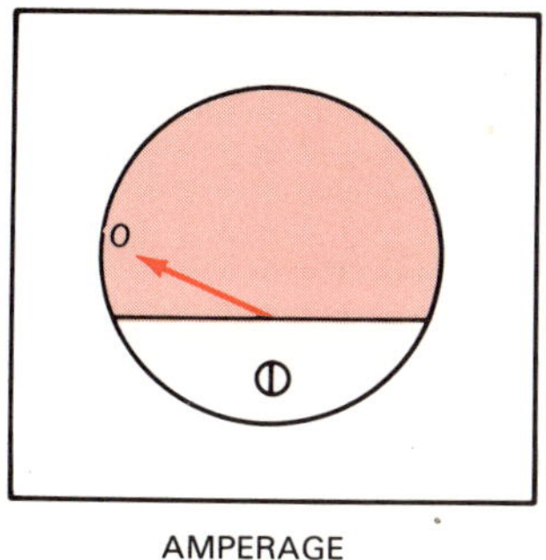


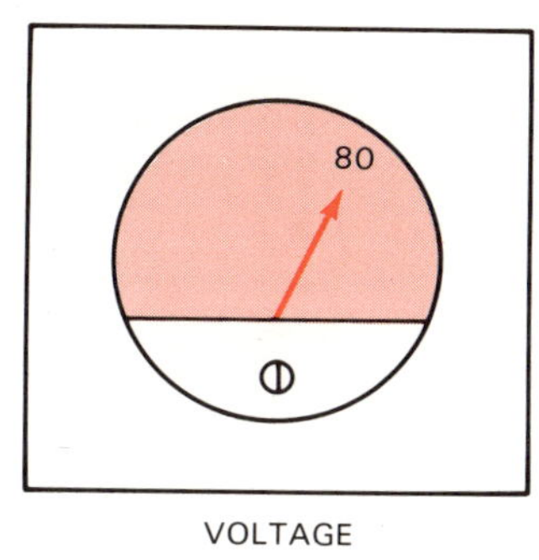


POWER SUPPLY — ON
NO OUTPUT (LOAD)
NO AMPERAGE FLOW
80 OPEN CIRCUIT VOLTAGE

Fig. 3-2. Welding machine operating without producing welding current. The voltmeter registers the maximum OCV allowed by NEMA.

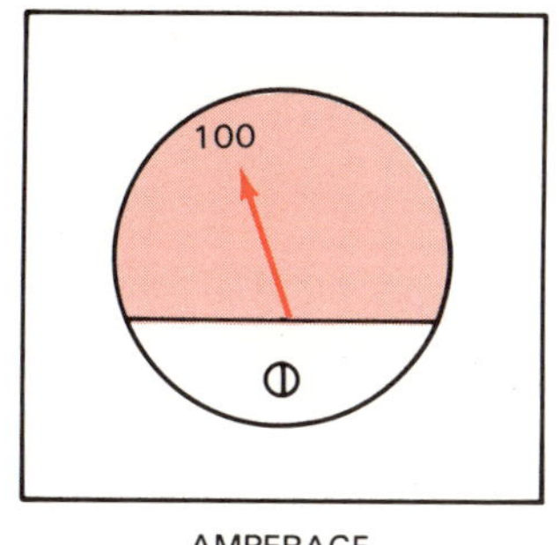


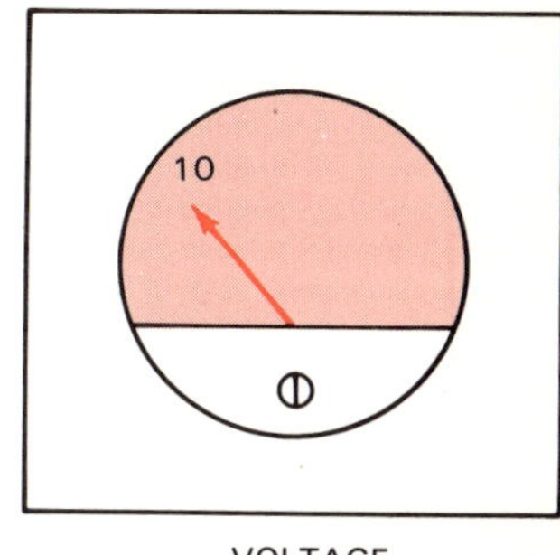


POWER SUPPLY — ON
MACHINE UNDER LOAD
AMPERAGE REGISTERS
ACTUAL ARC VOLTS REGISTERS

Fig. 3-3. Welding machine operating under load. The voltmeter now registers the actual arc voltage in the gap between the tungsten and the work.

without welding current being produced. Machines equipped with voltmeters show OCV when the machine is operating without welding current being produced, as shown in Fig. 3-2. When the arc is started, the voltmeter will register actual arc voltage. This is the voltage across the gap between the tungsten tip and the workpiece as shown in Fig. 3-3.

## Duty Cycle

The DUTY CYCLE RATING determines the maximum amperage output for a given period of time. Every type of power supply has an established duty cycle. Operation within these ratings will greatly extend the life of the equipment. The basic design of the power supply controls the amount of duty cycle. Design considerations include: 1. Internal components. 2. Size of wiring. 3. Insulation and amount of cooling.

Power supplies are manufactured with duty cycles having ranges of 20 to 100 percent. Each 10 percentage points represents one minute of operation in a 10 minute period. Fig. 3-4 shows the cycle and time period for the various types of GTAW power supplies. For example, by using the data in Fig. 3-4 to determine machine operation, a 300 ampere rated power supply with a 60 percent duty cycle could be used at 300 amperes for a 6 minute period. Then, the welder must allow 4 minutes for the machine to cool.

Many GTAW applications are for longer periods of time than the duty cycle, but with lower amperes than the rated load. In these cases, a power factor may be used to compute how long the machine may be used,

| DUTY CYCLE | NUMBER OF MINUTES MACHINE MAY BE OPERATED AT RATED LOAD IN A 10 MINUTE PERIOD |
|---|---|
| 100% | FULL TIME |
| 60% | 6 |
| 50% | 5 |
| 40% | 4 |
| 30% | 3 |
| 20% | 2 |

Fig. 3-4. Power supply duty cycles limit the number of minutes the unit may be operated at rated load.

| % Duty Cycle Of Power Supply | Factor No. X Rated Load | = Duty Cycle | Derated Duty Cycle For A/C Derated Amps X |
|---|---|---|---|
| 60 | 75 | 100% | 70 |
| 50 | 70 | 100% | 70 |
| 40 | 55 | 100% | 70 |
| 30 | 50 | 100% | 70 |
| 20 | 45 | 100% | 70 |

EXAMPLE NO. 1
PRESENT POWER SUPPLY RATED 300 AMPS — 60% DC
300 AMPERES
x .75 FACTOR NO.
=225 AMPERES 100% DUTY CYCLE

EXAMPLE NO. 2
SAME POWER SUPPLY
225 AMPERES
x .70 FACTOR NO.
=157 AMPERES 100% DUTY CYCLE

Fig. 3-5. Power factor numbers are used to compute various duty cycles at other than rated loads. Derating of the duty cycle is done on AC machines not specifically designed for GTAW.

and a 100 percent duty cycle may be figured at a lower amperage. Example 1 in Fig. 3-5 shows the factor number used for various duty cycles.

Some alternating current power supplies which are used for GTAW may not have been designed specifically for this purpose, and in these cases the duty cycle must be derated to protect the machine. Example 2 in Fig. 3-5 shows the percentage number which must be used to compute the new duty cycle.

### Volt Ampere Curve

The volt ampere curve is the operating characteristics of the power supply under load. It is established by the machine manufacturer. Fig. 3-6 shows a volt ampere curve for a typical CONSTANT CURRENT power supply using approximately 80 open circuit volts.

The term CONSTANT CURRENT is obtained from the intersection point of the volt ampere curve and the welding arc voltage. If the welder maintains a constant arc length, the welding current will remain the same. Should the welder raise or lower the torch height, a new voltage intersection line is obtained. This movement of the torch raises or lowers the welding power supply output current. Notice that the curve is not straight but drooping. Because of this characteristic, the machines may also be called DROOPERS.

As seen in Fig. 3-6, the welder has a potential spread of about 11 amperes in welding current by raising or lowering the torch height from the work. This allows the welder to raise or lower the welding current by changing the arc gap. This is very useful for changing the amount of current to bridge gaps, deepening penetration and to weld slower or faster.

GTAW power supplies which have a selection of current ranges or steps, change the volt ampere curve at each higher setting. This change makes the volt ampere curve flatter. The flatter the curve, the more the amperage will change as the welder varies the arc length.

To determine the actual variance in the amperage change, the machine instruction book should be consulted. In many instances, the manufacturer will have the volt ampere curve results determined by factory test and illustrated in the machine instruction book.

Because the welding arc voltage may be adjusted by the welder changing the arc gap during the welding operation, the power supply is also termed a VARIABLE VOLTAGE type power supply.

## ALTERNATING CURRENT TRANSFORMERS

Alternating current power sources are normally single phase transformers that use alternating current from the incoming (primary) power line. High voltage and low amperage current is then changed (transformed) into low open circuit voltage and high amperage currents for welding power. Several methods of controlling output current have been developed. Each method has a distinctive advantage and provides the welder with the needed controls for monitoring or controlling welding current.

The manual GTAW alternating current welding power supplies use one of three methods of control:

1. TAPPED SECONDARY COIL type machines may have a plug-in jack or rotary selector. These machines are usually the lowest in cost and produce a given amperage depending on the jack or selector used. The welding current leads are tapped into the secondary (output) windings of the transformer. Normally this type of machine is used in non-critical applications, as the jack or rotary selector position does not allow the possibility of mid-range adjustments. This control method is

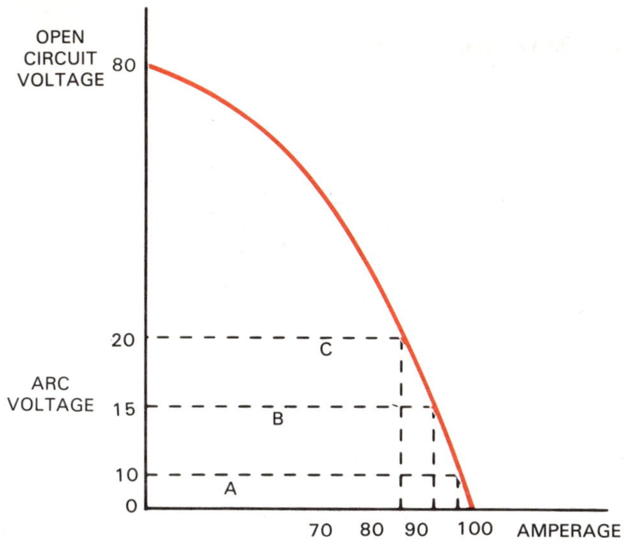


A. 10 ARC VOLTS (VERY CLOSE ARC GAP) 97 AMPERES
B. 15 ARC VOLTS (MEDIUM ARC GAP) 92 AMPERES
C. 20 ARC VOLTS (LONG ARC GAP) 86 AMPERES

Fig. 3-6. Volt ampere curve illustrates how the welder can select output amperage by adjusting the arc voltage (gap). Each type machine has its volt ampere curve developed by the manufacturer.

shown in Fig. 3-7.

2. The MOVABLE COIL type machine uses a primary and secondary coil with a method of moving the primary coil closer or further away from the secondary coil. Varying the distance between the coils regulates the amount of output current. The farther the two coils are apart, the more vertical the volt ampere curve. As the coils get closer together, the curve will be flatter. Fig. 3-8 shows a movable coil power supply set in both the minimum and maximum position. Fig. 3-9 shows a movable coil type power supply.

3. The SATURABLE REACTOR CONTROL type of power supply is usually a multi-range machine. An electrical low voltage, low amperage direct current is used to change the effective magnetic characteristics of the reactor cores. Often referred to as a magnetic amplifier, a small control power change will produce a sizable output power change. Changing ranges on the power supply will change the volt ampere curve. As the ranges increase, the volt ampere curve becomes flatter. Control of the welding output is simplified as a rheostat can be placed on the control panel for manual control. Remote hand or foot controls may also be inserted into the direct control circuit for complete control throughout the range selection. Fig. 3-10 shows a typical saturable reactor with direct current control.

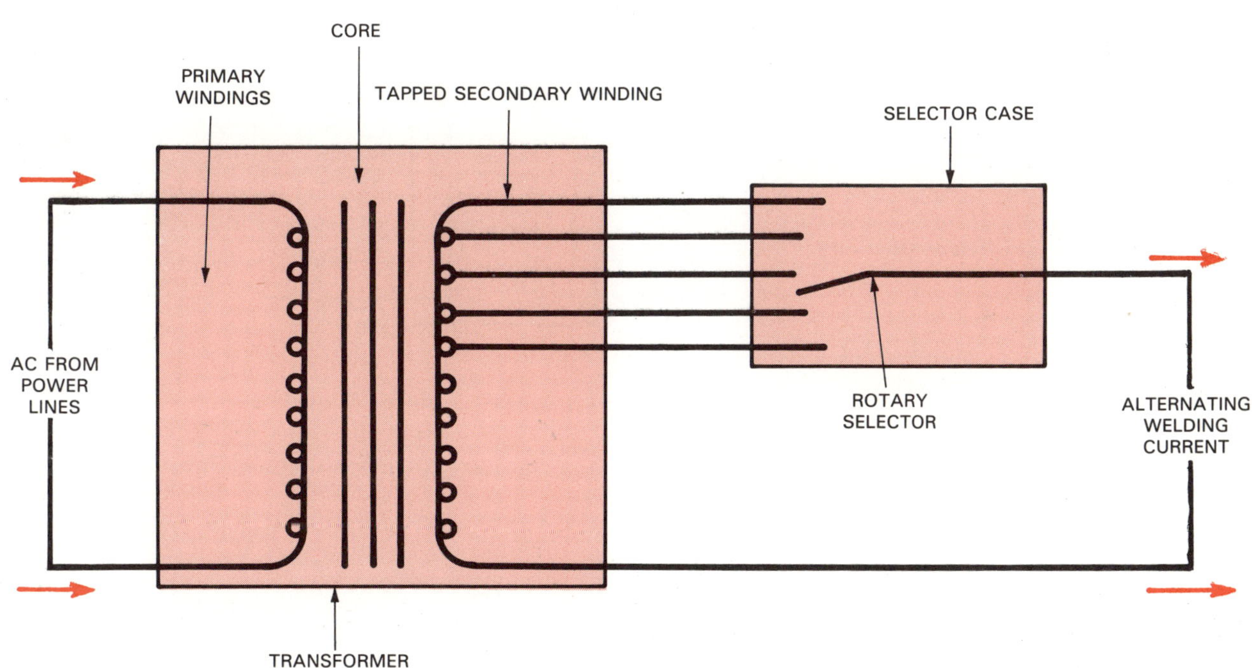


Fig. 3-7. A less expensive welding machine is produced from tapping various winding levels on the secondary transformer. Only when the selector is in contact with the tapped leads will the current flow to the torch.

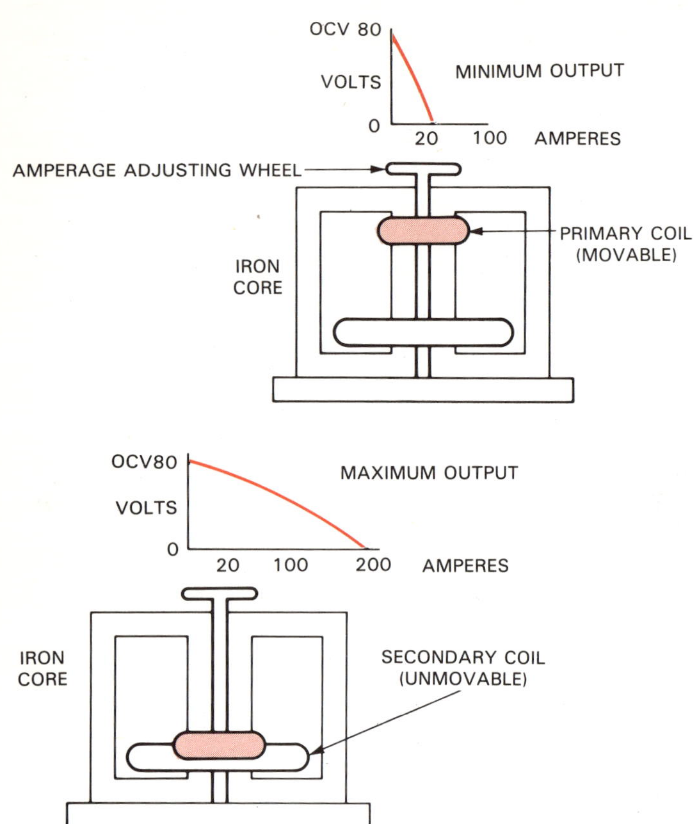


Fig. 3-8. Changing the output level of this transformer changes the shape of the volt ampere curve. This type machine can be adjusted within the operating range during welding.

ROTARY SELECTOR CRANK

Fig. 3-9. This AC transformer produces current in two ranges, 30-150 amperes and 40-225 amperes. The rotary selector crank allows many adjustments within each range. (Miller Electric Mfg., Co.)

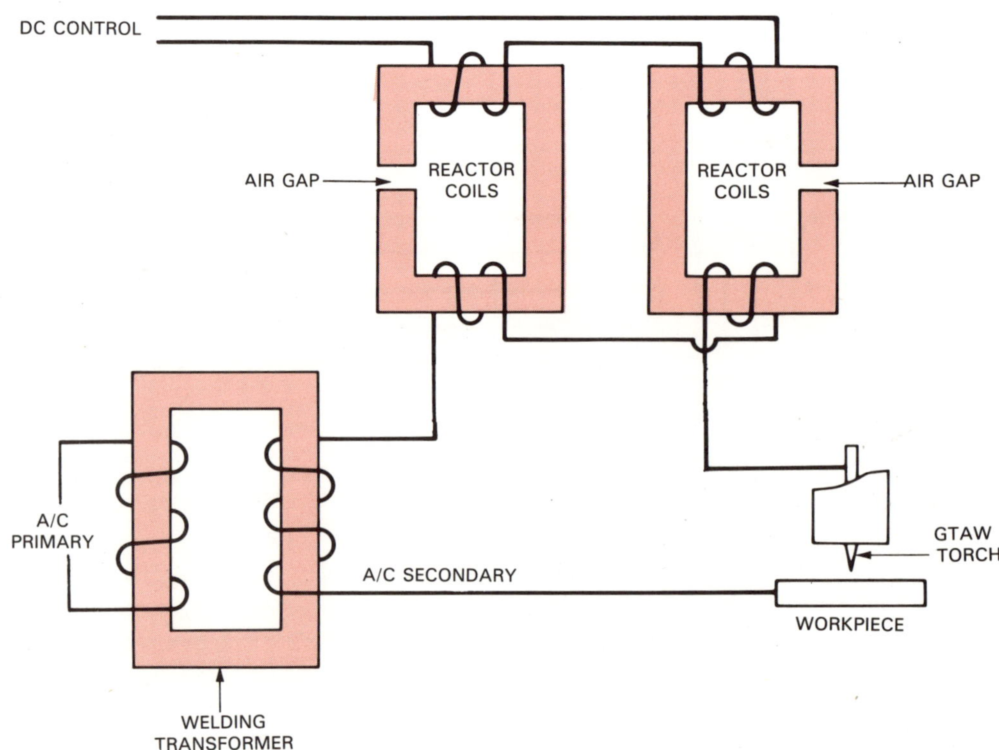


Fig. 3-10. Saturable reactors allow precise direct control or remote control by using a direct power control circuit.

**Balanced Wave Control**

The balanced wave control is used to change the pattern of the alternating current by adding capacitors in the output circuits. This may produce deeper penetration or better cleaning action. It is not used to select the amount of power supply output current. A rheostat located on the control panel enables the operator complete selection of the type of wave form desired. A balanced wave form and the results produced are shown in Fig. 3-11.

A recent addition to the wave balance circuit is the square wave pattern. With this action, the wave is squared instead of sloping from the zero amperage part of the cycle. Fig. 3-11 shows the operation of the balanced wave function with square wave.

The direct current is produced by rectifiers, which are commonly called SCR's (Silicon Controlled Rectifiers). An SCR is essentially an electrical gate which opens and closes to allow either straight or reverse polarity to pass through into the welding circuit. Timing of the SCR's operation also allows the use of the reverse polarity part of the cycle to be inserted into the straight polarity (or vice versa) part of the cycle, as shown in Fig. 3-12. This operation is called FULL WAVE RECTIFICATION. As seen in Fig. 3-12, the direct current output still rises to current level and drops to zero current level at the end of each half cycle. This type of output current cannot be used for welding as it is wavy or ripply. To reduce the ripple, inductor-capacitors are placed into the circuit as shown in Fig. 3-12. Current types and polarity changes are made by moving switch positions on the main power supply.

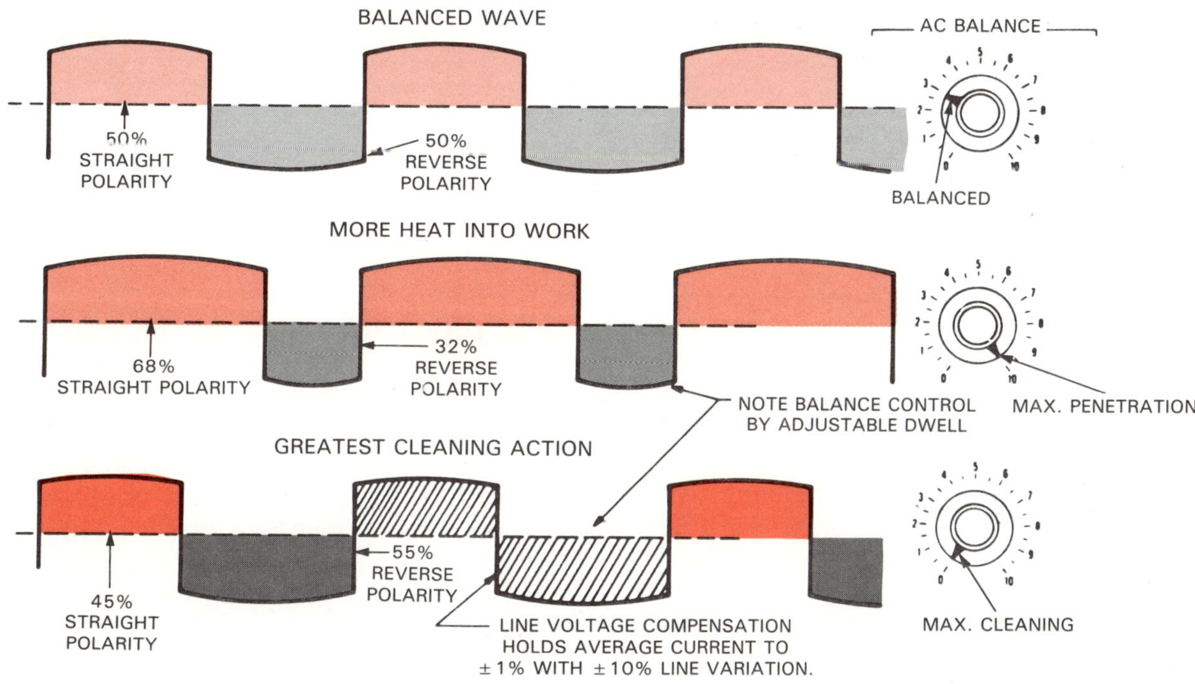


Fig. 3-11. Balanced wave allows the welder to achieve either deeper penetration or better cleaning of the oxide film from the surface of the material.

## ALTERNATING CURRENT TRANSFORMERS/ DIRECT CURRENT RECTIFIERS

The ALTERNATING CURRENT TRANSFORMER/DIRECT CURRENT RECTIFIER type of machine, commonly called an AC/DC WELDING POWER SUPPLY, is very useful in the welding industry because of the dual current selection from a single machine. The machine produces alternating current or direct current straight or reverse polarity. A single phase transformer with a saturable reactor is used to produce alternating current.

## DIRECT CURRENT RECTIFIERS

DIRECT CURRENT RECTIFIERS are machines which use three phase alternating current primary power. A transformer is used to convert high voltage electrical current to low voltage current, with rectifiers to produce direct current. With three phase input power, the cycle commences three times for each full

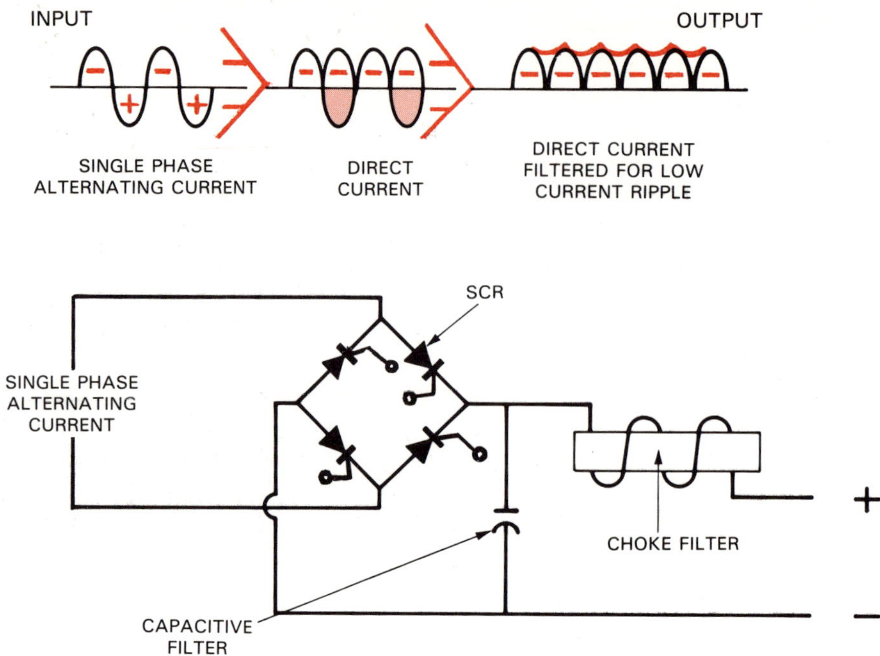


Fig. 3-12. SCR's insert the positive part of the AC cycle into the straight polarity part of the cycle. This operation is called rectification. The current flow is now in one direction and is called direct current. The filters smooth the ripple to make the current flow even.

cycle. The resulting straight polarity will have three full rises to current levels. Output current is extremely steady, with very little ripple and is used where precision control is required during welding. Fig. 3-13 shows a typical three phase SCR control power supply circuit.

Polarity changes are made by changing switches located on the main power supply control panel. Where switches are not available, the polarity is changed by reversing the power output leads.

## MOTOR GENERATORS/ALTERNATORS

Motor generators are generally used in an area away from a stationary power source. They may be driven from an electrical motor, gasoline or diesel engine. The gasoline or diesel unit is an ideal power supply for field work, as most units also provide 110 volts AC/DC power for using small power tools. There are two basic types of rotating power sources, the ALTERNATOR

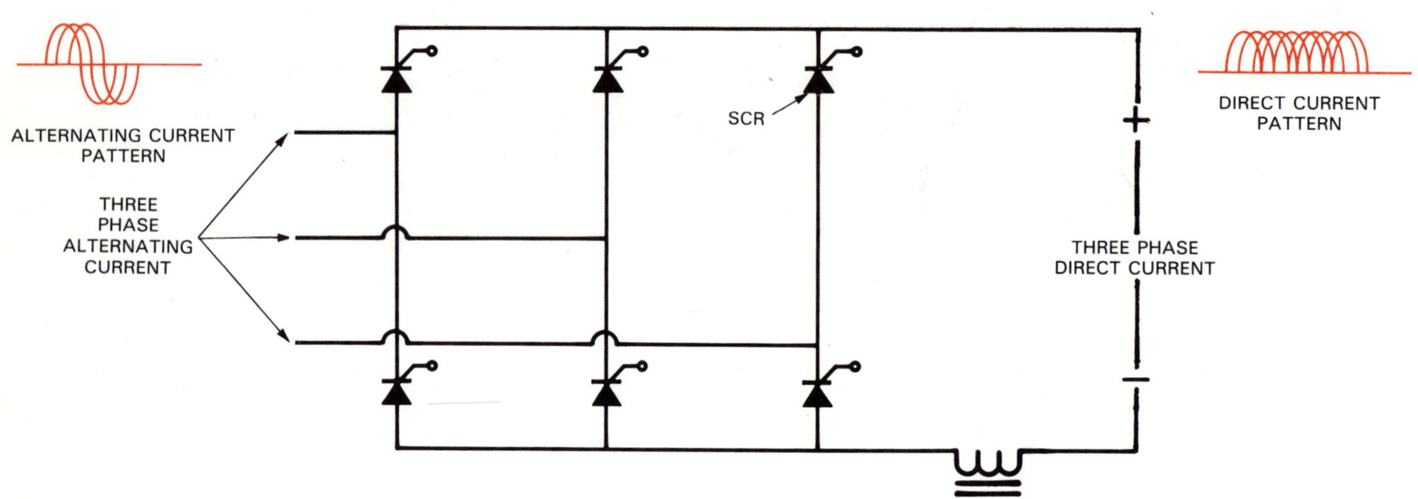


Fig. 3-13. Three phase alternating current input uses six SCR's to convert the AC to direct current. The filters are usually smaller, as the output ripple is much less. The DC output is extremely steady.

which produces alternating current, and the GENERATOR which produces direct current.

Some manufacturers produce power supplies which will produce both AC and DC from a single unit. Amperage control may be supplied in ranges, with fine adjustment control within the individual ranges. Some models allow adjustment of the Open Circuit Voltage (OCV) to give the welder complete control of the welding amperage. A motor generator with control of OCV and welding amperage is shown in Fig. 3-14. Typical engine driven portable AC/DC welders are shown in Figs. 3-15 and 3-16.

## MULTI-OPERATOR POWER SUPPLIES AND CONTROLS

Multi-operator power supplies are commonly used in construction sites, nuclear plants, etc., where large numbers of welders are employed. Rather than supply each welder with an individual power supply, a master power supply is used to supply individual controllers or grids, as diagramed in Fig. 3-17.

A typical main power supply is a silicon rectifier constant potential (voltage) power supply rated at 100 percent duty cycle, Fig. 3-18. Input voltage is 230/460 three phase when using utility power. When utility power is not available, the unit may be powered by a diesel engine.

Welding current is supplied from the main power supply to the individual controller (grid). Adjustments in welding current is made by the welder using switches or rheostats on his or her controller, Fig. 3-19.

Adjustments are usually made on the switch type in 1, 5, 10, 25 and 100 ampere increments, while the rheostat type is infinitely adjustable within a range. The welding current supplied by the individual controller is very steady. Regardless of the number of arcs operating on the system, each arc is independent of the

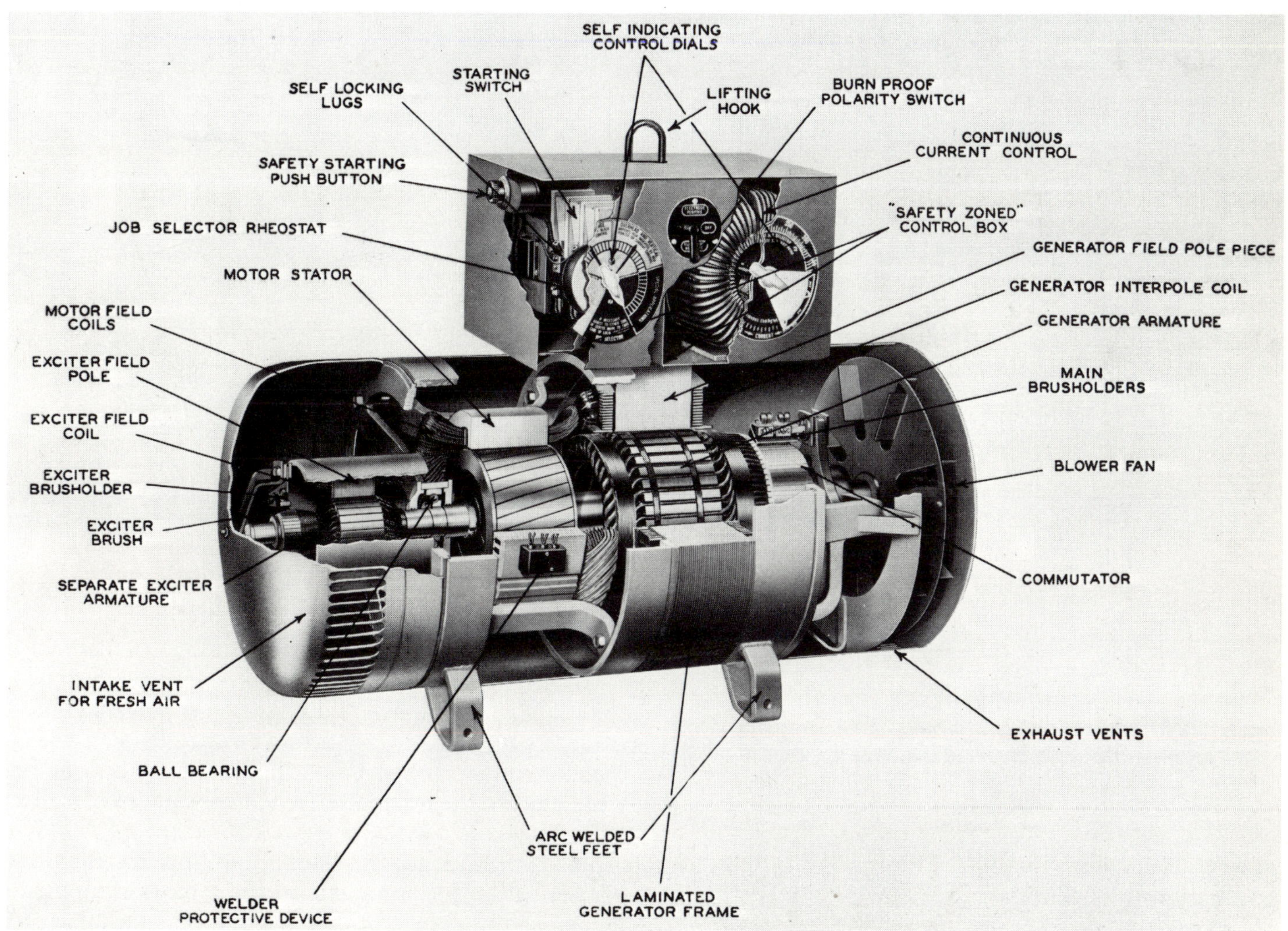


Fig. 3-14. This direct current motor generator has a rheostat to change the OCV from low to high. The welder can make almost any type of volt ampere curve desired. (Lincoln Electric Co.)

Fig. 3-15. This portable engine driven welder produces both AC and DC. 120 volt AC power is also available to operate grinders, drills, etc., while the machine is under load. (Miller Electric Mfg. Co.)

Fig. 3-16. This power supply has been mounted on a small cart for portability. (Lincoln Electric Co.)

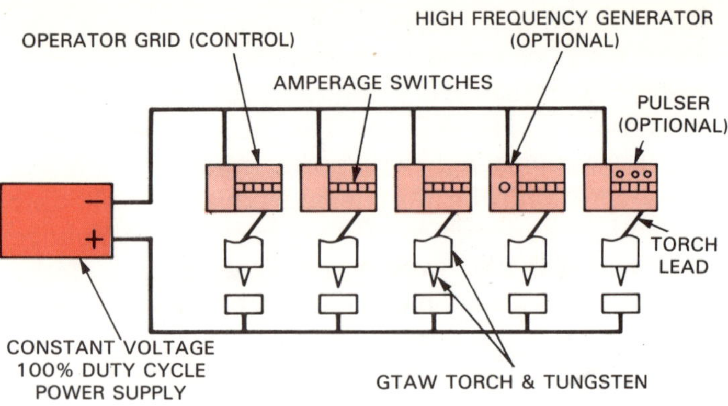


Fig. 3-17. Multiple operator systems allow the use of several separate grids with only one master power supply.

Fig. 3-18. This multi-operator master power supply provides 300 amperes to 12 individual grids. Since all of the grids will not be operating at full load, the power supply need not have the total ampere capacity. (Cooperheat Ltd.)

others. The main advantages of the multi-operator power supply system are:

1. Lower initial cost.
2. Lower operating cost.
3. Portability.
4. Lower maintenance cost.
5. Controllers may be removed from the system for repair without affecting remaining controllers.

Originally the multi-operator system was designed for shielded metal arc welding (SMAW or "stick electrode") which requires starting the arc by scratching the electrode to the work. Stopping the arc is done by withdrawing the electrode. When used in GTAW, the scratch start is used with a gas cooled torch. A valve is placed in the argon gas line to start and stop the flow of inert gas. Developments in the system equipment include:

1. Addition of automatic gas controls.
2. Water flow valves for increased duty cycle.
3. High frequency for starting the arc.
4. Slopers for starting and ending the welding current.

5. Pulsers to control the weld metal and joint penetration where the system is used for pipe welding.

Fig. 3-19. Individual switches are used on this type grid to select the amount of welding current. Note the system installation including: water cooling for torch, argon gas flowmeter and torch shutoff valve. Torch cables are protected by a vinyl cover. (Cooperheat Ltd.)

## POWER SUPPLY CONTROLS

GTAW power supplies have various controls to monitor the overall welding operation. More simple power supplies may only have jacks or screws for adjustment of the amperage. In these cases, the welder must adapt or adjust other requirements such as: start the gas flow, start the water flow, scratch the tungsten to start the arc, etc., to complete the weld operation. More complex machines may only require an initial setup and the sequence required for the operation to become semi or automatic. Manual GTAW AC/DC power supplies will usually have the following listed controls installed in the unit:

1. Off-on switch.
2. Range switch.
3. Current potentiometer with 100 percent control of each range.
4. High frequency for start only, continuous use for AC, or off.
5. High frequency intensity.
6. Stabilizer control to eliminate tungsten spitting on AC.
7. Wave balancer for penetration control.
8. Hot or soft starts for penetration control when starting AC welding.
9. Remote control for hand or foot control of current.
10. Argon post flow control to prevent contamination of the tungsten.
11. Water on or off when using machine solenoid control.
12. Argon on or off when using machine solenoid control.

Fig. 3-20 shows a typical industrial type power supply control panel used for manual welding.

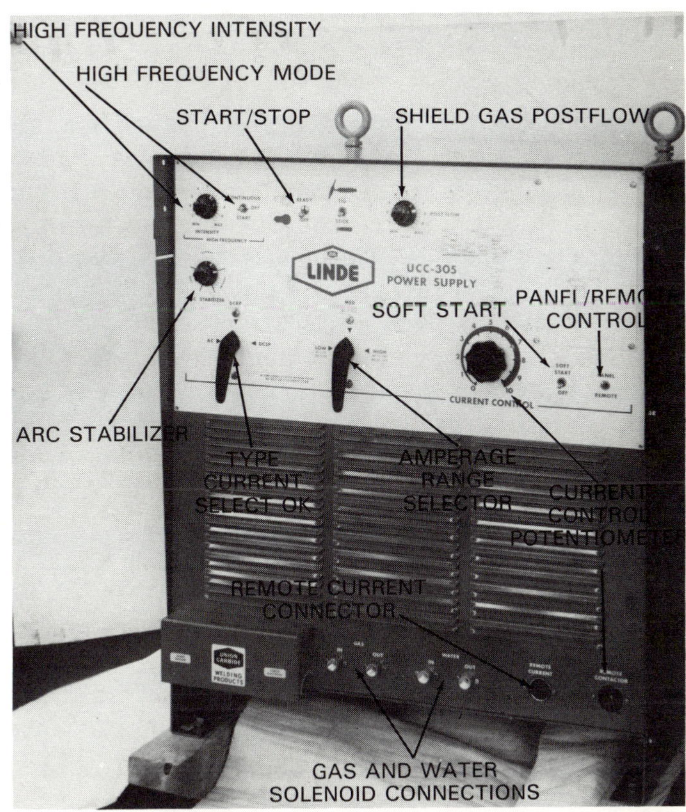


Fig. 3-20. Manual gas tungsten arc power supply. This machine is rated at 300 amperes with a 60 percent duty cycle. (Linde Co.)

## POWER SUPPLY SELECTION

Many factors influence the selection of a power supply for welding operations. Major points to consider are:

1. Ability of the machine to produce the type of current desired.
2. Machine rated load (amperage ranges).
3. Duty cycle.
4. Adequate control during the welding operation.

Comparison of various types of power supply types

may be made using the following factors.

### Current Type

A single type current machine is limited by the type of output to either ferrous or non-ferrous type metal welding, except in cases of automatic type applications. A single type current machine is very useful if the welding is for a long range project. Multiple type current machines are very useful where different types of metal are being welded with different types of current. When investing in new or used equipment, cost comparisons can be made considering single type power supplies versus multiple type current machines, and the amount of welding to be done with each type.

### Amperage Ranges

Tapped leads, selector switches, and adjustable coil machines are limited in their scope and usage. These machines deliver a set current output, and generally do not have sufficient control for fine adjustment of the machine output. These machines may be used where many parts are welded at a set amperage and do not require variations from a set current.

Industrial rated welding machines control the output of the machine in steps or ranges. They may use three or four steps for control of welding current. Machines of this type deliver a more precise amperage control with less drift or variation during the welding operation.

### Duty Cycle

Factors which may be used in the selection of a power supply for the duty cycle requirement are:

1. Lower amperage than the rated load. (This is figured by using the factor table shown in Fig. 3-5).
2. Higher duty cycle than required for normal use.
3. Welding time less than the duty cycle.
4. Irregular welding times such as:
   A. Changing tooling.
   B. Loading parts.
   C. Replacing tungstens.
   D. Cleaning operations during the welding cycle.

The basic duty cycle requirement consideration is not to exceed the manufacturer's design rating of the power supply. If this is done, the machine will not produce power at a stable rate. This will also cause overheating and malfunctions of the machine components.

Some power supplies have a thermal (heat) overload protection thermostat to protect the unit from overheating. When this occurs, the machine will automatically shut down the welding circuit until the machine cools to a lower internal temperature. Machines that do not have this device will overheat and this may cause damage to the unit.

### Open Circuit Voltage

Most power supplies designed for AC/DC operation operate near the maximum OCV (open circuit voltage) allowed by NEMA (National Electrical Manufacturer's Association). The maximum OCV for each type of machine is shown on the machine data label, as shown in Fig. 3-21. The high OCV is required to increase the response time or recovery voltage for arc ignition on the reverse polarity part of the AC cycle. These machines do not have any external method of changing the OCV.

Rectifier welding power supplies do not have the problem of recovery voltage, as they produce only direct current. However, some manufacturers do have a method of changing the OCV. By lowering the OCV, the volt ampere curve can be made flatter, which gives the welder additional current control of the arc.

Motor generators producing DC do not have a problem of recovery voltage since the current flow is either all straight polarity or reverse polarity. Some have an OCV control circuit in the system, which allows the welder to choose the type of arc control desired. Fig. 3-14 shows a welder with OCV control.

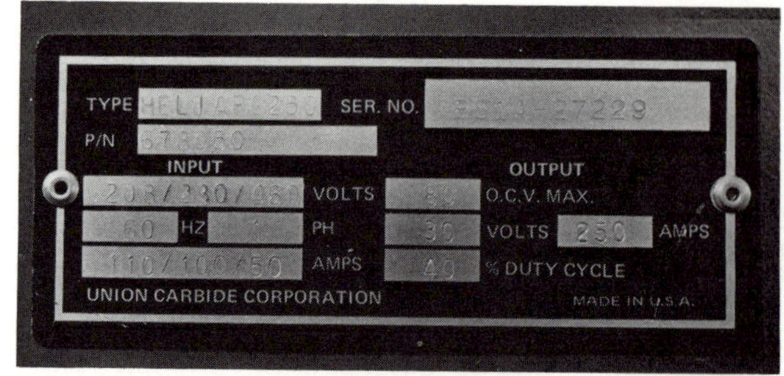


Fig. 3-21. Power supply data labels provide information for primary power and secondary power output. (Union Carbide Corp.)

### Amperage Controls

One of the two types of current control, either machine or remote, is selected depending on the mode in which the machine is to be used. Machines which use a preset selected amperage and are not changed during the welding process, can be controlled by the machine set-up. Where a welder is working away from the power supply and must make adjustments during the operation, a remote circuit must be included in the power supply.

Machine control of the amperage may be done by mechanically moving taps, jacks, or levers. Amperage may also be controlled by an electronic circuit. Electronic circuits are used to operate the main contactor, thus allowing control of the amperage by remote hand or foot controls. Remote electronic controls are used to:

1. Start the arc at very low levels of current.
2. Raise the amperage to the desired setting.
3. Reduce the current level to a low setting at the end of the weld.

The amperage control can also be used by the welder for:

1. Applying additional current for deeper penetration.
2. Reducing current level when bridging gaps.
3. Adjusting current flow for the addition of the filler wire.
4. Applying current slowly to prevent burn-through on thin materials.
5. Applying current slowly to prevent blasting the tungsten tip away from the electrode.

## Auxiliary Controls

Pulsers, slopers, timers, etc., are termed AUXILIARY CONTROLS. Many power supplies do not include these units in the basic power supply, or have the required circuitry planned so they may be added at a later time. When selecting a power supply, consideration should be given to the possible addition of these controls at a later time if they are not purchased with the basic power supply. The required circuitry may then be installed when the machine is assembled.

Many of the auxiliary controls are made today in modules. They are solid state in design. These auxiliary controls are either plug-in types or connect to a wiring strip board within the machine. In either case, installation is completed using the manufacturer's instructions. Fig. 3-22 shows a power supply with auxiliary controls. Fig. 3-23 shows an auxiliary control module.

## Power Supply Installation

Power supplies should be installed in an area that is free of dust, dirt, fumes and where the machine heat may be dissipated (may escape). Dirt and heat will cause a power supply to produce an output below it's rated load, and if not corrected, will ruin the machine. The area selected for installation should remain free of objects blocking the free flow of air into and out of the machine. The machine should not be exposed to moisture, as electronic controls may pick up moisture and may fail.

Utility supplied power supplies operate on 208, 220, and 440 volts AC power, single and three phase power. They operate on 60 cycles unless otherwise specified. Machines using other than 60 cycle power are especially made at the factory for this requirement. Fig. 3-24 shows a primary power connection board inside of the power supply. The required fuse size for the incoming power is always shown on the data label, see Fig. 3-21. The fuse panels should be close to the power supply for safety reasons. A typical power supply installation is shown in Fig. 3-25.

Machines that do not match the utility power voltage may be used if a step up or step down transformer is

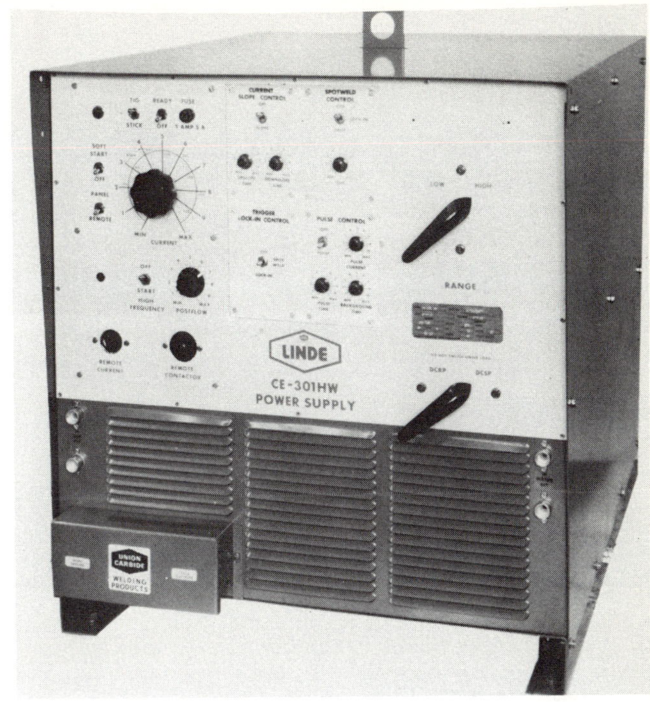

Fig. 3-22. Auxiliary control modules have been added to this power supply for slope control, spot weld control, lock-in and pulse control. (Linde Co.)

Fig. 3-23. Typical module which may be added to many machines, providing that required circuitry is available on the power supply. (Union Carbide Corp.)

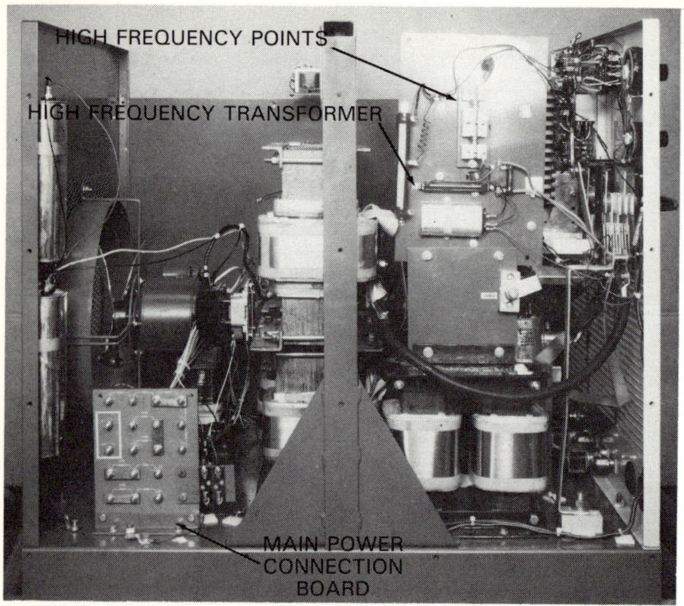


Fig. 3-24. Primary power connections are made to boards within the power supply. The high frequency spark is generated in the unit shown in the upper right section of the photograph. (Linde Co.)

used. Fig. 3-26 shows a stepdown transformer installed for this use. The manufacturer's instruction manual should always be contacted for information regarding installation of machines where utility voltages do not match the machine requirement.

Manufacturers have excellent warranties on their products. Using a power supply voltage other than the one stated on the power supply data label may nullify the warranty.

Primary power which is supplied to residential homes for electrical power for stoves, dryers, etc., is not suitable for welding machine primary power as it is only 110 volts. The electrical power utility or supplier should be contacted in all cases regarding power requirements before installation of a welding power supply.

## POWER SUPPLY MAINTENANCE

Given reasonable care and routine maintenance, a welding machine will operate for many satisfactory hours before repairs are required. Modern methods of insulating transformers, using solid state designed components, and using good basic design have extended

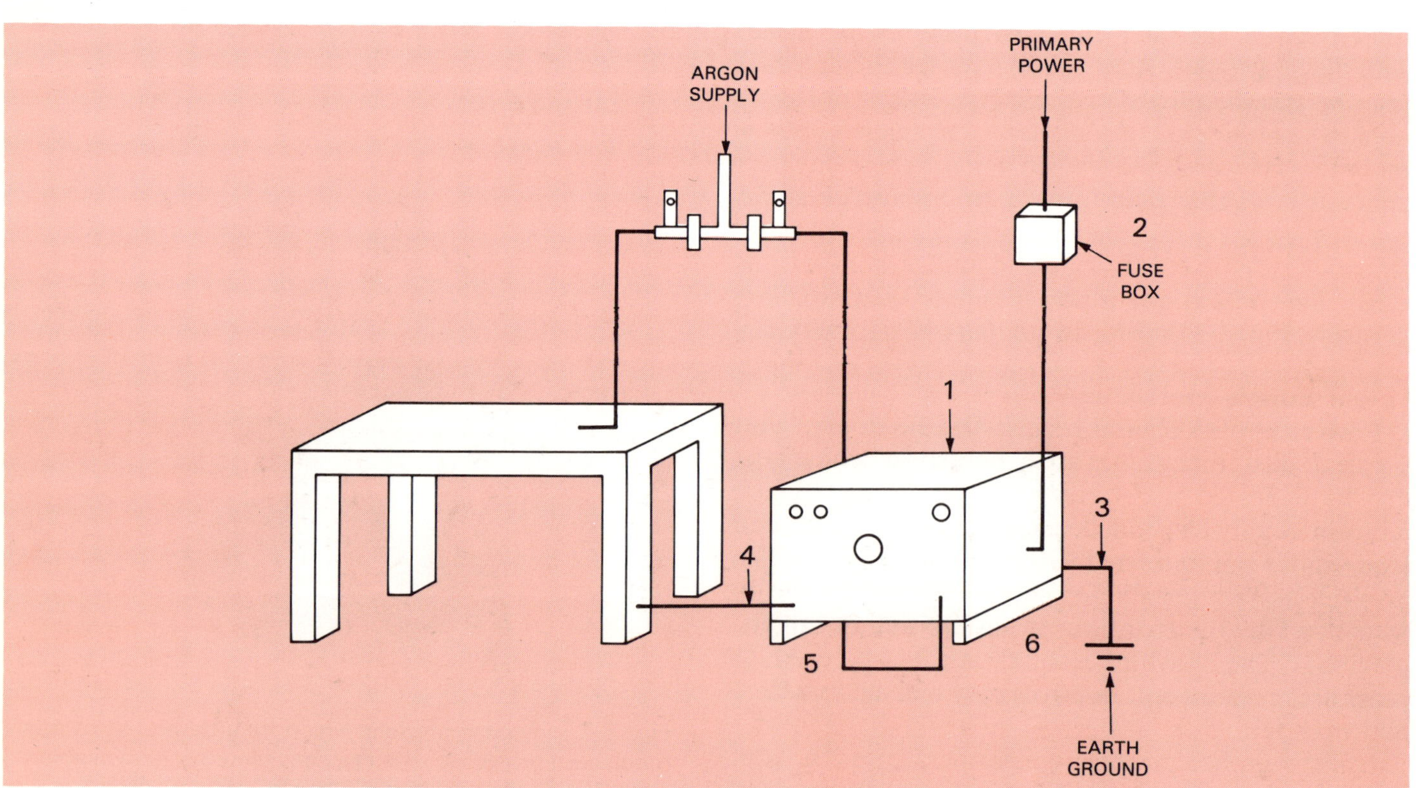


1. LOCATE POWER SUPPLY AWAY FROM WALLS FOR PROPER AIR FLOW.
2. LOCATE PRIMARY FUSE BOX NEAR POWER SUPPLY.
3. CONNECT POWER SUPPLY FRAME TO EARTH GROUND.
4. CONNECT WORK GROUND WITH 2/0 MIN-4/0 MAX CABLE.
5. KEEP POWER SUPPLY DRY.
6. KEEP AREA NEAR POWER SUPPLY CLEAN.

Fig. 3-25. Typical power supply installation includes a fuse box mounted near the welder. An earth ground from the power supply protects the welder from primary line high voltage.

Fig. 3-26. These transformers change incoming primary power to match the power supply requirements.

the hours of operation of a modern power supply. The manufacturer's instructions should always be followed for periodic inspection of the power supply. The following points should be observed:

1. Cleaning or blowing out the unit should be done on a periodic basis. Use only dry, filtered, compressed air, nitrogen gas, or an electrical nonconducting cleaner.
2. Check all wiring terminals within the power supply on a regular basis.
3. High frequency points should be checked every six months, or when operation is faulty, for excessive pitting and for proper gap. Replace pitted points if required. Regap points according to the manufacturer's instructions. A typical high frequency assembly is located in the power supply as shown in Fig. 3-24. The gap check area is shown in Fig. 3-27. Too close a gap will cause improper operation of the high frequency. Too large a gap will cause high frequency radiation loss, which results in radio and television interference.
4. Lubrication of the fan and motor bearings on most AC/DC machines is not required as they are sealed units. Older units may still require lubrication, but be careful to not over-oil. Over-oiling will cause bearing failure.
5. Check mechanical arm and switches for freedom of movement. Mechanical joints may be lightly greased if required.
6. Terminal blocks for cable connections should be tight and clean. If corroded, they will restrict the flow of current. They may be cleaned by brushing or wiping with a cloth.
7. Motor generator brushes should be checked and replaced when worn beyond the manufacturer's tolerance. Worn brushes wear armatures and worn armatures must be machined for proper operation.
8. Lubrication, inspection, and adjustment of portable gasoline and diesel power supplies requires close attention to the manufacturer's instructions. This applies to both the engine and the power supply. Operation of these units usually occurs under very adverse conditions, and improper maintenance will shorten the capacity and operation of the unit.
9. USE EXTREME CAUTION WHEN WORKING WITH PRIMARY POWER.
10. Do not work inside a power supply that is operating.
11. Be careful when fueling a gasoline or diesel engine.
12. Use only authorized replacement parts.

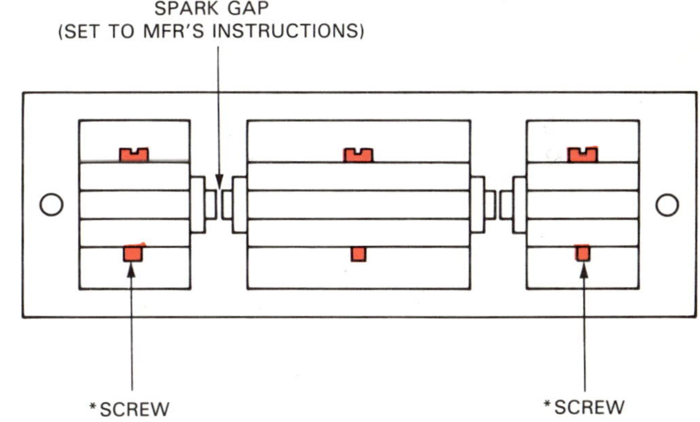


Fig. 3-27. Always maintain proper gap setting on the high frequency points. Improper settings affect the arc stability.

## REVIEW QUESTIONS

1. The GTAW power supply provides electrical current which is also called _______, and is used to provide _______.
2. The utility company delivers _______ voltage and _______ amperage to the power supply.
3. Transformers within the power supply convert the incoming power to _______ voltage and _______ amperage.
4. NEMA specifications limit the maximum _______ that the machine may produce on the secondary side of the power supply.
5. Duty cycle ratings for power supplies are controlled by _______, _______, and _______.

6. Power supplies are made with ______, ______, ______, ______, ______, and ______, percent duty cycles.
7. The duty cycle period of time is measured in a 10 ______ period.
8. Where machine loading is for more or less time or amperage than the specified duty cycle a ______ number may be used.
9. Power supplies for GTAW may also be termed ______, ______ current, and ______ voltage.
10. The welder may change the ______ being delivered to the workpiece by changing the arc length.
11. The gap between the tungsten and the workpiece is called ______ ______.
12. Alternating current contains a half cycle of ______ polarity and a half cycle of ______ polarity.
13. Jack or selector type alternating current power supplies (do/do not) have control of mid-range amperage valves.
14. What is the volt ampere curve?
15. How is the volt ampere curve used?
16. Do all machines have a volt ampere curve?
17. Do all machines have a volt ampere curve controller?
18. What type of current is used to change the output current of a magnetic amplifier power supply?
19. What does a balanced wave term mean?
20. What do the letters SCR mean?
21. What are SCR's? How do they work?
22. SCR's produce ______ ______.
23. ______ ______ are used in power supplies to reduce the amount of output current ripple.
24. What is the multi-operator (M.O.) system?
25. List five advantages of a multi-operator power supply system.
26. List four major considerations to be used when selecting a power supply.
27. What is a thermal overload protection switch?
28. Does the data label on the machine list the incoming power fuse limits?
29. List three operations the remote control circuit may be used for.
30. What equipment is used to raise or lower primary power to match the machine requirements?
31. List five general maintenance areas for servicing a power supply.

# Chapter 4

# AUXILIARY EQUIPMENT AND SYSTEMS

This chapter details the various types of equipment used in GTAW to:

1. Assist in starting the arc.
2. Increase control of welding current.
3. Apply and control filler metal.
4. Maintain arc gap.
5. Move the arc across the puddle (oscillate).
6. Pulse the weld current to aid penetration.
7. Time the weld cycle.
8. Program the weld operation.

This equipment may or may not be included in the basic power supply. Most of the equipment can be added to the basic power supply or system as an auxiliary item.

Welding torches, tungstens and some special welding systems are termed auxiliary equipment, even though they are a basic part of the system as they attach to the basic power supply.

Gas supply and regulation are an integral part of the system and will be explained in Chapter 5.

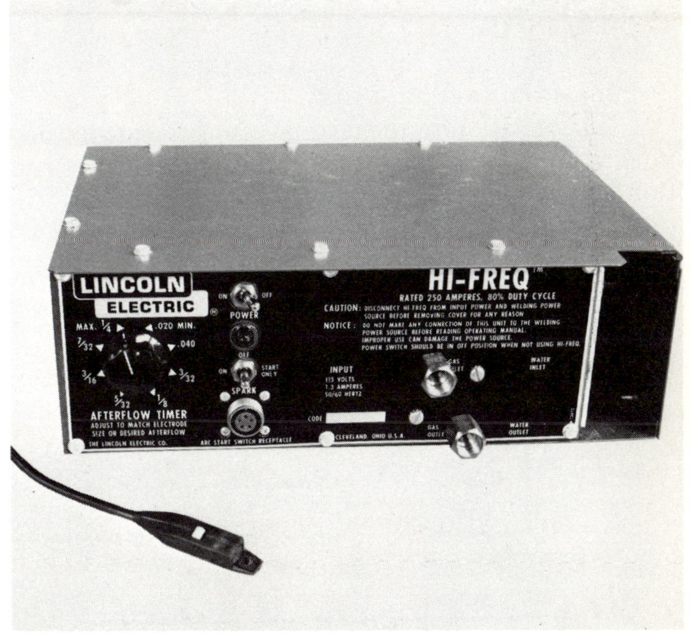


Fig. 4-1. Typical auxiliary high frequency generator. The timer matches the diameter of the tungsten to control the afterflow (post flow) of gas. (Lincoln Electric Co.)

### Auxiliary High Frequency Generator

The AUXILIARY HIGH FREQUENCY GENERATOR is designed for use on power supplies that do not have a built-in high frequency generator. Some auxiliary spark generator units have only a high frequency spark control. Others may have a combination of high frequency and gas or water controls.

Fig. 4-1 shows an auxiliary high frequency generator with gas control. It also has a timer for controlling torch gas post flow to protect the tungsten from contamination. Referring to Fig. 4-1, the controls for the high frequency operation are:

1. OFF. The unit is not operating. Most units may be left in the system when not in use and the unit will not interfere with the normal operation of the power supply.
2. ON or CONTINUOUS. The unit is on continuous operation for welding on AC. Generally used for welding of aluminum or magnesium.
3. START ONLY. The unit is on only to start the arc on DC using either straight or reverse polarity. Generally used for welding of materials other than aluminum or magnesium.
4. SPARK INTENSITY (optional on some units). This control adjusts the amount of spark from minimum to maximum. A maximum spark is sometimes used for heavily oxidized material to assure positive arc starting.
5. AFTERFLOW TIMER. Afterflow timers allow the shielding gas to flow for an adjustable period of time after the arc is broken. This shields the tungsten from the air and contamination. The afterflow timer is variable depending on the size of the tungsten. To prevent contamination, the

tungsten should always be a silver color before the gas is shut off. A blue or grey tungsten color indicates insufficient length of time on the afterflow timer.

6. REMOTE STARTING OF HIGH FREQUENCY. The remote arc start switch is attached to the torch or foot control. By pressing the remote control switch, the control sequence starts. Once the arc is established, the switch can be released and the arc will continue until the arc is withdrawn from the work.
7. COOLING WATER CONTROL (when used with a water cooled torch). The cooling water will flow when the high frequency (H.F.) start switch is depressed and will flow for the length of time the afterflow timer is set to operate. Cooling water will flow for the same period of time as the gas flows. Fig. 4-2 shows the proper installation of a H.F. unit in a welding system.

The installation of the auxiliary high frequency unit must be done according to the manufacturer's instructions for radiation protection and for the safety of the operator.

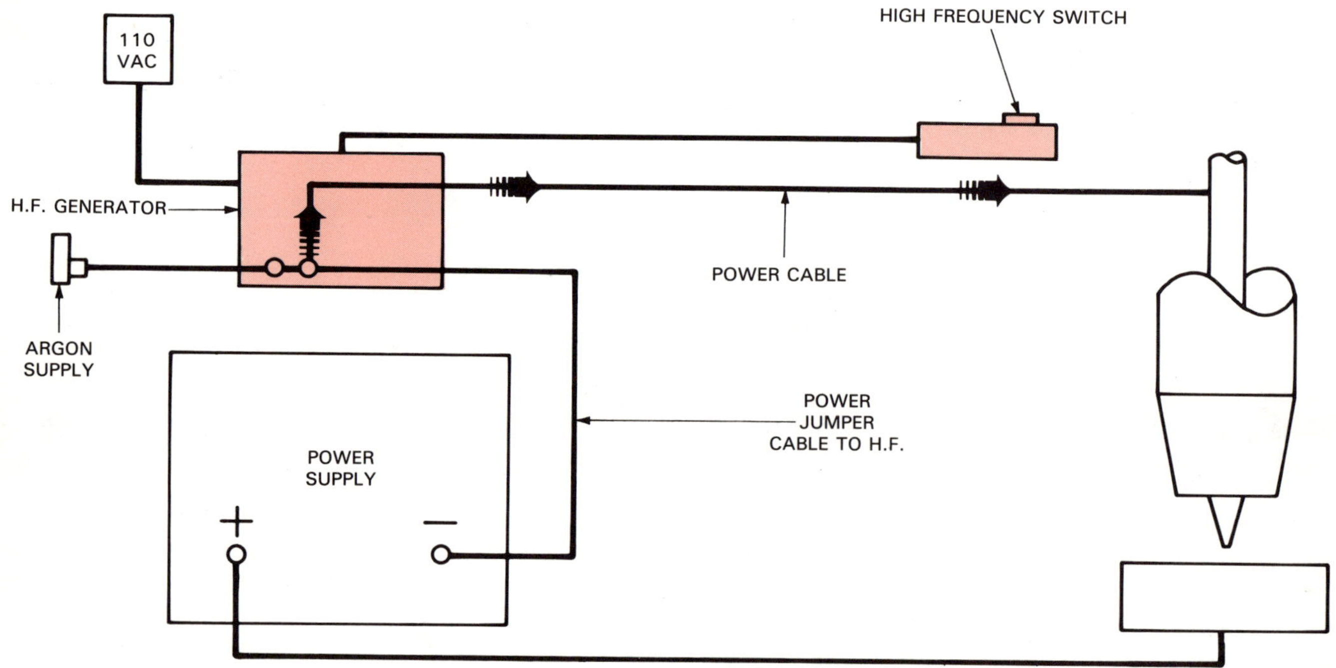


Fig. 4-2. High frequency generators must be installed properly to prevent interference with electronic components and to insure operator safety.

## HIGH FREQUENCY INTERFERENCE PROTECTION

The spark gap oscillator in the high frequency generator unit is similar to a radio transmitter. Improper installation can result in radio and TV interference or problems with nearby electronic equipment. Radiated interference can develop in the following four ways:

1. Direct interference from the high frequency unit.
2. Direct interference radiated from the welding leads.
3. Direct interference radiated from feedback into the power lines.
4. Interference from reradiation of "pick-up" by underground metallic objects.

Keeping these four contributing factors in mind, installing equipment according to the following instructions should minimize problems:

1. Keep the power supply lines as short as possible. Completely enclose the power supply lines in rigid metallic conduit or equivalent shielding for a minimum distance of 50 feet. There should be a good electrical contact between this conduit and the welding power supply. Both ends of the conduit should be connected to a solid electrical ground and the entire length should be continuous.
2. Keep the work lead and electrode lead as short as possible and as close together as possible. Usual lengths should not exceed 25 feet. Tape the leads together when practical.
3. Be sure the rubber covering of torch and work cables are free of cuts and cracks that allow high

frequency leaking. Cables with high natural rubber content resist high frequency leakage better than neoprene and other synthetic rubber insulated cables.

4. Keep the torch in good repair. Keep all connections tight to retard high frequency leakage.
5. The work terminal must be connected to a solid ground at the welding power supply, or to a water pipe that enters the ground within 10 feet of the welding machine. Ground the connection using cable of the same size as the work cable or larger. Grounding to the building frame or a long pipe system can result in reradiation, effectively making these members radiating antennas. A solid ground is one that is driven into the ground or earth.
6. When the high frequency is in operation, keep all panels and covers securely fastened in place to minimize radiated interference.
7. All electrical conductors within 50 feet of the welder should be enclosed in grounded rigid metallic conduit or equivalent shielding. Flexible helically-wrapped conduit is generally not suitable.
8. If the welding operation is in a metallic building, several good electrical driven grounds from the metallic walls all around the building are recommended. When operating the high frequency, the spark intensity should be kept at the lowest setting possible consistant with satisfactory performance. This will minimize the level of radiated interference.

Failure to observe these recommended installation procedures can cause radio or TV interference problems and result in unsatisfactory welding performance resulting from lost high frequency power.

### Pulsers

Pulsing of weld current from low to high levels in GTAW is a method developed to:

1. Control heat input.
2. Control penetration.
3. Control bead shape.
4. Refine grain structure.
5. Increase agitation of molten metal for "outgassing."

Two basic systems of pulsing have been developed for GTAW use. One uses a low frequency switching of weld current while the other uses a very high frequency switching weld current. For the manual welding operation, the low frequency model is generally used. The automatic welding system uses a low or high frequency system or both.

The sequence of the pulsing operation is to apply weld current for a sufficient period of time to form a weld with controlled penetration through the joint. The weld current is then lowered to a point where the weld metal cools to partially solidify. The sequence then repeats until the weld is complete. The finished weld bead will actually be an OVERLAPPING SPOT WELD.

Pulsing systems may be used on both alternating current and direct current. They may be installed as a plug-in module on some power supplies, or may be added to other units as an auxiliary control. When used on manual welding power supplies, the maximum current level is generally set using the remote amperage control with the power supply fine current potentiometer adjusted to the desired maximum current level. Low current level (background current) is set on the pulser. The foot control can also be used to start the arc, upslope to the desired current level and then as a downslope at the end of the weld to eliminate craters. Regardless of the type used, the system should have at a minimum, the following controls:

1. Cycle start.
2. Amplitude (maximum current).
3. Time at amplitude.
4. Background (low) current level.
5. Time at background current.
6. Downslope current decay.

Fig. 4-3 shows a typical auxiliary pulser unit designed for manual GTAW operation. Fig. 4-4 shows a pulser installed as a module unit in a GTAW power supply programmer.

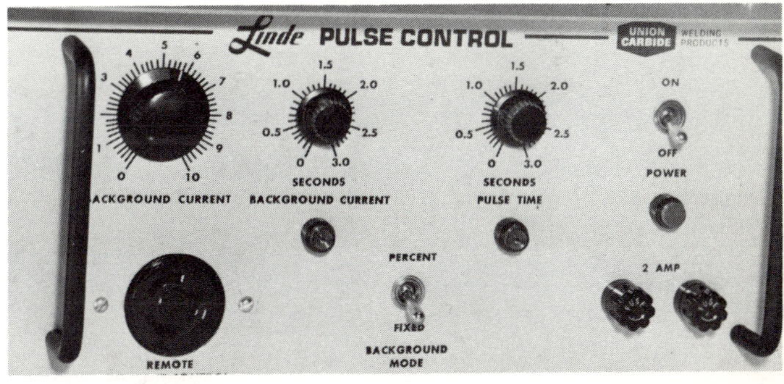


Fig. 4-3. Pulsers control the low (background) current level and the high current level time. The high current level is controlled by the remote control potentiometer setting. (Union Carbide Corp.)

Programs are established where the background current is fixed at a specified amperage, as shown in Fig. 4-5. The peak current may vary, but the background current will not. This type of program normally operates on materials other than aluminum and magnesium.

A program set up where the background current varies in a preset ratio to the peak current level changes

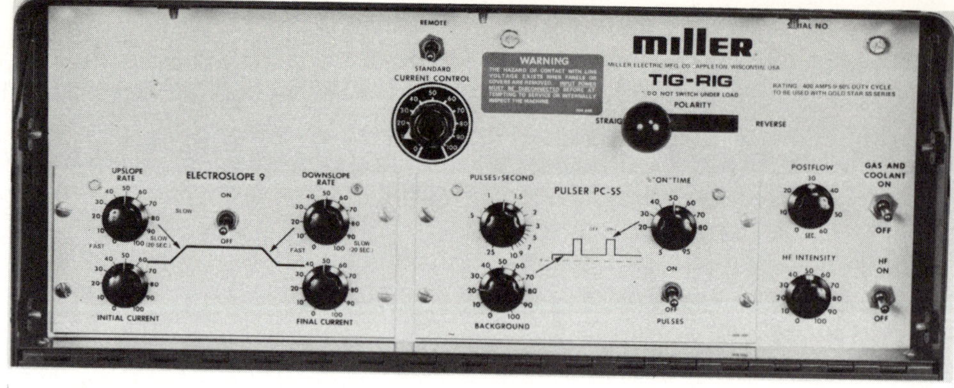


Fig. 4-4. This machine was designed with the circuitry required to accept a pulser and a sloper control module. (Miller Electric Co.)

is shown in Fig. 4-6. This type of program normally operates on aluminum and magnesium to give a rapid pulsing to agitate the molten metal rapidly for outgassing.

Automatic welding power supplies with weld current high pulsing features are being used in many applications. Generally, this type of pulser has additional controls such as pulsing of the filler wire and sequencers for programming the various functions into the proper sequence during the welding operation. Where high quality repetitive welds are mandatory, these systems are very effective. Constructed using solid state circuitry, operating on 110 volts AC, 60 Hz, they are relatively trouble free.

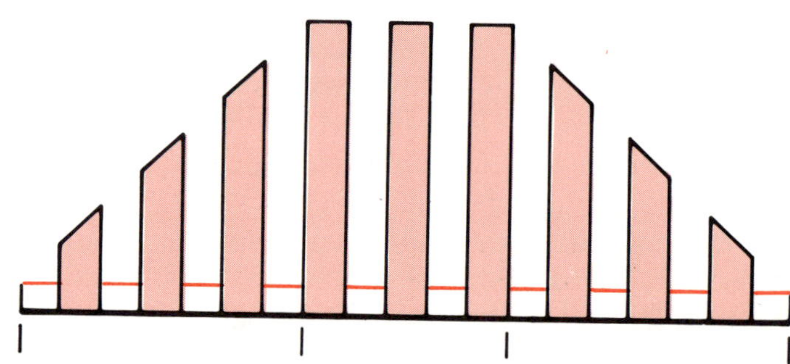


Fig. 4-5. Upper limit control by the operator allows changing maximum amperage levels to allow for thickness changes, gaps, etc. The background current level remains constant.

## Slope Controllers

SLOPE CONTROLLERS are used in both manual and automatic GTAW. They control the rise and fall of the welding current at the start and end of the weld operation. They may be called a SLOPER, PROGRAMMER, ELECTROSLOPE, or SEQUENCER. A typical sloper is shown in Fig. 4-4. The units are designed to perform the following functions:

1. Control the amount of initial current. Also may control this current level for the prescribed period of time.
2. Control the upslope time rate (initial weld current to weld current).
3. Control the length of main current welding time (optional).
4. Control the downslope time rate (welding current to final weld current).
5. Control the final weld current for a prescribed period of time.

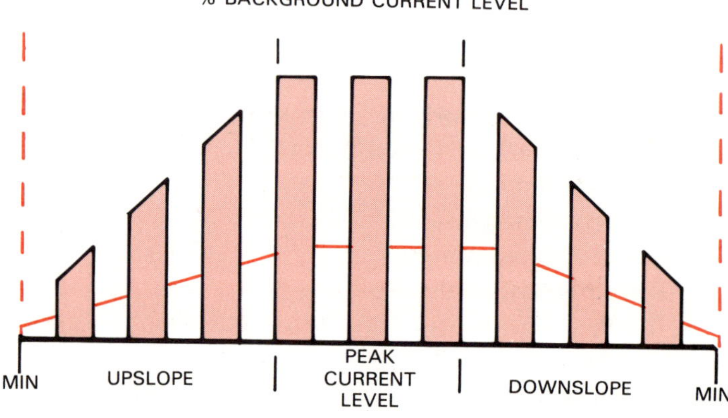


Fig. 4-6. Porosity (hydrogen gas) has the opportunity to rise from the melt during the pulse agitation, when the background current changes.

The INITIAL WELD CURRENT and the INITIAL TIME PERIOD is generally set at levels where burnthrough on thin materials is prevented. This also gives the welder time for location and final adjustment of the welding torch before full current is applied. Filler wire, if required, may be brought into the proper position at this time.

The UPSLOPE TIMER determines the period of elapsed time from the start (initial) current to the weld

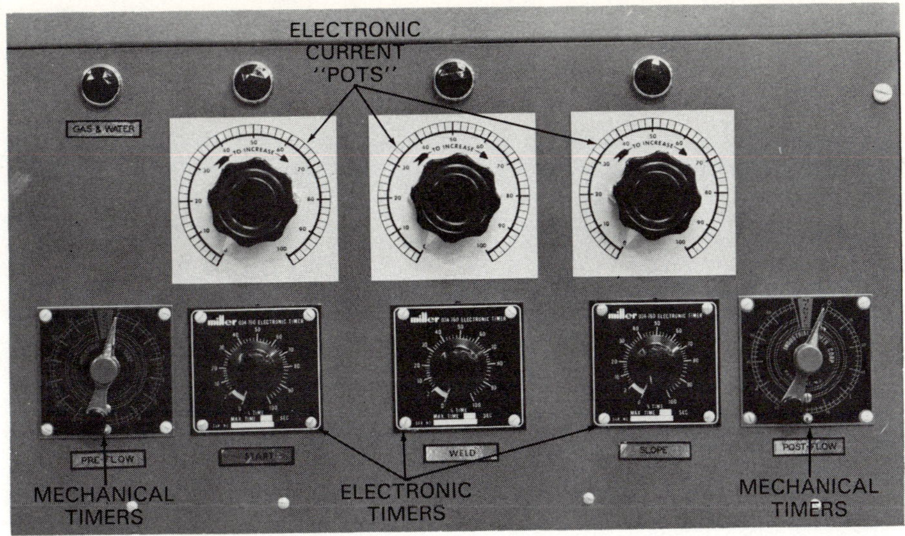


Fig. 4-7. Mechanical timers and electronic timers are both used on this timer panel. (Miller Electric Co.)

current. Note that during this time period, some welders obtain penetration through the base material. Filler wire, if required, is added when the upslope times out.

MAIN WELDING CURRENT TIMER is set for the length of time the weld current is desired.

The DOWNSLOPE TIMER determines the period of elapsed time from weld current to final current. This is called a TAPERING DOWN TIME and is set for a period of time for the weld wire to be stopped (if used) gradually.

FINAL WELD CURRENT is set at a low amperage to prevent craters at the end of the weld.

### Timers

Timers used in sequencing circuits fall into three types of categories:

1. ELECTRICAL-MECHANICAL: These units are electrically operated after mechanical setting for the period of time desired. When the circuit is activated, the timer will automatically reset to the time period desired and operate for the set length of time. Timers of this type are generally used to control the pre-flow and post-flow of gas and water. Fig. 4-7 shows a sequencer using mechanical and electronic timers to control gas and water flow.
2. ELECTRONIC-POTENTIOMETER: Commonly called "Pots," these units are used to control upslope times, weld times, downslope times. Time periods are set for seconds, minutes or cycles. They are simple to set and are easily adjustable through the entire range. When replacing these units, be sure to use an exact replacement as each type of potentiometer has a separate range of operation within the circuit.

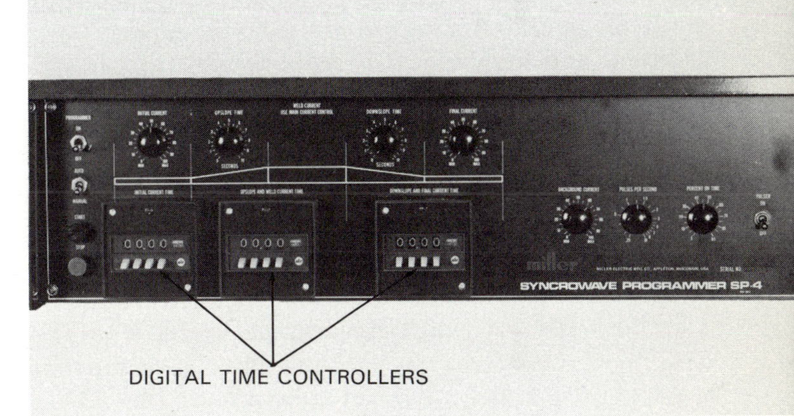


Fig. 4-8. Electronic timers and digital timers are both used on this programmer panel. (Miller Electric Co.)

In this style "Pot," the knobs can loosen and calibration or original setting may be lost. Usually, the "Pot" will overrun the numbers marked on the panel. If this happens after a procedure is set up, the original setting is often difficult to find. Calibrate the potentiometer in this manner:

A. Turn the potentiometer shaft as far left as possible or to minimum position with the knob removed.

B. Install and adjust the knob until the marker is aligned on the "O" mark or the first mark on the case.

C. Tighten the knob holding screw.

The potentiometer is now calibrated and should the knob come loose at a later date, follow the steps outlined above to recalibrate the unit.

Fig. 4-8 shows a programmer and pulser with electronic potentiometers. Note: This unit is designed to operate only with specific power supplies. The unit may be purchased as an accessory. However, it must be installed by an authorized technician for the warranty to be valid.

3. ELECTRONIC-DIGITAL: This timer is the newest addition to the field of weld time control. It has several advantages over the older types of timers. The main advantages include:
   A. Can be set to a desired time.
   B. The time is accurate.
   C. Can be seen as individual numbers.
   D. No calibration is required.
   E. The numbers can be reset by pressing the tabs until the proper number is showing.

   Digital time controllers are shown in Fig. 4-8. Fig. 4-9 shows a sequencer installed on a power supply.

Fig. 4-9. This sequencer (programmer) gives the operator complete control over the entire welding operation. (Miller Electric Co.)

**Remote Controls**

Control of welding sequences and welding current is often accomplished by remote controls, such as foot or hand operated switches and potentiometer. Foot controls and hand controls are designed for specific machines and in most cases are not usuable with other types and models. The basic types and uses of the foot control are:

1. Foot control with micro switch only.
   A. Start the high frequency only.
   B. Start high frequency, gas and water flow only.
   C. Energize the weld current contactor only.
2. Foot control with potentiometer only.
   A. Potentiometer is used to control weld current within the range established on the power supply.
3. Foot control with micro switch and potentiometer.
   A. Switch is used to start high frequency, gas and water flow, and energize weld current contactor. The potentiometer is then used to upslope to weld current level and down to final current level. Release of the foot control reactivates the micro switch and starts the post-flow gas and water circuit.

A basic foot control is shown in Fig. 4-10. This unit has two circuit cables connecting the micro switch for operation of the contactor and one for current control.

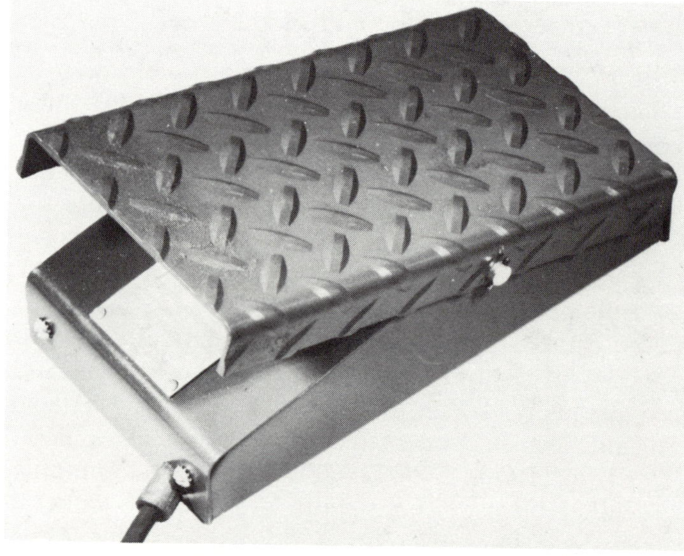

Fig. 4-10. Depressing this foot control energizes all of the welding sequence circuits. The operator raises or lowers the pedal for more or less welding current. Releasing the foot control energizes the afterflow circuits. (Lincoln Electric Co.)

Hand controls operate in the same manner as foot controls. They may include a micro switch, potentiometer, or both. Mounted on the torch for accessability, they can be located to the user's preference and are usually taped or bolted on. Some require constant pressure on the switch for continuous operation. Others may and can be released during welding, then reoperated to energize the post-flow operation. Fig. 4-1 shows a typical torch micro switch. In this operation, the switch is used to energize a high frequency operation. Fig. 4-11 shows a torch mounted control which contains a micro switch and a potentiometer. A hand control in operation is shown in Figs. 4-12 and 4-13.

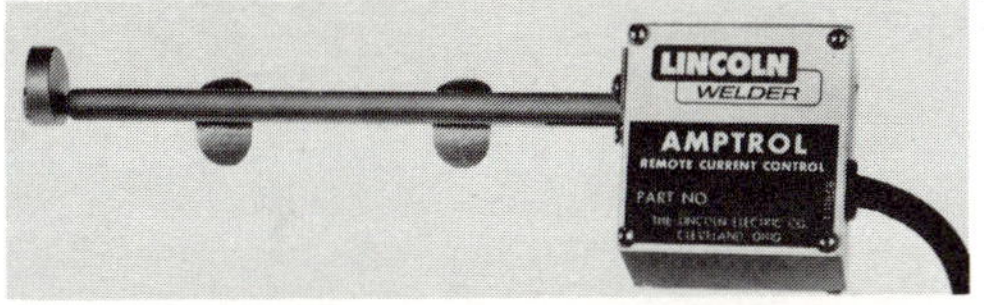


Fig. 4-11. Hand controls of this type operate by turning the thumb button on the end of the long shaft. The controls operate the same as a foot control.
(Lincoln Electric Co.)

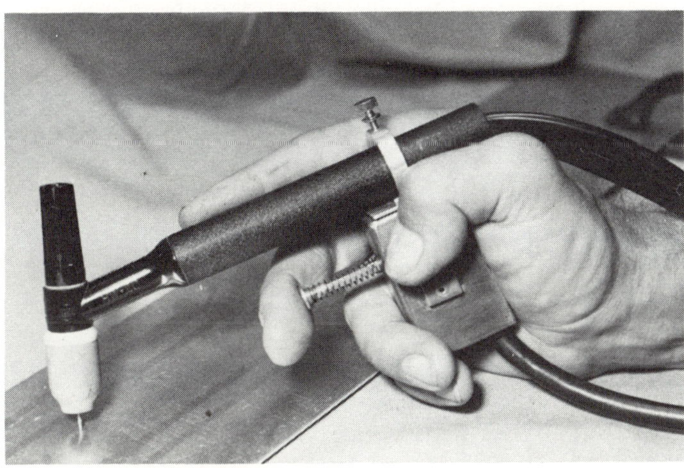

Fig. 4-12. This model of hand control operates in the same manner as the unit in Fig. 4-10, however, the unit operates by depressing a spring loaded shaft.
(J & K Welding Co.)

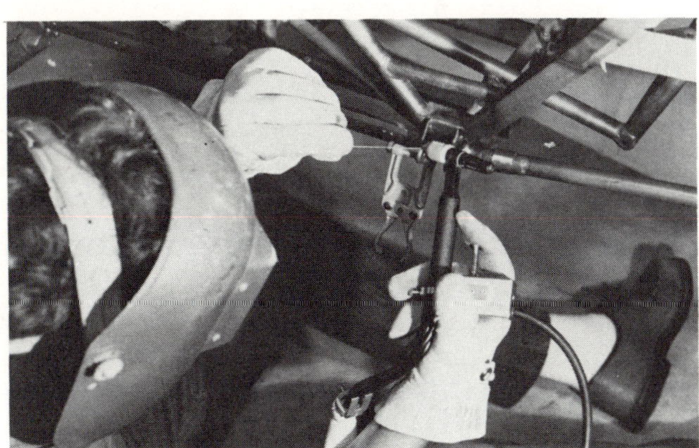

Fig. 4-13. In this operation, a hand control was selected because a foot control could not be used due to the position of the weld. (J & K Welding Co.)

The foot control is normally used when working on a table or in an area where the welder can sit down. Since one foot is required to operate the unit, out of position and long welds become very difficult as the welder tires from standing on one foot. For welds of this type, a hand operated control allows the welder more freedom of movement and still offers complete control of the welding operation.

## COLD WIRE FEEDERS

COLD WIRE FEEDERS are used in manual, semi, and fully automatic welding operations where filler wire application is required. They are adaptable to both hard (steels, stainless steel, etc.) and soft (aluminum, magnesium, etc.) wires. Various types of drive mechanisms and guides adapt the units to different diameters of filler wire. The equipment includes:

1. Filler wire (spooled) supply mounting bracket.
2. Controller to meter the amount of wire required.
3. Wire feeder drive motor and drive rollers.
4. Conduit assembly.
5. Wire guide manipulator.

Fig. 4-14 shows a complete wire supply and drive system ready for installation on a semi or automatic welding unit.

The wire feed drive controller, shown in Fig. 4-15, is motor powered by 110 volt AC. It may be installed for manual start and stop operation by the operator or for automatic sequencing during the weld operation.

Solid state controllers using digital readouts are able to control wire speed with reliability. Older units which do not have digital control may be updated by adding

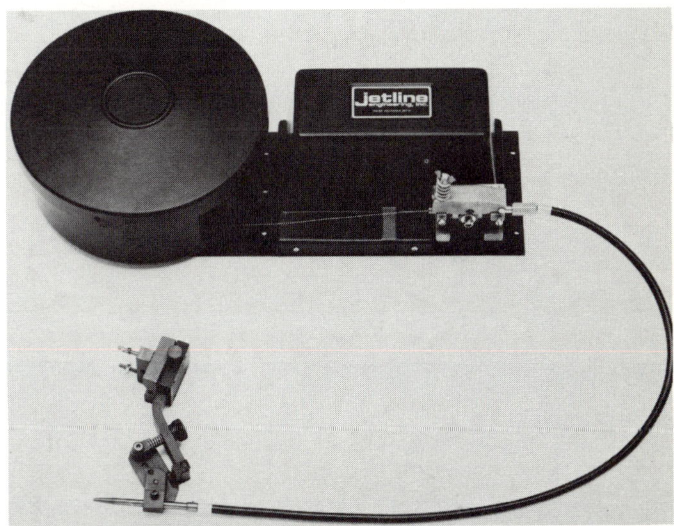


Fig. 4-14. Wire supply, drive motor, drive rollers, conduit assembly and wire manipulator assembled into a single package for installation.
(Jetline Engineering Co.)

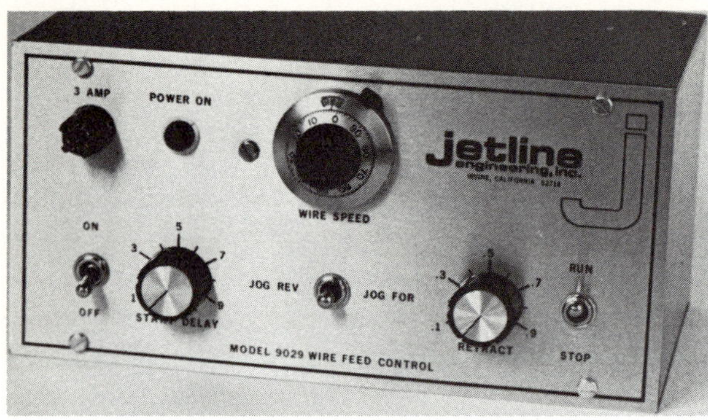


Fig. 4-15. Wire feed control has a retract timer to draw wire out of molten pool after welding ceases. Retraction of the wire prevents the wire from freezing into the weld metal. (Jetline Engineering Co.)

Fig. 4-17. Wire feeders may be used for manual or automatic feeding of the filler wire. All of the control and drive components are assembled in one unit. (Conley & Kleppen Co.)

a tachometer to give accurate inches per minute readout. Fig. 4-16 shows a filler wire drive unit using an auxiliary tachometer to register wire feed in inches per minute.

Another type of wire feeder, shown in Fig. 4-17, may be used for machine welding or adapted to manual welding. Drive rollers on the feeder shown in Fig. 4-17 are usually designed for one diameter of either a hard or soft wire. Hard wires use V-groove rollers while soft wires require U-groove rollers. This requires changing rollers when changing types or diameters of wire, as shown in Fig. 4-18. Conduits with varying diameter liners are used to feed the wire from the feeder to the

Fig. 4-16. Tachometers give instant readouts of wire feed in inches per minute. (Jetline Engineering Co.)

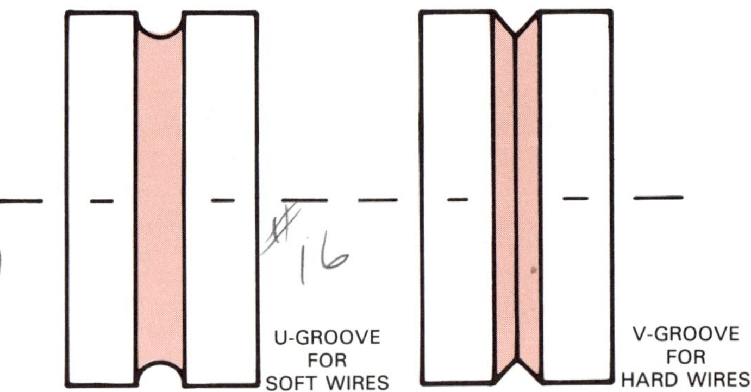


Fig. 4-18. Wire drive rollers have different design grooves for hard or soft wires. Using incorrect grooves causes wire feeding problems.

manipulator. Conduits require replacement when changing wire diameters or when worn. Conduits should not be installed with severe bends, as they cause the liner to wear very rapidly and they restrict the flow of wire.

Adjustment of the filler wire into the weld pool is accomplished during semi and automatic welding by turning screws in the manipulator. The manipulator is preset prior to starting the weld to feed the wire in the proper position. It is adjusted as required by the operator during the feeding sequence. A manipulator and the proper placement of the wire is shown in Fig. 4-19.

When wire feeders are used for manual welding, the wire manipulators are held by the welder, as shown in

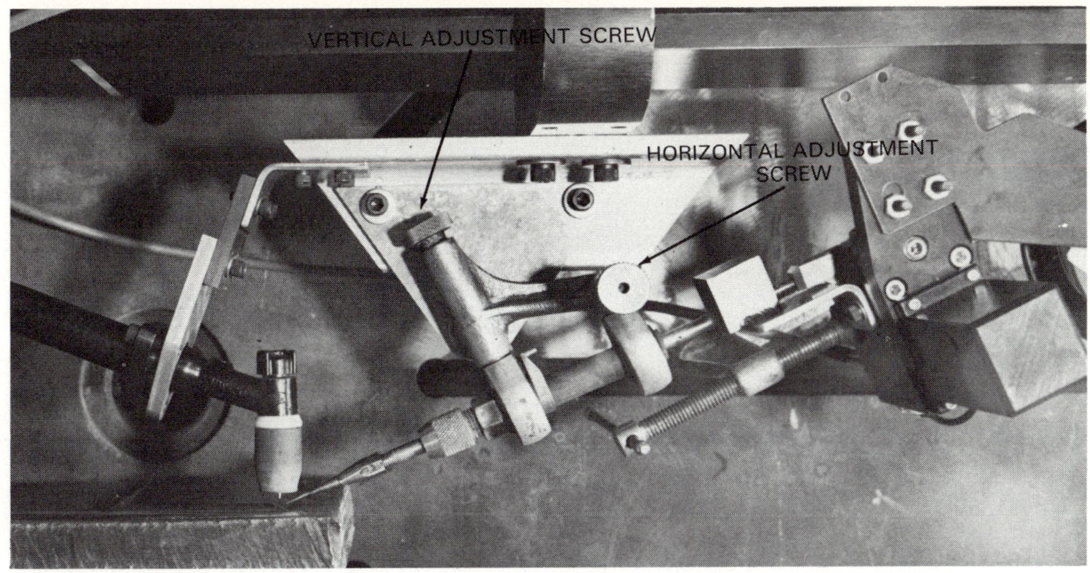


Fig. 4-19. Turning the adjusting screws on the wire manipulator adjusts the wire entry into the arc stream. (Airco)

Fig. 4-20. The wire feeder, in manual operation, is started and stopped by operating the button switch mounted on the manipulator body.

Wire tip guides in the manipulators are made of copper or an alloy material and fit a specific wire diameter. Tip guide exit holes wear considerably when using hard wires. Therefore, they must be checked often and replaced when worn.

## HOT WIRE FEEDERS

The basic HOT WIRE FEEDER SYSTEM is similar to the cold wire feeder system except that the wire is electrically resistance heated to the melting point prior to entering the weld pool. The system is used only on semi and automatic operations. The system has two major advantages over the cold wire feed unit:

1. High filler wire deposition rate. The increase in filler wire deposition rates lowers the cost of fabrication. This makes the system compatible with other high deposition processes such as GMAW (Gas Metal Arc Welding) and SAW (Submerged Arc Welding).
2. Excellent weld quality. By heating the filler wire to the melting point before inserting the wire into the molten pool, any surface contaminates, such as oil, grease and drawing lubricants, are driven off. As a result of the heating, the source of hydrogen porosity is removed and the completed welds are virtually porosity clear.

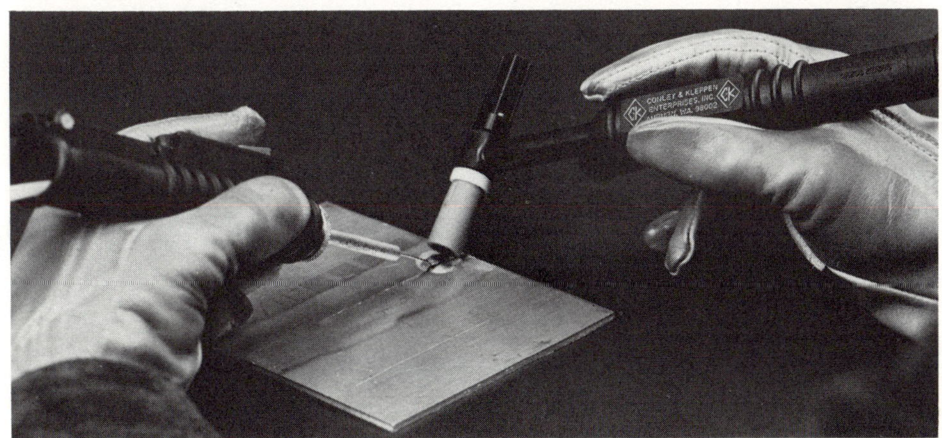

Fig. 4-20. The wire is fed into the puddle by energizing the drive motor on the wire feeder. This eliminates the welder feeding straight length wire as normally used in manual welding. (Conley & Kleppen Co.)

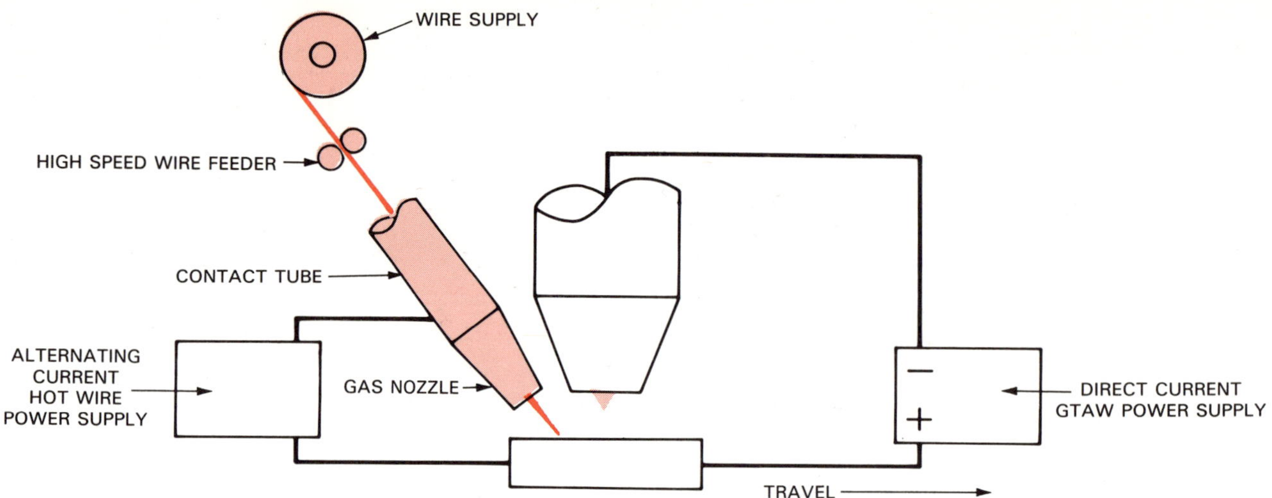


Fig. 4-21. Hot wire systems have a very high deposition rate. They are used only in semi and automatic welding systems.

Equipment for the hot wire feed system include:

1. High speed filler wire feeder.
2. Contact tube and holder.
3. AC constant potential power supply for melting wire. (Direct current power supply cannot be used.)

A basic hot wire system is shown in Fig. 4-21.

Operation of the system requires feeding the wire through a contact tube for resistance heating as the wire contacts the workpiece and melts. The speed of the wire and the amount of heat required to melt the wire is controlled by an AC power source with a single control. As the operator increases the wire speed, the electrical current is automatically increased. Lowering the wire speed automatically lowers the melting current.

With this hot wire feed system, welding current is not used to melt the filler wire as in cold wire feed.

Fig. 4-22. This automatic welding system contains hot wire system (the power supply for the system is not shown). Also included is an automatic voltage control head, oscillator, and automatic carriage for longseam travel. (Union Carbide Corp.)

The arc energy is used only to melt the parent metal resulting in a highly efficient arc pattern. As the filler wire enters the rear of the molten pool (cold wire enters from front), the entire composition of the filler metal is fed directly into the weld. (In cold wire feed some elements of the wire may burn-out by passing under the arc stream.)

Gas nozzles feed an inert gas supply to protect the heated wire as it passes through the space from the contact tube to the weld pool, thus, protecting the wire from atmospheric contamination.

The systems are normally used for welding of carbon steel, low alloy steels, stainless steels, copper alloys, and nickel. The use of this system is not recommended for copper and aluminum as the low resistance of these filler wires requires high heating currents which result in arc blow and uneven melting. Fig. 4-22 shows an automatic GTAW welding system with a filler wire drive unit and the hot wire unit attached to a machine welding torch.

## OSCILLATORS

OSCILLATORS are used on semi and automatic torches to control the motion of the welding arc and the molten pool. The advantages of oscillation include:

1. Control heat distribution and reduce weld shrinkage.
2. Minimize undercut.
3. Reduce porosity.
4. Improve penetration.
5. Improve overall integrity of the weld.
6. Improve weld contour.

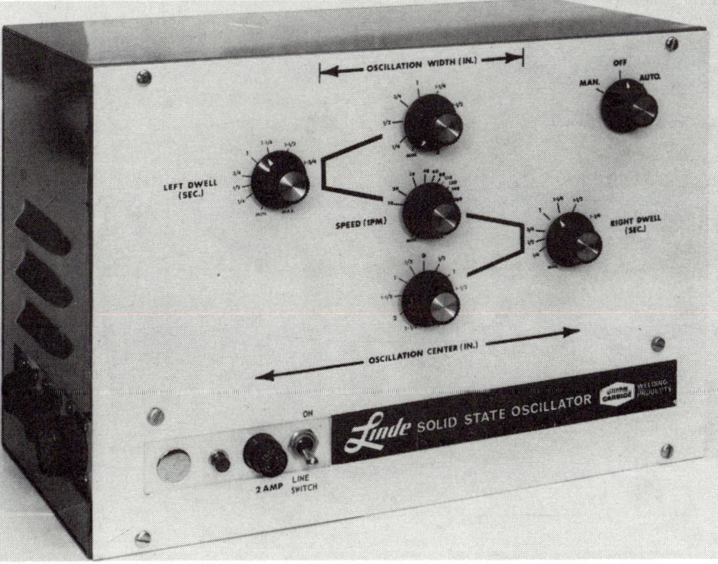


Fig. 4-23. Oscillator controls offer many types of oscillation patterns. (Union Carbide Corp.)

Oscillators may be used on all types of materials and may be either mechanical or magnetic. The MECHANICAL TYPE OSCILLATORS operate with a solid state controller as shown in Fig. 4-23. Step motors and drive screws oscillate the welding torch across the weld area. Fig. 4-22 shows an oscillator mounted on an automatic welding system.

MAGNETIC TYPE OSCILLATORS operate by generating a controlled magnetic field around the welding arc. This direct current magnetic field interacts with and effects the arc itself. Fig. 4-24 shows a magnetic oscillator and a solid state controller for controlling the oscillation program. A weld with and without oscillation is shown in Fig. 4-25.

Fig. 4-26 shows some of the areas where oscillation is used and the various types of oscillation patterns. Controllers operate on 110 volt AC and potentiometers are either solid state electronic or digital for control.

Mechanical oscillators require a movable platform on which the weld torch is mounted. This requires a rigid base for the equipment.

Magnetic oscillators move only the arc itself. The only additional equipment to be mounted directly on the welding torch is the probe. Heavy duty magnetic probes must be water cooled.

## AUTOMATIC VOLTAGE CONTROL SYSTEMS

AUTOMATIC VOLTAGE CONTROL SYSTEMS are commonly called AVC systems. They operate on semi and automatic welding operations to control the arc voltage (tungsten to work gap) to a preset condition. Normal operation is on DCSP (Direct Current Straight Polarity), however they can be used on AC with a special converter. The equipment includes:

1. Control unit.
2. Drive unit on which the machine welding torch is mounted.

Fig. 4-27 shows an AVC head and controller.

An AVC head is a must for high quality welds where arc voltage is a critical factor, and tooling or driving mechanisms cannot be set with precision. The major advantages of the system are:

1. Welding can be done at higher speeds.
2. Welds can be made on uneven surfaces.
3. Set-up time is minimized.
4. Less operator skill is involved.
5. Electrode deterioration is automatically compensated.
6. High quality welds can be obtained.

Another advantage of the system is that when using cold wire feed addition, the operator can set the wire feed manipulator to the desired height and the wire will stay in the proper relationship to the tungsten and the molten pool. With manual adjustment of the voltage, changes of the arc length cause variations in the wire

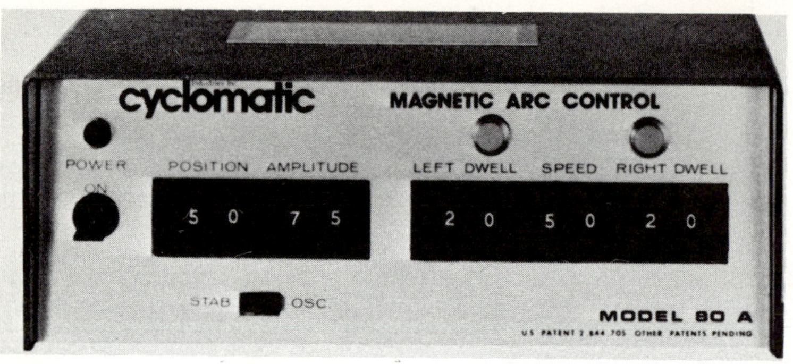


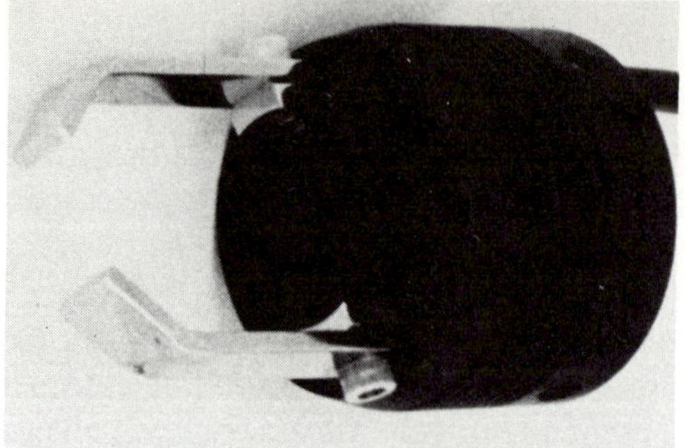

Fig. 4-24. Magnetic oscillators attach directly to the welding torch. The probe emits a magnetic field established on the programmer to deflect the arc to the desired pattern. (Cyclomatic Ind., Inc.)

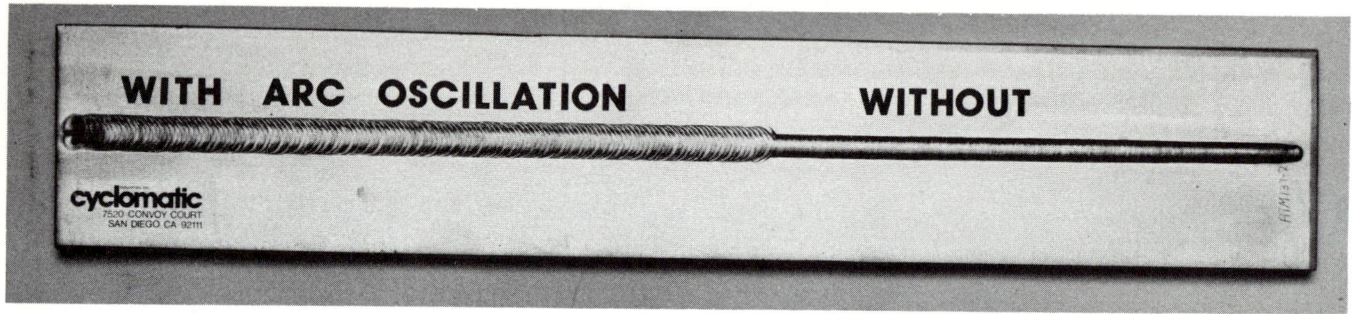


Fig. 4-25. Oscillators flow weld metal to improve surface contours. (Cyclomatic Industries, Inc.)

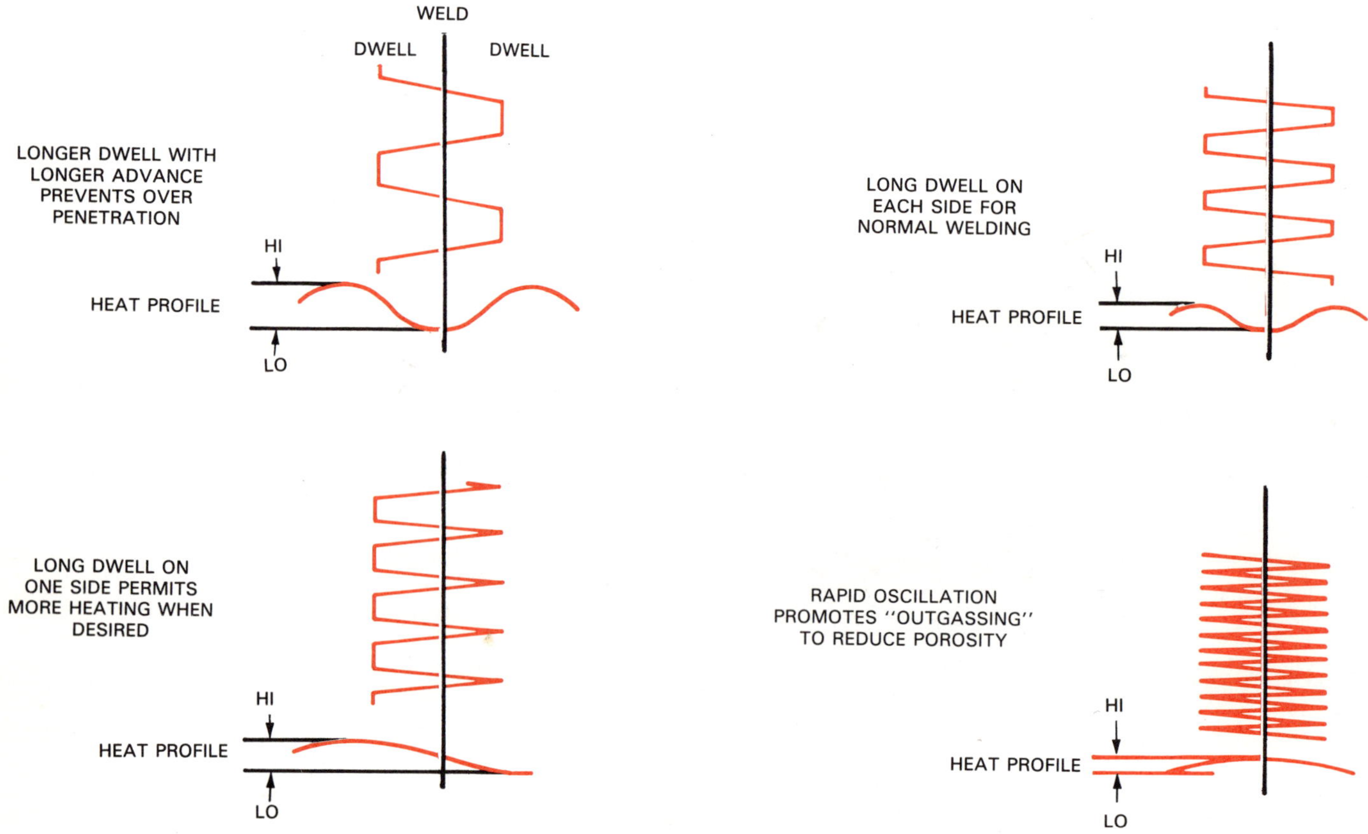


Fig. 4-26. Oscillation patterns are designed to achieve specific results.

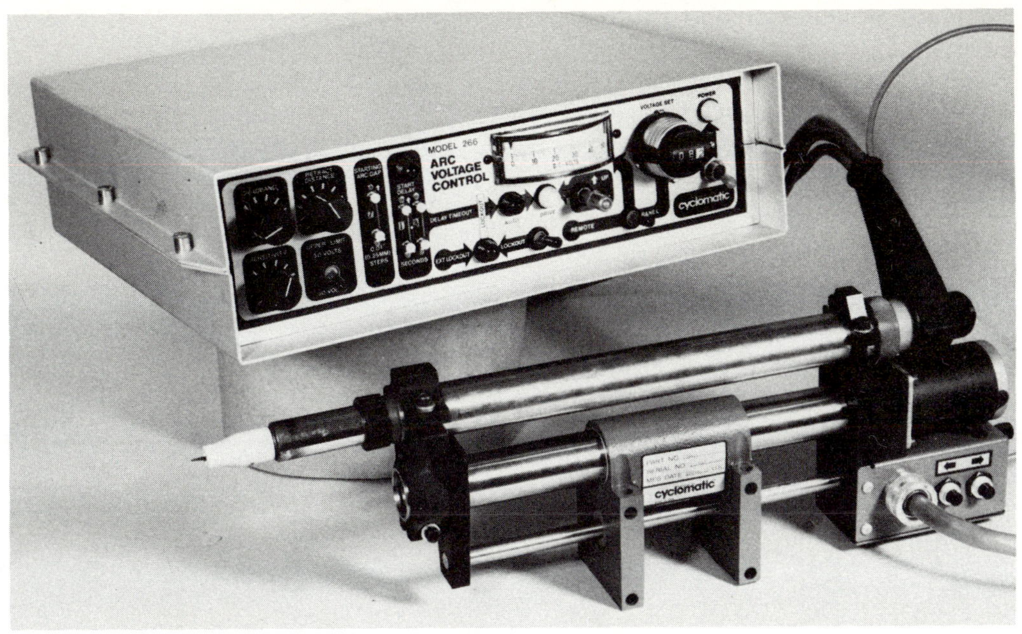


Fig. 4-27. Automatic voltage control system, drive unit, and a machine welding torch mounted on the drive unit. (Cyclomatic Ind., Inc.)

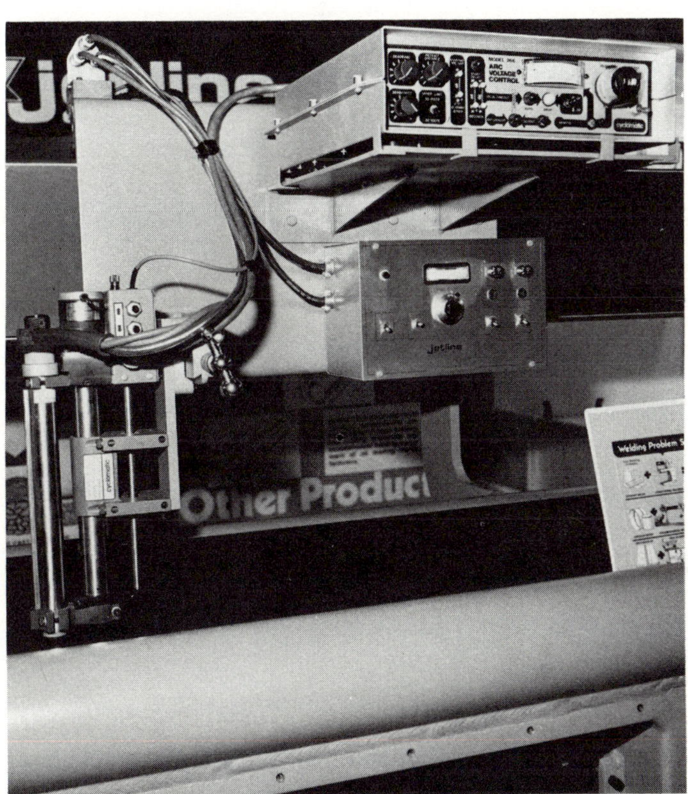


Fig. 4-28. The automatic voltage control system maintains a preset arc voltage regardless of the weld joint contour. (Cyclomatic Industries, Inc.)

feed into the molten pool. This requires constant attention to voltage fluctuations and wire feed manipulator changes.

Operation of the system requires:

1. Setting a reference arc voltage on the controller.
2. Set up of the drive unit sensitivity regarding response time for changing the arc length.

The major sequence of events in an AVC welding operation include:

1. Start weld operation.
2. Drive motor moves the head to the work at a preselected speed.
3. Tungsten touches workpiece and retracts.
4. High frequency fires and starts arc.
5. Voltage control sets the tungsten to work gap and maintains preset voltage during weld operation.
6. Stop weld operation.
7. Drive motor automatically retracts a set distance after arc goes out.

The control unit and drive unit operate on a 110 volt power. They are usually solid state and are relatively trouble free. Fig. 4-28 shows an AVC head and controller mounted on a seamwelder.

Fig. 4-22 shows an AVC and controller mounted on an automatic welding system.

## GTA SPOT WELDING SYSTEMS

GTA spot welding may be used for tackwelding components together for assembly of parts, or as the main fabrication welding process. The process may be done manually, semi, or automatic depending on the application. Simple systems employ a special torch, a timer and a power supply. Semi and automatic systems require additional equipment to that noted above for control of welding current and program sequence. The process may be done on all types of materials, however it is limited as to the thickness which may be welded. A typical GTA spot weld application is shown in Fig. 4-29.

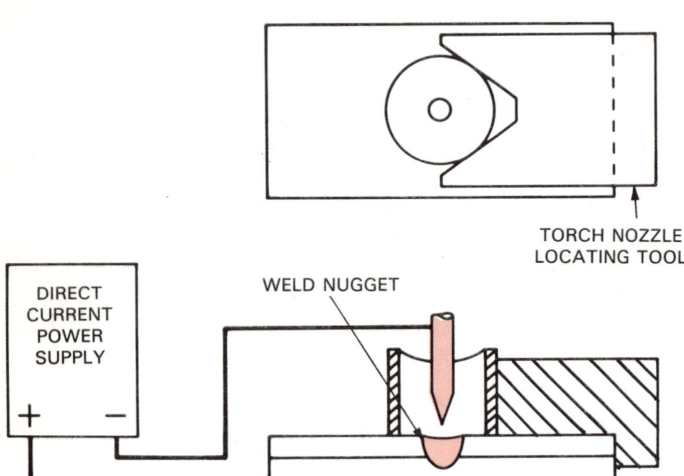


Fig. 4-29. Gas tungsten arc spot welding requires that the two pieces to be welded must be in intimate contact with each other. An air gap will reduce the heat flow and a nugget will not form.

The type of equipment used will be governed by the number of weld nuggets desired and the quality of the spot weld. Fig. 4-30 shows a typical manual GTA spot weld torch and controller. The torch may have a tungsten that is fixed in relation to the nozzle to work distance, or it may have a mechanism which automatically moves the tungsten to the workpiece until the arc is started. The tungsten then automatically retracts to the preset arc gap. Placement of the tungsten is critical as varying the height of the tungsten will change the amount of weld amperage.

A simple system with the required controls is shown in Fig. 4-31 and includes:

1. Power supply.
2. GTA spot welding torch.
3. Timer control.
4. Inert gas supply.
5. Controls for gas, water flow and high frequency.

The type of program produced by this system is diagramed in Fig. 4-31.

Complex systems required for many materials is shown in Fig. 4-32 and include:

1. Power supply.
2. GTA spot welding torch.
3. Pulser.
4. Sloper.

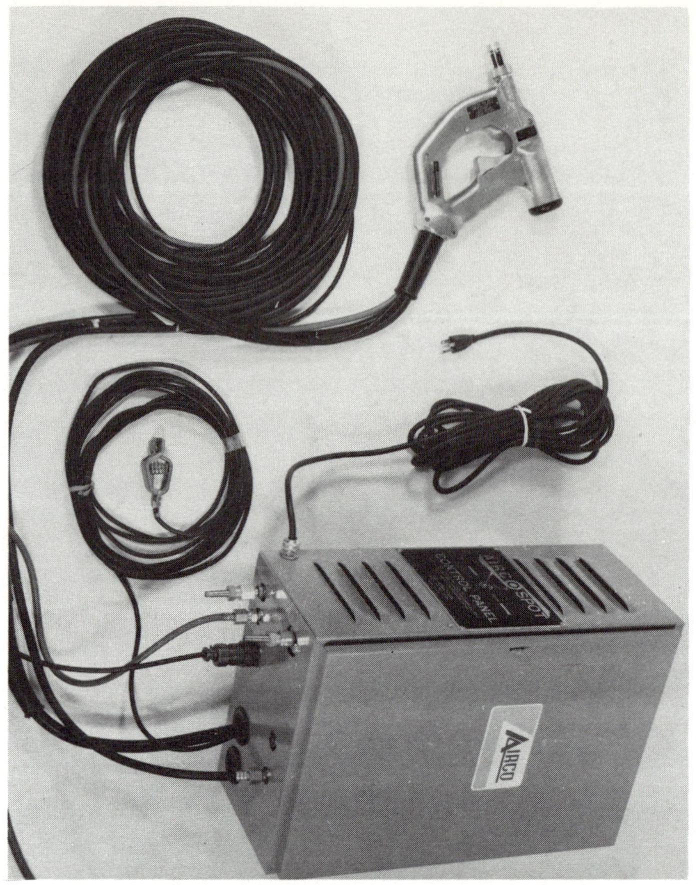

Fig. 4-30. The gas tungsten arc spot welding torch is made so that slight pressure may be applied to hold the proper alignment with the base material. The gas nozzle is slotted so that the argon gas envelope may vent to the air. (Airco)

5. Cycle timer.
6. Controls for gas, water flow and high frequency.
7. Pilot arc (if used).
8. Inert gas supply.

The type of program produced by this sytem is diagramed in Fig. 4-32.

The pilot arc system may not be required if the power supply has a high frequency unit or if an auxiliary unit is in the system. However, in some cases where a long arc voltage is used, the high frequency spark may not be dependable.

A PILOT ARC SYSTEM is used to provide positive arc starts. Since the tungsten is hot, electrons are flowing in the arc zone and the main arc is immediately established when the power circuit is energized.

Operation of the spot welding system consists of energizing the proper circuits for a set period of time with the correct parameters to form a weld nugget with the desired properties. To establish the correct procedure, test must first be made on material simulating the parts to be welded. Each test is then destroyed by

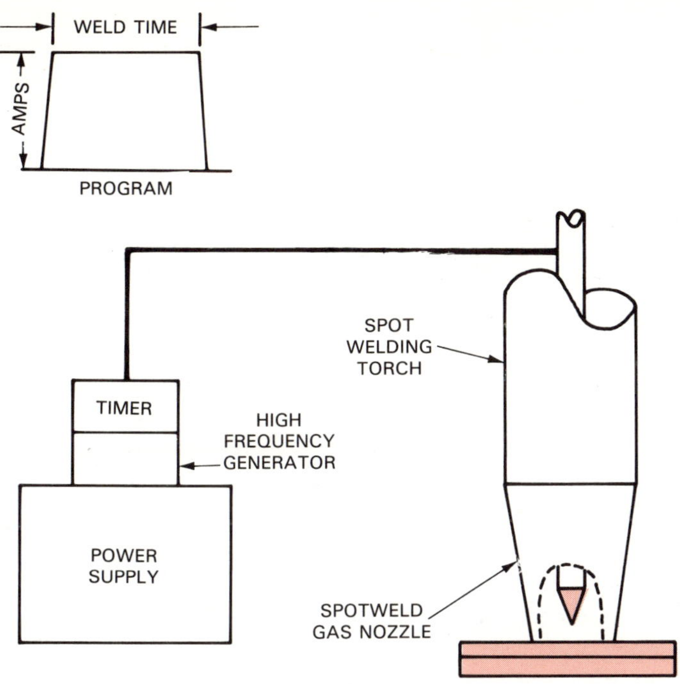


Fig. 4-31. Simple gas tungsten arc spot welding systems are often used for tackwelding assembly of parts.

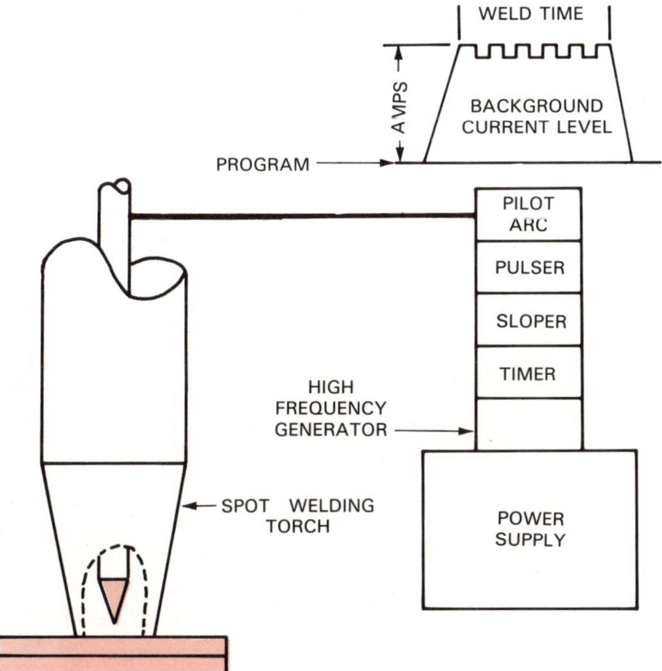


Fig. 4-32. Rigid welding specifications require complex systems of spot welding for reliability.

tensile test and/or macro (cross section of the weld) test to determine the nugget quality. The major areas of inspection of the nugget quality include:

1. Tensile strength.
2. Depth of penetration into the lower piece of metal.
3. Diameter of the nugget width.
4. Surface crater depression and crater cracks.

Areas that affect spot weld quality include:

1. Torch nozzle must be flat against the workpiece.
2. The thinnest material must always be on the top.
3. The material must fit together without a gap.

The problem areas in GTA spot welding include:

1. Tungsten placement in the torch is a major parameter and affects both amperage and voltage. The tungsten must be precision ground to a predetermined taper and point dimension. The tungsten must be set into the torch with a gauge to an exact height to the nozzle tip. Fig. 4-33 shows a torch with a tungsten presetting tool in place.
2. Tungsten deterioration will occur as the number of spot welds progress. Metallic vapors, spatter, etc., will cause the tungsten shape to change which requires frequent inspections of the tungsten condition.
3. Nozzles that are not water cooled may break down after a number of welds. They should be checked often and cleaned of spatter as required. Nozzles that are cracked or beyond cleaning should be replaced.

## TUBE TO TUBE WELDING

Tube welding systems are designed to weld around tube material while the torch is stationary. With systems

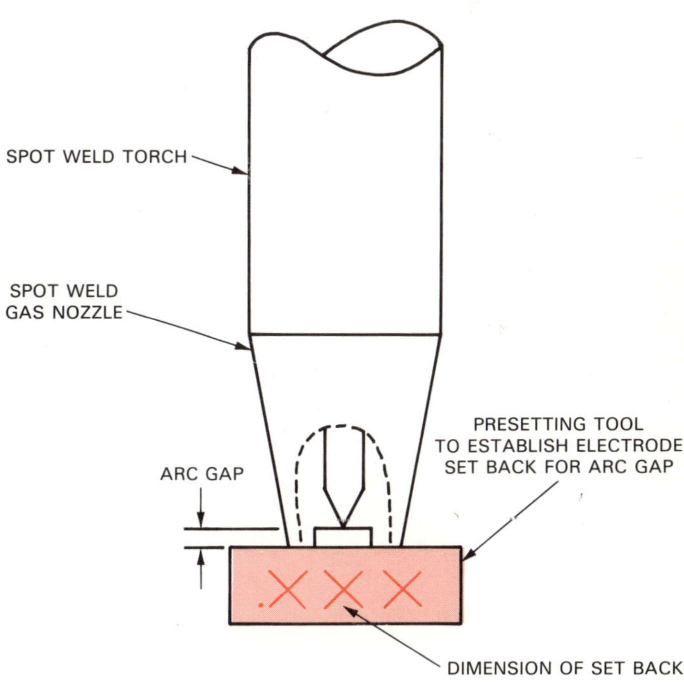


Fig. 4-33. Tungstens can be set in a minimum amount of time with a presetting tool.

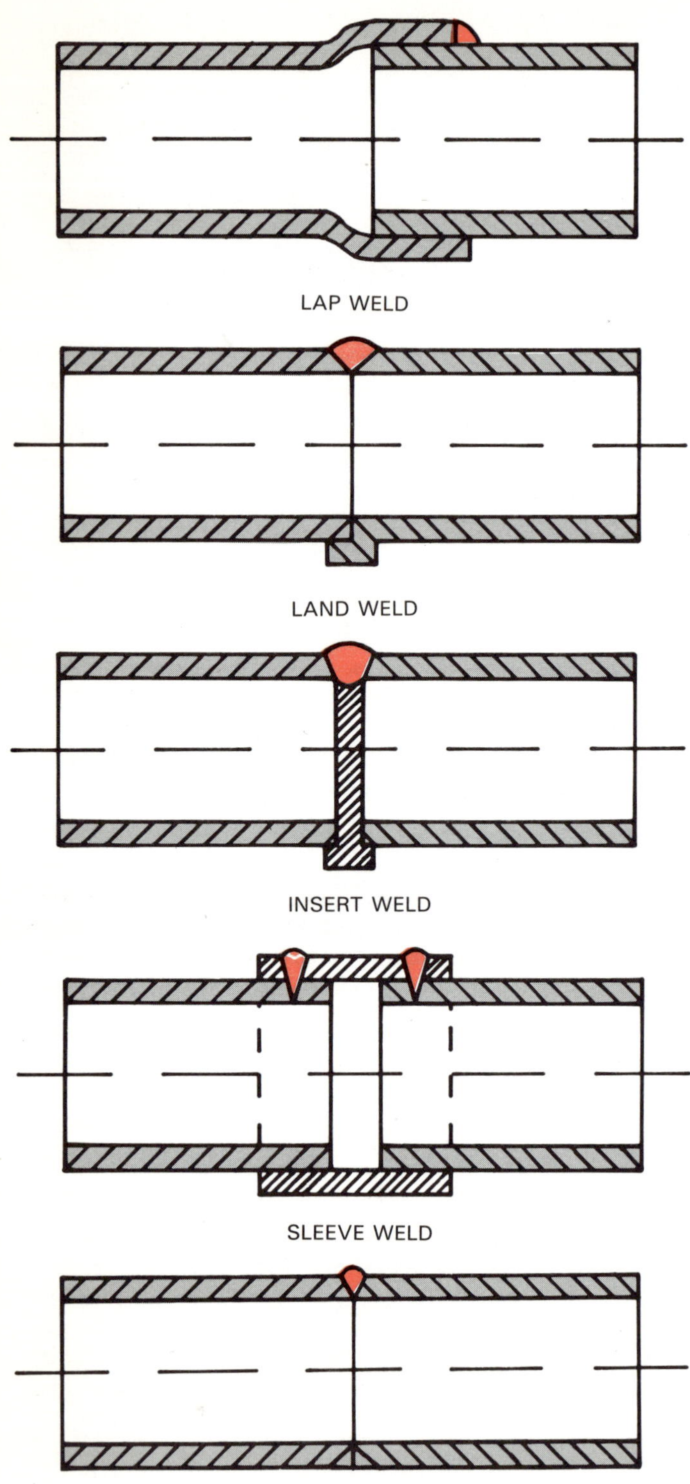


Fig. 4-34. Various types of tube weld designs may be used with tube welding heads.

of this design, the material can be prepared, assembled, and welded in place without turning equipment. Material can be repaired in place by cutting out a section. Replacing sections or components of new configurations can be added without completely removing the old system. In many areas of construction, mechanical connections cannot be tolerated and welding must be accomplished regardless of the location of the weld joint.

The systems operate on all types and varying thicknesses of material, however each weld must be assembled to a specific tolerance. Welding parameters must be established in the weld programmer prior to operation.

Some of the typical joint designs for tubing welds are shown in Fig. 4-34. Fig. 4-35 shows an actual tube weld being made in place in an aircraft.

The equipment designed for tube welding is made into systems for a specific task. In many cases, the equipment is interchangeable for various sizes of tubing and can be quickly adapted without major changes. A block diagram of a tube welding system is shown in Fig. 4-36.

Major equipment components for tube welding include; power supply and programmer, shown in Fig. 4-37 and welding head, shown in Fig. 4-38.

To accomplish the weld cycle:

1. Weld parameters are entered into the programmer.
2. The torch is mounted over the tube weld joint.
3. Welding current levels are established in the power supply.
4. Starting the cycle automatically sequences all of the events in the programmer and power supply required for the weld.
5. The unit shuts down when the sequencer times out.

Fig. 4-35. This tube welder is making a weld in a confined area. Since the tungsten revolves around the tube for welding (the torch does not move), the unit is very useful for welding in restricted areas. (Astro-Arc Co.)

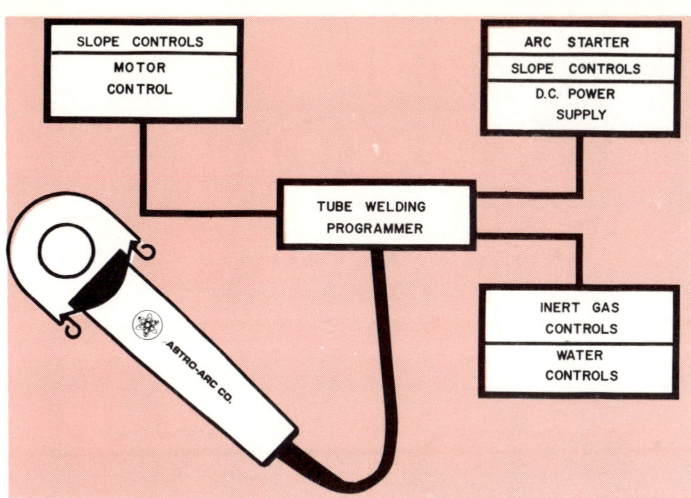


Fig. 4-36. Equipment required for a tube welding system is shown in the block diagram. (Astro-Arc Co.)

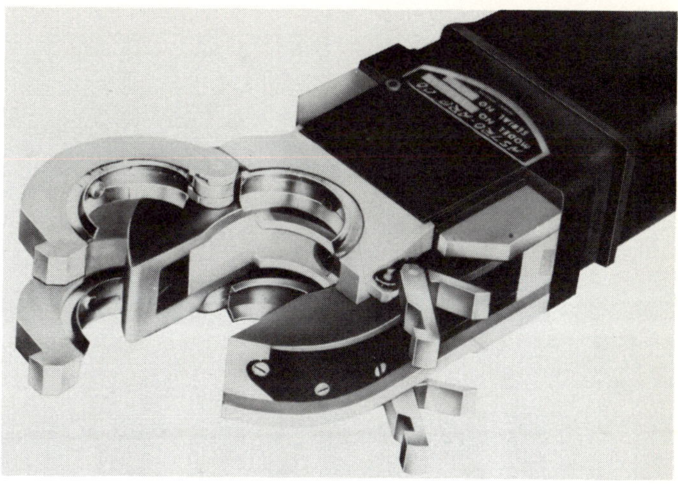

Fig. 4-38. The welding torch clamps to the tube for welding. The tungsten is rotated around the tube by a drive motor located in the handle. This unit is water cooled for heavy duty work. (Astro-Arc Co.)

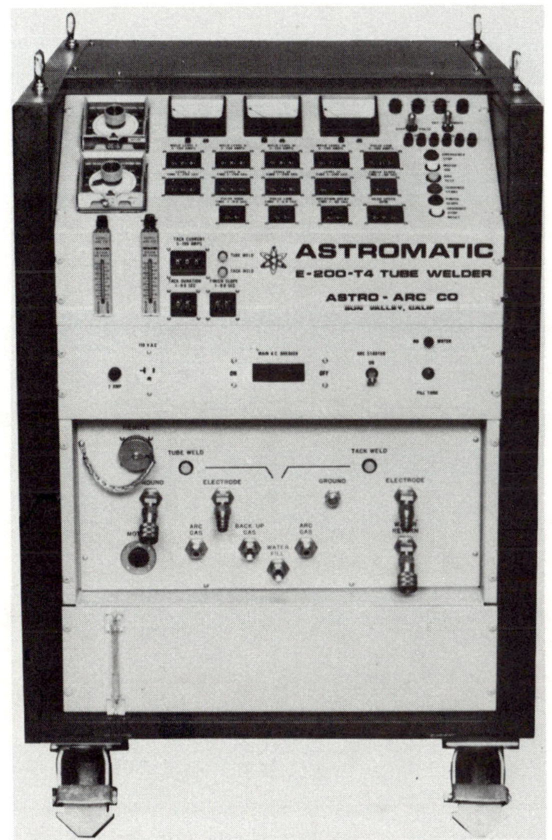


Fig. 4-37. Tube welding power supply contains a power source and programmer for sequencing the entire operation. (Astro-Arc Co.)

## TUBE TO TUBE SHEET WELDING

Boilers, condensers and heat exchangers require the welding of tube to tube sheet as the major method of fabrication. As each tube weld is critical in thc opcration of the unit, each weld must be made with the highest quality possible. Weld joint design and machining tolerances must be established to accomplish this task. The equipment must be capable of making very high quality welds with 100 percent repeatability. Fig. 4-39 shows some common tube to tube sheet weld designs.

Power supplies shown in Fig. 4-37 contains all the needed controls for making tube to tube sheet welds. The welding head used for this operation is shown in Fig. 4-40. The two types of tube to tube sheet welds shown in Fig. 4-41 have been sectioned after welding to determine the weld quality.

## PIPE TO PIPE WELDING

The weld joint design and the equipment for a pipe welding system is much more complex than tube welding as the thickness of the weld joint changes. Several pipe weld joint designs are shown in Fig. 4-42. In addition, the weld is rotated around the pipe in a 360 degree rotation. The molten puddle must be controlled to prevent sagging non-uniform contour. To prevent sagging and to provide uniform contours, AVC (automatic voltage control) torches to maintain arc gap, pulsers, and an oscillator with dwell control are used. The pulsers may control the pulse of the current, the wire feed, and the travel speed. They may be used in any combination in the system.

The major equipment components for pipe welding include: power supply and programmer, shown in Fig. 4-43 and the welding head shown in Fig. 4-44.

Computer numerical control techniques (CNCT) are

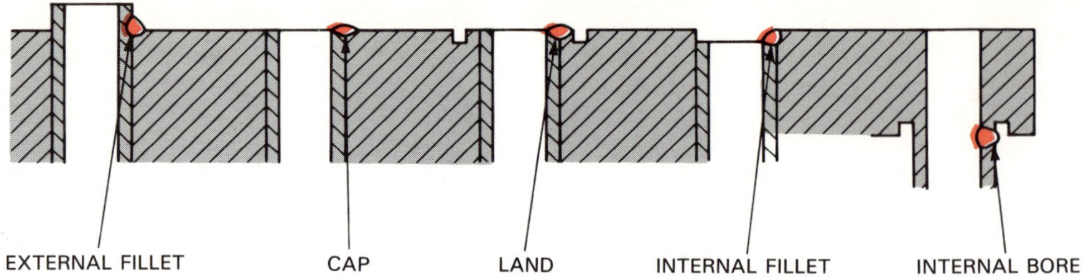


Fig. 4-39. Tube to tube sheet weld joint designs.

Fig. 4-40. Tube to tube steel welding head can be set up to weld internal or external weld joints. (Astro-Arc Co.)

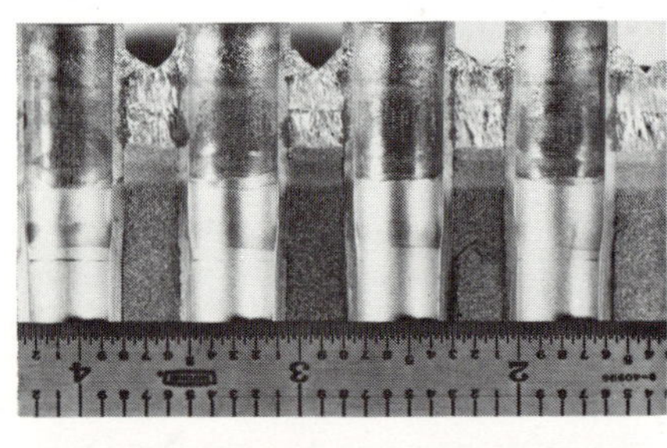

Fig. 4-41. Above — Fillet welds were used to join the extended tubes to the tube sheet. Below — This joint design extended the tube slightly above the tube sheet. The extended edge, when melted, became the filler material for the weld crown.

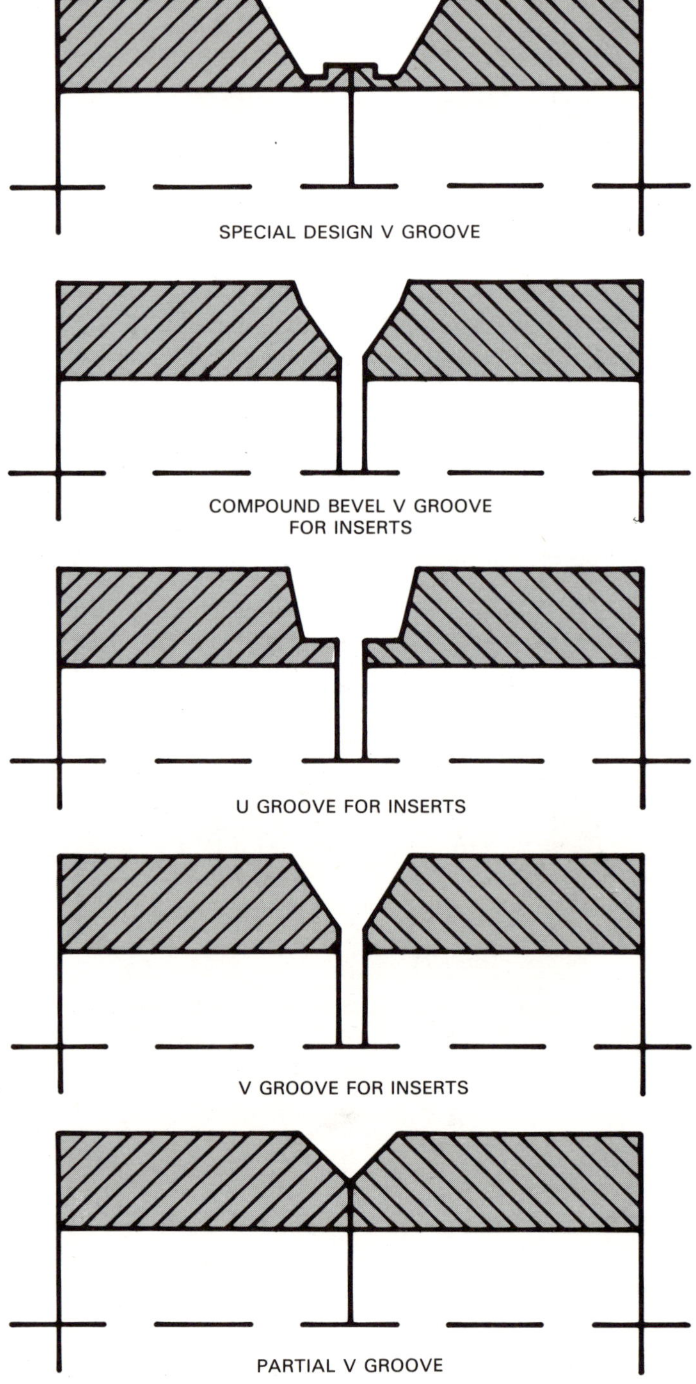


Fig. 4-42. Standard pipe weld joint designs.

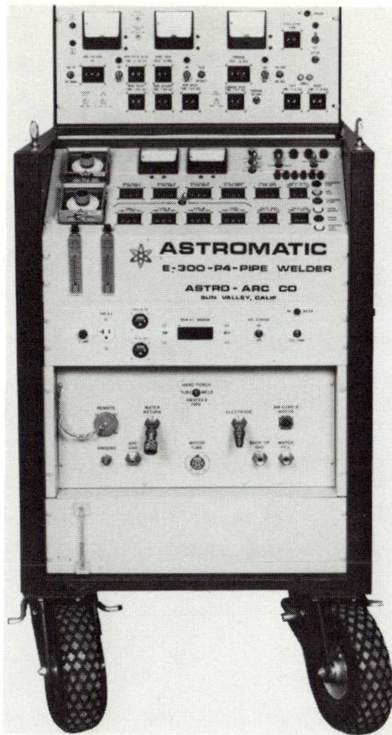


Fig. 4-43. Pipe welding power supplies require a higher duty cycle than tube weld power supplies as they carry higher amperages for longer periods of time. The programmer also controls more functions such as wire feeding, oscillators, and pulsers. (Astro-Arc Co.)

often added to control of all the welding variables. These computer controls will:

1. Improve weld quality.
2. Improve reliability.
3. Reduce material requirements.
4. Increase productivity and adaptability.
5. Simplify equipment operation.

A complete pipe welding system with computer numerical control is shown in Fig. 4-45. Step pulsing of filler wire, welding current and oscillation in combination during the welding is diagramed in Fig. 4-46.

With the improved designs of computers, programmers, and sequencers, welding variables are controlling every aspect of the weld operation. However, the most modern and expensive equipment made cannot overcome poor fit up of the detail parts. The system may have all of the latest technology available, but it will only operate to include a specific tolerance. To expect the system to operate beyond these tolerances is very costly. When making procedures for tube and piping welds, always use the extreme tolerances that will be encountered on the actual production part. Procedures established under these conditions can then be modified for the fit up condition below these tolerances.

Fig. 4-44. Pipe welding head travels around the pipe. All of the equipment required to align and move the torch in the pipe groove and the wire supply is contained in this unit. (Astro-Arc Co.)

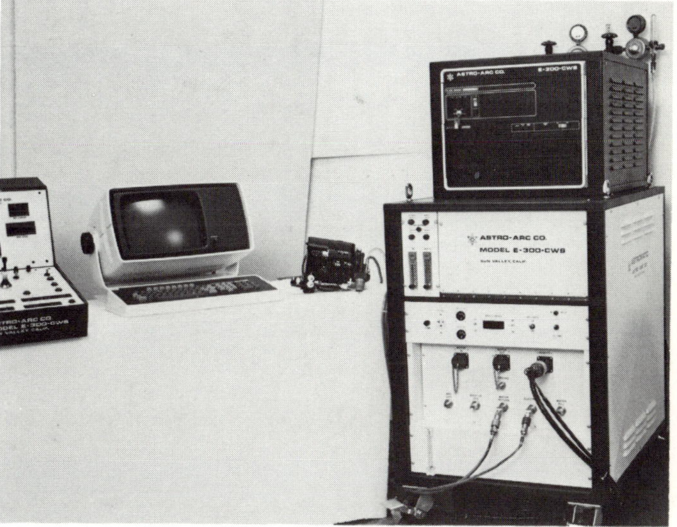


Fig. 4-45. A computer numerical control pipe welding system can be used to make the entire weld automatically. (Astro-Arc Co.)

## GTAW TORCHES

Electrode holders are commonly called TORCHES. They are available in many styles and types for many varied GTAW welding tasks. The basic torch functions are:

1. To grip and hold the electrode.
2. To provide electrical current to the electrode.
3. To conduct heat away from the electrode.
4. To provide inert gas to the weld area.
5. To protect the welder from heat and electrical shock.

The torches are made from high-conductivity copper or brass. The internal connections are silver brazed. The outer insulation covering is either phenolic or silicone rubber. Torches are designed for either manual or machine welding, and may be gas or water cooled.

All GTAW torches are rated for MAXIMUM CONTINUOUS AMPERAGE CAPACITY (DUTY CYCLE). Manufacturers have designed accessories such as valves, nozzles, and adapter accessories for each type. Accessories from one manufacture may not be adaptable to another manufacturer's torch.

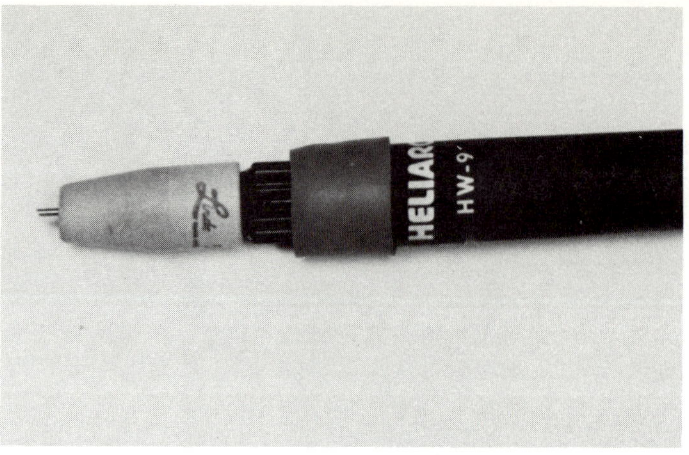

Fig. 4-47. Pencil type gas cooled manual torch used for light duty work. (Linde Co.)

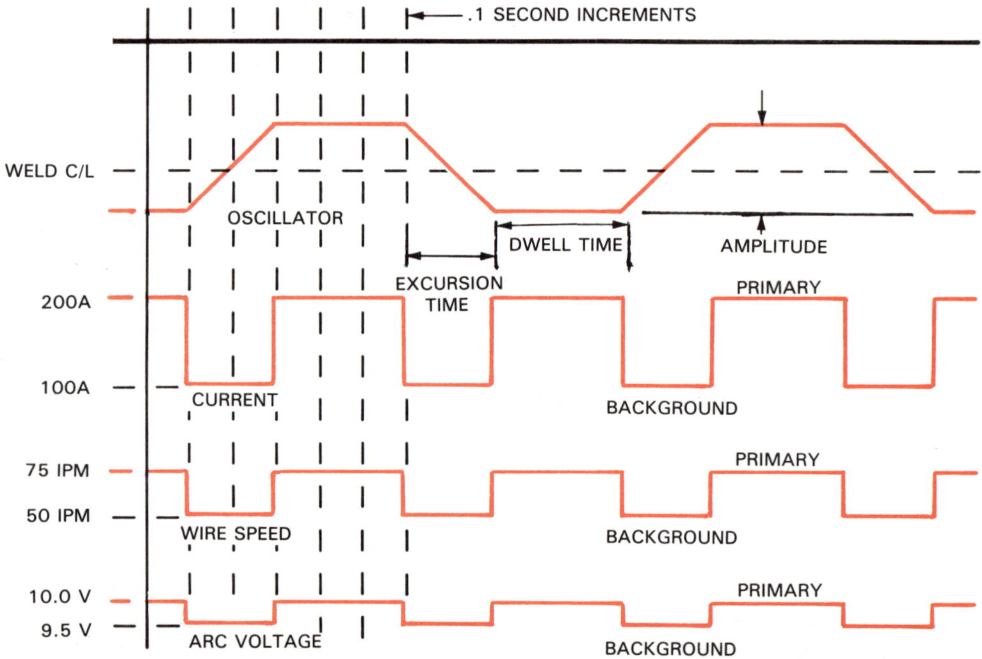


Fig. 4-46. Typical weld schedule or program sequences each operation in proper relationship with other operations.

Various types of GTAW torches are illustrated: Fig. 4-47 – pencil type gas cooled torch; Fig. 4-48 – angle head gas cooled torch; Fig. 4-49 – angle head gas or water cooled swivel head torch; Fig. 4-50 – angle head water cooled torch; and Fig. 4-51 – machine water cooled torch.

GAS COOLED TORCHES (commonly called AIR COOLED) are cooled by the passage of the inert shielding gas through the power cable vinyl covering tube. The gas is used to carry away heat generated in the power cable. These types of torches are generally

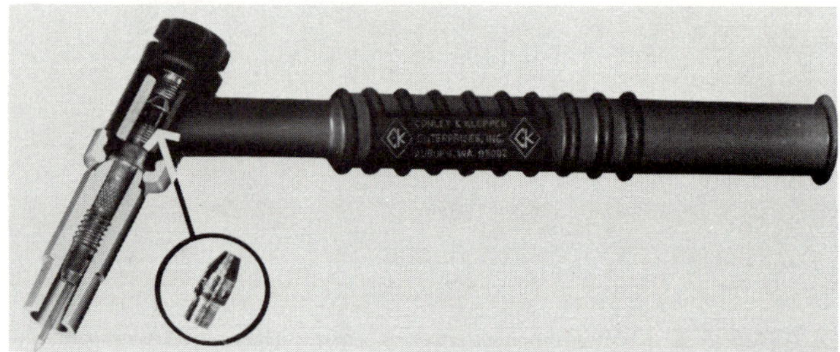

Fig. 4-48. Angle head gas cooled manual torch shows arrangement of collet and collet body. (Conley & Kleppen Co.)

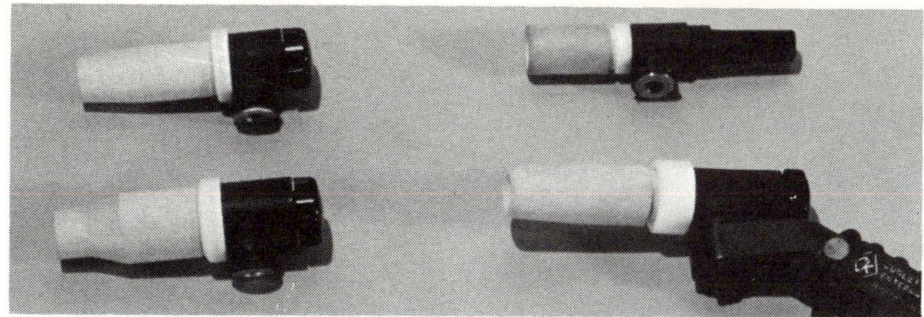

Fig. 4-49. Swivel head gas or water cooled manual torch with various nozzles and caps for working in confined locations. (Conley & Kleppen Co.)

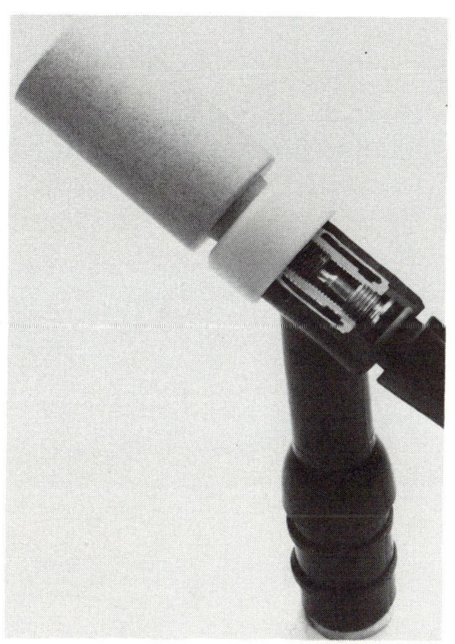

Fig. 4-50. Cooling water circulates through passages in head to remove heat. Duty cycle is greatly increased over gas cooled torches. (Conley & Kleppen Co.)

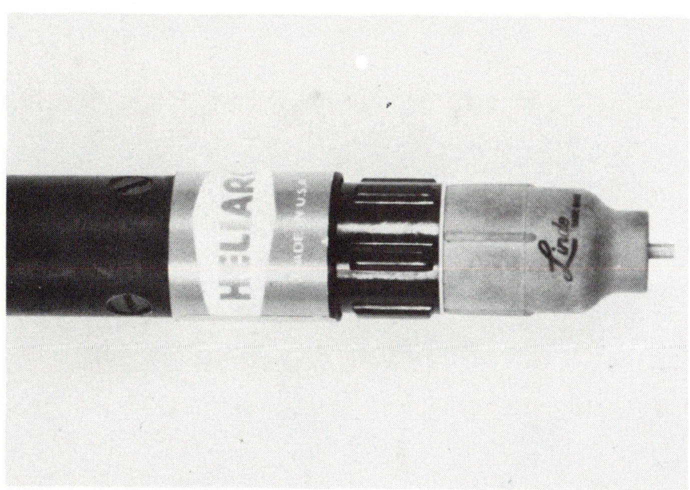


Fig. 4-51. Machine welding torch has a long barrel or tube for mounting in the torch clamp. This torch is water cooled for a high duty cycle. (Linde Co.)

used for light duty and field work, where cooling water is not available. A gas cooled power cable connection arrangement to the torch is shown in Fig. 4-52. Notice that this torch has a gas shut off valve located in the torch handle.

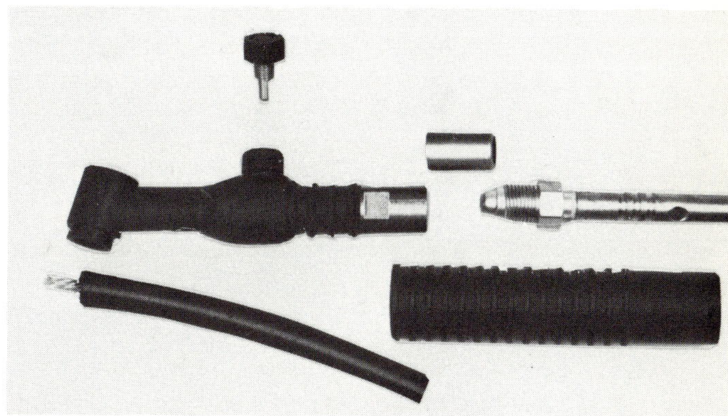

Fig. 4-52. Gas cooled manual torch components and arrangements. (Ceramic Nozzles Inc.)

WATER COOLED TORCHES are cooled by the passage of water through the torch head. The water then exits the system through the power cable. Fig. 4-53 shows a water cooled torch with connections for gas inlet, water inlet, and power cable with water outlet.

The main components of a GTAW torch consists of:

1. Torch body.
2. Collet body.
3. Collet.
4. Cap or cover for tungsten.
5. Nozzle.

Depending on the manufacturer's design, torches may have "O" rings to seal the tungsten caps. Nylon or teflon insulators are used to insulate the nozzle end of the torch body as shown in Fig. 4-54.

The COLLET and COLLET BODY are sets which

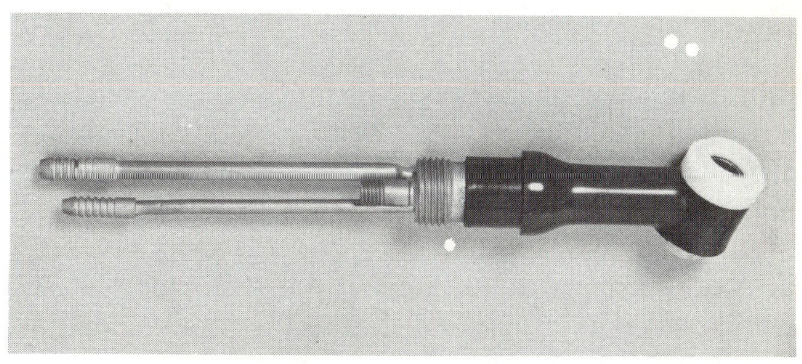

Fig. 4-53. Connections for gas, water out, and power are silver brazed into the manual torch body. (Ceramic Nozzles Inc.)

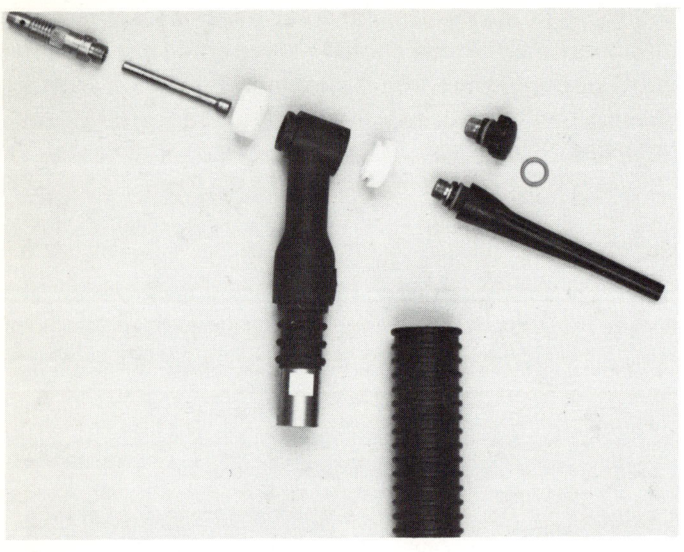

Fig. 4-54. Nylon or teflon insulators shield the high frequency spark and the "O" ring seal is used to prevent gas leakage. (Ceramic Nozzles Inc.)

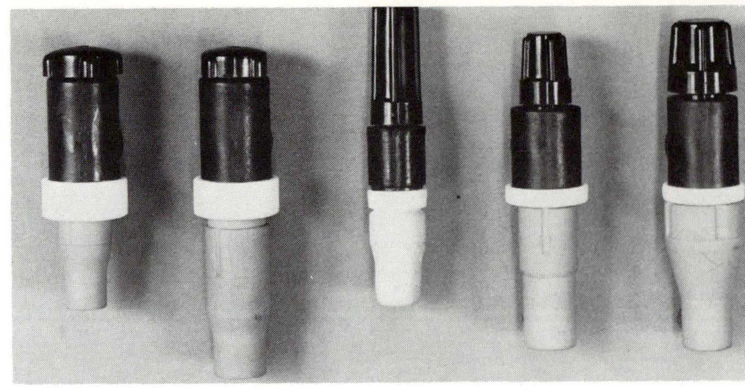

Fig. 4-55. Nozzle configurations. (Ceramic Nozzles Inc.)

are designed to grip one diameter size of tungsten and cannot be used with any other size. Using the wrong size will result in possible damage to the torch and lack of proper gripping of the tungsten. Collet bodies should always be installed and tightened by hand or a special chuck wrench. Failure to properly tighten the collet body will damage the end cap threads when attempting to grip the tungsten.

END CAPS are designed in varying lengths to match standard tungsten lengths and adapt the torch for welding in small areas. Turning the end cap into the torch head forces the collet into the collet body base. This grips and locates the tungsten in the position desired. Never use pliers to tighten the end cap. If the tungsten will not stay in adjustment, check the following:

1. Cap threads may be stripped.
2. Wrong size collet or collet body may exist.
3. Collet may be distorted or bent.
4. Torch threads may be damaged where collet or collet body attaches.
5. Collet body may be loose.

GAS NOZZLES are usually designed for installation into a particular type of torch and generally do not adapt to another make or model. They come in all sizes, shapes, and materials. Gas nozzles are reasonable in cost, therefore they should be replaced when they become unusable. A nozzle which has chips or cracks or a metal build up on the outlet end should be discarded. These types of defects can alter the gas flow pattern from the nozzle and cause contamination of the weld metal. Typical nozzle configurations are shown in Fig. 4-55.

Nozzles are identified by the size of the orifice (opening) and the length of the nozzle as shown in Fig. 4-56. Each torch manufacturer assigns part numbers to the various nozzles for individual type torches and these must be used when ordering replacement nozzles.

| SIZE | | NO. |
|---|---|---|
| 3/16 | = | NO. 3 |
| 4/16 | = | NO. 4 |
| 5/16 | = | NO. 5 |
| 6/16 | = | NO. 6 |
| 7/16 | = | NO. 7 |
| 8/16 | = | NO. 8 |
| 9/16 | = | NO. 9 |
| 10/16 | = | NO. 10 |
| 11/16 | = | NO. 11 |
| 12/16 | = | NO. 12 |

LENGTHS*
SHORT
REGULAR
LONG
EXTRA LONG
SPECIAL

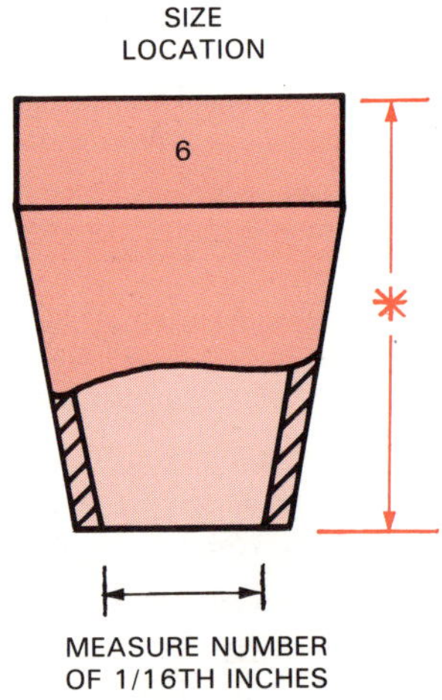


Fig. 4-56. Standard nozzle diameters and lengths.

Fig. 4-57. Gas lens use a stainless steel screen to establish a column of shielding gas rather than a blanket of gas. (Linde Co.)

## Gas Lens

The GAS LENS device allows the welder to use a longer electrode extension than with a standard nozzle. The device uses a series of stainless steel wire mesh screens to "firm up" the argon gas column. This aids in maintaining a blanket of inert gas around the tungsten and over the weld area. This is very helpful when wind drafts are present, or the tungsten must be extended due to the location of the weld area. Use of the gas lens requires a special gas lens collet body and gas nozzle. A gas lens device is shown in Fig. 4-57.

Fig. 4-58. Power cable for the torch is contained inside of a vinyl tube. Gas or water passes around the cable for cooling. (Conley & Kleppen Co.)

## Cables, Hoses, Connectors, and Valves

Power cables for GTAW torches are usually made of copper wire enclosed in a vinyl tube as shown in Fig. 4-58. Either gas or water is used to carry away the heat generated in the torch and the power cable. Special connectors are used at each end of the cable assembly for the gas or water to flow through. Fig. 4-59 shows typical connectors for attachment to the power supply.

Some torches available today use a separate power cable and do not use a coolant to carry away heat as shown in Fig. 4-60. With this type assembly, repairs to the torch cable are easily made and generally are less expensive. However, the cable assembly is heavier due to the added weight of the electrical cable and insulation.

Standard cable lengths are 12 1/2 feet and 25 feet. Couplings may be used to increase the length beyond these dimensions.

Manual shielding gas shut off valves are mounted on the torch body or on the inlet gas hose. These are generally used in field type work or where timers, circuits or solenoids are not included in the system. Several types are in use and are shown: Fig. 4-52 – Threaded screw valve in body type; Fig. 4-61 – Push-pull valve in body type; and Fig. 4-62 – Push-pull valve in shielding gas line.

Remember, these valves operate manually. If the valve is left open after the operation is completed, the shielding gas will continue to flow as long as the gas is available from the main supply.

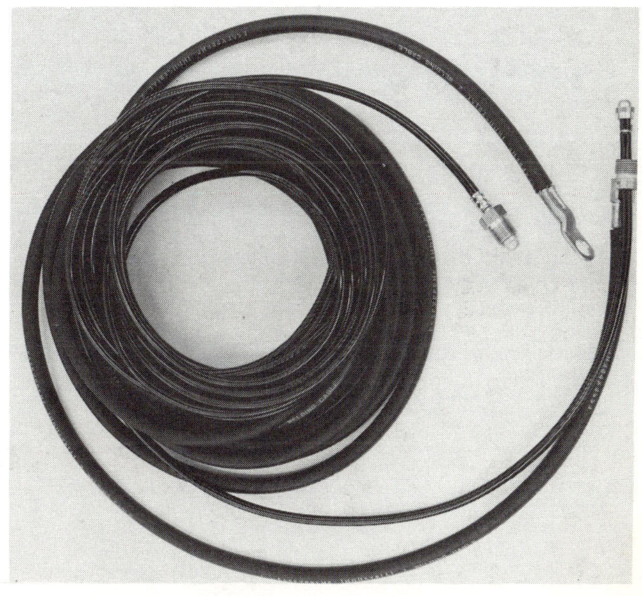

Fig. 4-60. External type power cables do not require cooling. (Conley & Kleppen Co.)

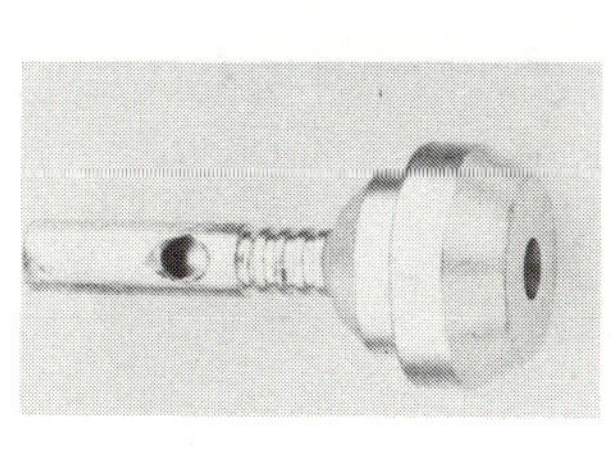

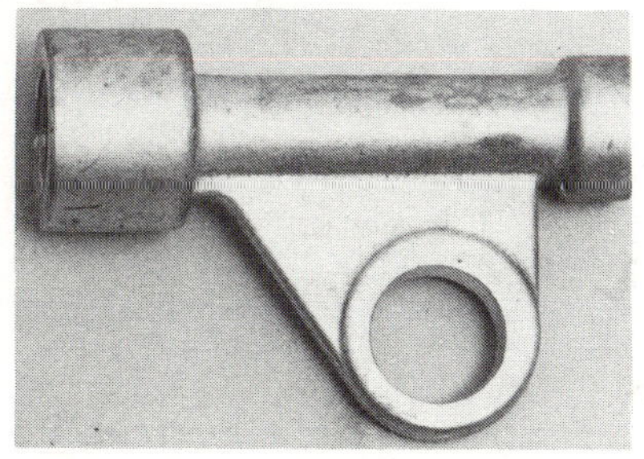

Fig. 4-59. Power cable adapters are used for connecting the cable to water or gas and welding machine terminals. (Linde Co.)

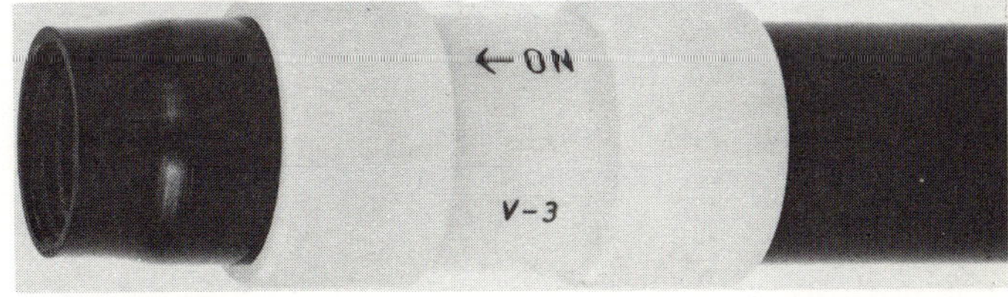


Fig. 4-61. Gas shut-off valve attaches to manual torch body and is a push-pull operation for on or off of the gas flow. (Ceramic Nozzles Inc.)

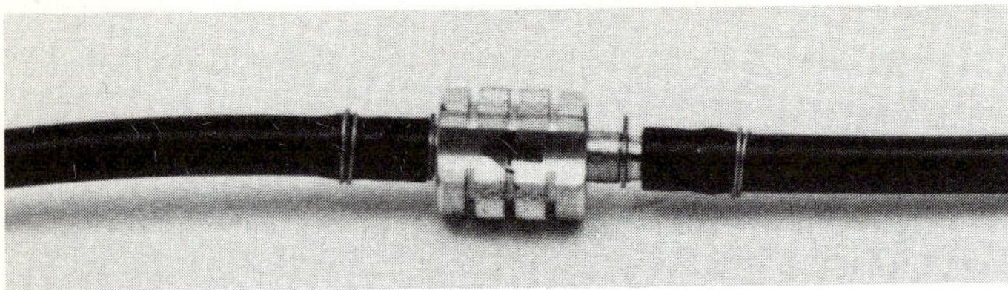

Fig. 4-62. This push-pull valve for gas flow or shut-off mounts in the gas inlet hose. This valve operates only on torches which have a separate gas supply hose. (Linde Co.)

### Torch Maintenance

Maintenance of torches and components should be carried out on a daily basis to:

1. Reduce the possibility of contaminating welds.
2. Reduce the cost of shielding gas.
3. Increase welder safety.

Areas of maintenance inspection include:

1. Replace burned or cracked insulators.
2. Replace burned or cracked "O" rings.
3. Test all water and gas lines daily for leaks.
4. Repair or replace leaky tubing.
5. Keep all connections tight.
6. Keep coolant water clean.
7. Use recommended coolant water pressure levels.

### Torch Installation

For a GTAW torch to operate properly at it's rated capacity it must be installed properly. Always read the manufacturer's instructions and carry them out fully. If you do not understand the instructions, stop and contact the manufacturer or representative for additional information. Be safe rather than sorry.

Fig. 4-63 shows a GTAW gas cooled torch installation using machine solenoid valve.

Fig. 4-64 shows a GTAW gas cooled torch installation using a torch gas shut off valve.

Fig. 4-65 shows a GTAW water cooled torch installation using city water.

Fig. 4-66 shows a GTAW water cooled torch installation using a hydrotank.

Fig. 4-67 shows a GTAW water cooled torch installation using a water tank and cooler unit.

Fig. 4-68 shows a typical hydrotank unit with running gear for portability.

Fig. 4-69 shows a water tank and cooler unit.

Fig. 4-70 shows a GTAW torch installation using solid wire power cable installation.

Installation notes for GTAW torches:

1. Always use a filter on all water lines into the torch.
2. Use a regulator and pressure gauge to determine pressure into the torch. Never exceed manufacturer's instructions.
3. When using a hydromount or water cooler, check

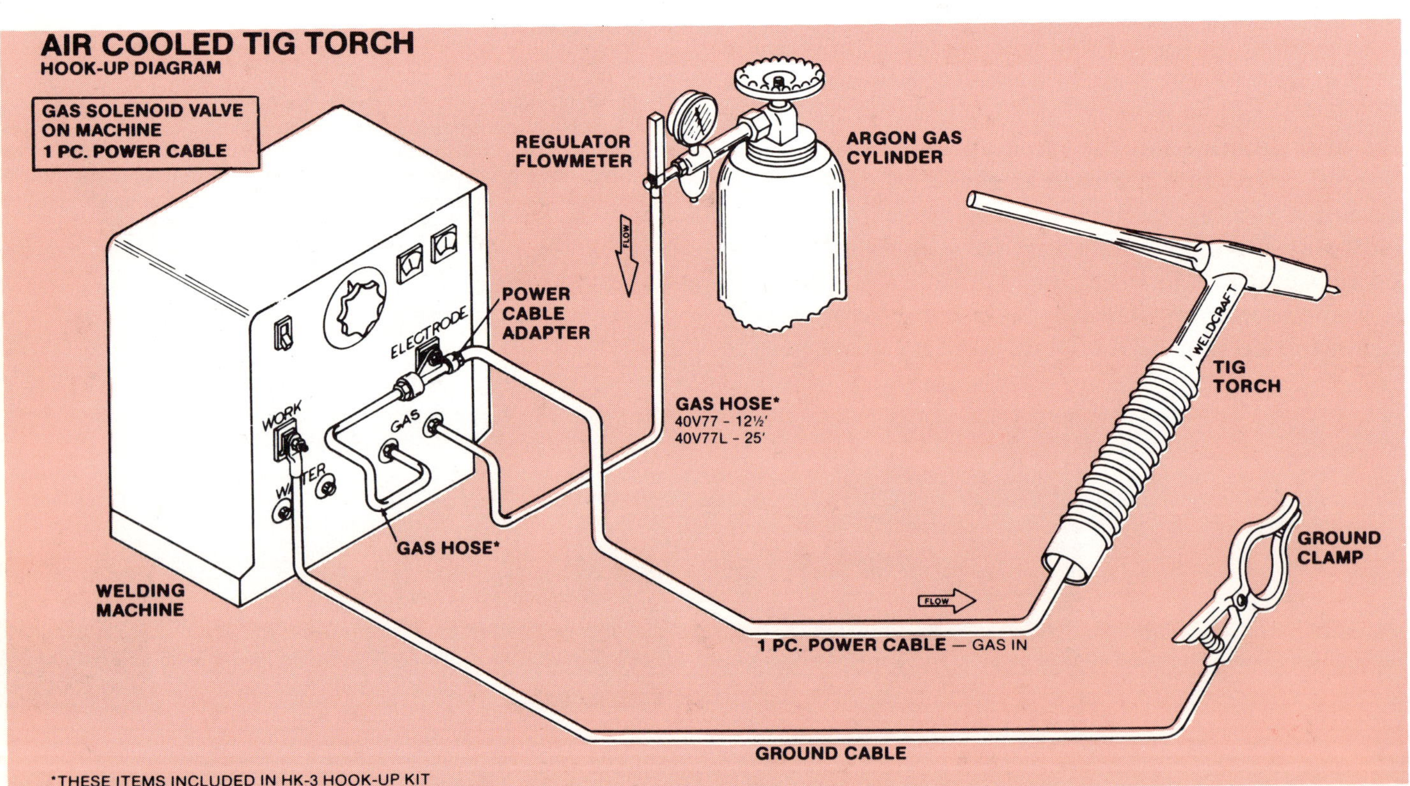


Fig. 4-63. Gas cooled manual torch installation requires a switch on either the torch or foot control to start and stop the gas solenoid control. (Weldcraft Co.)

the flow rate at the water return line outlet.

4. Use a rust inhibitor (colored) in hydromounts or water cooled systems.

5. Always use special left hand nuts and adapters for

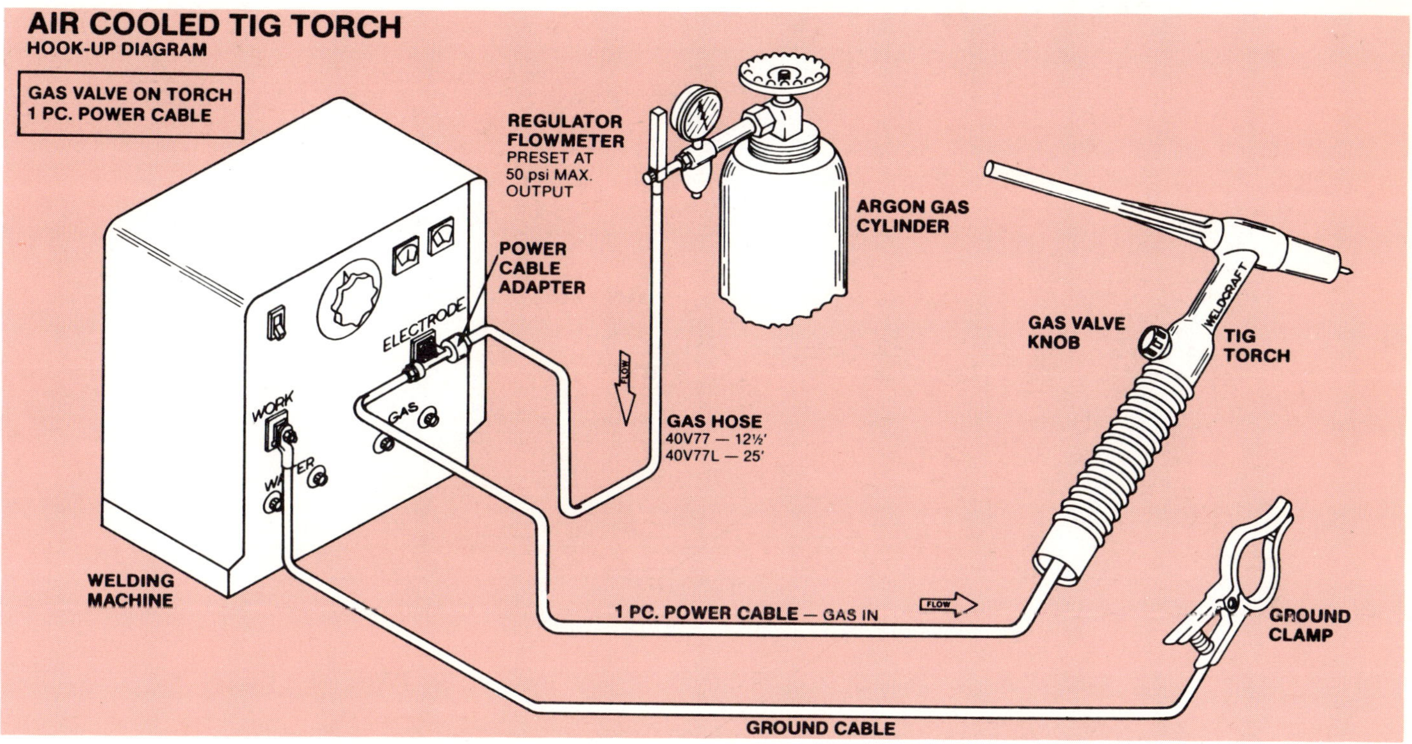


Fig. 4-64. Gas cooled manual torch installation uses a torch mounted valve to start and stop gas flow. (Weldcraft Co.)

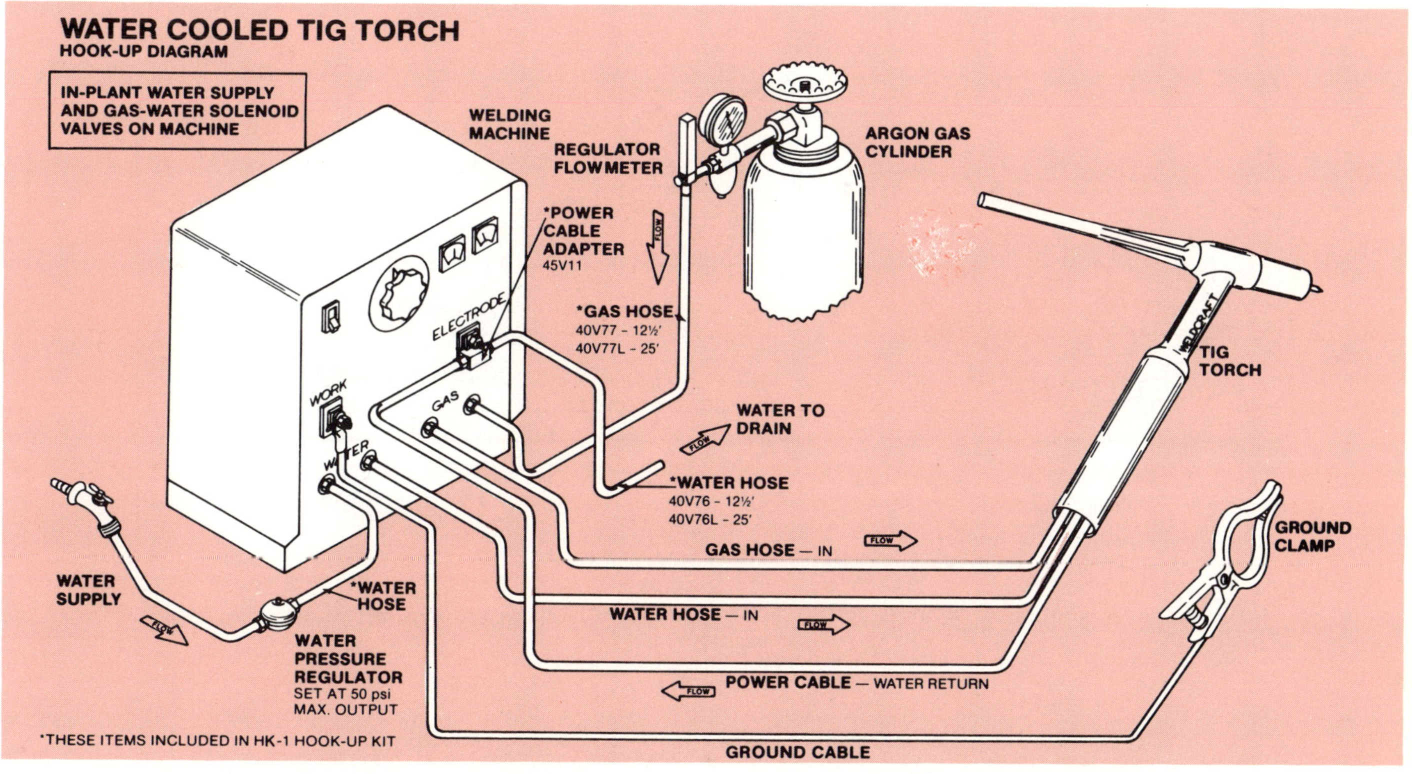


Fig. 4-65. City water cooled manual torch installation. (Weldcraft Co.)

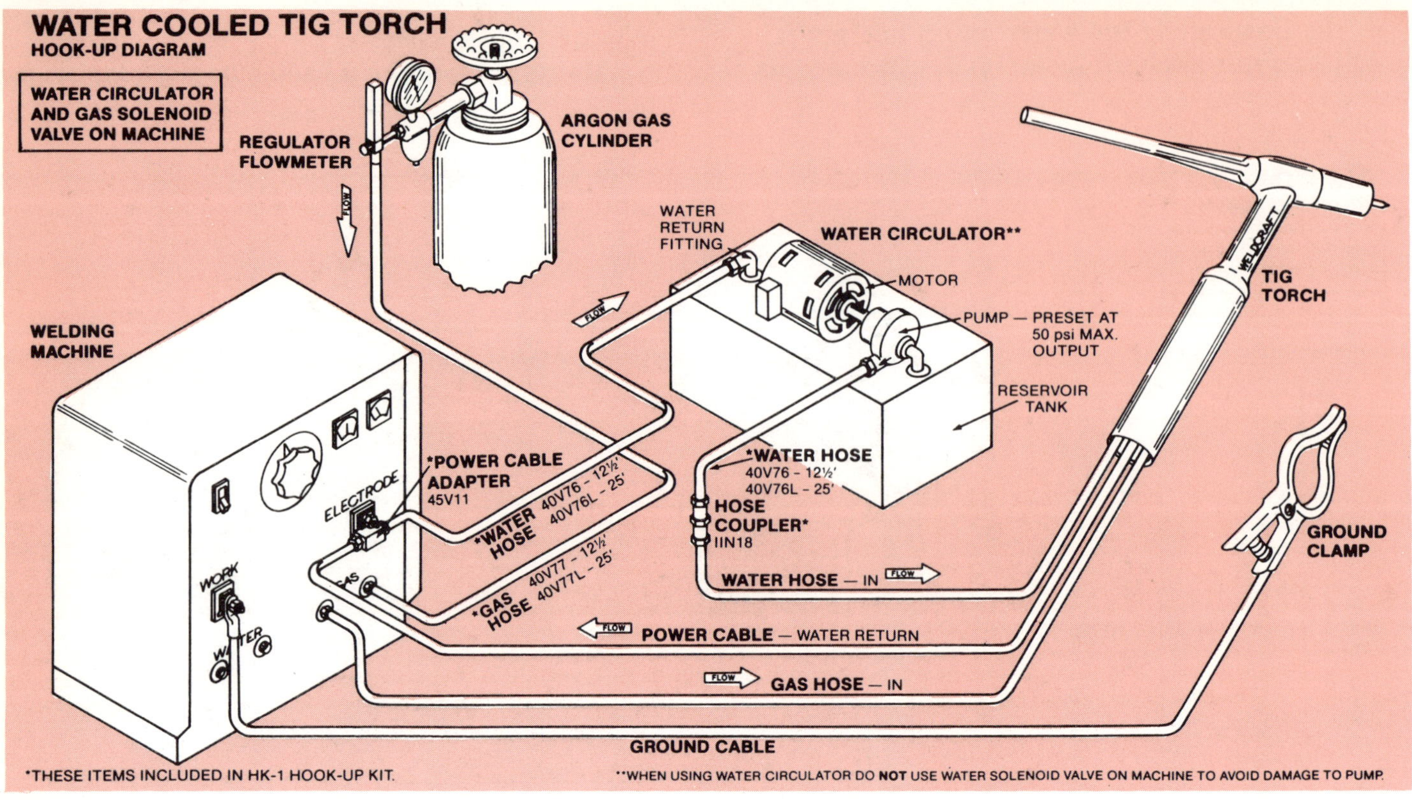


Fig. 4-66. Recirculated water cooled manual torch installation. (Weldcraft Co.)

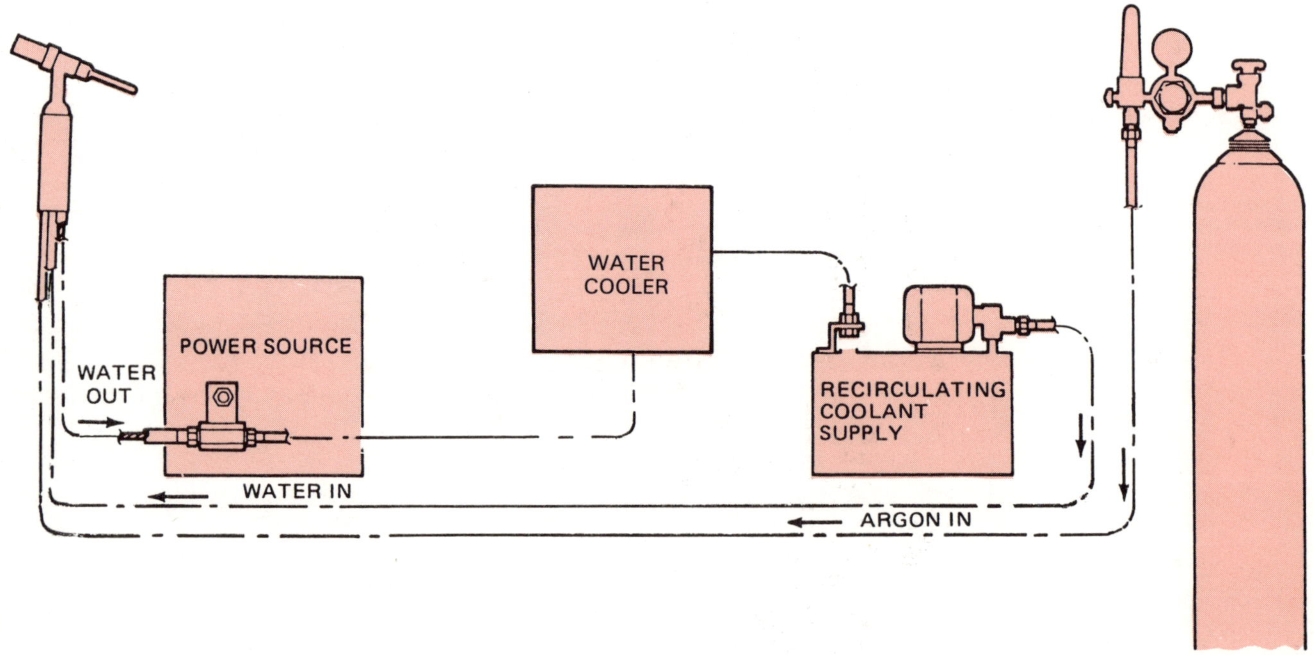


Fig. 4-67. Recirculated water system with a water cooler installed in the system. (Conley & Kleppen Co.)

water connections. Gas fittings use right hand threads on the nuts and adapters.

6. Never use vinyl tubing that has contained water for argon gas lines.
7. Always connect water lines to the torch so that the water will enter to the torch and exit from the

Fig. 4-68. Hydro (water) tank reservoir for use with recirculated water. (Linde Co.)

Fig. 4-69. Water coolers lower temperature of the water, therefore smaller reservoirs of water are required. (Linde Co.)

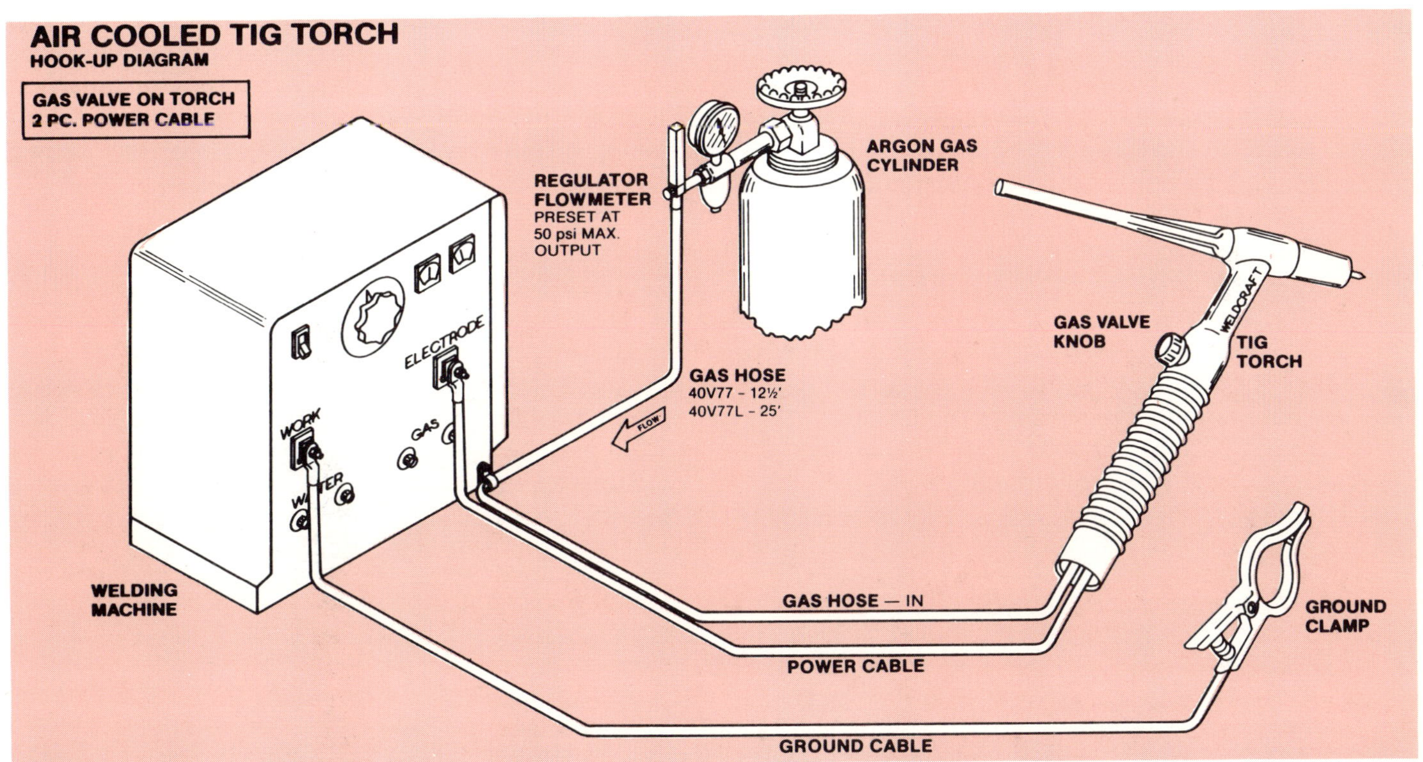


Fig. 4-70. Solid wire (power cable) manual torch installation. (Weldcraft Co.)

power cable.

After the torch installation is complete and the system is checked for proper operation, a torch cable cover jacket may be installed to protect the vinyl tubing from heat, abrasion, puncture from sharp objects, etc. These covers are available in asbestos, leather, and heavy vinyl and are made in standard torch cable lengths. A typical cover jacket is shown in Fig. 4-71.

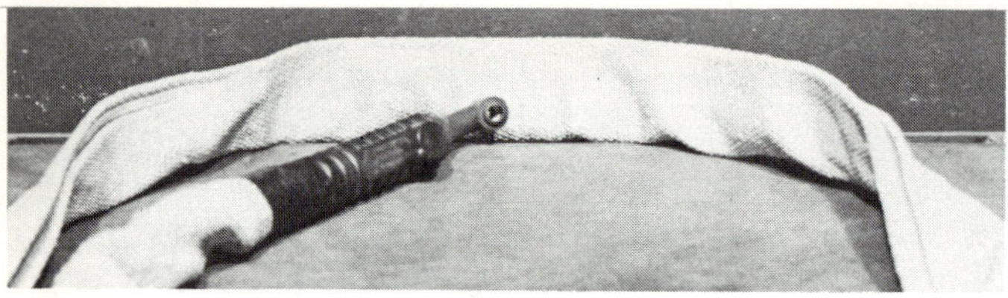

Fig. 4-71. Cable covers protect the torch cables from damage.

## TUNGSTEN ELECTRODES

Electrodes used in GTAW are manufactured to meet the requirements of the American Welding Society specification A5-12. The chemical requirements for the various types of electrodes are shown in Fig. 4-72. The different types of electrodes used are:

1. PURE TUNGSTEN. It achieves good arc stability and a clean balled end for AC welding with reasonably good resistance to contamination.
2. THORIATED TUNGSTEN. Thorium is added to tungsten for the following reasons:
   a. Arc starting is easier.
   b. Wider current ranges may be used.
   c. Tip does not become molten.
   d. The tungsten exhibits less tendency to stick or freeze to the work.
   e. Has a high resistance to contamination on both AC and DC when properly used.
   f. Will not cause contamination if accidentally touched to the work during the welding operation.

   Both one and two percent thoriated electrodes have a longer life due to the extra thoria content while still maintaining the capability of easier arc starting, high current capability, and increased resistance to weld pool contamination.
3. ZIRCONIATED TUNGSTEN. This combines the desirable characteristics of a pure tungsten electrode with the capability and starting characteristics of a thoriated tungsten electrode. Normally used on AC welding when tungsten inclusions are not tolerated.

### Electrode Finish

Tungsten electrodes are manufactured with a surface free of impurities, undesirable films, foreign inclusions, pipes, seams or silvers in order to assure correct operation of welding equipment with no adverse effect on the weld deposit. They are available in a "clean" or "ground" finish. The "ground" finish electrode is the more expensive.

### Electrode Color Codes

Each type of electrode is identified by a different color code which is painted on one end of the tungsten. Some manufacturers use the same color as the code for the electrode type for the electrodes package. Electrode color codes are shown in Fig. 4-72.

### Electrode Sizes, Tapering, and Balling

Electrodes are made in 3 1/2 inch, 7 inch, and longer lengths. Normal practice is to order the 7 inch length and use a long cap on the torch for the tungsten cover. The tungsten will shorten when grinding or tapering. Electrode sizes and amperage ranges for the various diameters are shown in Fig. 4-73. The values shown are for the full diameter of the tungsten.

### Electrode Preparation

In normal operation on DC current, the tungsten is ground to a taper point for directing and controlling the arc stream. When the electrode is prepared in this manner, a lower amperage rating must be used to compensate for the metal removed to make the taper. Electrodes may be tapered by using an electrode grinding machine as shown in Fig. 4-74, sanding belts, or a chemical powder. Shop belt sanders, as shown in Fig.

CHEMICAL REQUIREMENTS

| AWS-Classification | Tungsten, min, percent (by difference) | Thoria, percent | Zirconium, percent | Total Other Elements, max, percent | Color |
|---|---|---|---|---|---|
| EWP . . . . . . . . . . | 99.5 | – | – | 0.5 | Green |
| EWTh-1 . . . . . . . | 98.5 | 0.8 to 1.2 | – | 0.5 | Yellow |
| EWTh-2 . . . . . . . | 97.5 | 1.7 to 2.2 | – | 0.5 | Red |
| EWTh-3 . . . . . . . | 98.95 | 0.35 to 0.55 | – | 0.5 | Blue |
| EWZr . . . . . . . . . | 99.2 | – | 0.15 to 0.40 | 0.5 | Brown |

| Electrode Type | Color |
|---|---|
| EWP | Green |
| EWTh-1 | Yellow |
| EWTh-2 | Red |
| EWTh-3 | Blue |
| EWZr | Brown |

Fig. 4-72. Tungsten electrode chemical requirements and color codes.

TYPICAL CURRENT RANGES FOR TUNGSTEN ELECTRODES.a

| Electrode Dia, In. | Straight Polarity Direct Current, amps | Reverse Polarity Direct Current, amps | High-Frequency Unbalanced Wave AC, amps | | | High-Frequency Balanced Wave AC, amps | | |
|---|---|---|---|---|---|---|---|---|
| | EWP, EWTh-1, EWTh-2, EWTh-3 | EWP, EWTh-1, EWTh-2, EWTh-3 | EWP | EWTh-1, EWTh-2, EWZr | EWTh-3 | EWP | EWTh-1, EWTh-2, EWZr | EWTh-3 |
| 0.010 . . . . . | up to 15 | b | up to 15 | up to 15 | b | up to 15 | up to 15 | b |
| 0.020 . . . . . | 5-20 | b | 5-15 | 5-20 | b | 10-20 | 5-20 | 10-20 |
| 0.040 . . . . . | 15-80 | b | 10-60 | 15-80 | 10-80 | 20-30 | 20-60 | 20-60 |
| 1/16 . . . . . . | 70-150 | 10-20 | 50-100 | 70-150 | 50-150 | 30-80 | 60-120 | 30-120 |
| 3/32 . . . . . . | 150-250 | 15-30 | 100-160 | 140-235 | 100-235 | 60-130 | 100-180 | 60-180 |
| 1/8 . . . . . . . | 250-400 | 25-40 | 150-210 | 225-325 | 150-325 | 100-180 | 160-250 | 100-250 |
| 5/32 . . . . . . | 400-500 | 40-55 | 200-275 | 300-400 | 200-400 | 160-240 | 200-320 | 160-320 |
| 3/16 . . . . . . | 500-750 | 55-80 | 250-350 | 400-500 | 250-500 | 190-300 | 290-390 | 190-390 |
| 1/4 . . . . . . . | 750-1000 | 80-125 | 325-450 | 500-630 | 325-630 | 250-400 | 340-525 | 250-525 |

a All values are based on the use of argon as the shielding gas. Other current values may be employed depending on the shielding gas, type of equipment, and application.
b These particular combinations are not commonly used.

Fig. 4-73. Current ranges for tungsten electrodes.

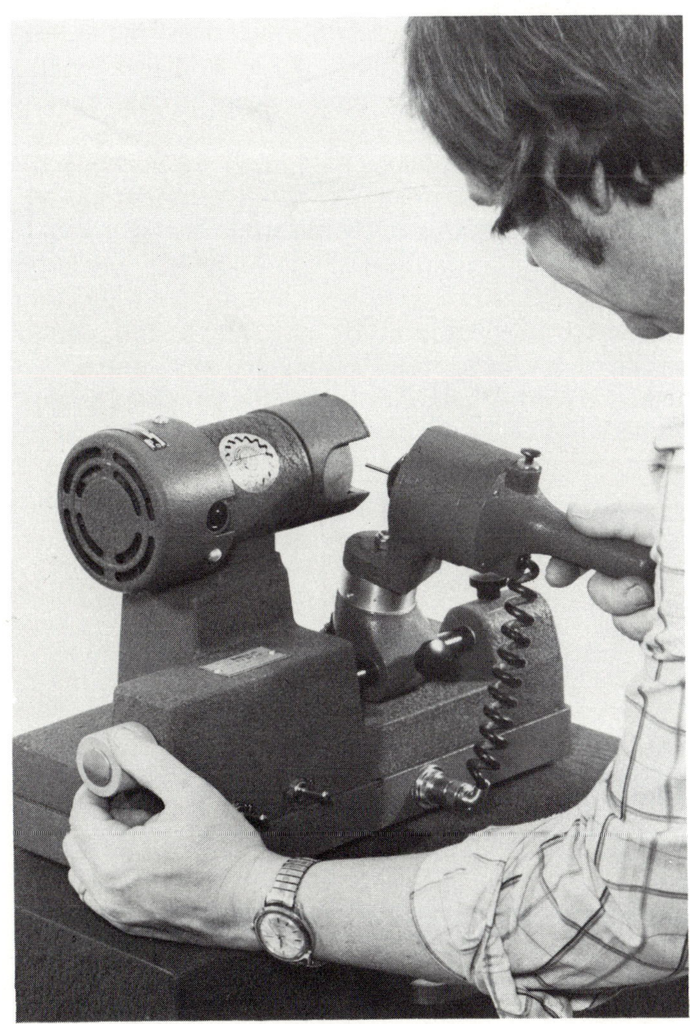

Fig. 4-74. Special tungsten grinding machines prepare precision ground tungstens. (The Weldma Co.)

4-75, are sometimes used, however the belt must have a very fine grit or the tungsten will break during the tapering operation. When using a chemical powder as shown in Fig. 4-76, the electrode must be dipped into the powder when hot. Powders produce a very good taper angle.

The method used to prepare the tungsten is not as critical for manual welding applications as the preparation of a tungsten used for automatic welding. In manual welding, the welder can change heat inputs, torch angle, travel speed, etc., as required for the operation. In automatic welding the parameters are established, thus the tungsten tapering preparation becomes a critical part of the operation. Fig. 4-77 shows some

Fig. 4-75. The tungsten is rotated during pointing when using a belt sander.

common electrode taper angles. The 30 degrees taper is most often used by manual welders, however the point is the most fragile of all the angles and resharpen-

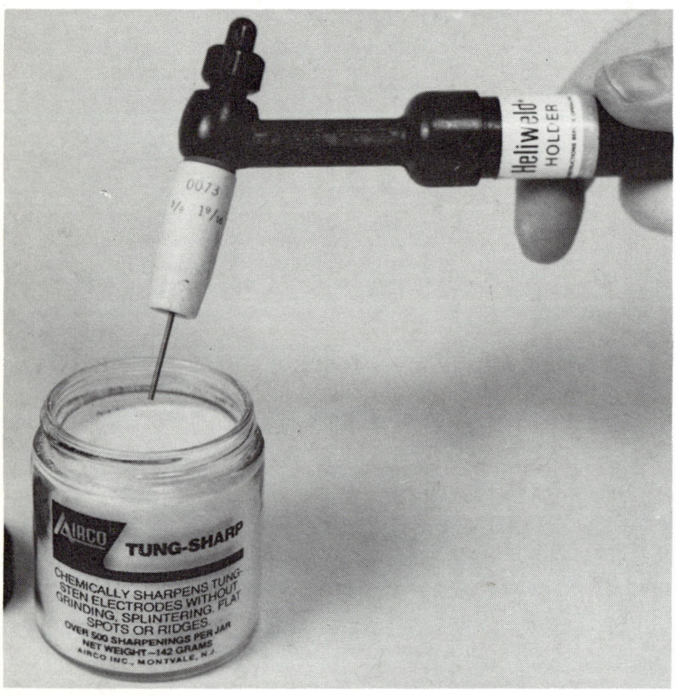


Fig. 4-76. Chemical powders produce sharp points on tungstens. The tungsten must be dipped into the powder while red hot. (Airco)

ing is often required during the operation.

Balling the electrode for AC operation can be readily accomplished by grinding a radius on the end of the electrode. Start an AC arc on a test piece of steel or copper with a low current, then increase the current to form the desired ball diameter. Application of too much current during balling or the welding operation will cause the ball to dance with the possibility of the tungsten ball separating from the electrode. This causes contamination of the weld metal. If the ball dances during welding, change to a larger diameter electrode which will carry more amperage.

Contamination of the electrode can be caused by:

1. Contact of the electrode to the weld metal or filler metal.
2. Lack of shielding gas during or after the weld operation.
3. Expelled metal from the molten pool attaching to the electrode tip.

Should CONTAMINATION occur, remove the electrode and retaper or ball before reusing. Foreign material which is not removed from the electrode tip will burn away as the tungsten becomes hot. This residue will cause smoke fumes to further contaminate the molten metal.

EXCESSIVE SPITTING from the electrode taper point is usually caused by electrode contamination. The electrode should be removed from the torch and reground before using.

WHISKERS forming near the electrode tip indicates a possible moisture contamination in the shielding gas and may cause arc wandering. The shielding gas supply, torch, and all connections should be checked for leaks of moisture or air. When using helium or helium-argon mixtures, whiskering may also develop. When this occurs, the tungsten electrode must be removed and reground.

DIFFICULTY IN STARTING an arc on DCSP (DCEN) can occur when using a pure tungsten electrode due to the low electron emissivity. (Emissivity is the relative power of a surface to emit heat by radiating electrons.) Usually a scratch start is required to start the arc. If this blunts the tip, then a thoriated electrode must be used. Another method used is to hold the electrode against the work. Start the arc at a very low current which will heat the electrode, shut off the current, and then restart while the electrode is hot. Should continuous high frequency be available and not detrimental to the weld operation, this may be used as an arc starter.

POINT DETERIORATION may be caused by excessive currents, previous metal contamination, splintering or cracking of the electrode during preparation, or by using continuous high frequency operation on AC or DC. Always remove all contamination from the electrode when grinding new tapers, and inspect the point for cracks and splinters.

ARC WANDERING is usually caused by poor preparation of the electrode taper, or the electrode may

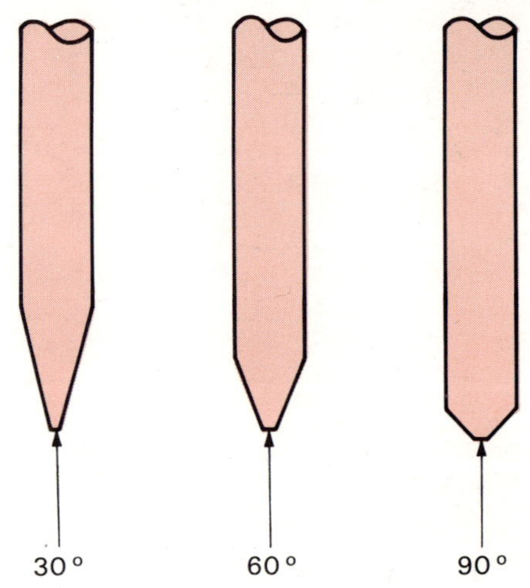


Fig. 4-77. Electrode tapers used in manual and automatic gas tungsten arc welding.

be bent or not centered in the gas nozzle. The taper preparation should be carefully ground so that the tip is in the center line of the electrode. Incorrect grinding of the taper angle is shown in Fig. 4-78. Electrodes that are bent should have the bent section removed or discard the electrode.

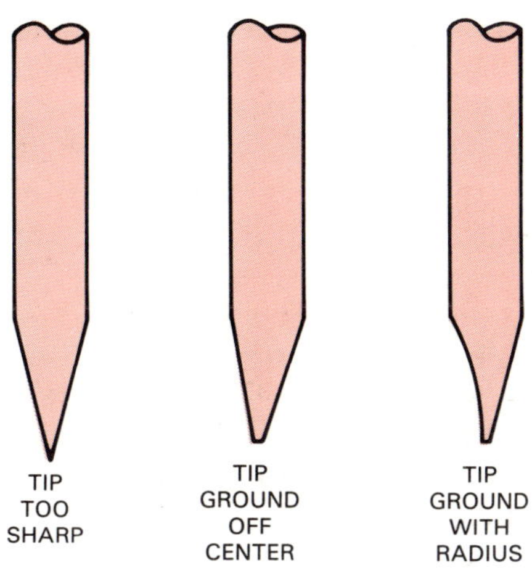


Fig. 4-78. Incorrectly ground electrodes cause many problems during the welding cycle.

Arc wandering is sometimes caused by sanding or grinding the tungsten as shown in Fig. 4-75 as the grinding marks are across the tungsten. To prevent this condition, grind the tungsten as shown in Fig. 4-79.

Fig. 4-79. Grind the tungsten with the grind marks parallel with the tungsten length.

Since the tungsten is very brittle, extreme care must be taken when grinding a new taper or the electrode may break. Use a fine grit wheel or belt, and use only moderate pressure on the tungsten point. Always rotate the tungsten evenly during manual grinding to maintain the tip in the center line of the tungsten.

To prevent loss of identification of the tungsten type, do not grind the painted end. Once the painted end is ground and the paint removed, identification of the tungsten type is very difficult.

Always handle tungstens carefully to prevent breakage. Wear clean gloves to prevent contamination with oil or grease. If the tungsten is contaminated with oil or grease, remove the contamination with a degreaser before using.

Contaminated electrodes introduce contamination into the molten weld metal. This action may cause porosity, cracking or the entrance of foreign material. Fig. 4-80 shows some common types of contamination.

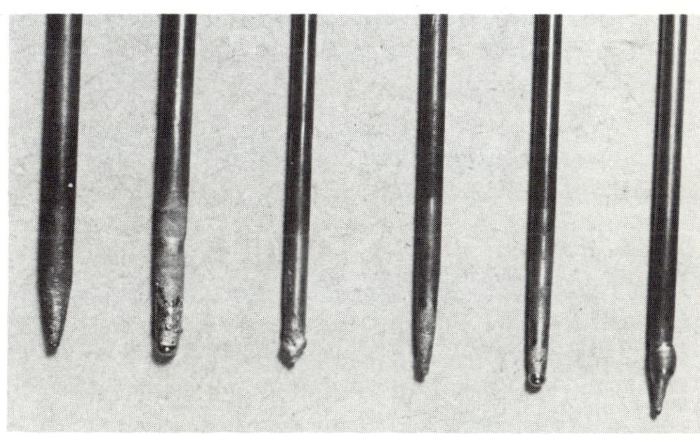

Fig. 4-80. These electrodes have been contaminated by metal pickup or lack of inert gas shielding during the cooling period after the arc was extinguished.

## REVIEW QUESTIONS

1. When welding on aluminum with alternating current, the high frequency spark control is set to ________.
2. When welding on ferrous materials with direct current, the high frequency spark control is set to ________ ________ type use.
3. Where is the intensity control on the high frequency generator used for maximum spark?
4. The afterflow timer control on the high frequency generator is adjusted for a period of time to prevent contamination of the ________.
5. If the cooling water is controlled by the afterflow timer, how long does it flow after the arc goes out?
6. Improper installation of the high frequency

generator can result in ________ and ________ interference.

7. List five benefits derived from the use of pulsers in the weld operation.
8. Pulsers actually create ________ welds.
9. Can pulsers be used on both alternating current and direct current?
10. What type of a pulsing program is used when welding aluminum? Why?
11. Where is a slope controller used?
12. What other names are used to identify a slope controller?
13. List the three types of timers used in the control of welding operations.
14. The two basic types of remote controls are the ________ and the ________ operated switches.
15. Cold wire feeders may be used in both ________ and automatic welding.
16. What type of groove design is used on cold wire feeder drive rolls for soft wires? For hard wires?
17. What is the piece of equipment called that controls the wire feed into the weld puddle? What connects this piece of equipment to the drive rollers?
18. A cold wire feeder will feed wire under the tungsten and into the ________.
19. How does a hot wire feeder place feed wire into the molten pool differently than a cold wire feeder?
20. How is the welding wire heated when fed through a hot wire feeder?
21. Hot wire feeders are not normally used to weld ________ and aluminum metals.
22. What are the main advantages of oscillators?
23. Name two types of oscillation equipment.
24. Automatic voltage control welding is used where arc voltage is a ________ factor and driving mechanisms be ________ set.
25. What is the basic use of a pilot arc system?
26. Why is the placement of the tungsten so important in the spot welding operation?
27. What are the main functions of a welding torch?
28. What two materials are used to insulate welding torches from the welder?
29. How are welding torches rated for use?
30. Gas cooled torches are sometimes called ________ cooled even though they are cooled by the shielding gas.
31. What does a gas lens do?
32. The standard gas tungsten arc torch cable lengths are ________ and ________ feet long.
33. When connecting argon gas fitting onto a machine or torch, the threaded connection is always a ________ hand thread.
34. What type of threads do the water connections have?
35. Name the three basic types of tungstens.
36. What current types are used with the three basic types of tungstens?
37. List the three types of tungstens and the color codes used to identify each type.
38. Which of the electrodes has difficulty in starting the arc when using DCSP? Why?
39. What types of problems are encountered when using contaminated electrodes?

# Chapter 5

# SHIELDING GASES AND REGULATION EQUIPMENT

## SHIELDING GASES

Shielding gases are used in gas tungsten arc welding to:

1. Shield the tungsten electrode from the atmosphere.
2. Shield the molten metal from the atmosphere.
3. Transfer heat from the tungsten to the metal.
4. Assist in starting the arc.
5. Stabilize the arc.
6. Aid in controlling bead contour and penetration.

The gases used include argon, helium, hydrogen, and nitrogen. Argon and helium are inert gases (will not combine with any material) which are pure, colorless, odorless, tasteless, and are used as shielding gases. Inert gases will not contaminate the weld. Hydrogen is fuel gas that burns, is not inert, and has limited applications in welding. Nitrogen is not an inert gas and is limited to special purging applications.

### Argon

Argon is the most commonly used shielding gas for gas tungsten arc welding. It is separated from the atmosphere during the production of oxygen and is readily available at reasonable cost.

Argon, of all the shielding gases, gives the welder the greatest tolerance for arc gap variation. It's lower arc voltage tolerance permits arc length gap variations with less influence on the arc power and bead shape. In addition, argon provides superior arc starting over helium. For alternating current welding use, argon is better than helium due to it's superior cleaning action, the final weld appearance, and the weld metal quality.

Argon, being heavier than air, forms a blanket of gas above the weld while helium, being lighter than air, floats (rises) above the weld. Helium therefore requires higher flow rates than argon to properly shield the weld.

### Helium

Helium gas is used primarily for four applications:

1. Assisting in obtaining deep penetration into the weld joint.
2. Increasing welding speed in machine welding.
3. Welding of high heat conductivity metals (copper, aluminum, and the refractory metals).
4. Machine welding of aluminum with DCSP.

Helium is separated from natural gas produced from wells. It is not as available as argon and therefore costs more than argon. Usage costs are also higher as helium is lighter than argon, so the higher flow rate required for proper shielding will increase the cost of this welding operation.

Starting the arc in a pure helium shield is very difficult. Therefore many machine welding applications use an argon gas starting system. After the arc is started using argon, the operation shifts over to using helium shield gas. Manual welding with pure helium is seldom done as the arc length (gap) tolerance is so critical that the welder cannot usually maintain the arc.

### Argon-Helium Mixtures

Argon-helium mixtures are used to increase control of the voltage and heat input of the arc, while maintaining the favorable characteristics of argon. Helium rich mixtures are preferred in order to achieve significant benefits from the helium. The most commonly used mixture, He-75, contains 75 percent helium and 25 percent argon. Fig. 5-1 shows the substantial increase in heat input and weld penetration afforded by He-75 over argon. Additions of less than 50 percent helium have little influence on the arc characteristics. The common argon-helium mixes and argon-hydrogen mixes are as described below:

He-75 — The speed or quality of AC welding in aluminum can be improved with this mixture. Cleaning action is almost as good as with pure argon. The mixture is sometimes used for manual welding of aluminum pipe and mechanized welding of butt joints in aluminum sheet and plate.

He-90 — This contains 90 percent helium and 10

percent argon. This mixture is used occasionally for DC welding when it is desirable to obtain heat input characteristics approaching helium but maintaining the good starting behavior of argon. The arc in argon mixtures is much easier to start than an arc in helium when using a high frequency start. In some cases, an argon mixture is used for igniting the arc, then admitting the helium rich mixture to make the weld.

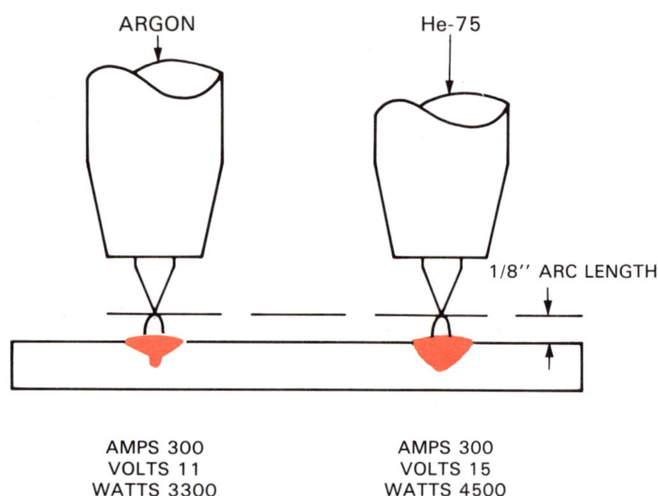


Fig. 5-1. Thermal conductivity of helium is much higher than argon. Helium transfers more heat from the arc region to the puddle.

### Argon-Hydrogen Mixtures

Argon-hydrogen mixtures are used in special cases, such as mechanized welding of light gauge stainless steel. Faster speeds can be obtained by increasing the amount of hydrogen added to the argon. However, excessive hydrogen will cause porosity. The 15 percent hydrogen mixture commonly called H-15 is the most popular argon-hydrogen for gas tungsten arc welding. It is used most often for welding tight butt joints in stainless steel up to 1/16 inch thick at speeds comparable to helium. This mixture also allows a 50 percent speed increase over argon. A hydrogen content of 5 percent is sometimes preferred for manual welding to obtain cleaner appearing welds. Fig. 5-2 shows some typical gas selections.

### Nitrogen

Nitrogen gas cannot be used in the torch as a shielding gas. It can be used as an inexpensive purge gas for initial purging of tanks, vessels, and pipe lines prior to the admission of the inert gas for welding. The gas can also be used to protect the back side of fillet welds where the joint design does not allow admittance of nitrogen into the torch shielding gas. Nitrogen is a contaminate in argon gas. The uses of nitrogen gas as a purging gas or backing gas is shown in Fig. 5-3.

## GAS PURITY

Shielding gases for welding are refined to high purity specifications. Cylinder argon has a minimum purity of 99.996 percent and contains a maximum of about 15 parts per million moisture (a dew point temperature of −73 degrees F. maximum). Driox (Linde Co.) argon has a minimum purity of 99.998 percent and a moisture content less than 6 parts per million. Helium is produced to a minimum purity of 99.995 percent and generally contains less than 15 parts per million of moisture. At these purity levels, differences in the kind and amount of impurities usually cannot be detected during welding.

Steels and copper alloys have high tolerances for various amounts of contaminates. Aluminum and magnesium are sensitive to the gas purity level and will have severe porosity if contaminated gas is used.

Still other metals such as titanium and zirconium have extremely low tolerances for foreign contaminates in the inert gases. Therefore, high purity standards are maintained by gas suppliers to insure the shielding gases used will be more than adequate for the most severe application.

## GAS SUPPLY

Shielding gases are supplied in cylinders of various sizes for shop use. Dewars (liquified gas container) as shown in Fig. 5-4 and high pressure gas truck or trailer mounted cylinders are used where high volumes of gases are required.

In all cases, the gas is sold through the distributor by the total number of cubic feet of each type of gas. As the gas is distributed in cylinders, the distributor will charge a rental fee (demurrage) on each cylinder used. In areas where large amounts of gases are used, liquified gases are the most economical as a smaller number of cylinders will be required. Various types of gas cylinders are shown in Fig. 5-5.

## STORAGE

Storage of gas cylinders and containers should be rigidly controlled to prevent the incorrect use of the shielding gases. The gas cylinder or container should always be stored in an outside or well ventilated area in an upright position and properly identified as to the type or mixture. All safety precautions should be followed to avoid injury to the user. Remember, inert gases do not contain oxygen, and therefore, will not support life. You cannot see, smell, or taste the inert gases.

| Metal | Shielding Gas | | Advantages |
|---|---|---|---|
| ALUMINUM | Manual Welding | Argon | Better arc starting, cleaning action and weld quality; lower gas consumption |
| | | Helium | High welding speeds possible |
| | Machine Welding | Argon-Helium | Better weld quality, lower gas flow than required with straight helium |
| MAGNESIUM | 0-1/16″ | Helium | Controlled penetration |
| | 1/16″ + | Argon | Excellent cleaning, ease of puddle manipulation low gas flows |
| MILD STEEL | 0-1/8″ | Argon | East of manipulation, freedom from overheating |
| | Spot Welding | Argon | Generally preferred for longer electrode life Better weld nugget contour Ease of starting, lower gas flows |
| | Depending on method of joint preparation | Argon-Helium | Helium addition improves penetration on heavy gage metal |
| | Manual Welding | Argon | Better puddle control, especially for position welding |
| STAINLESS STEEL | | Argon | Permits controlled penetration on thin gage material (up to 14 gage) |
| | Machine-Welding | Argon-Helium | Higher heat input, higher welding speeds possible on heavier gages |
| | | Argon-Hydrogen (65%-35%) | Prevents undercutting, produces desirable weld contour at low current levels, requires lower gas flows |
| | | Helium | Provides highest heat input and deepest penetration |
| COPPER & NICKEL Cu-Ni ALLOYS (MONEL & INCONEL) | | Argon | East of obtaining puddle control, penetration, and bead contour on thin gage metal |
| | | Argon-Helium | Higher heat input to offset high heat conductivity of heavier gages |
| | | Helium | Highest heat input for high welding speed on heavy metal sections |
| TITANIUM | | Argon | Low gas flow rate minimizes turbulence and air contamination of weld; improved metal transfer; improved heat affected zone |
| | | Helium | Better penetration for manual welding of thick sections (inert gas backing required to shield back of weld against contamination) |
| SILICON BRONZE | | Argon | Reduces cracking of this 'hot short' metal |
| ALUMINUM BRONZE | | Argon | Less penetration of base metal |

Fig. 5-2. Gases commonly used with gas tungsten arc welding.

## GAS DISTRIBUTION

Distribution of the gases to the welding area may be done in several ways. Fig. 5-6 shows a bank of cylinders that have been connected together. They may be located in a convenient area and the gas is then piped to the welding area. A single Dewar is shown in Fig. 5-7 where a manifold is used to connect up to 6 stations. Fig. 5-8 shows a commercial manifold connector for connecting several banks of cylinders together. With this arrangement, one bank may be used until empty, then the other is placed into use. The empty cylinders may then be replaced without affecting the bank in use.

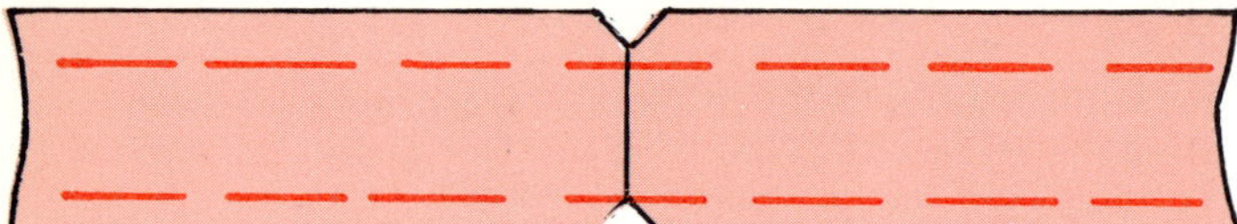


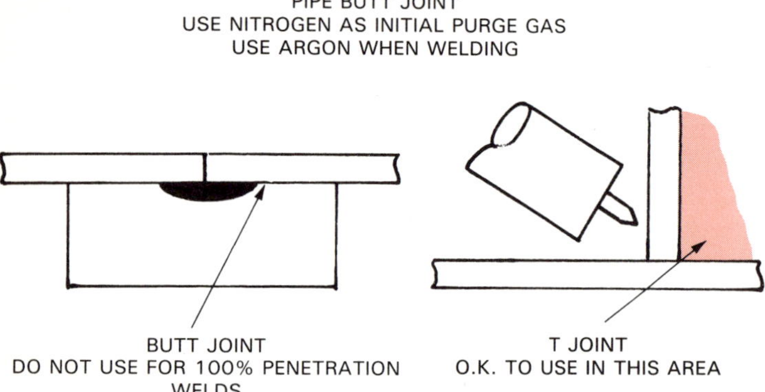


Fig. 5-3. Nitrogen used as a purge gas (ferrous metals only.)

Fig. 5-4. Two argon Dewars are connected to a pressure regulator and a switching valve. (Airco)

The distribution system must be leak free to maintain the purity and inertness of the gas being used. Therefore, all of the cylinder fittings must be cleaned before installation and seated into the regulators properly. High pressure connectors, tubing and pipe connectors must be protected to prevent entry of foreign materials, water, and oil when not in use. Plastic thread covers or tape may be placed over any unused openings for this purpose.

| CYLINDER STYLE | CONTENTS CUBIC FT. | FULL PRESSURE OF CYLINDER AT 70 °F | HEIGHT | O.D. |
|---|---|---|---|---|
| AS | 78 | 2200 | 35 | 7 1/8 |
| S | 150 | 2200 | 51 | 7 3/8 |
| T | 330 | 2640 | 60 | 9 1/4 |
| LC-3 | 2900 | 55 | 58 | 20 |

Fig. 5-5. Types and capacities of cylinders and Dewars supplied to industry.

Fig. 5-6. Individual cylinders are connected to a supply manifold by high pressure copper tubing, often referred to as pigtails. (Airco)

Fig. 5-7. Dewars supply gas at approximately 50 psi. A 6 station manifold is attached to this Dewar. (Distribution Designs, Inc.)

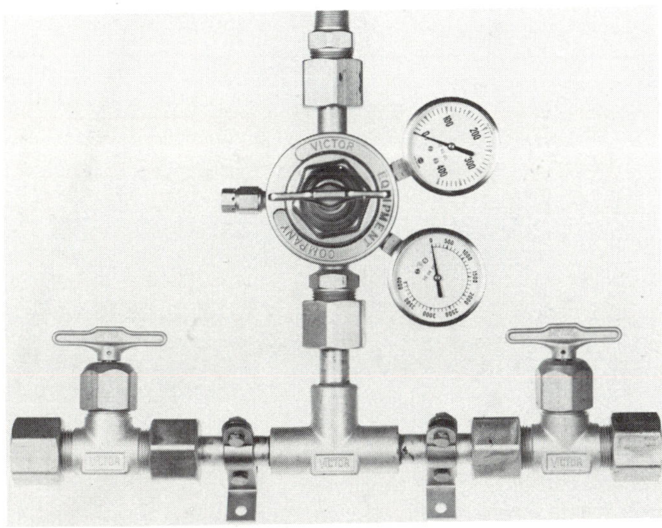

Fig. 5-8. This manifold is used for high pressure gases. A regulator is used to adjust desired manifold pressure. (Victor Equipment Co.)

## Manifolds

Manifolds or pipelines are often used to distribute gases to the welding area from the supply area. This reduces the number of individual cylinders required at the welding station. More than one station can be supplied from a single cylinder or a bank of cylinders. Rigid copper pipe is generally used in the construction of manifolds. The joints are either soft soldered or silver brazed. The joints must be sound and the manifold tested for leaks at a pressure above the normal operating pressure before placing the system in use.

This may be done by applying pressure to the system for several hours and noting the test pressure. A drop in pressure indicates leaks are present and they must be repaired.

Prior to placing the manifold system in use, the lines must be purged to remove any flux vapors and moisture. This can be done using nitrogen gas. Connect the nitrogen gas to the manifold, set the flow at the other end of the manifold for approximately 5 CFH and purge until analyzer test shows the system is clear of oxygen. Another test is to make a weld on a piece of titanium and hold the gas flow over the weld until the part cools below approximately 700 degrees F. If the test weld is silver in color, the system is clear. If a blue or grey color is present, the system is still contaminated and further purging is required.

Should the system require repair, disassembly or modification, the entire system should be repurged and tested prior to use.

## GAS REGULATION

Regulation of shielding gases is accomplished by several types of special equipment. Gases distributed through a manifold require regulators, as shown in Fig. 5-8, to reduce the pressure from the cylinder to the desired manifold pressure level. This pressure is usually set from 20 to 50 psi. A flowmeter as shown in Fig. 5-9 is then used at the welding station to regulate the flow to the welding torch.

Where a cylinder is used at the welding station, the regulator/flowmeter, as shown in Figs. 5-10 and 5-11,

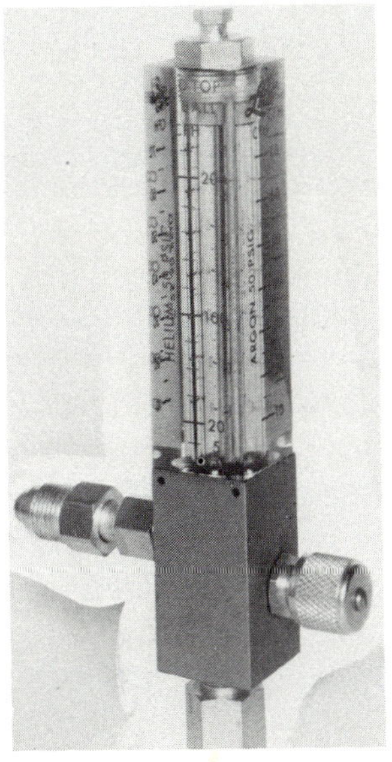

Fig. 5-9. Station flowmeters of this type operate from manifolds at approximately 50 psi. (Linde Co.)

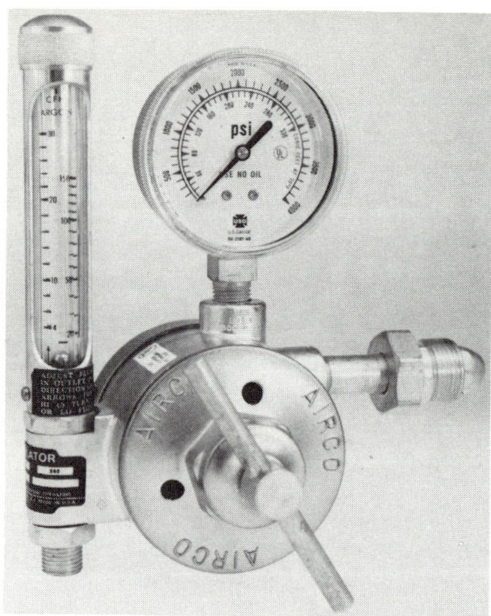


Fig. 5-10. Single cylinder regulator/flowmeters have a gauge that shows cylinder pressure. Gas flow (cubic feet per hour) to the torch is adjusted by turning the adjustment knob and reading the ball position in the vertical tube. (Airco)

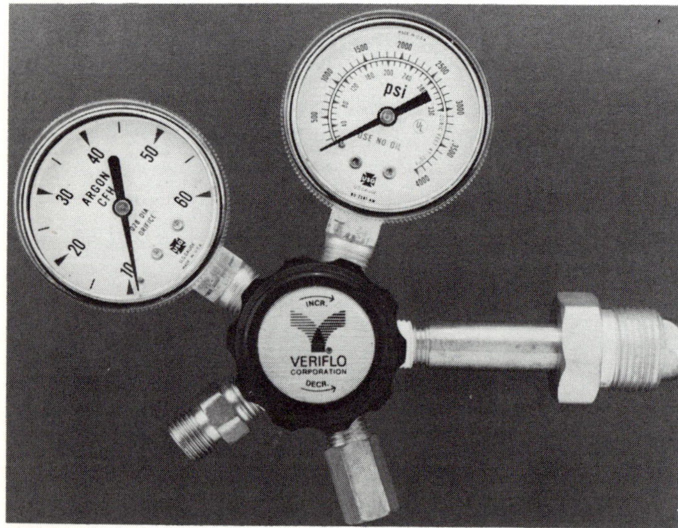


Fig. 5-11. A dial gauge is used with this regulator to read gas flow in cubic feet per hour. The desired gas flow is obtained by turning the adjustment knob on the front of the regulator (Veriflow Co.)

is used to reduce the pressure from the cylinder and regulate the flow. When installing a regulator/flowmeter or flowmeter with a ball tube, the ball tube should always be in the vertical position for proper operation. The amount of flow is always indicated at the TOP of the ball unless otherwise indicated.

Regardless of the type of gas supply (cylinder, Dewar, manifold), when the gas flow valve is opened for welding, a surge of gas will exit from the gas nozzle. This is due to the pressure build-up when the gas is not flowing. This surge of gas will last for several seconds until the excess pressure is reduced. To eliminate this condition, a special designed surge check valve may be used, as shown in Fig. 5-12.

Fig. 5-12. Surge check valves eliminate surging of gas from the nozzle during the start of the operation. By eliminating the surge, they readily pay for themselves in gas savings. (Weld World Co.)

## Gas Mixing

Mixing of inert gases can be done at the manifold, as shown in Fig. 5-13. This type of mixer is used where large volumes of gases are used. The gas mixers shown in Figs. 5-14 and 5-15 are often used in single station operations. Another type of mixer shown in Fig. 5-16 may also be used in manual applications with the gas mixed within the Y valve. These Y valves are installed on the outlet side of the flowmeter with the gas metered by the two separate gas flowmeters.

To prevent back flow of the gases and improper mixing, a flow check valve, Fig. 5-17 should be installed between the flowmeter and the Y valve. Fig. 5-18 shows a back flow check valve installed on a flowmeter.

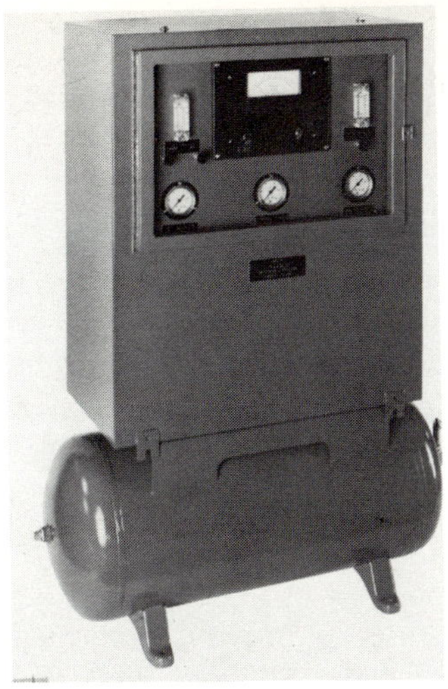

Fig. 5-13. The large tank on the bottom of the mixer serves as a mixing and storage chamber. This assures sufficient volumes of mixed gases for large users. (Thermco Instrument Corp.)

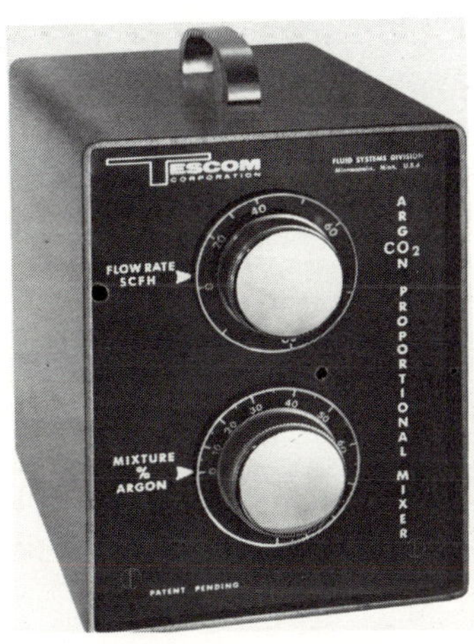


Fig. 5-14. Small proportional mixers of this type are used for individual station mixing. (Tescom Corp.)

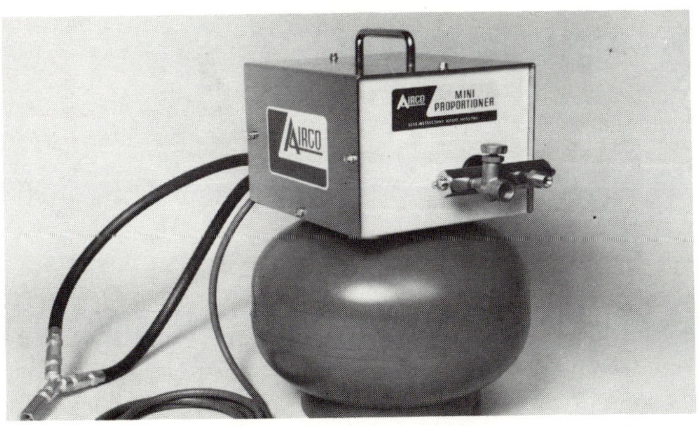


Fig. 5-15. Mini-proportioner with mixing and storage tank for multiple stations. (Airco)

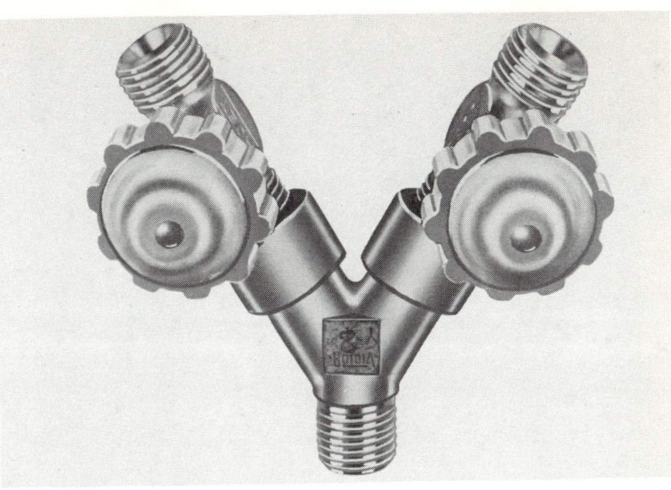

Fig. 5-16. Y valves have shut-off valves for single flow of single gas or gas mix of two gases. (Victor Equipment Co.)

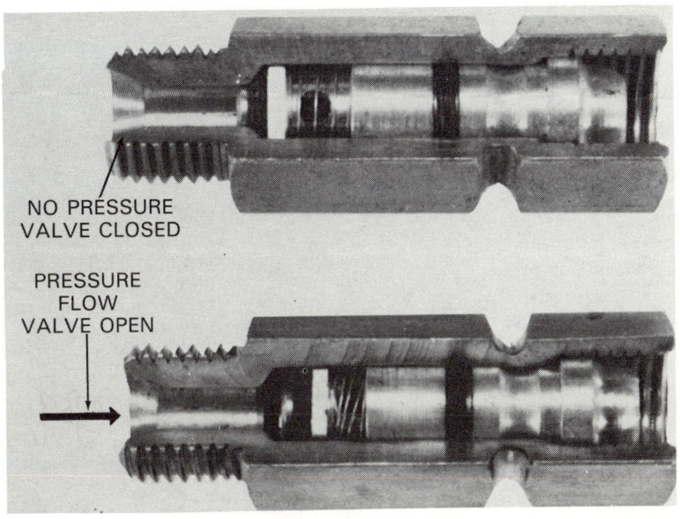


Fig. 5-17. Reverse flow check valves prevent mixing of gases in supply line. The upper valve is shown closed. The bottom valve is shown open. (Airco)

## Purity Testing

Testing for proper mixes at the welding stations can be accomplished by gas analyzers, Fig. 5-19. These instruments can also be used for manifold leak testing and testing for adequate purging of pipes, vessels, etc., prior to the weld. The instrument shown in Fig. 5-20 is used also for testing purged vessels, chambers, and pipes to determine if the purge is adequate. A sample of the purge gas is introduced into the light bulb and current is turned on to heat the bulb filament. After the current is turned off, a silver filament in the bulb indicates the purge is satisfactory. Blue or grey filament colors indicate contamination and the purge is not satisfactory.

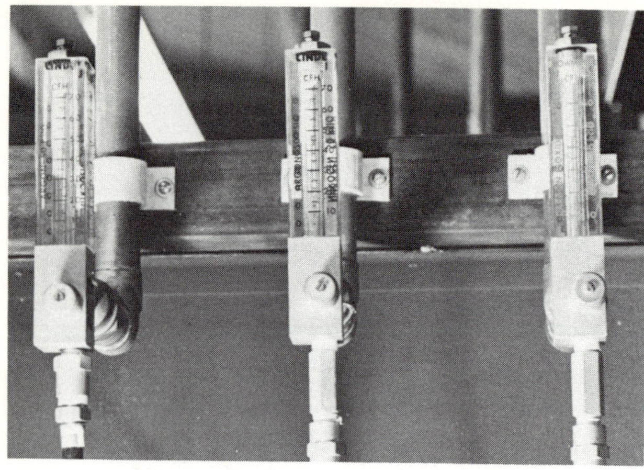

Fig. 5-18. Reverse flow check valves are installed on the outlet side of the flowmeter.

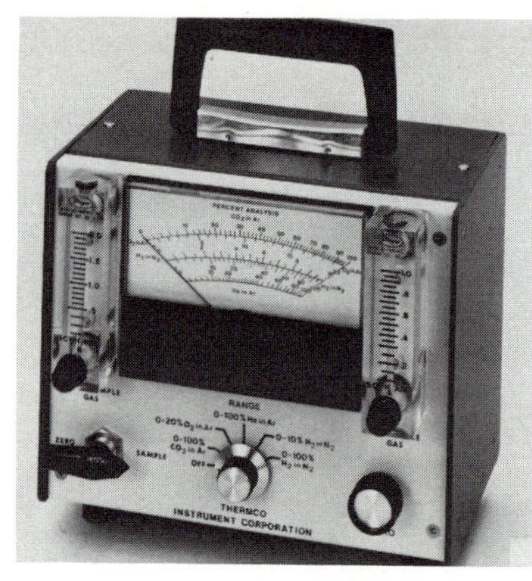


Fig. 5-19. Portable analyzers of this type measure the percentages of the individual gases. (Thermco Instrument Corp.)

Fig. 5-20. Portable analyzers using a filament bulb are often used to determine the condition of purging gases of pipes, tubes, vessels, and chambers prior to welding. (Jetline Engineering Inc.)

**Precaution Notes**

1. Use only cylinders that have a content identification.
2. Use only high pressure equipment on high pressure gas cylinders.
3. Always clean cylinder outlet port before installing regulator.
4. Make sure regulator is firmly attached before opening cylinder valve.
5. Never stand in front of regulator when opening cylinder valve.
6. Open cylinder valve slowly, then open valve completely.
7. When cylinder is empty, close the valve.
8. Never use water, oil, vaseline, etc., when installing tubing or connectors in any part of the system.
9. Never use rubber hoses for inert gases.
10. Never use a tube that has contained water, air, or any fluid.
11. Never try to repair a tubing leak with tape or a patch. Discard the tubing, and replace.
12. Always close flowmeter valves when done.
13. Keep vinyl tubing away from heat and off of the floor. Use covers to protect tubing.
14. Do not allow tubing to kink and shut off flow.
15. Maintain an efficient supply of inert gas by checking the system often and carefully.

## REVIEW QUESTIONS

1. Why are shielding gases used in GTA welding? Give six reasons.
2. List two gases used for arc shielding gases.
3. Why are the arc shielding gases inert?
4. What are the main characteristics of a shielding inert gas?
5. Where is a nitrogen gas used?
6. Does argon gas or helium gas have the same arc starting characteristics?
7. Why does helium gas require higher flow rates than argon?
8. Where does helium gas perform better than argon?
9. Which argon-helium gas mixture is commonly used in GTAW for welding aluminum on AC?
10. Which argon-helium gas mixture is commonly used in GTAW welding with DC?
11. Where is an argon-hydrogen gas mixture used? Why?
12. Does titanium require a very high purity shielding gas? Why?
13. Liquified gas cylinders are commonly called ______.
14. Rental rates for gas cylinders are often called ______ rates.
15. ______ gas is used to purge manifolds, vessels, and pipes before welding.

16. Why are regulators used on gas cylinders?
17. Manifolds usually have a pressure range of ___20___ to ___50___ psi.
18. ___FLOWMETER___ are used to control the flow of gases from a cylinder to the work area.
19. A special design valve may be used to prevent ___SURGES___ from the shielding nozzle when starting the gas flow.
20. ___CHECK___ valves are used to prevent back flow of the shielding gases when mixing different gases.
21. What type of equipment should be used to protect vinyl hoses? LEATHER COVERS

Single gas cylinders are connected to a single manifold. Any number of cylinders may be used with this arrangement. (Airco)

# Chapter 6

# FILLER MATERIALS

FILLER MATERIALS used in the GTAW process must be of the highest quality possible to make acceptable welds. For this purpose, manufacturers of the material use specialized machines, processes and inspections. The filler material is furnished in many forms.

## Manufacturing

Material to be made into filler material may be selected based on several factors:

1. Chemical composition.
2. Mechanical properties.
3. Notch toughness.
4. Grain size.
5. Internal defects.
6. Impurity level limits.

Material selected at the primary mill is hot drawn through reducing dies to a pre-determined size and cleaned. Further reduction of the wire to the desired size is then done cold. Annealing and cleaning operations are performed as the wire is further reduced in size.

Various types of lubricants are used in the draw dies during the drawing process. These lubricants are used to decrease the wear of the dies, to decrease the friction between the wire and the die, and to carry away the heat generated in the die.

During the drawing process several types of defects can occur which will effect the quality of the filler material. These defects include:

1. Overlapping.
2. Splitting.
3. Cracking.
4. Seaming.
5. Oxide formations.

Examples of these defects are shown in Fig. 6-1. Filler materials with these defects should generally not be used.

At the completion of the drawing process, the wire is then cleaned of all surface impurities and processed for shipment to the user.

## Specifications

Many quality specifications are used for the manufacture, testing, inspection, and packaging of the various filler material. Fig. 6-2 lists some of the common filler metal specifications. A typical specification may include the following requirements:

1. Scope.
2. Classification and usability.
3. Manufacturing methods.
4. Acceptability test.
5. Chemical composition.
6. Usability test and results.
7. Standard sizes and lengths.

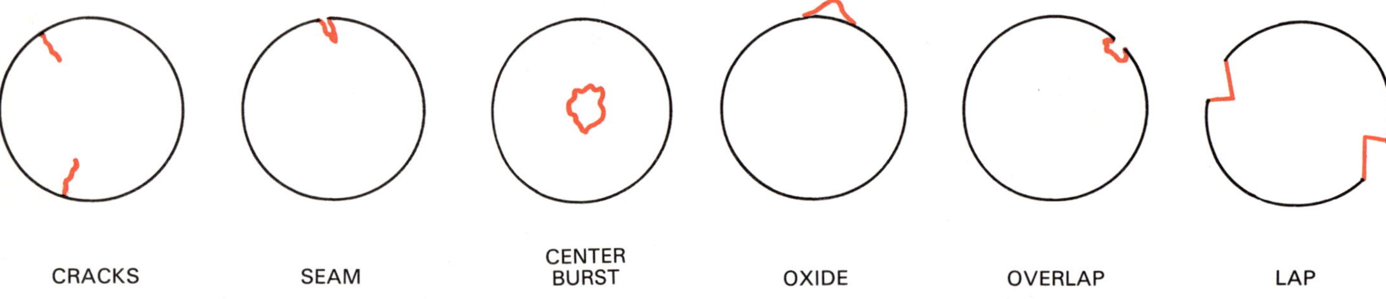


Fig. 6-1. Typical defects that can occur in the wire during the drawing operation.

| Specifications | Material Types |
|---|---|
| American Welding Society | All Types |
| American Society for Testing Materials | All Types |
| Mil-R-5632 | Steel |
| Mil-R-5031 | Stainless Steel |
| Mil-E-23765 | Steel |
| Mil-E-16053 | Aluminum |
| QQ-R-566 | Aluminum |
| Aeronautical Material Specifications | All Types |

Fig. 6-2. Common welding wire specifications used throughout the welding industry to establish quality.

8. Finish and temper.
9. Spool and winding requirements.
10. Packaging.
11. Marking of packages.
12. Guarantee.

In some cases, the user may add some additional requirements to the basic specification such as chemistry or weld testing, identification, packaging, etc. Each additional requirement will add cost to the filler material.

Manufacturers of filler materials only guarantee their product to meet the requirements of the specification and replacement of material. They do not guarantee that the material will make welds with acceptable results, as they cannot govern the welding application.

Filler wire manufacturers make welding material for three major areas:

1. GENERAL USE wire which will meet a specification requirement. However, no record of chemical composition, strength level, etc., is submitted to the user.
2. Fabrication specifications that require more rigid control over the filler metal may require the user to obtain a certificate of conformance with the purchase of the material. A CERTIFICATE OF CONFORMANCE as shown in Fig. 6-3, is a statement that the filler materials meet all the requirements of the material specifications. All of the material will be identified by heat numbers, lot numbers, or code numbers located on the package. On some work, the buyer may require that these numbers are recorded wherever the material is

CERTIFICATION OF QUALITY CONFORMANCE TESTS

Manufacturer or Distributor ______ Customer's Name ______
Address ______
Date ______ Customer's Order No. ______
Specification MIL- ______
Type MIL- ______ Core Wire Heat No. ______
Diameter & Length ______ Lot Identification MIL- Para. ______
Inspection Level ______
Lot No. ______ Wet Batch No. ______

Chemical Analysis (Complete): Carbon ______ Manganese ______ Silicon ______ Phosphorus ______ Sulfur ______ Chromium ______ Nickel ______ Molybdenum ______ Vanadium ______

Mechanical Test — As-Welded / Stress-Relieved: Yield Strength (0.2% offset method); Tensile Strength; Elongation (%); Reduction of Area (%)

Charpy Impacts: 1. 2. 3. 4. 5. (As-Welded) 1. 2. 3. 4. 5. (Stress-Relieved)

Chemistry was taken from: Chem Pad ☐ Groove Weld ☐
X-Ray Results ______
Concentricity (%) ______
Covering Moisture ______
Grinding During 8a Test Plate Preparation

Groove Weld Test: Test No. 3 8a Chem Pad; Amperage ______; Operator Error (Layer Nos.) ______

We hereby certify that the above material has been tested in accordance with the listed specification and is in conformance with all requirements.

Fig. 6-3. Certificate of Conformance form. (Techalloy Maryland, Inc.)

Reid - Avery
Dundalk, Baltimore, Md. 21222

QUALITY ASSURANCE TEST REPORT

DATE: ______
SOLD TO: SHIPPED TO: DATE SHIPPED: ______
P.O. NO.:- P.O. NO.:-

SPECIFICATION:

| ITEM | POUNDS | SIZE | TYPE | LOT NO. | HEAT NO. |
|---|---|---|---|---|---|
| 1. | | | | | |
| 2. | | | | | |
| 3. | | | | | |

ACTUAL CHEMICAL ANALYSIS OF WIRE OR WELD METAL

| ITEM | C | Mn | P | S | Si | Cr | Ni | Mo | | | | |
|---|---|---|---|---|---|---|---|---|---|---|---|---|
| 1. | | | | | | | | | | | | |
| 2. | | | | | | | | | | | | |
| 3. | | | | | | | | | | | | |

ADDITIONAL TEST RESULTS

TESTS PERFORMED WITH: ______
X-RAY: ______
STRESS RELIEVED ______ HRS. @ ______ °F.
AS WELDED PLATE NO. ______ STRESS RELIEVED PLATE NO. ______
YIELD POINT, PSI
TENSILE STRENGTH, PSI
ELONGATION IN 2", %
REDUCTION OF AREA, %
IMPACTS PLATE NO. ______ @ ______ °F ______ ft. lbs. PLATE NO. ______ @ ______ °F ______ ft. lbs.

State of ______
City of ______
Subscribed and sworn to before me this ______ day of ______ 19 ______.
Notary Public
My commission expires ______

I certify the chemical analysis and physical or mechanical test results reported above are correct as contained in the records of the company.

QUALITY ASSURANCE DEPARTMENT

Fig. 6-4. Certified chemical analysis form. (Techalloy Maryland, Inc.)

used in welding applications.

3. A CERTIFIED CHEMICAL ANALYSIS report as shown in Fig. 6-4 is the actual chemical analysis of the individual heat or lot of material. The test is made on a spectrometer machine shown in Fig. 6-5. Critical welding operations on missiles, nuclear reactors, and pressure vessels usually require very close control of the actual filler material chemistry. Records are maintained during the fabrication cycle wherever the specific materials are used. In case of a joint failure, either because of the filler material or base metal, other joints in the system can be located and evaluated for possible replacement.

Fig. 6-5. Actual chemical analyses are obtained from a wire sample on this machine. (Techalloy Maryland, Inc.)

### Filler Material Forms

Filler materials are furnished in various forms and sizes for GTA welding and these include:

1. SPOOL OR COIL. Spools or coils are used in semi and automatic welding for both cold and hot wire operations. The spools or coils adapt to a standard hub attachment and are made of plastic or wood pulp. They are disposable after use. The filler materials are wound on the spool either layer wound, as shown in Fig. 6-6, or level layer wound, as shown in Fig. 6-7. Depending on the type and size of the material, they contain varying amounts of wire.

Fig. 6-6. Welding wire on this spool is layer wound.

Fig. 6-7. The aluminum wire on this spool is level layer wound. Note that the outer surface is flat.

To prevent wire feeding problems in automatic welding, all spool wire must meet cast and helix requirements. CAST is the diameter of one complete circle of wire as it lies on a flat surface. HELIX is the maximum height of any point of this circle of wire above the flat surface. Fig. 6-8 indicates these tolerances. Wire that has incorrect cast and/or helix will cause excessive wear of the conduit liner and the guide tip. If the wire is used with an automatic voltage control head, the welding voltage cannot be maintained properly.

2. STRAIGHT LENGTH ROD. These rods are used in manual welding where the welder feeds the material into the pool as required. They are made

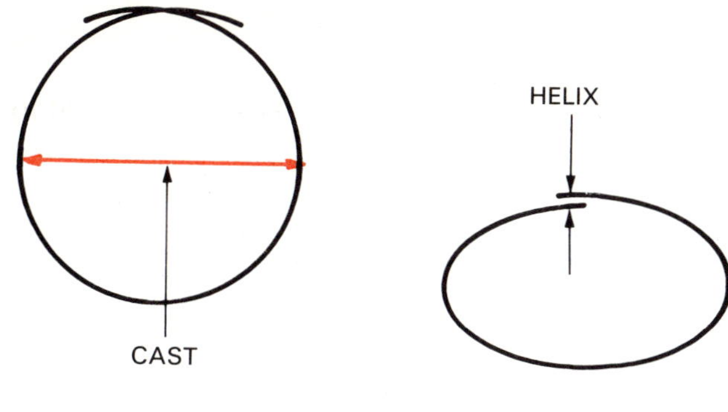


| | | Cast | | | | Helix | |
|---|---|---|---|---|---|---|---|
| Spool Size | Wire Types | Min. in | mm | Max. in | mm | Max. in | mm |
| 4-in (100 mm) | Low-Alloy, Stainless, and Nickel Alloy | 6* | 150 | 9 | 230 | ½ | 13 |
| | Aluminum | † | | 6 | 150 | 1 | 25 |
| 8-in (200 mm) | Low-Alloy, Stainless, and Nickel Alloy | 15* | 380 | 30 | 760 | 1 | 25 |
| 12-in (300 mm) | Low-Alloy, Stainless, and Nickel Alloy | 15* | 380 | 30 | 760 | 1 | 25 |
| | Aluminum | † | | 15 | 380 | 1 | 25 |

Standards for Cast and Helix

* Measured on outside strand of full spool
† Diameter of wire level from which sample is taken

Fig. 6-8. Cast and helix. These terms define the characteristics of any form of continuous wire as it comes from the spool or coil. Cast is the diameter of one complete circle of wire as it lies on a flat surface. Helix is the maximum height of any point of this circle of wire above the flat surface.

in the drawing mill and furnished in varying sizes and lengths. The standard length is 36 inches.

3. INSERTS. Inserts are generally used in manual, semi, and automatic welding of pipe. The standard designs are shown in Fig. 6-9.

### Identification

Identification of spools or coils of wire is done with the use of labels or tapes on the outside of the spool or flange. A spool with a label identifying the material is shown in Fig. 6-10.

Straight length rods are identified by flag-tags as shown in Fig. 6-11, or by embossing as shown in Fig. 6-12. Inserts are identified by ink, as shown in Fig. 6-13.

### Filler Material Packaging

Manufacturers of filler materials use a variety of packaging methods to protect the material while be-

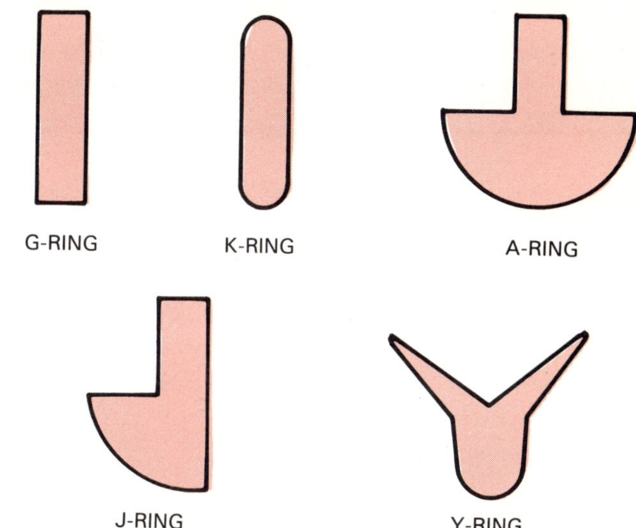


Fig. 6-9. Consumable inserts for pipe welding are formed from welding wire.

Fig. 6-10. The label identifies this particular material as: 4043 aluminum. HQ letters identify the material as a special processed high quality material. The Linde stock number is 5202F07. The Linde control number is 707. The actual heat number is 00350. The wire is 3/64 inch diameter. The weight of the welding wire on the spool is 14.3 pounds. (Linde Co.)

Fig. 6-11. Flag-tags are attached to the end of the rod to identify the material. (Techalloy Maryland, Inc.)

Fig. 6-12. Embossing the rod identification is usually done on soft materials such as aluminum, magnesium, and copper.

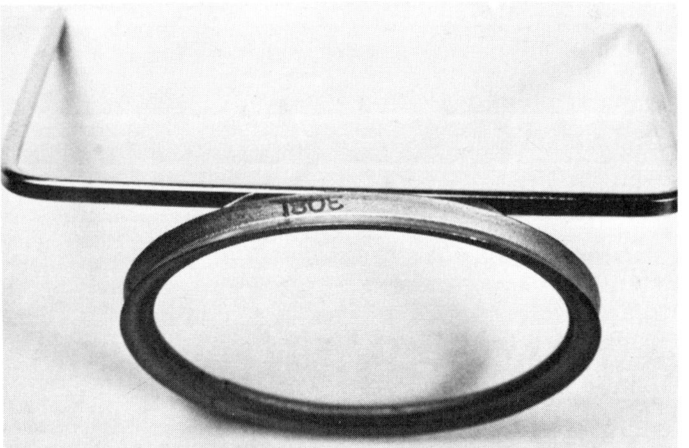

Fig. 6-13. Inserts are ink stamped only for identification. As the insert is completely consumed during the welding operation, contamination from stamping or embossing is avoided.

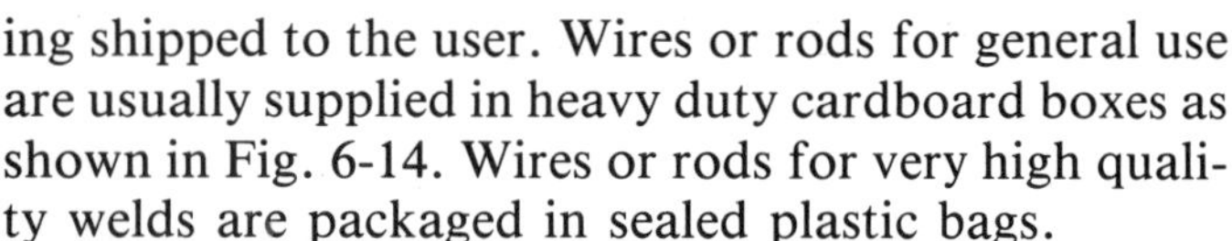

ing shipped to the user. Wires or rods for general use are usually supplied in heavy duty cardboard boxes as shown in Fig. 6-14. Wires or rods for very high quality welds are packaged in sealed plastic bags.

Spooled or coiled wire for very high quality welds are packaged as shown in Fig. 6-15.

## Filler Material Use in the Welding Shop

Filler materials are shipped to the user packaged to prevent contamination of the material. Preventing contamination of the material after the package is opened is the responsibility of the user.

Filler materials are easily contaminated by oil, moisture, grease, smoke, soot, and salt. Salts and oils on the hand, fingers, or gloves will also contaminate the materials. Dirty work areas, tables, rags, etc. readily contaminate the filler material when they come into contact. These foreign materials often cause defects in the weld metal in the form of porosity and cracks. Rejection of the defective weld which requires rework to eliminate the defect adds cost to the job.

The simplest method of avoiding contamination of the filler metal is to keep it clean. This is done by:

1. Keep the material packaged for as long as possible.

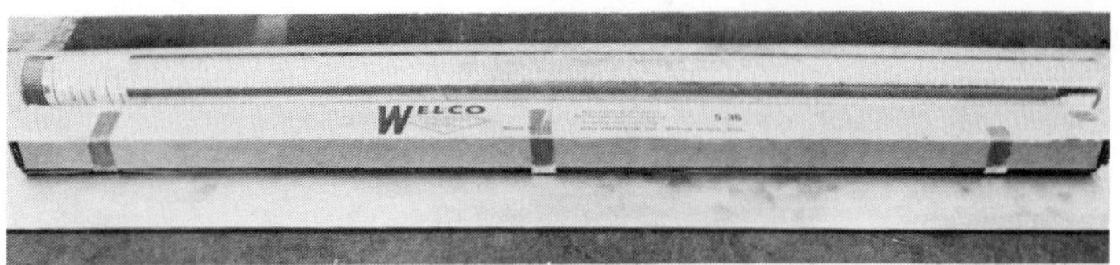

Fig. 6-14. General use rod is packaged in sturdy cardboard packages. Plastic bags give additional protection for the wire within the cardboard box. Plastic bags seal out any contamination from the outside environment. The bag should remain sealed until the rod is to be used.

Fig. 6-15. Spooled wire in this package will store idefinitely if the package is not broken.

2. Work in a clean and dry area.
3. Store unsealed filler materials in a heated cabinet.

4. Handle material as little as possible, then only with clean gloves.
5. Spooled filler material mounted on automatic welding machines should always be covered, as shown in Fig. 4-14.
6. Clean filler materials prior to use by wiping with a clean rag soaked in alcohol, acetone, or a similar degreaser.
7. Contamination on the ends of filler wire should be snipped prior to re-use, as shown in Fig. 6-16. Welding fillers removed from the inert gas shield during the welding operation are hot and readily absorb oxygen, nitrogen, and hydrogen.

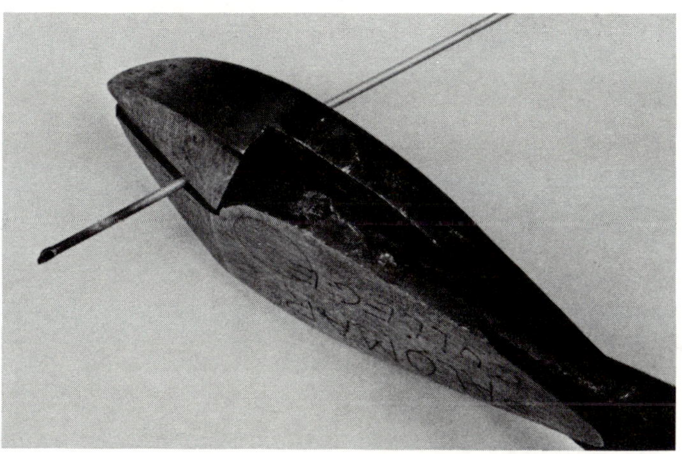

Fig. 6-16. The dark, discolored area on the end of the wire indicates contamination. The area at the end should be cut off before re-use.

Contaminated filler materials cannot be tolerated where high quality welds are required. Requiring the filler wire manufacturer to clean, inspect, and package the material to a specification does little good if the material is contaminated prior to use by improper handling of the material. Quality welds cannot be made using a contaminated filler material.

## REVIEW QUESTIONS

1. What are the six major factors used in selecting welding wire?
2. To draw the wire to the proper diameter, both ________ and ________ drawing methods are used.
3. List five types of defects that may occur during the drawing operation.
4. Specifications for filler materials establish specific requirements for the ________ of the material.
5. Does the manufacturer of the wire guarantee the product to make a good weld?
6. A certified chemical analysis is an actual analysis of a specific ________ or ________ of material.
7. Spooled wire may be supplied either ________ or ________ wound.
8. What two requirements are placed on welding wire to be used for automatic welding?
9. What are welding inserts made from?
10. How are straight length rods identified?
11. List five ways in which filler wires may become contaminated.
12. Name two typical degreasers used to clean filler wires before use.

# Chapter 7

# WELD JOINTS AND WELD TYPES

## JOINT TYPES

The American Welding Society defines a JOINT as the manner in which materials fit together. There are five basic types of joints. They include the following joints.

1. Butt joint.
2. T joint.
3. Lap joint.
4. Corner joint.
5. Edge joint.

The basic types of joints are shown in Fig. 7-1.

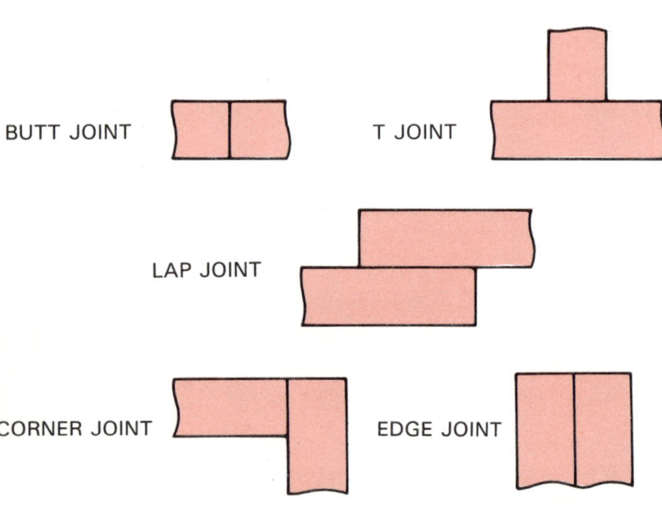


Fig. 7-1. Basic types of joints.

## WELD JOINT PREPARATION

Weld joints may be initially prepared in a number of ways. These include:

1. Shearing.
2. Casting.
3. Forging.
4. Machining.
5. Stamping.
6. Filing.
7. Etching.
8. Grinding.
9. Routing.
10. Oxygen-acetylene burning-cutting.
11. Plasma arc cutting.

Final preparation of the joint prior to welding will be covered in another chapter which details the welding of a particular material.

## WELD TYPES

Various types of welds can be made in each of the basic joints. They include the following types.

1. Butt joint, Fig. 7-2.
   - A. Square groove butt weld.
   - B. Bevel groove butt weld.
   - C. V groove butt weld.
   - D. J groove butt weld.
   - E. U groove butt weld.
   - F. Flare V Groove butt weld.
   - G. Flare Bevel Groove butt weld.

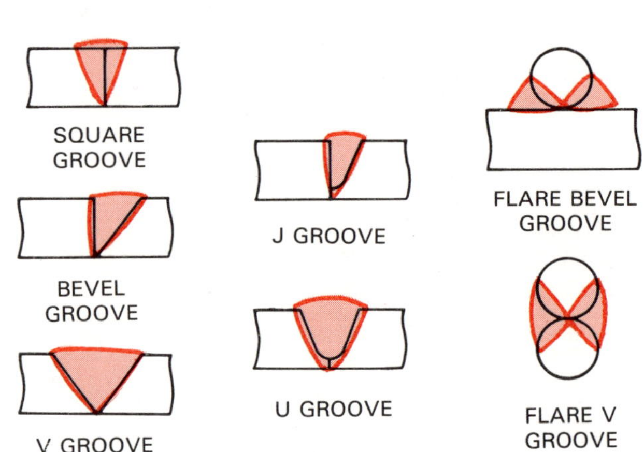


Fig. 7-2. Types of welds which may be made with a basic butt joint.

2. T joint, Fig. 7-3.
   A. Fillet weld.
   B. Plug weld.
   C. Slot weld.
   D. Bevel groove weld.
   E. J groove weld.
   F. Flare Bevel groove weld.
   G. Melt thru weld.

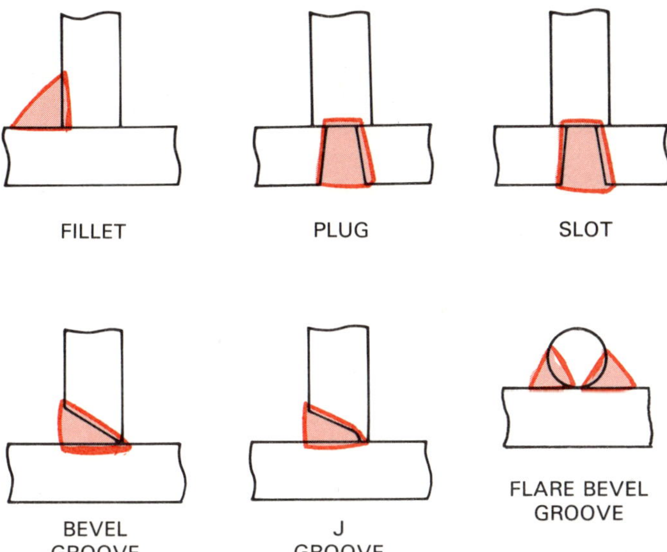


Fig. 7-3. Types of welds which may be made with a basic T joint.

3. Lap joint, Fig. 7-4.
   A. Fillet weld.
   B. Plug weld.
   C. Slot weld.
   D. Spot weld.
   E. Bevel groove weld.
   F. J groove weld.
   G. Flare Bevel groove weld.
4. Corner joints, Fig. 7-5.
   A. Fillet weld.
   B. Spot weld.
   C. Square groove weld or butt weld.
   D. V groove weld.
   E. Bevel groove weld.
   F. U groove weld.
   G. J groove weld.
   H. Flare V groove weld.
   I. Edge weld.
   J. Corner flange weld.
5. Edge joint, Fig. 7-6.
   A. Square groove weld or butt weld.
   B. Bevel groove weld.
   C. V groove weld.

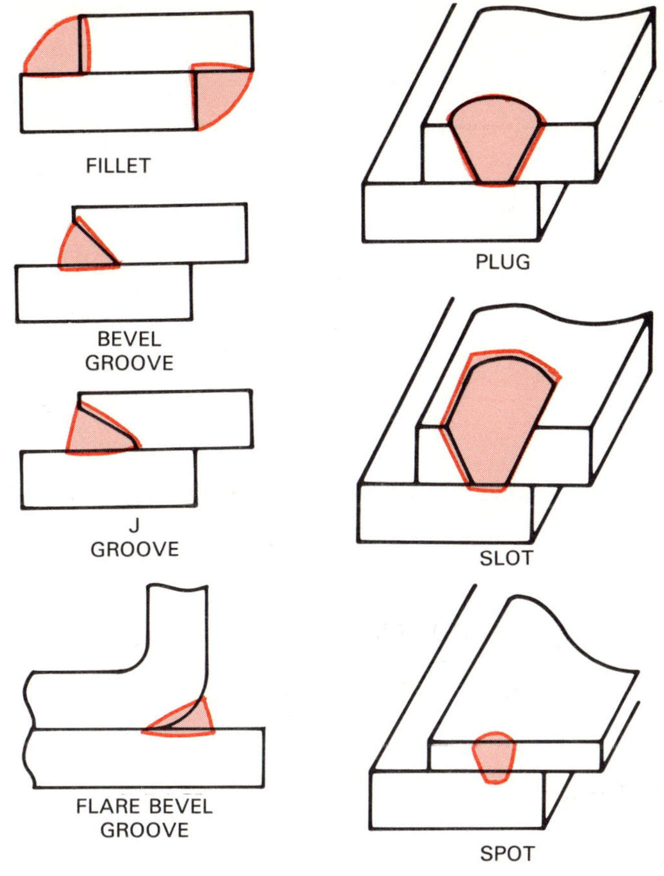


Fig. 7-4. Types of welds which may be made with a basic lap joint.

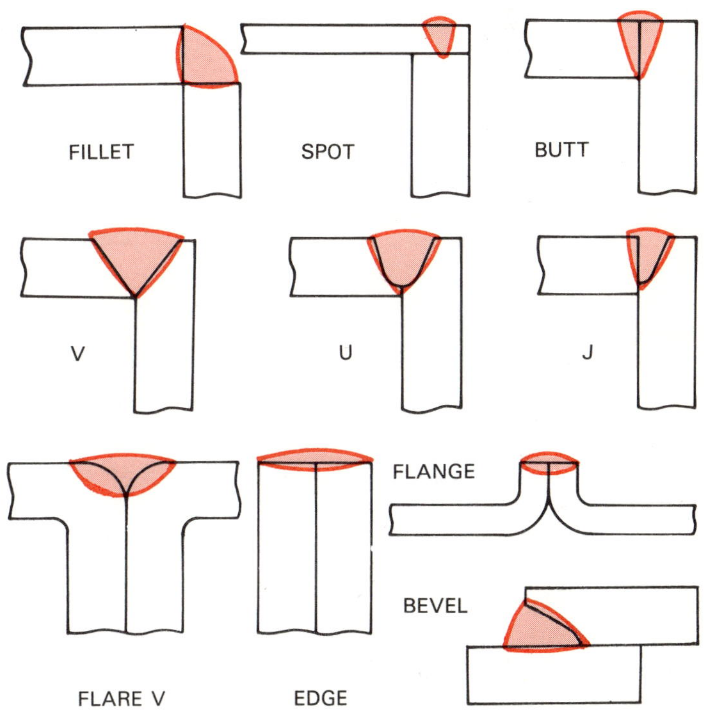


Fig. 7-5. Types of welds which may be made with a basic corner joint.

D. J groove weld.
E. U groove weld.
F. Edge flange weld.
G. Corner flange weld.

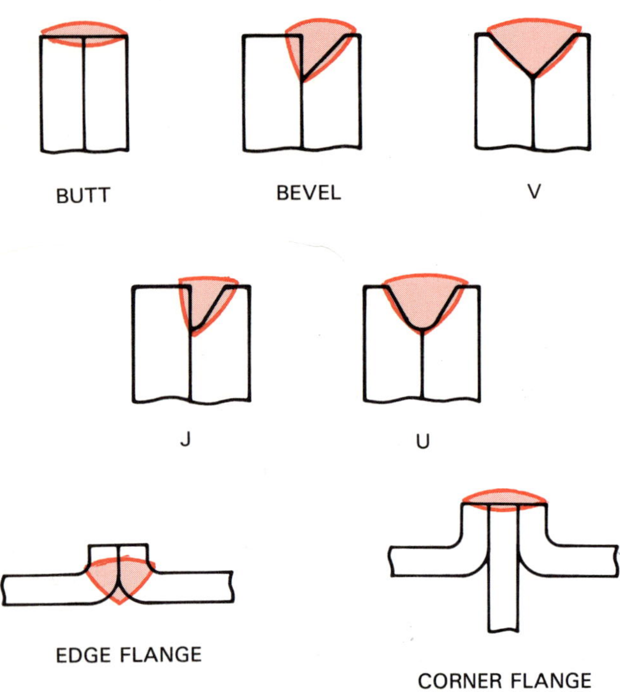


Fig. 7-6. Types of welds which may be made with a basic edge joint.

## DOUBLE WELDS

In some cases the weld cannot be made from one side of the joint. The weld must be made from both sides and these welds are identified as DOUBLE WELDS. Fig. 7-7 shows some common applications of double welds on the basic joint designs.

## WELDMENT CONFIGURATIONS

Often times the basic joint is modified to assist in assembly of the components, to gain access to the weld joint area, to change the metallurgical properties of the weld, etc. Some common weldment configuration designs are described here. Joggle type joints are used in cylinder and head assemblies where back-up bars or tooling cannot be used, Fig. 7-8. Tubing to heavy wall tube with a built-in back-up bar are used where sufficient material is available for machining the required back-up, Fig. 7-9. Groove welds with a fabricated back-up bar are fitted to the joint very tightly, Fig. 7-10. Loose fitting bars cause problems concerning heat flow and penetration. Special designed weld joints for a controlled penetration joint are used where excess penetration would cause a problem with assembly or liquid flow, Fig. 7-11. Series of bead welds overlayed on the face of a joint is called BUTTERING, Fig. 7-12. Buttered welds are often used to join dissimilar metals. Series of overlayed welds on the surface of a part to form a weld joint or to protect the base material is called SURFACING or CLADDING, Fig. 7-13. Tube to tube or tube to flange weld have

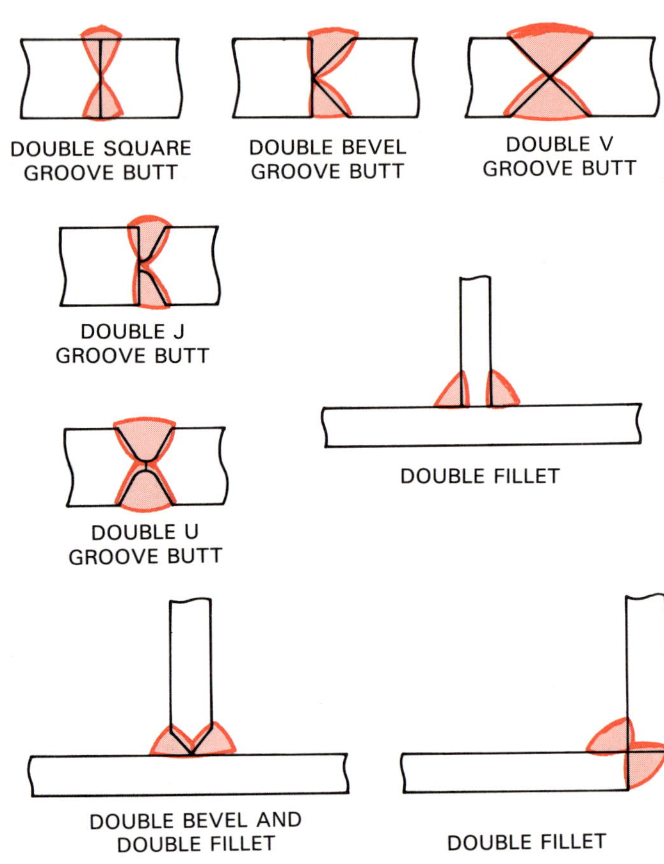


Fig. 7-7. Applications of double welds.

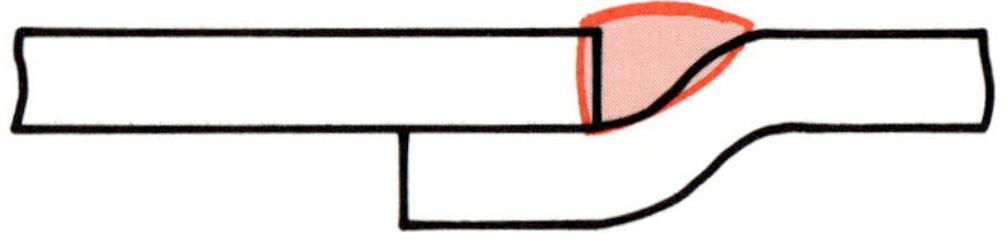

Fig. 7-8. Joggle type joint.

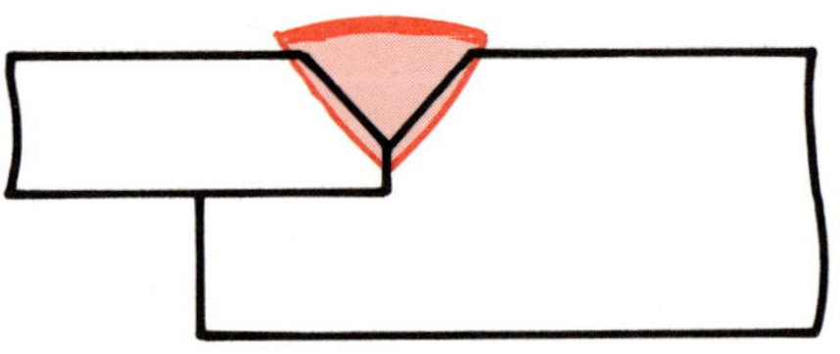

Fig. 7-9. Built-in back-up bar joint.

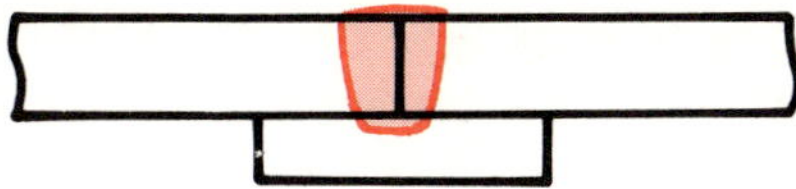

Fig. 7-10. Fabricated back-up bar configuration.

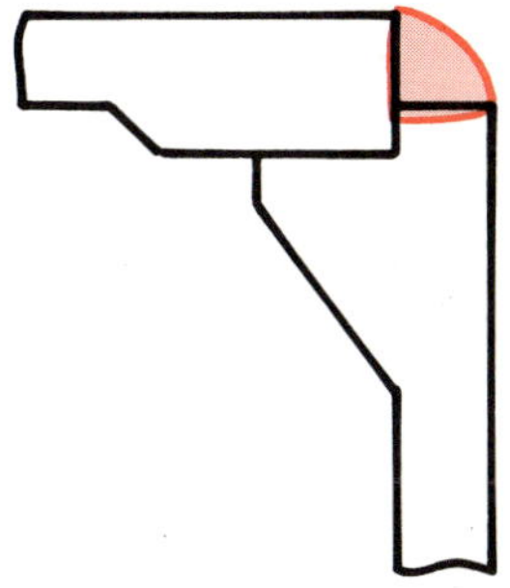

Fig. 7-11. Controlled penetration joint.

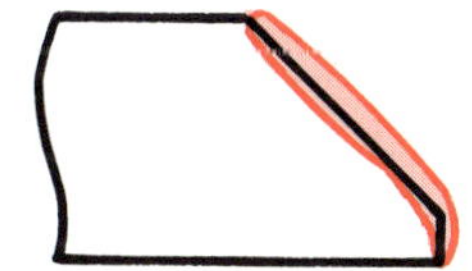

Fig. 7-12. Buttered welds.

Fig. 7-13. Overlay weld.

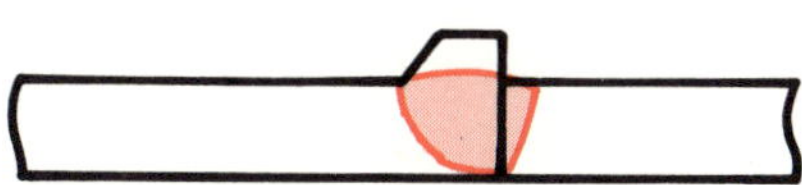

Fig. 7-14. Tube to tube or tube to flange weld.

the required filler material machined into the joint, Fig. 7-14.

Tubing weld joints are often modified by adding a sleeve either on the inside or outside of the joint. This type of joint allows the use of tubes without an end to end fit as shown in Fig. 7-15.

Pipe weld joints are often modified to fit a special type of insert or back-up ring. Several types of back-up rings are shown in Fig. 7-16.

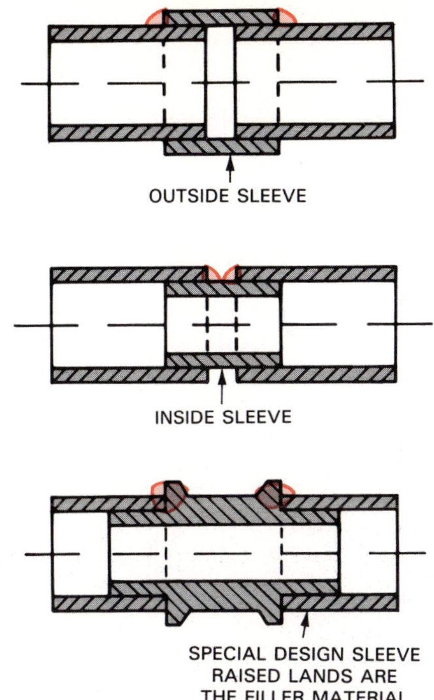


Fig. 7-15. Sleeve type joints are commonly used where the tubing is restrained and cannot shrink during the welding.

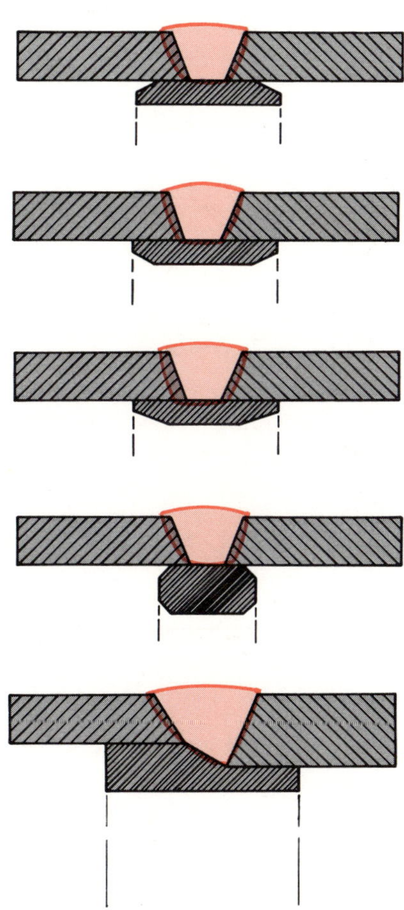

Fig. 7-16. Back-up rings for pipe joints.

### Welding Terms and Symbols

Some of the common terms used to describe the weld joint are shown in Fig. 7-17. Common terms used to describe the weld and the weld area are given in Fig. 7-18.

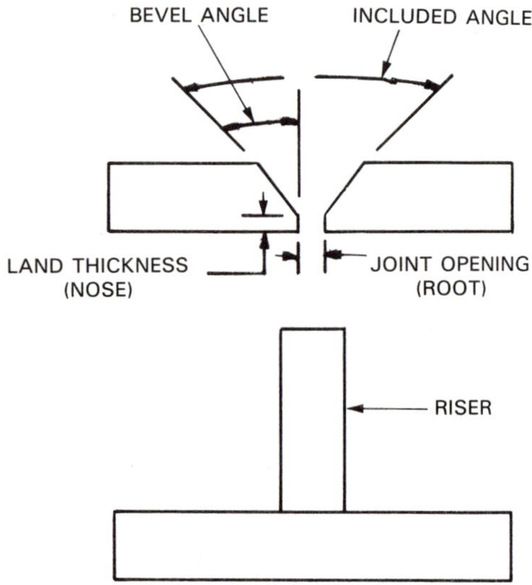


Fig. 7-17. Weld joint terms.

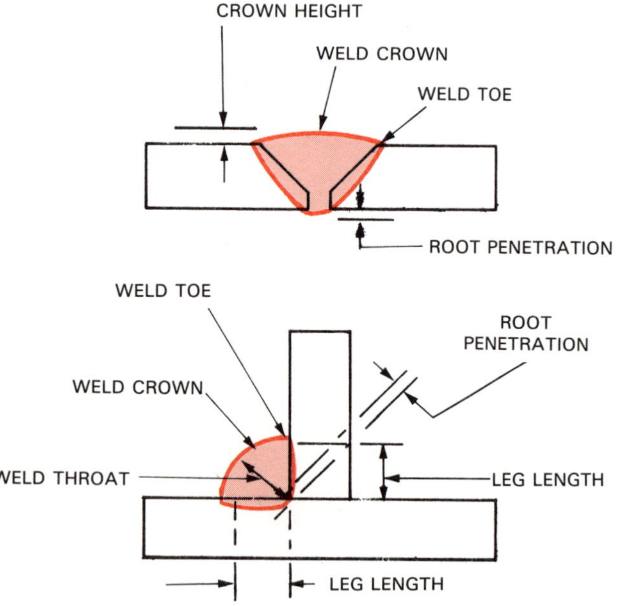


Fig. 7-18. Weld and weld area terms.

The welding symbol shown in Fig. 7-19 has been adopted by the American Welding Society. This symbol is used on drawings to indicate the weld to be made. The symbol in Fig. 7-19 includes the location of information to be used in making the proper weldment. Drawings show how the surfaces and edges of parts fit together, how the parts will be assembled, and the type of weld joint to be made.

The complete weld symbol gives the welder the instructions to prepare the base metal, the welding process to use, and the finish to add to the completed weld. Fig. 7-20 contains the "basic weld symbols" which direct the welder to select the proper type of weld joint. By combining the basic weld symbols and the welding symbol, the designer or drafter can communicate any welding operation to the welder.

It is important to study and understand each part of the welding symbol. The arrow indicates and marks the point at which the weld is to be made. The line to the arrow is always at an angle to the reference line. Whenever the basic weld symbol is placed below the reference line, the weld is to be made at that point, as shown in Fig. 7-21. Whenever the basic symbol is placed above the reference line, the weld is to be made on the other side of the joint, as shown in Fig. 7-22. By placing dimensions on the symbol and drawing, the exact size of the weld may be indicated. Study the examples of typical weld symbols and weldments shown in Fig. 7-23.

Separate classes are often offered which provide advanced study in the area of blueprint reading for welders. By taking such a class, the welder can improve his or her ability to read and interpret welding drawings. Another method of gaining ability to read prints is by reading and studying texts on the topic of blueprint reading.

## DESIGN CONSIDERATIONS

Design of the weld type and the weld joint to be used are of prime importance if the weldment is to do the job intended. The weld should be made with reasonable cost. Several factors concerning the weld design must be considered. Evaluations should be made concerning the areas of:

1. Material type and condition.
2. Service conditions.
3. Physical and mechanical properties of the completed weld and heat affected zone.
4. Preparation and welding cost.
5. Assembly configuration and weld access.
6. Equipment and tooling.

### Butt Joints and Welds

Butt joints are used where high strength is required. They have good reliability and withstand any type of stress better than any other weld type. To achieve full stress value, the weld must have 100 percent penetration through the joint. This can be done by either welding from one side or both sides with the welds joining in the center.

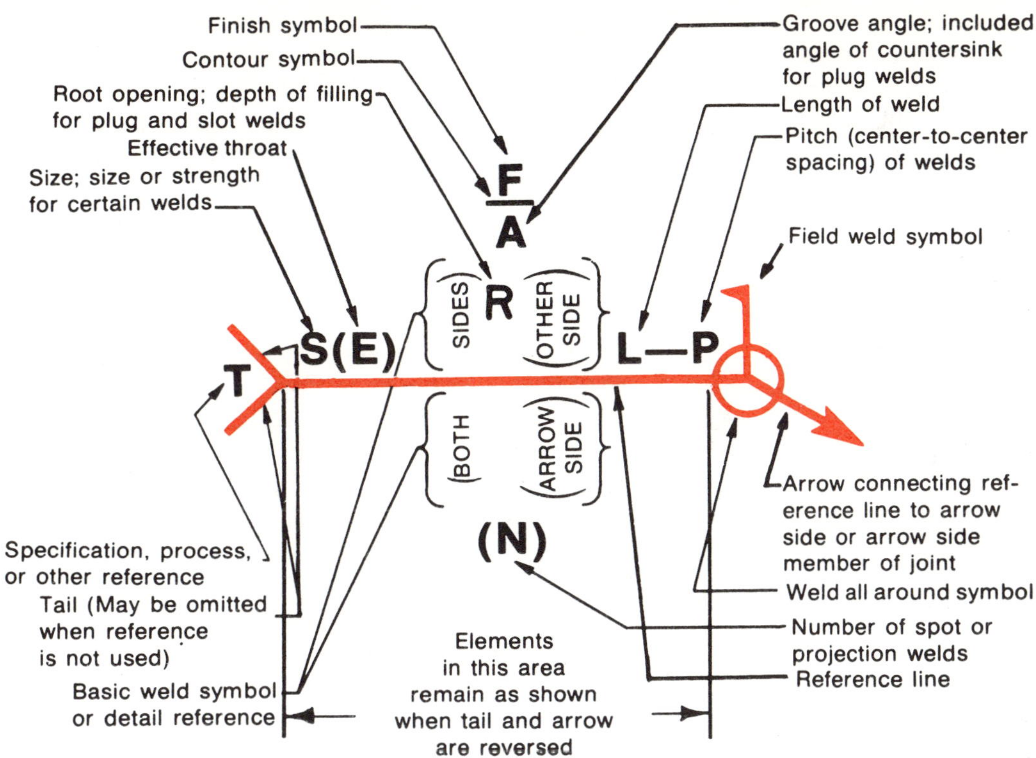


Fig. 7-19. The welding symbol approved by AWS gives complete and specific welding information to the welder.

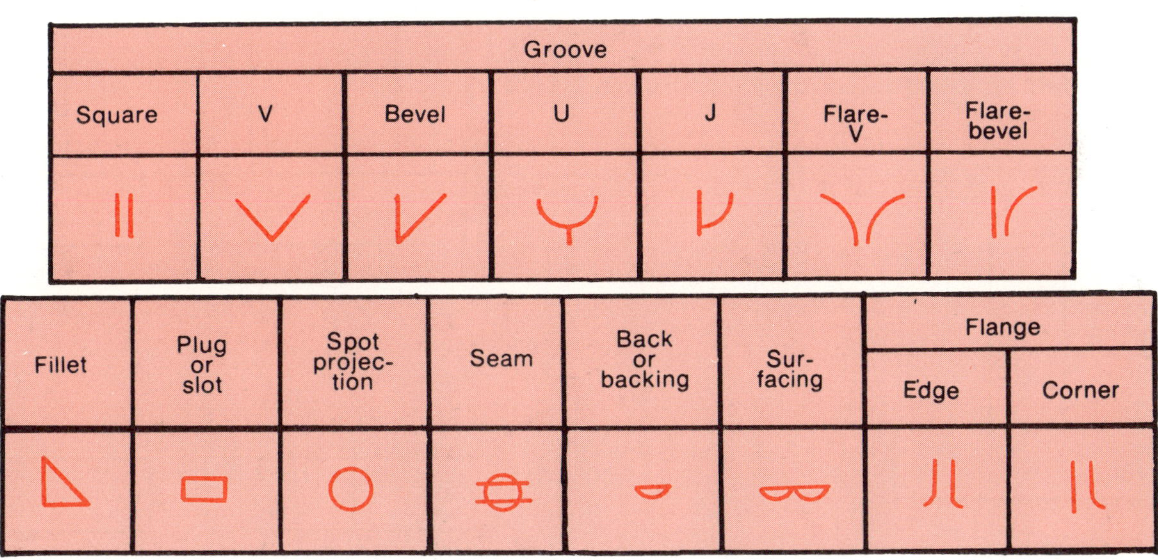


Fig. 7-20. Basic welding symbols.

Thinner gauge metals are more difficult to fit-up for welding. The thinner gauge metals also require more costly tooling to maintain the proper joint configuration. Tack welding may be used as a method of holding the components during assembly. However, tack welds present many problems when compared with good tooling, including:

1. The conflict with the final weld penetration into the weld joint.
2. Add to the crown dimension (height).
3. Tack welds often crack during welding, due to the heat and expansion of the joint during the joint welding operation.

The expansion of the base metal during welding will

often cause a condition known as MISMATCH, as shown in Fig. 7-24. When the mismatch occurs, the weld generally will not penetrate completely through the joint. Many specifications limit highly stressed butt joints to a 10 percent maximum mismatch of the joint thickness.

ARROW SIDE
DESIRED WELD
FILLET WELD SYMBOL

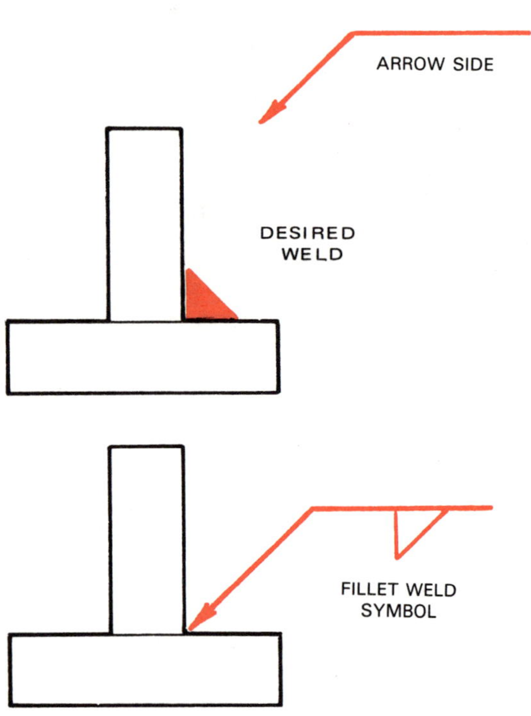

Fig. 7-21. Fillet weld symbol shown on the arrow side and toward the reader.

OTHER SIDE
DESIRED WELD
FILLET WELD SYMBOL

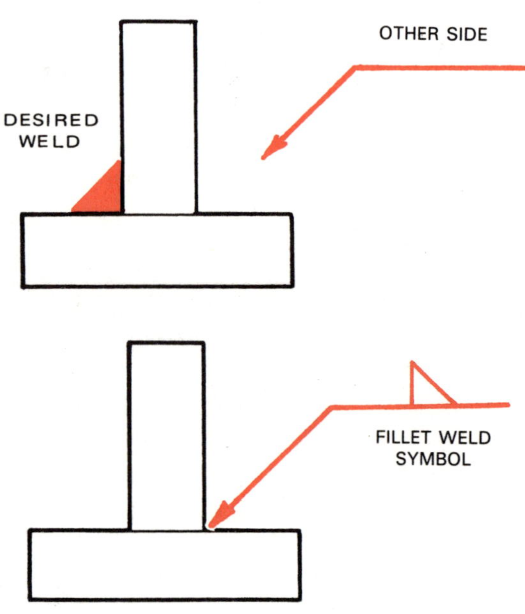

Fig. 7-22. Fillet weld symbol on the other side, away from the arrow. Note the resulting weld.

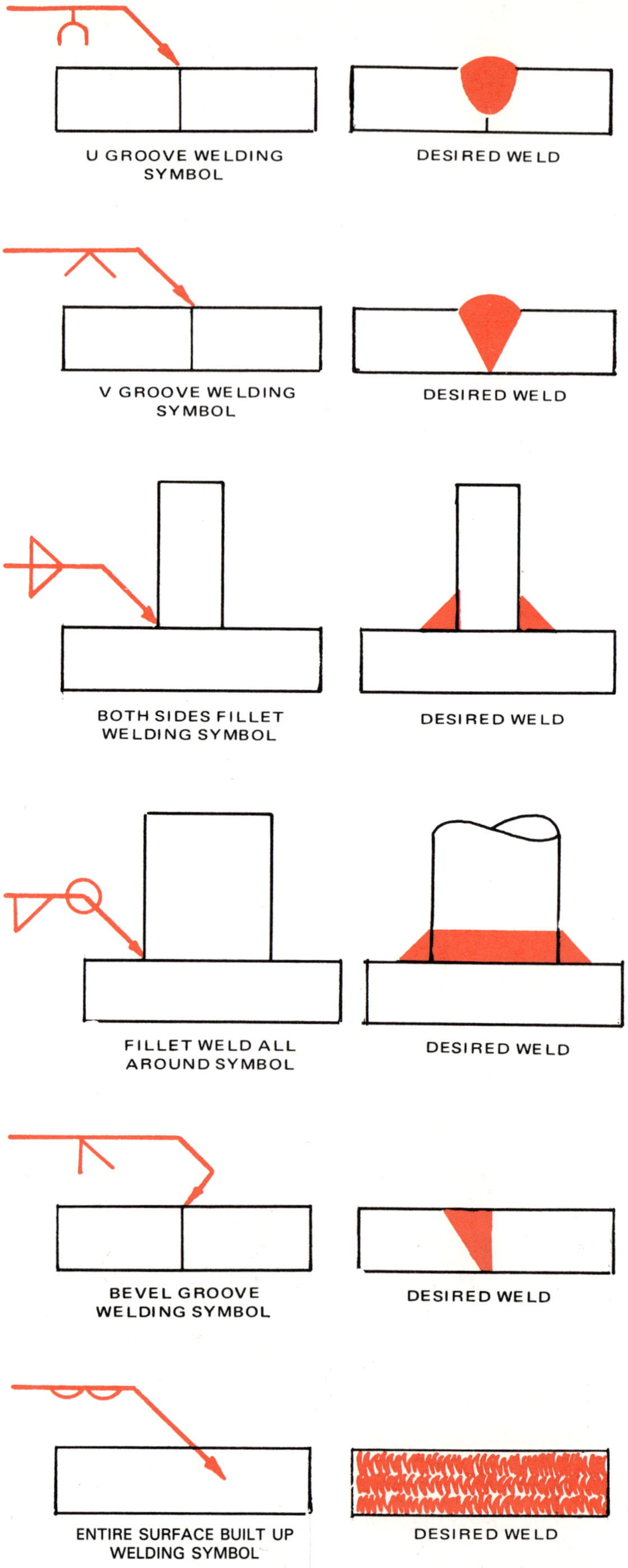

U GROOVE WELDING SYMBOL
DESIRED WELD
V GROOVE WELDING SYMBOL
DESIRED WELD
BOTH SIDES FILLET WELDING SYMBOL
DESIRED WELD
FILLET WELD ALL AROUND SYMBOL
DESIRED WELD
BEVEL GROOVE WELDING SYMBOL
DESIRED WELD
ENTIRE SURFACE BUILT UP WELDING SYMBOL
DESIRED WELD

Fig. 7-23. Typical weld symbols and their weldments.

Whenever possible, butt joints should mate at the bottom of the joint as shown in Fig. 7-25. Joints of unequal thicknesses should be tapered in the weld area to prevent incomplete or inadequate fusion as shown in Fig. 7-26.

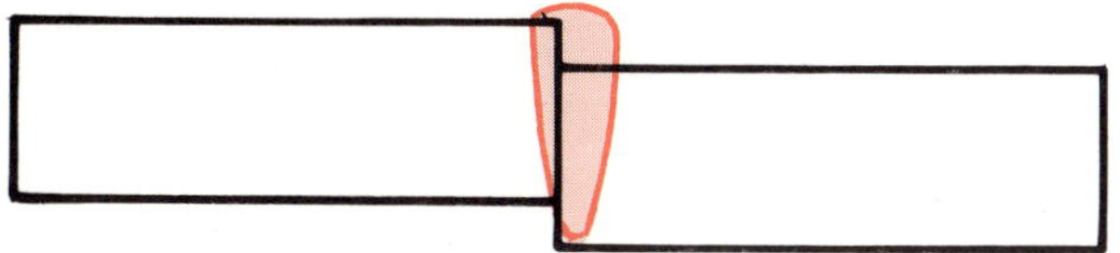

Fig. 7-24. Welds made on mismatched joints often fail far below the rated load when stressed.

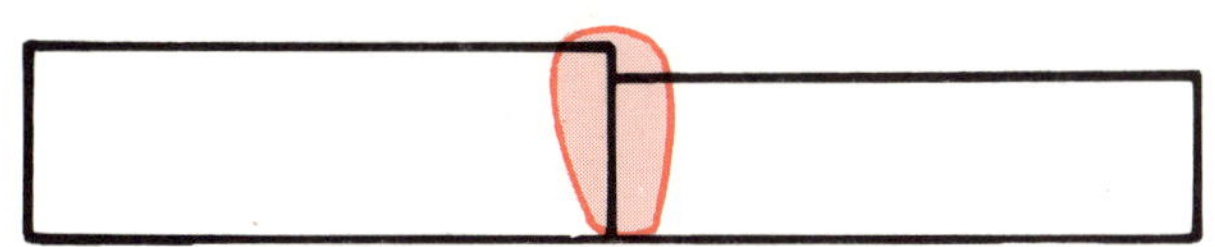

Fig. 7-25. Mating the joint at the bottom equalizes the load during stress.

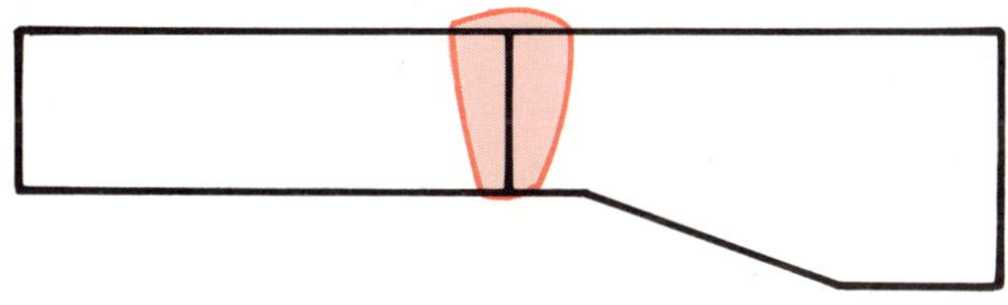

Fig. 7-26. Joints of unequal thickness absorb different amounts of heat and expand at different ratios. Equalize the heat flow by tapering the heavier material to the thinner material thickness.

Butt welds always shrink across (transverse) the joint during welding. For this reason a shrinkage allowance must be made if the after welding part dimensions have a small tolerance. Butt welds in pipe, tubing, and cylinders also shrink on the diameter of the material. This shrinkage is shown in Fig. 7-27. In areas where these dimensions must be held, a shrinkage test must be made to develop the amount of shrinkage. Fig. 7-28 shows how the test is made. Heavier materials will shrink more than thinner materials. Double groove welds will shrink less than single groove welds since less welding is involved and less filler metal is used.

### Lap Joints and Welds

Lap joints may be either single, double fillet, or spot welded. They require very little joint preparation and are generally used in static load designs. Where corrosive liquids are used, both edges of the joint must

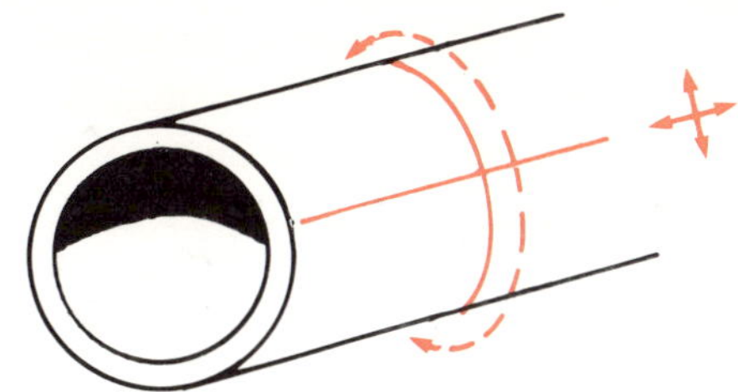

Fig. 7-27. Butt welds shrink during welding in both transverse and longitudinal directions.

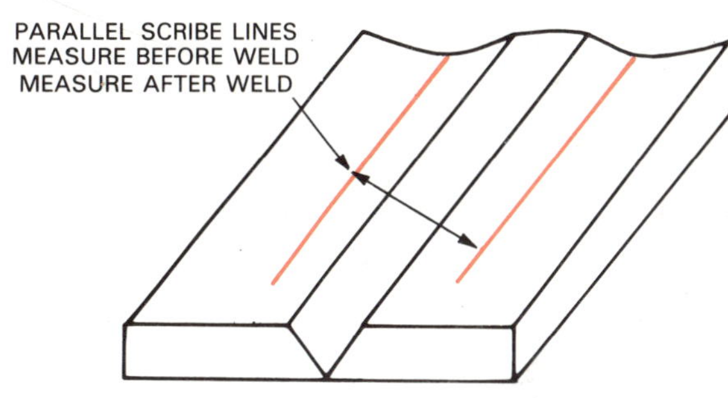


Fig. 7-28. By determing the amount of shrinkage across the joint additional metal may be added to the components for dimensional tolerances.

be welded as shown in Fig. 7-29. One of the major problems with this joint design is shown in Fig. 7-30 where the component parts are not in intimate contact. A bridging fillet weld must then be made which leads to incomplete fusion at the root of the weld and oversize fillet weld dimensions.

When using this type of design in sheet or plate material, adequate clamps or tooling must be used to maintain contact of the material at the weld joint.

Cylindrical components may be assembled as shown

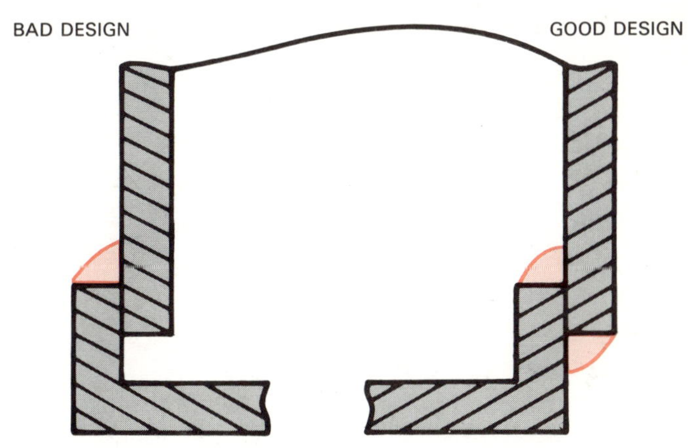


Fig. 7-29. Corrosive liquids must not be allowed to enter the penetration side of the joint.

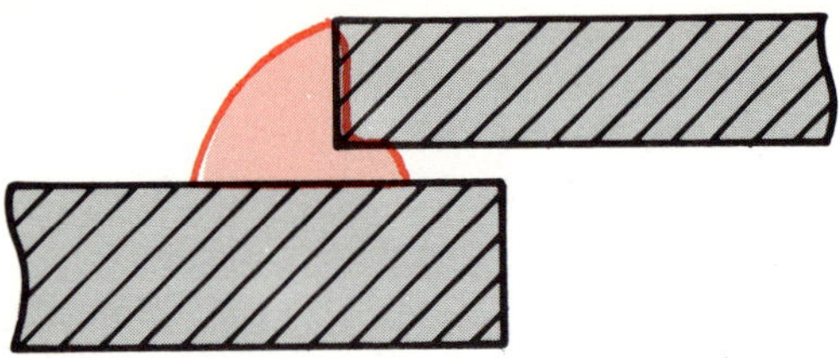

Fig. 7-30. Lap joint problem areas.

in Fig. 7-31, using an interference fit to eliminate this problem. The outer part inside diameter (ID) dimension is made several thousands of an inch smaller than the inner part outside diameter (OD). When ready for assembly, the outer part is heated enough for the part to expand larger than the inner part. The outer part is then assembled and allowed to cool in place. The outer part when cool, will be locked tightly into place.

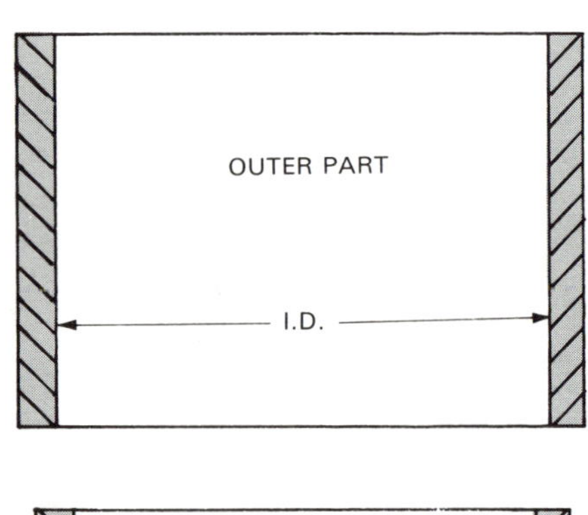


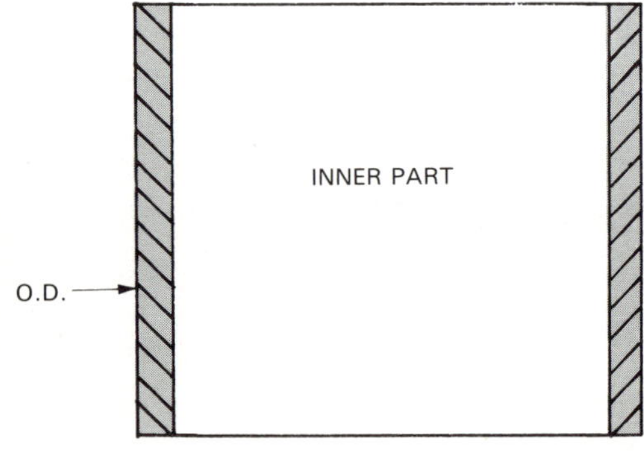


Fig. 7-31. The diameters of the component parts to be assembled may be determined by using a "Pi" tape around the inner and outer cylinder components. The tape measures in thousandths of an inch and full inches.

### T Joints And Welds

T weld joint designs are used for joining components at angles to each other. Depending on the use of the joint, they may be made with a single fillet, double fillet, or a groove and fillet weld combination. Fig. 7-32 shows how these welds may be used.

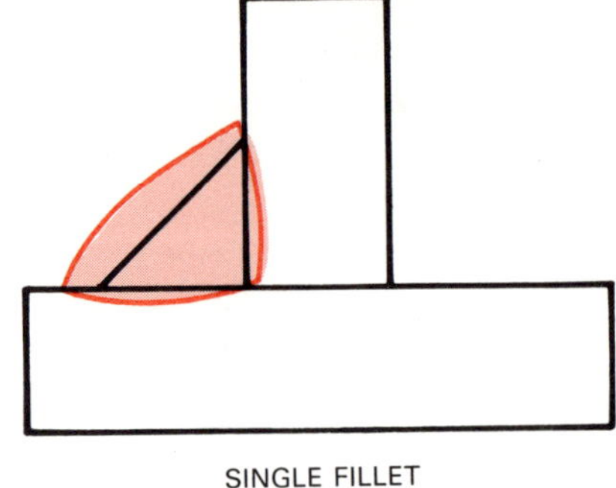


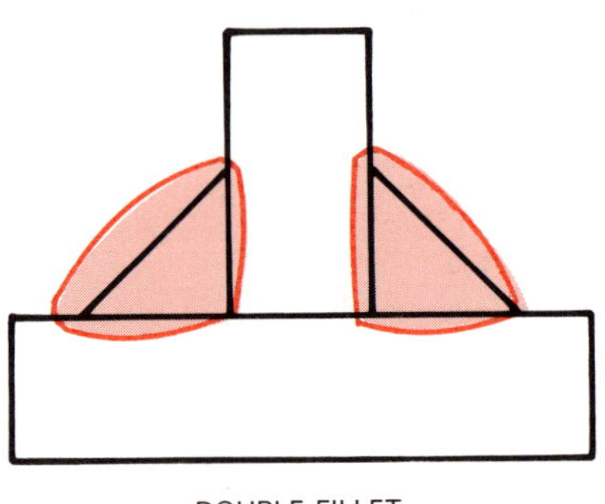


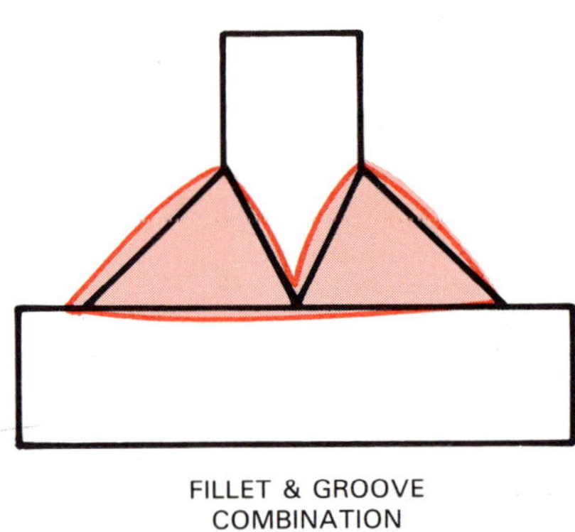


Fig. 7-32. Various types of T joints and welds.

Fillet welds are made to specific sizes as determined by the allowable design load, and are measured as shown in Fig. 7-33. Where design loads are not known, the "rule of thumb" may be used for determining the fillet size. In these cases, the fillet leg lengths must equal the thickness of the thinner material.

The main problems encountered in the T joint fillet welds is the size of the fillet throat and lack of penetration into the mating corners. Gas tungsten arc fillet welds are usually concave rather than convex. Therefore, to achieve the proper size, the leg length must be lengthened as shown in Fig. 7-33 to increase the throat dimension.

The lack of penetration problem is caused by insufficient heat at the root of the joint. The adjoining materials absorb the heat before it gets into the corner. With the addition of filler material the problem

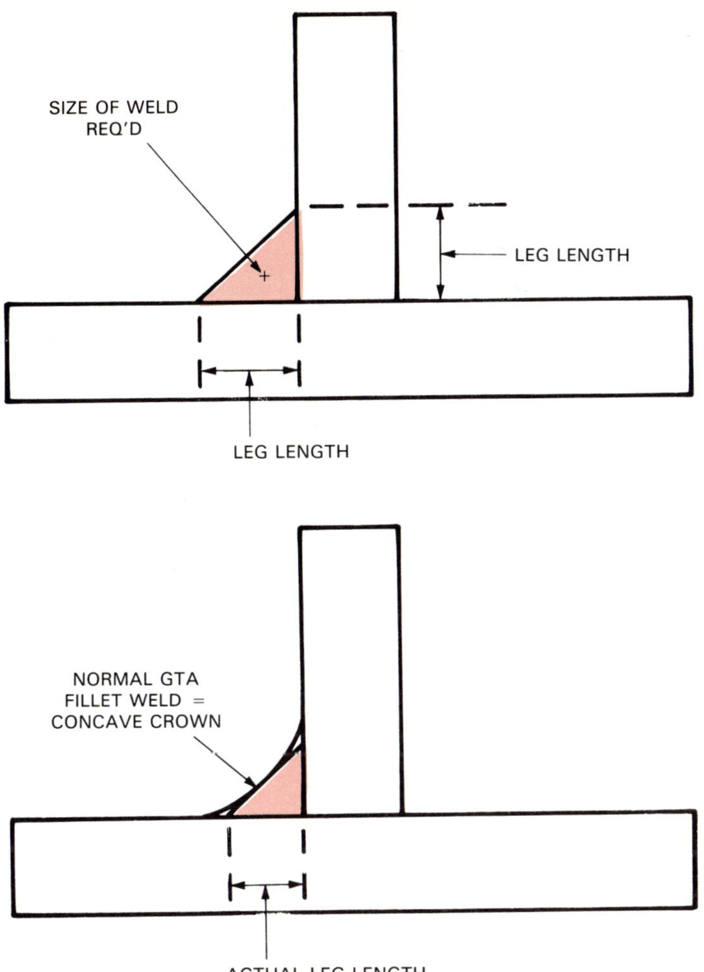


Fig. 7-33. Fillet weld leg lengths should be equal distance from the root of the joint. Unequal leg length, unless otherwise specified, will not carry the designed load and may fail under load.

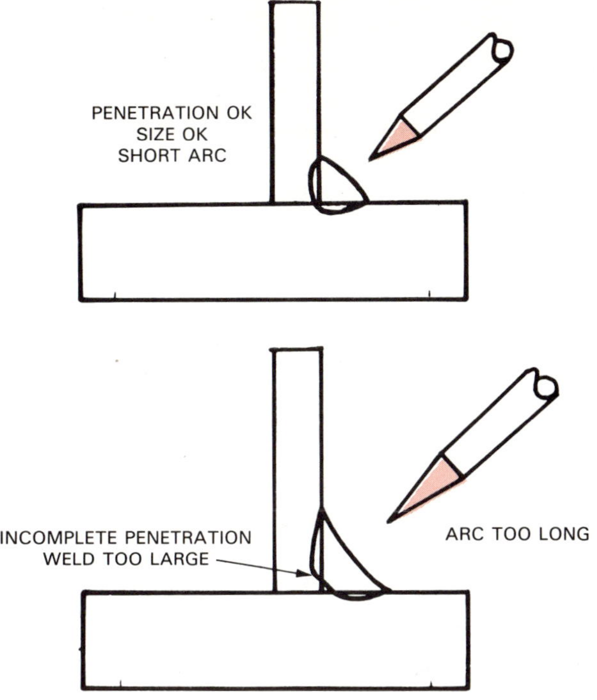


Fig. 7-34. Lack of penetration into the root of the joint may cause failure of the weld under load.

gets worse as more arc energy is directed at melting the filler material. The weld must penetrate beyond the mating point as shown in Fig. 7-34 to be effective. To eliminate this condition, a small diameter filler material should be used and large welds should be avoided.

## Corner Joints and Welds

Corner welds are very similar to T joints as they consist of sheets or plates mating at an angle to one another. They are usually used in conjunction with groove welds and fillet welds. Many different designs may be used, some of which are shown in Fig. 7-35 (edge to edge, half overlap, flush corner). When using thinner gauges, assembly of the component parts may be difficult without the proper tooling. Tack welding and welding heat causes distortion of thinner gauges very readily. For the most part, this type of joint design should be limited to heavier type materials in structural assemblies.

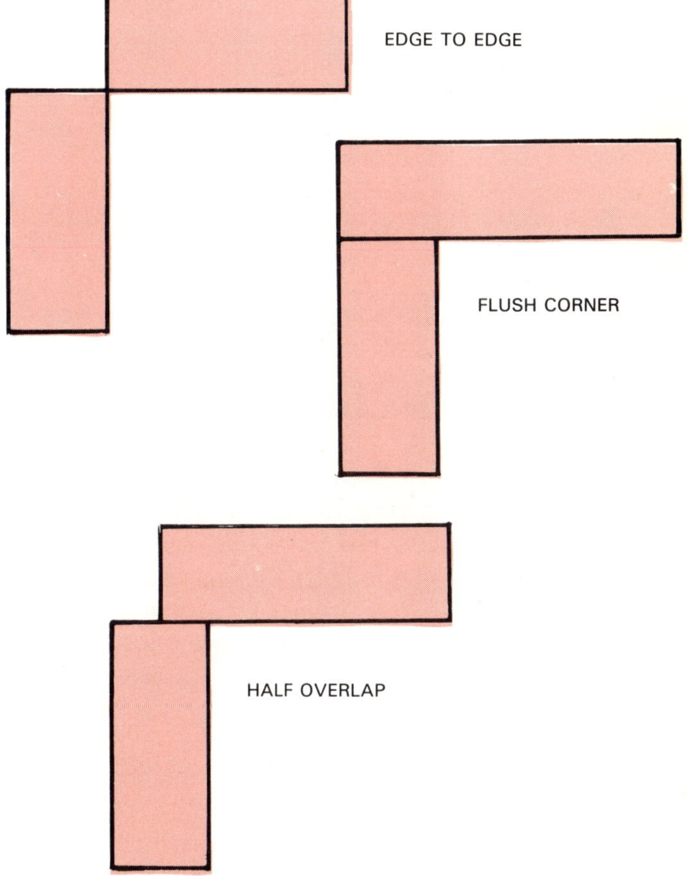


Fig. 7-35. Common corner weld joint designs which may be used in fabrication of component parts.

### Edge Joint and Welds

Edge welds are used where the edges of two sheets or plates are adjacent and in approximately parallel planes at the point of welding. Fig. 7-36 shows several types of edge weld designs. These designs are common in structural applications only, since the weld does not penetrate completely through the joint thickness it should not be used in stress or pressure applications. The main problem area in this design concerns the melt down of the flanged material. In many cases, the base material cannot be welded to itself because of cracking tendancies and a different filler material must be used. In these areas, a filler wire must be added to make a sound weld.

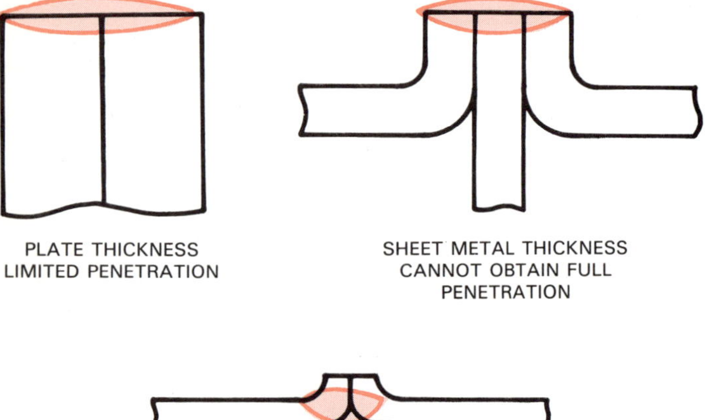


Fig. 7-36. Common edge weld joint designs which may be used in fabrication of component parts.

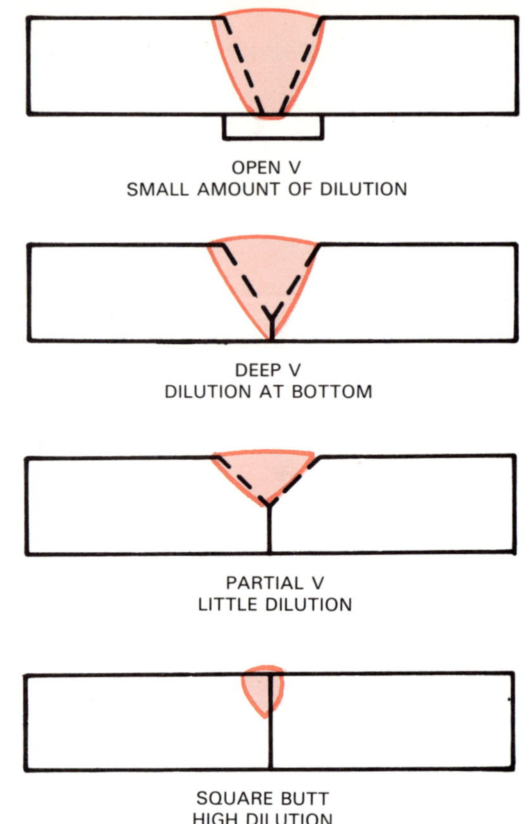


Fig. 7-37. Weld grain structures are obtained by cross sectioning the weld, polishing the section and inspecting with magnification.

## WELD AND HEAT AFFECTED ZONE GRAIN STRUCTURES

The design of the weld joint and the amount of welding done has a very distinct effect on the properties of the weldment. HEAT SOAK describes the extent of the heat absorbed into the weld and weld area.

The amount of heat, type of filler wire, and the number of passes (layers) used in the joint determine the dilution of the metal and the grain structure. DILUTION, in welding terminology, describes the change from the original chemical composition to another chemical composition with different physical or mechanical properties. GRAIN STRUCTURES describe the final alignment of the individual grains or types of grains after all of the heating, melting, and cooling cycles are completed. Fig. 7-37 shows several types of welds and how the grain structures are obtained.

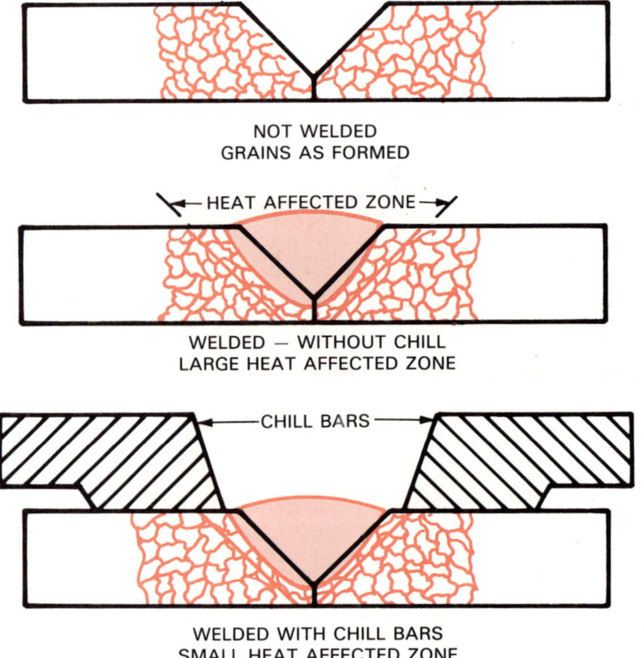


Fig. 7-38. Chill bars and tooling are used to remove heat from the weld area and to restrict heat flow into the material.

The amount of heat, and the number of passes used in the joint also determines the extent of the heat soak into the parent metal. Multi-pass groove or fillet welding adds considerable heat input to the weld area. Where this heat flow is not controlled by tooling, the grains will enlarge to a degree which may affect the mechanical properties. Fig. 7-38 shows how the heat flow affects the parent metal grain structure.

## SPECIAL DESIGNS AND PROCEDURES

Special designs are often used in the fabrication of a weldment where:

1. Welds cannot be thermally treated after welding because of configuration and or size. A typical joint design as shown in Fig. 7-39 is used to achieve full mechanical values required in the weld.

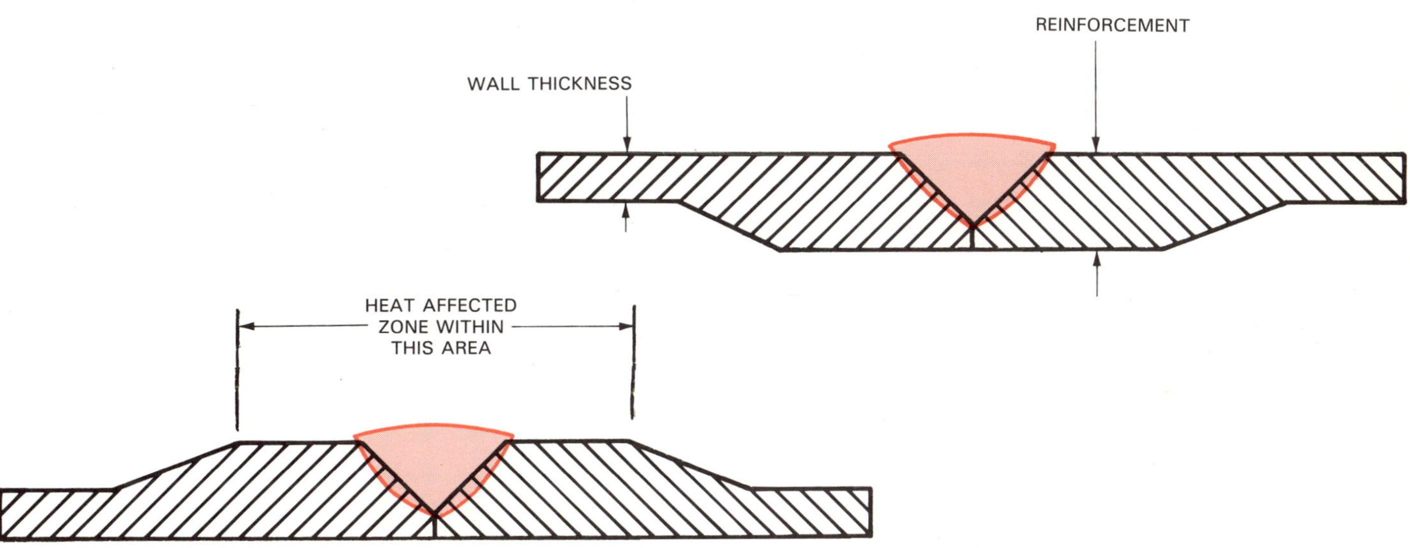


Fig. 7-39. The joint thickness and the filler material tensile strength is equivalent to the strength of the base material in this design.

2. Joining of dissimilar materials can be done by buttering the face of one material to match the other as shown in Fig. 7-40.

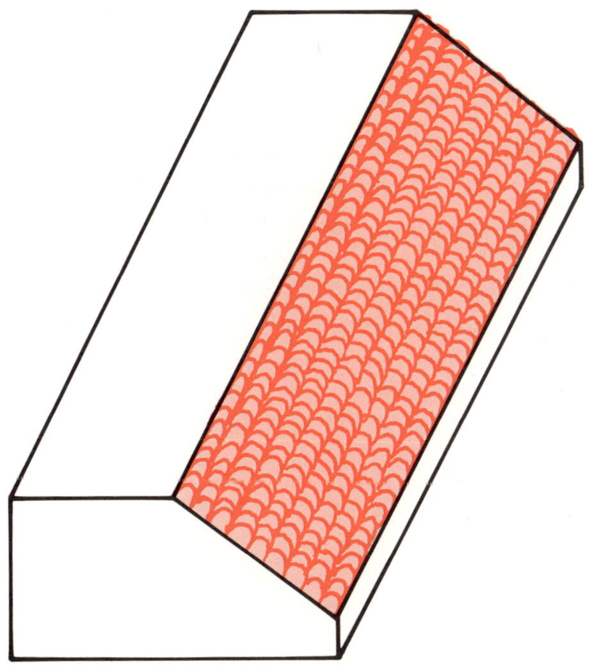

Fig. 7-40. Buttering techniques are commonly used to adapt dissimilar metals for welding.

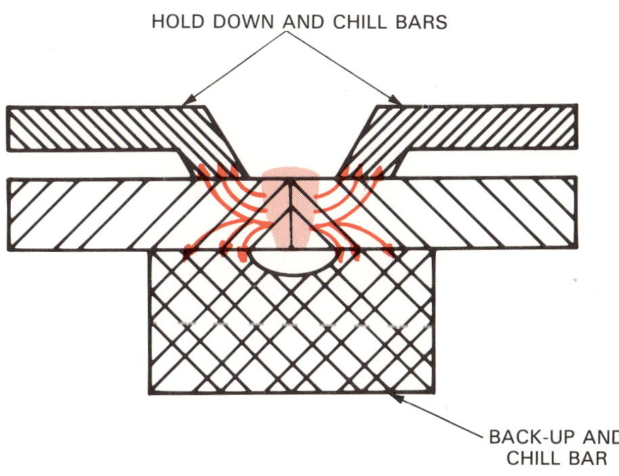


Fig. 7-41. Tooling and bars used to remove heat from the weld and resist heat flow into the base material are called CHILL BARS.

3. Special procedures and tooling may be used to provide a preheating, interpass and postheating operation to control grain size. Preheating is generally used to slow down the cooling rate of the weld to prevent cracking. Interpass is the period of time on a multi-pass weld after a weld pass is completed and a new weld pass is started.
4. Special procedures, tooling, and chill bars may be used to localize and remove welding heat during the welding application. Fig. 7-41 shows an application of tooling used to remove heat from the part.

## WELDING POSITIONS

In addition to the flat position, it is often necessary to make welds in various other positions. The American Welding Society has defined the positions of welding to include:

1. Flat.
2. Horizontal.
3. Vertical.
4. Overhead.

Fig. 7-42 shows the four positions for fillet welds, grooved butt welds, and pipe welds.

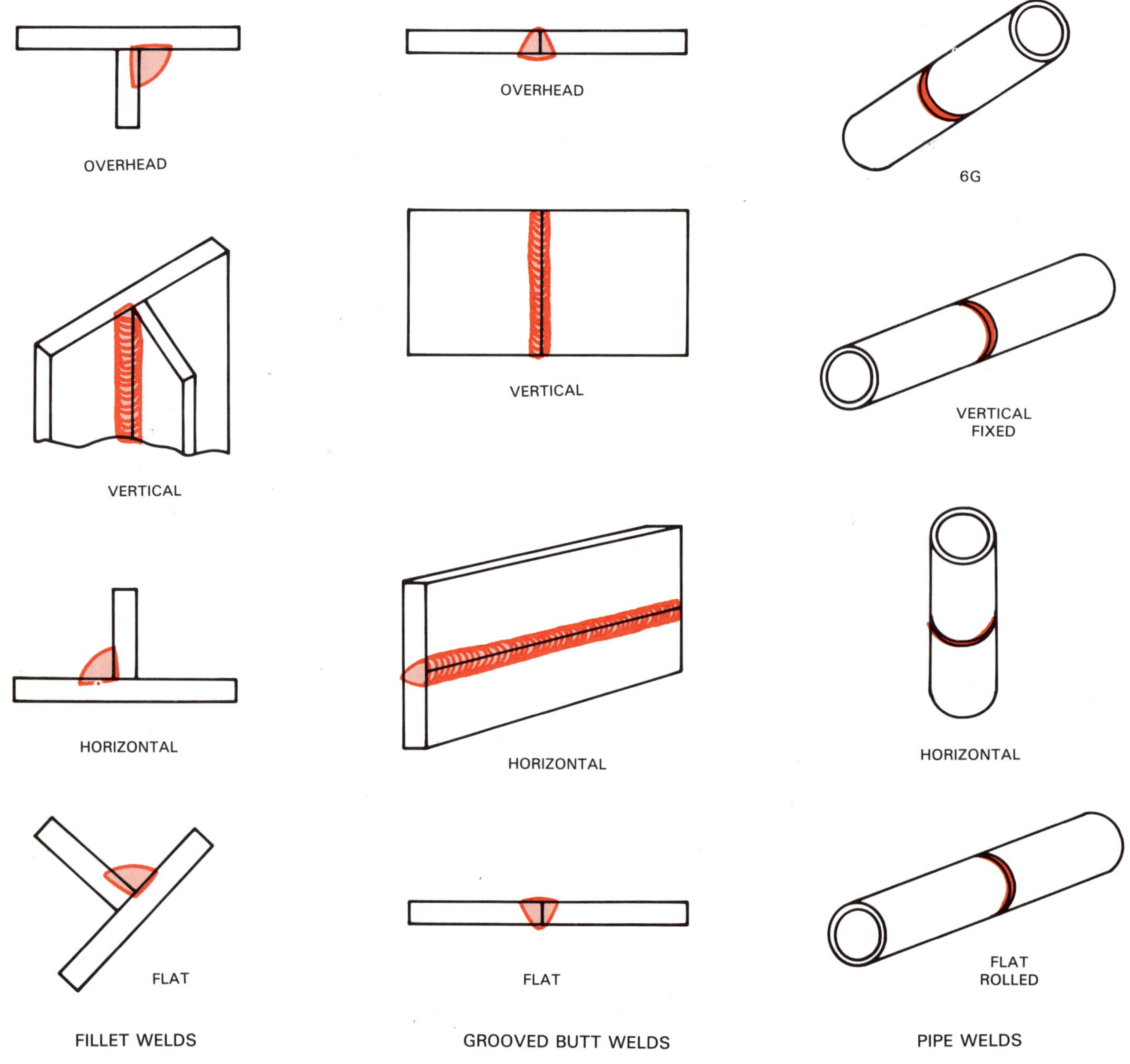


Fig. 7-42. American Welding Society definition of welding positions.

## REVIEW QUESTIONS

1. What are the five basic types of joints?
2. Can more than one type of weld be made in each of the basic joints?
3. Where the weld is made from both sides of the joint, it is called a ________ ________ .
4. Where are joggle type joints often used?
5. An operation known as ________ is often used to build up a weld joint for joining dissimilar metals.
6. A series of overlayed welds on the surface of a part to form a weld joint is called ________
7. Back-up rings, made to fit into a pipe joint, are also called ________.
8. Name two types of thermal cutting used to prepare weld joints.
9. Butt joints are normally used where ________ ________ is required.
10. Full penetration through the weld joint is considered to be ________ percent penetration.
11. List three problem areas encountered when using tack welds during assembly.
12. When two materials do not mate at the weld joint in the same plane this condition is called ________.
13. What two problems may occur when the materials do not mate at the bottom of the joint during welding?
14. Double groove welds shrink more or less than single groove welds?
15. Where are lap joints usually used?
16. List two problem areas encountered in welding lap joints.
17. What type of measuring tool is often used to determine the diameter of cylindrical components?
18. How is the "rule of thumb" used to determine fillet weld leg length?
19. Are GTA fillet welds usually concave or convex?
20. What is the main problem encountered in making a GTA T joint fillet weld?
21. The area in the parent metal next to the weld where grain structure changes occur is called the ________ ________ ________.
22. Tooling and bars used to control heat flow from the weld into the parent metal are called ________ bars.

# Chapter 8

# TOOLING

## DESIGN

Tooling for GTAW is a very important part of the manufacturing cycle of the weldment. If the tooling is not designed and constructed properly, serious problems can result, affecting both quality and cost in the production of the weldment. The major tooling items to be considered include:

1. Alignment of the components to be welded.
2. Heat control of the weld zone.
3. Positioning of the joint for welding.
4. Assembly (loading) and disassembly (unloading) of the components.
5. Providing atmosphere to prevent contamination of the weld.
6. Accessability for the weld torch and filler material.

Some of the other factors which may be considered in the tool design include: Dimensional tolerances; type of material to be welded; complexity of the weld; cost of the tool; number of parts to be made; quality of the weld; welding conditions.

## DESIGN CONCEPTS

Major tools designed for assembly of component parts and welding are considered HARD TOOLING. They are generally used where large numbers of items are welded. Fig. 8-1 shows one type of hard tooling.

Fig. 8-1. Large tools align component parts at assembly and hold the parts during the welding operation. (Aerojet-General Corp.)

Where a small number of parts are to be made, SHOP AIDS are generally used. These tools may use the weldment as part of the tooling or they may be a complete tool. Fig. 8-2 shows a simple Shop Aid tool used to align a butt joint for tackwelding and welding. The angle is used to center the two pieces of material. The part is rotated by hand as each segment of the weld is completed.

The tool shown in Fig. 8-3 is a location and holding tool used to locate the component parts for tackwelding. The tackwelded assembly is removed from the tool for the final welding operation.

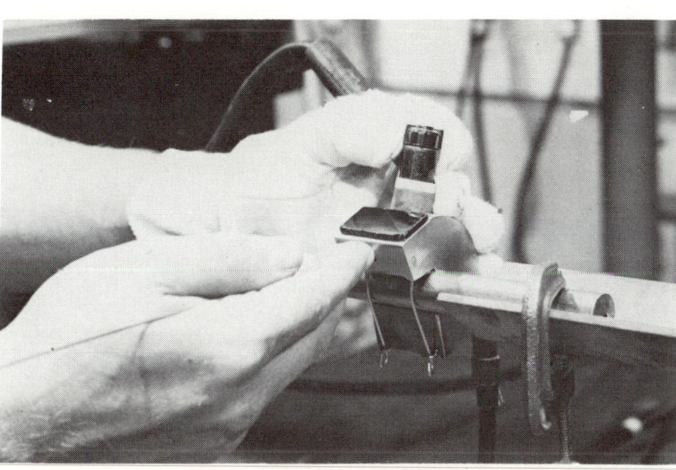

Fig. 8-2. Shop aid type tooling. (Aerojet-General Corp.)

Fig. 8-3. Holding fixtures are very rigid to maintain accurate alignment of the component parts. (Aerojet-General Corp.)

## TYPES OF TOOLS

Basic tool designs include several types and these are:

1. INTERNAL TOOLING – the weld is made from the outside of the joint. A typical tool is shown in Fig. 8-4.
2. EXTERNAL TOOLING – the weld is made from the inside of the joint. A typical tool is shown in Fig. 8-5.

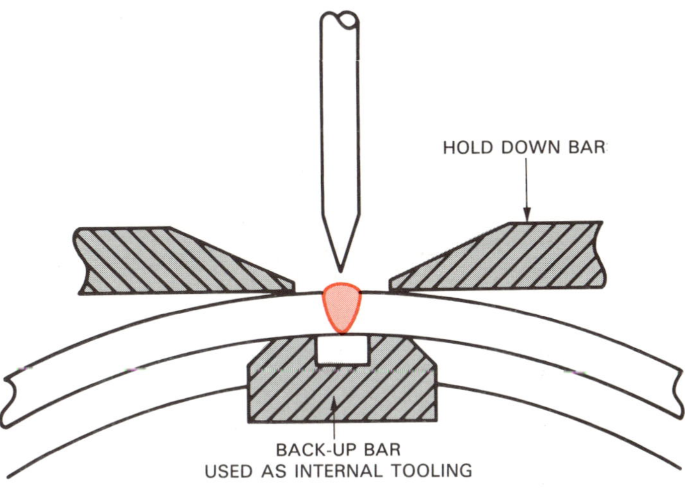


Fig. 8-4. Air pressure is applied to the hold down fingers to firmly hold the cylinder longseam joint to the back-up bar for welding when using internal tooling.

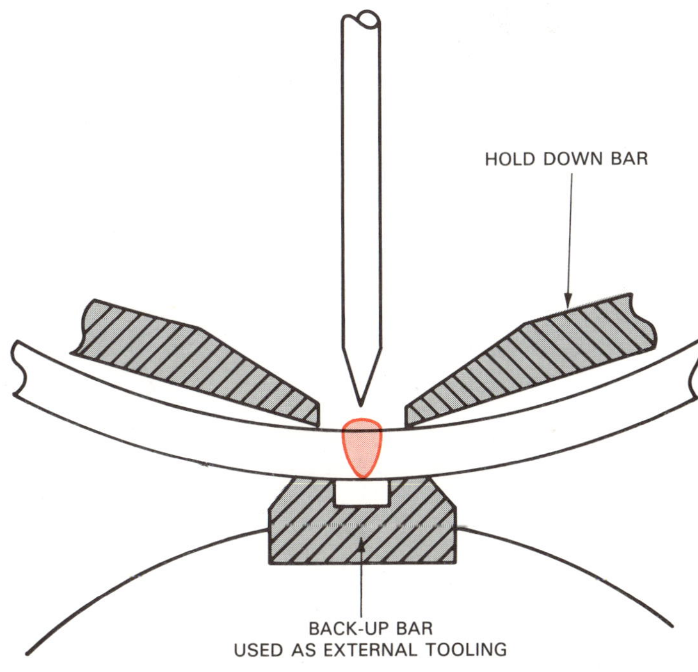


Fig. 8-5. The inside diameter of the cylinder must be sufficient to accept the welding torch and component parts of the fixture in external tooling.

3. COMBINATION ASSEMBLY AND WELDING TOOLING – the weld may be made from inside or outside or the part may become part of the tool. Often the tooling is removed after tackwelding. A typical tool is shown in Fig. 8-6.
4. HEAT SINK TOOLING – the tool is used only to remove heat from the part during the welding operation. This reduces excessive penetration and distortion. A typical heat sink tool is shown in Fig. 8-7.

Since the weld and the weld areas are hot and expand during welding, the tool must not lock the parts in place. The tool must allow for this expansion, as well as the shrinkage during cooling. If the parts cannot shrink as they cool, the weld joint stresses may exceed the strength of the weld. This may result in weld center line cracking either in the form of visual cracks on the surfaces of the weld, or micro cracking in the weld grain structure.

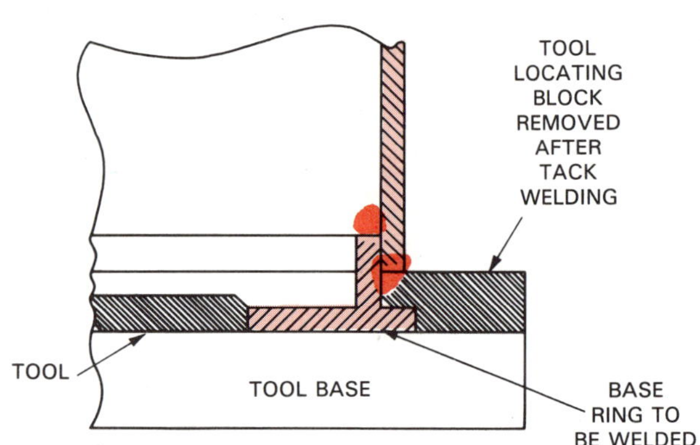


Fig. 8-6. This circumferential tool accurately locates and holds the base ring. Locating blocks space the outer ring during assembly. The blocks are removed after tack welding the assembly for welding the inner or outer weld.

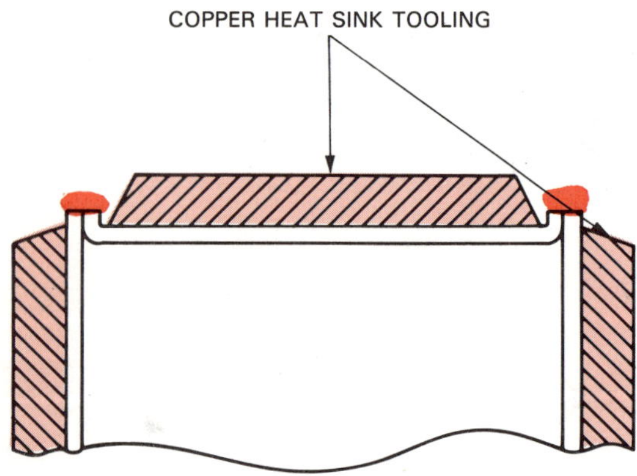


Fig. 8-7. Copper heat sink is placed on both sides of the weld joint to remove the heat generated during the welding cycle.

## APPLICATION AND OPERATION OF TOOLING

The application of the tool in producing the weldment or in aligning the component parts usually consists of two parts. A fixed member is used for location of the part and a movable member is used for clamping purposes. In some cases, when using simple tools, a movable member may not even be required. In these cases, the part may be clamped directly to the fixed tool as shown in Fig. 8-8.

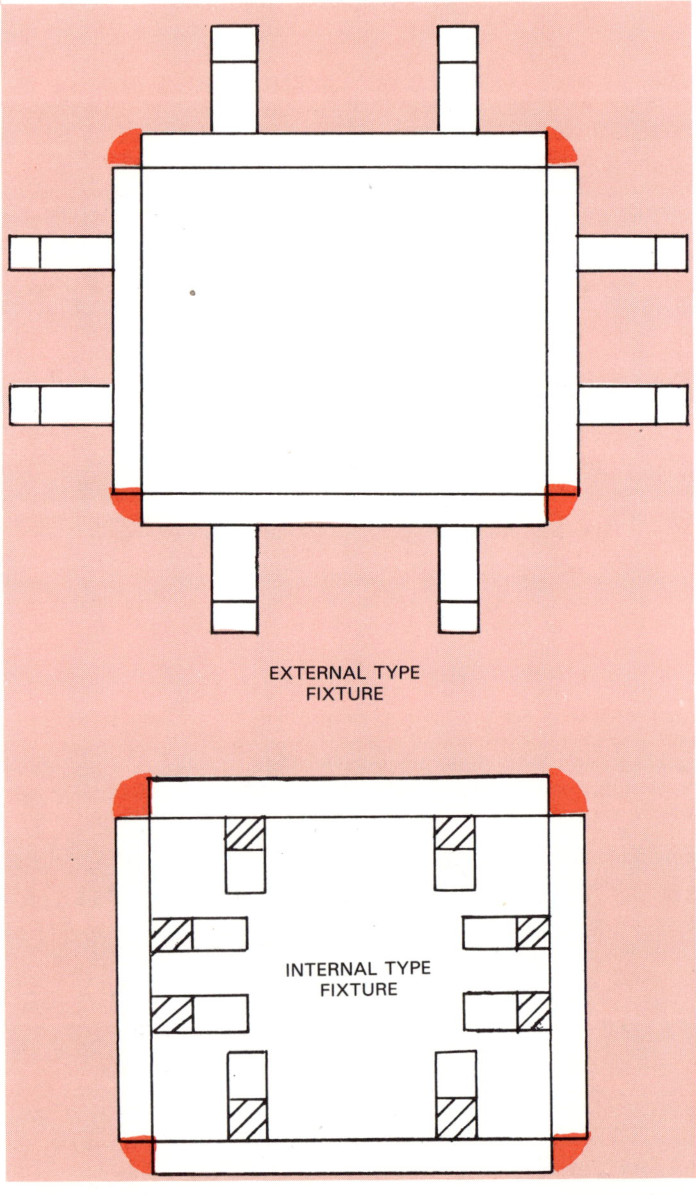


Fig. 8-8. In the external type fixture, location points are located on the outside of the frame. Since the welds will shrink, the part inside the location points will not become wedged in the fixture. In the internal type fixture, location points are located on the inside of the frame. Shim blocks are then used with the location points for assembly. They are removed before welding to allow for any possible shrinkage.

Mechanical tooling may utilize a number of methods for the holding or clamping operations. An example of mechanical clamping is shown in Fig. 8-9. The use of these types of clamps and dogs often require excessive amounts of time in loading and unloading of the tool. Other types of clamping forces such as air pressure, hydraulic and air-hydraulic may be used to shorten this time period. Air pressure and hydraulic systems operate very rapidly. They are reliable and far more efficient than mechanical types where these systems may be used. Various systems are shown in Figs. 8-10 and 8-11.

## MATERIALS

The materials used in the main structure or the base of the tool must be capable of securely holding the component parts. Equipment for clamping and the actual weld zone tooling components must operate consistently after repeated use. Fig. 8-12 shows a typical tool base designed in this manner. Square tubing, pipe, and box-type frames offer considerable rigidity and are used extensively in the manufacture of welding tooling.

Materials used in the weld joint area present many problems to the designer. Ferrous materials such as iron

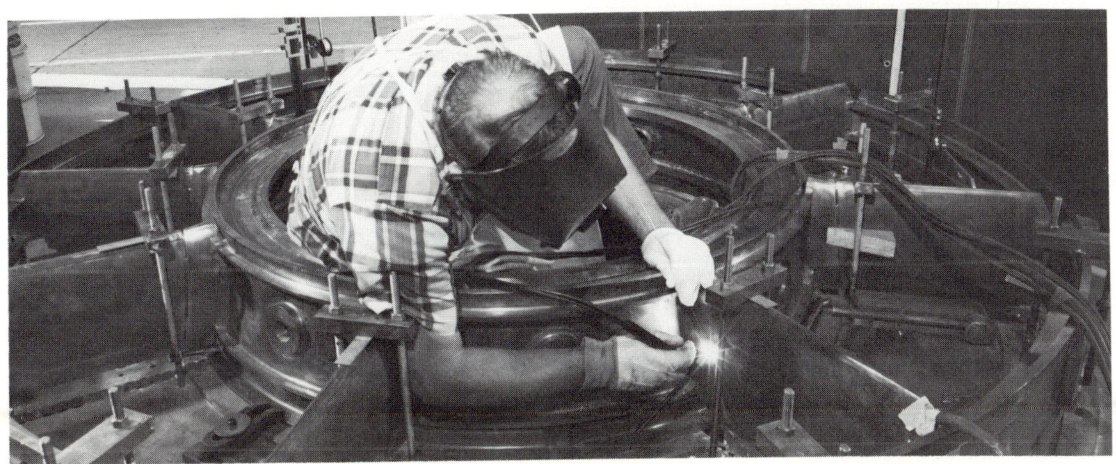

Fig. 8-9. Commercial type clamps and hold down "dogs" are often used for mechanical clamping operations. (Aerojet-General Corp.)

Fig. 8-10. Air pressure systems are often used for rapid loading and unloading of component parts. (Aerojet-General Corp.)

Fig. 8-11. Hydraulic pressure is used to expand the cylinder outward into the clamping rings for proper joint alignment. (Parker Hannifin Corp.)

Fig. 8-12. This clamping tool has been strengthened by the use of gussets (triangular inserts) and stiffener rings. (Aerojet-General Corp.)

and steel are magnetic and in some cases may cause arc blow (magnetic influence on the arc direction) when using direct current straight polarity. The non-ferrous materials such as aluminum and copper are non-magnetic, however they are soft. The tool is then readily deformed by the welding heat and pressures used for clamping purposes. Stainless steels in the 300 series are often used for tooling in the weld area as they are non-magnetic and not as soft as aluminum or copper. Combinations of these materials are often used to diminish these problems, as shown in Fig. 8-13. The use of material combinations also retains the tool rigidity and lowers the maintenance cost.

## BACK-UP BAR DESIGN

The backing bar for full penetration groove welds must have sufficient rigidity to support the component parts and still allow for penetration of the weld metal on the root of the weld. The groove that is cut into the backing bar may be of various designs. Normally when welding aluminum a shallow radius groove is used for forming the penetration on the root of the weld. The welding industry often refers to this type of forming or molding of the penetration into a grooved back-up bar as "casting the penetration." A deeper radius or square groove is used when an inert gas backing is required. Fig. 8-14 shows the common type grooves used.

Fig. 8-15 shows back-up groove dimensions. Larger radius and square groove designs are used where surface tension of the weld penetration controls the amount of penetration, rather than casting or molding.

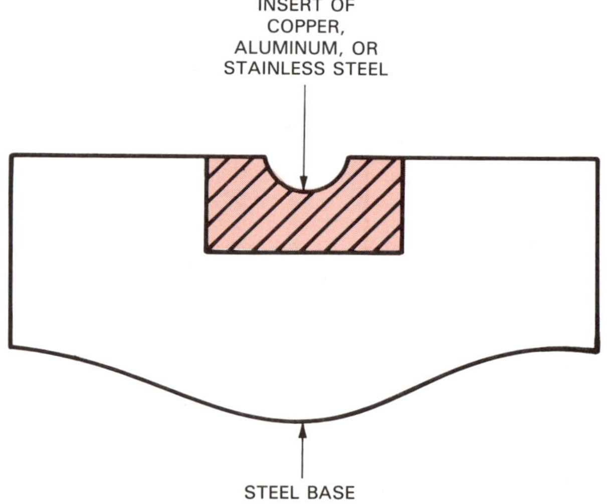


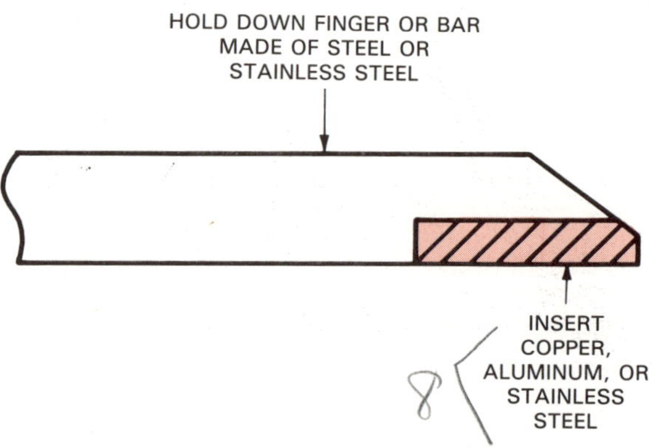


Fig. 8-13. The major part of the tool is made of steel for rigidity. The area in contact with the weld area is stainless steel, aluminum, or copper.

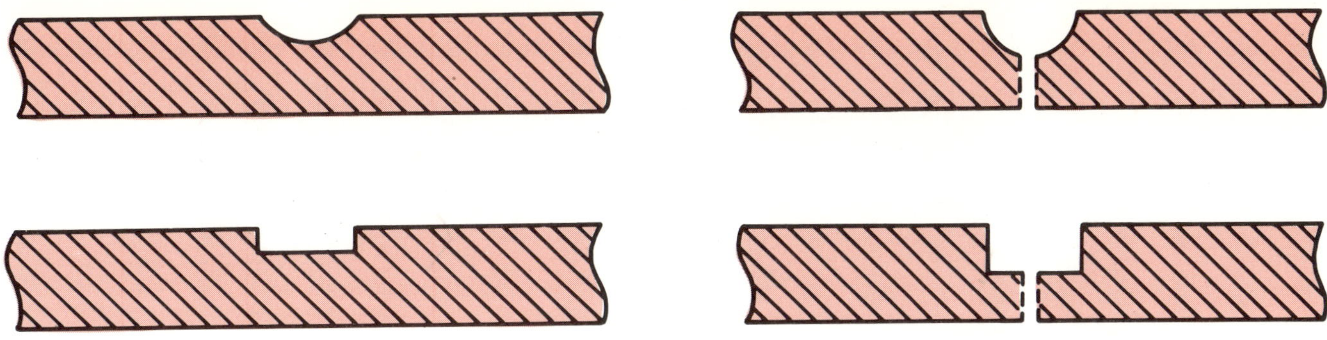

Fig. 8-14. Types of weld backing grooves for full penetration groove welds.

| Metal Thickness | Weld Type | Groove Dimensions For Casting Penetration | Groove Dimensions For Gas Back-Up Penetration |
|---|---|---|---|
| .005–.012 | FUSION | 040W010D | 040W125D |
| | FILLER | — | |
| .013–.020 | FUSION | 063W010D | 125W100D |
| | FILLER | — | |
| .021–.032 | FUSION | 093W010D | 187W100D |
| | FILLER | 125W020D | |
| .033–.040 | FUSION | 125W020D | |
| | FILLER | 187W025D | |
| .041–.050 | FUSION | 125W020D | |
| | FILLER | 187W025D | |
| .051–.062 | FUSION | 187W020D | |
| | FILLER | 250W040D | |
| .063–.072 | FUSION | 187W020D | 250W100D |
| | FILLER | 250W040D | |
| .073–.125 | FUSION | 250W020D | |
| | FILLER | 312W040D | |
| .126–.250 | FUSION | 312W020D | 312W100D |
| | FILLER | 375W050D | |
| .251–.375 | FUSION | — | |
| | FILLER | — | |

NUMBER REFLECTS WIDTH & DEPTH
E.G. 040W010D IS .040 WIDE & .010 DEEP

Fig. 8-15. Back-up bar groove dimensions for full penetration welds.

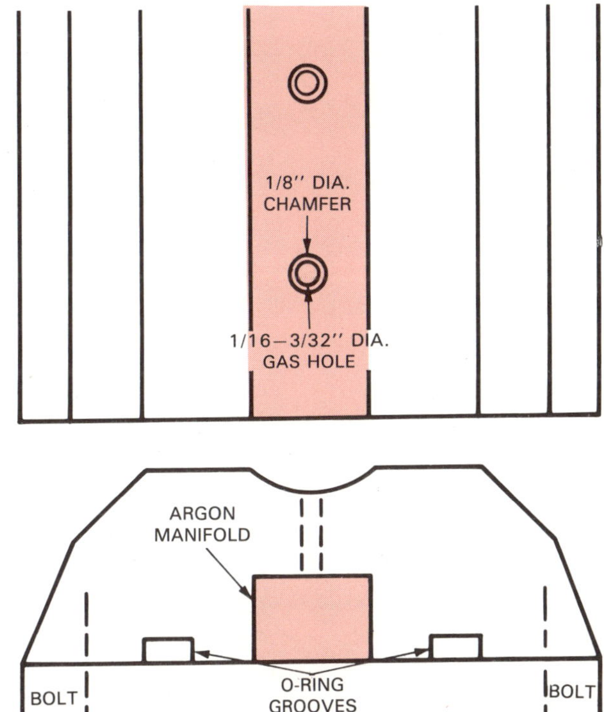


Fig. 8-16. Common back-up bar designs for admitting gas to the penetration side of the weld joint.

Where inert gas back-up is used, the gas must be admitted into the groove with very little pressure. This is done by drilling holes into a gas manifold in the base of the back-up bar. The holes may be placed on or off center and chamfered at the top to prevent jetting of the gas onto the weld penetration. Fig. 8-16 shows a typical back-up bar groove with holes in place.

Manifolds for distribution of the gas can be placed in or below the grooved bar as shown in Fig. 8-17. In all cases, the manifold should be as large as possible to provide the inert gas in sufficient volume with low pressure to the weld area.

Some welding operations require preheat and interpass temperature control of these components prior to and during the welding operations. To control these temperatures, a back-up bar may be used. Two common methods are heated liquids such as water and oil, or electric strips or rods. Thermostats are used to control the temperature. The heat is applied into the part to be welded in the actual weld zone. The major benefit from this type of heating is the very close control of the desired temperatures of the weld area throughout the welding operation. Fig. 8-18 shows a typical back-up bar designed for electric strip heaters.

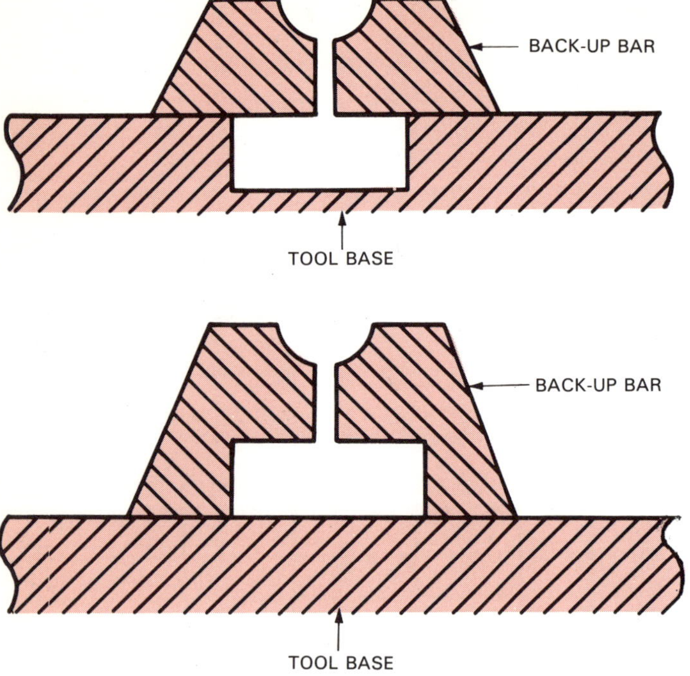


Fig. 8-17. Back-up bars designed with manifolds are costly to make and maintain. The smaller bar shown is less costly to replace.

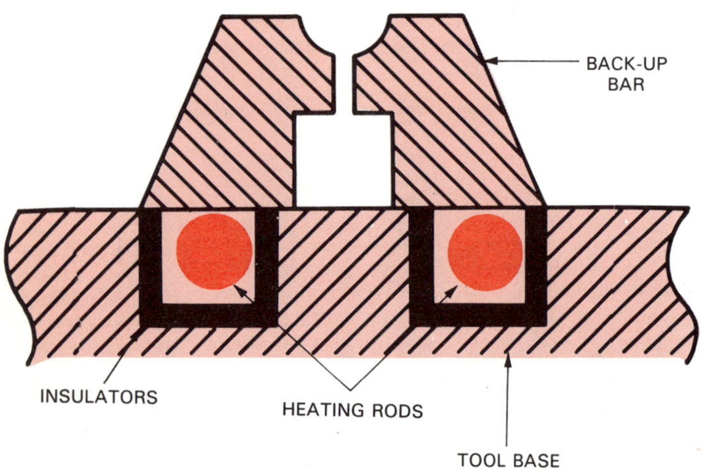


Fig. 8-18. Back-up bars which use strip heaters must be of sufficient thickness to accept the strip heater and insulators.

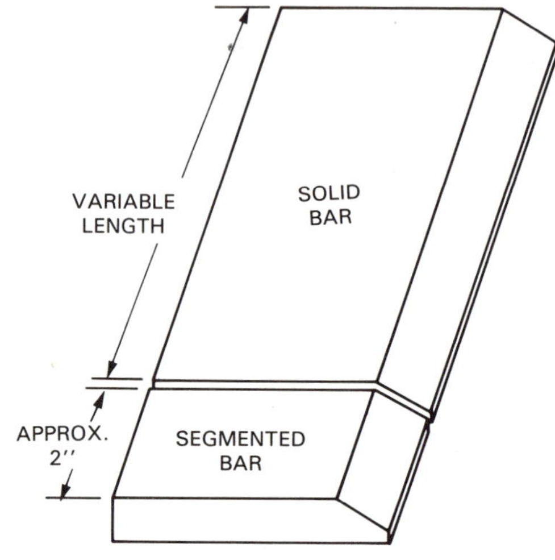


Fig. 8-19. Solid and segmented type hold down bar design.

## HOLD DOWN BAR DESIGN

The function of the SOLID HOLD DOWN BAR or SEGMENTED FINGER is to hold the components in place and to control the heat flow from the weld joint. The hold down bar may be a solid bar type or segmented finger type. Solid bars are generally used for smaller or short welds. Finger bars are used for longer welds. The type of basic tool design may dictate one type over the other. Longer solid hold down bars distort more readily and replacement cost is higher than segmented hold downs. Finger bars are cheaper to make and replace. Fig. 8-19 shows the two types of designs.

Fig. 8-20 shows some of the common designs in use. The sharp nose bar is fragile regardless of the material used in construction and may require frequent maintenance to keep the bar or finger in intimate contact with the workpiece. The flat nose design affords

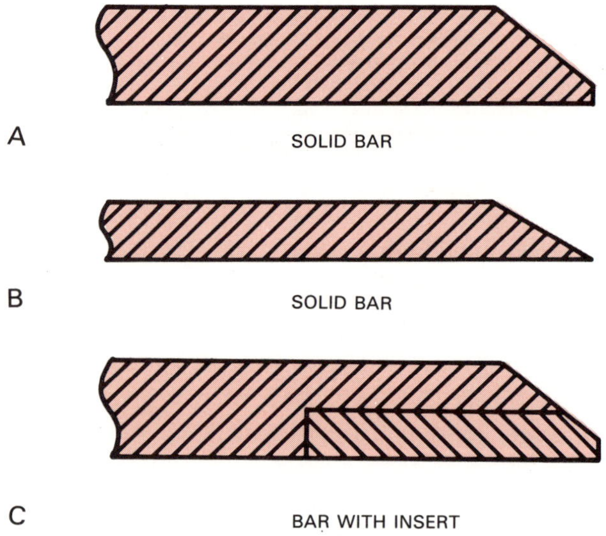

A SOLID BAR

B SOLID BAR

C BAR WITH INSERT

Fig. 8-20. Design and use of finger bar contact area. A is used where thicker welds are made and good heat transfer is required from the weld joint. B is used for thin gauge material with minimum hold down pressure. C is used where considerable pressure is used for holding part in place.

greater rigidity than the sharp nose design. However, they generally are placed further back from the joint to give the operator more visibility. This also allows for space for the torch and the wire feeder manipulator if used in the operation.

The RELIEF STEP DESIGN with a flat nose, as shown in Fig. 8-21, has several advantages over the other designs. However the overall cost of the part and replacement parts is higher due to the added machining required for the relief step. The main advantage of the relief step design is that the part may be used on slightly warped parts. The main bearing area is small compared to a solid finger or bar in intimate bearing over the entire area of the part. This assures intimate contact with the part in the weld area. Fig. 8-22 shows relief step finger design in use.

Fig. 8-21. Relief step design hold down bar requires a thicker cross section.

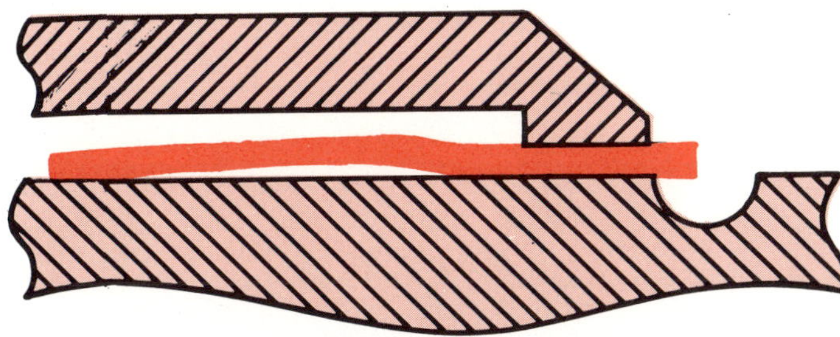

Fig. 8-22. Relief step design assures positive clamping of material and good heat transfer characteristics at the weld joint.

## MAINTENANCE

Welding tooling undergoes tremendous stresses and strains during welding operations. Application of pressure and absorbtion of heat during the tool's operation requires close inspection and maintenance to produce consistant welds. Even the most rigid tool will lose alignment after constant use. If the component parts are made to a close tolerance, the tool should be inspected quite often for tolerance deviation.

Back-up bars and chill rings should be checked for burrs and nicks which prevent close contact with the component part. Without the proper chill of the weld zone, a consistent weld cannot be made, and weld defects (low welds, no penetration) will occur. Fig. 8-23 shows how this condition may be detected.

Inert gas leaks may cause contamination of the weld penetration with oxygen, hydrogen and nitrogen which in many welds, would cause rejection. In some materials, the lack of gas coverage prevents weld root formation and the penetration is uneven and does not blend into the parent metal.

Cleanliness of the tool is a must. If the tool is to be used for an inert back-up gas atmosphere, moisture and grease will have an adverse effect on the purity of the inert gas.

Back-up bars and chill rings should be stored in a clean, dry area and cleaned again prior to use. Steel fixtures absorb moisture when cool. If used near the weld, any moisture pick-up will effect the gas purity.

Any portion of the tool's inert gas area which may be contaminated during storage should be disassembled and thoroughly cleaned before use. This should also be done before the tool is initially used, as oil and grease may accumulate during manufacture of the tool.

Mechanical/manual areas of the tool should be checked often for proper operation, loose bolts, nuts, screws, toggles, cams, etc. These are all part of the tooling operation and should be replaced or repaired whenever required.

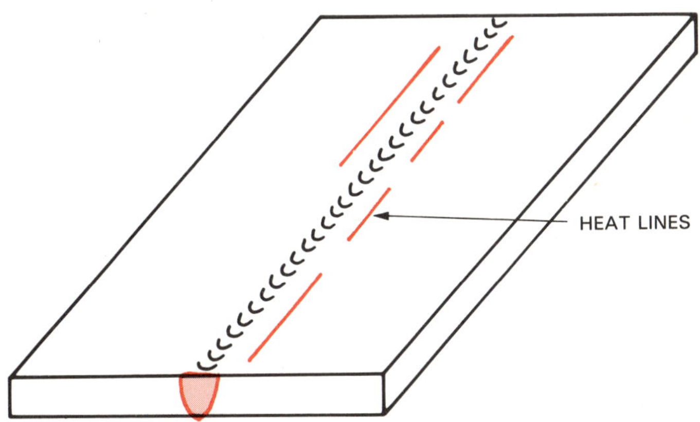


Fig. 8-23. Inspection of the weld area after the weld is completed will indicate whether or not the finger bar was in proper contact. Poor contact will be indicated by a heat line along the contact point. Where this line varies or becomes larger, the bar did not transfer heat properly.

## TEST TOOLING

TEST TOOLING is often made to prove tooling design concepts, make test welds, and establish weld joint shrinkage dimensions. In some cases, the training and certification of welders can be done on test tooling prior to the production task. Tools used in testing can be made rather quickly and modified as required at a much lower cost than changing the production tool. A simple test tool is shown in Fig. 8-24. Modifications of the tool could include the addition of heater strips or rods for preheating and interpass temperature control. A very important factor when making butt welds is the amount of shrinkage of the components at the weld joint. This can be determined by making lines or marks on each component prior to the weld. Measuring the distance between the lines, complete the weld, then measure between the lines again, as shown in Fig. 7-28. Information such as this is very important in making components to exact dimensions. It will allow sufficient machining stock for overall dimensions after welding.

Fig. 8-24. Test tools of this type are often used to determine the final design of the production tool. (Aerojet-General Corp.)

## AUXILIARY TOOLING

Auxiliary tooling includes Seamers, Manipulators, Positioners and Lathes, Planishers, Gas Atmosphere Welding Chambers, and Seam Trackers.

### Seamers

SEAMERS are used for making longseam internal or external welds in flat or cylindrical stock. They offer an excellent method of holding the material in place, provide the proper type backing bar and hold down finger bar design, and operate consistantly. They may vary in size from a bench model to a large unit capable of welding cylinders several feet in diameter.

Side beam tracks and carriages may be added to the seamer which contains the welding torch and equipment for movement of the torch. Units of this type can be aligned to a very close tolerance for tracking the weld seam. Weld repeatablity is very good. Fig. 8-25 shows a seamer with a mechanized carriage and the welding equipment mounted over the welding area. The welding equipment in Fig. 8-25 is also used for welding components on the adjoining welding lathe.

Fig. 8-25. A commercial seamer used to make longseam welds in tubing, pipes, cylinders, and flat materials. The lower bar, or mandrel as it is sometimes called, can be removed and replaced with different designs for almost any type of longitudinal welds. (JetLine Engineering, Inc.)

Fig. 8-26. Technician adjusting a manipulator which does not move during the welding operation. The weld joint is moved by the rotating positioner. (Parker-Hannifin Corp.)

## Manipulators

MANIPULATORS are often used to locate the welding head at various locations and heights. In some cases, the main base may be mounted on rails to enable movement throughout the welding area. They are made in many different designs. When used only for the location of the weld when the head is in a stationary position, as shown in Fig. 8-26, the accuracy and speed of the manipulator is not critical.

If the manipulator is used for longseam welding, the accuracy and speed are very critical areas of concern. Fig. 8-27 shows a manipulator which may be used for longseam welding and for varying locations and heights. Notice the welding head on one end and some of the welding equipment mounted on the other end.

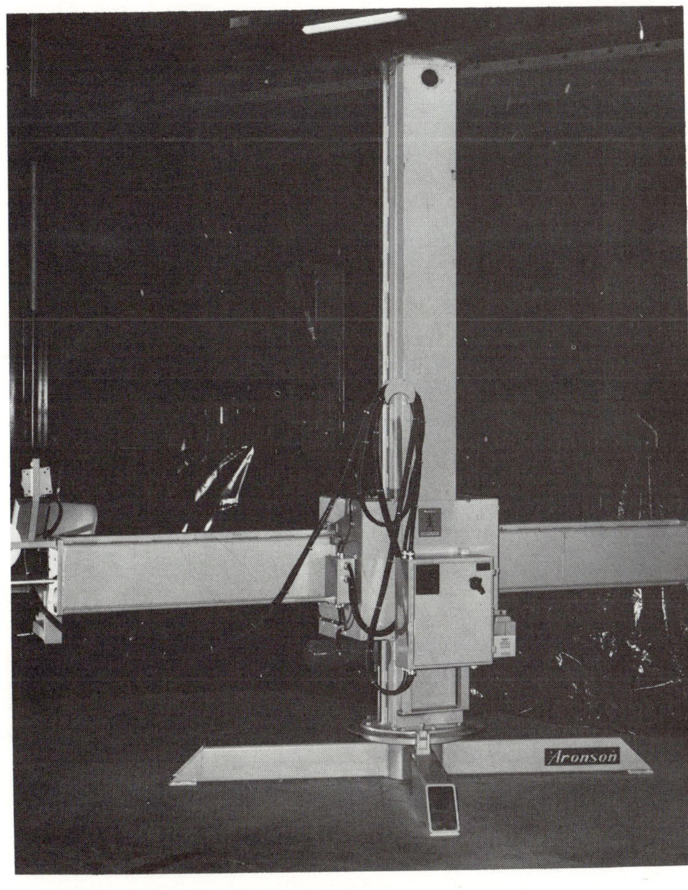


Fig. 8-27. This manipulator has a 360 degree rotational base and a ram drive for longitudinal welds. (Aronson Machine Co.)

## Positioners and Lathes

Equipment such as POSITIONERS and LATHES are used in many GTAW circular type welds. With modern electronic components such as SCR feedback and dynamic braking systems, they can be adapted to almost any task with a very high repeatability of the operation. Fig. 8-28 shows a small positioner and Fig. 8-29 shows a welding lathe.

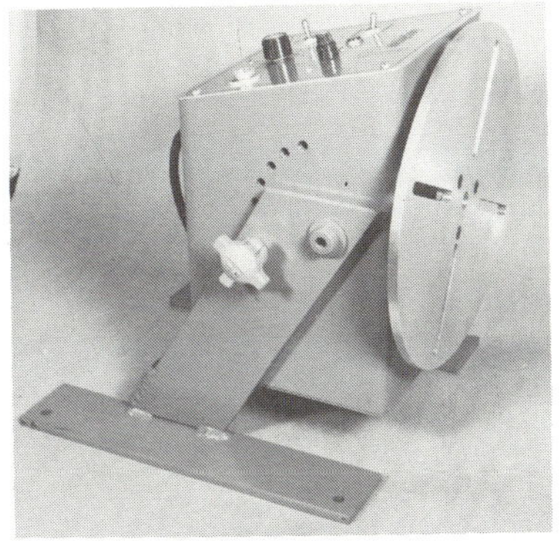

Fig. 8-28. Small welding positioners of this type uses a DC motor for very accurate rotational drive and are very useful in the welding shop. (JetLine Engineering, Inc.)

Fig. 8-29. The small positioner is being used as a headstock for a welding lathe. The tail stock may be moved for adapting to long or short parts. (JetLine Engineering, Inc.)

When considering the design of these systems or individual components, the capacity of the equipment should be thoroughly investigated. For example, one positioner may turn from .5 to 10 rpm, while another positioner may turn from .1 to 3 rpm. Therefore, the diameter of the weldment and the welding speed must be considered when selecting the unit for use.

Another example might be a 200 pound positioner that when welding with the faceplate flat, the unit will accept a 200 pound load. However, when the faceplate is vertical and the load is in the horizontal position, the maximum load is 200 pounds with the center of gravity 4 inches from the plate. With this in mind, also consider the added weight of the part tooling and the possibility of overloading. See Fig. 8-30.

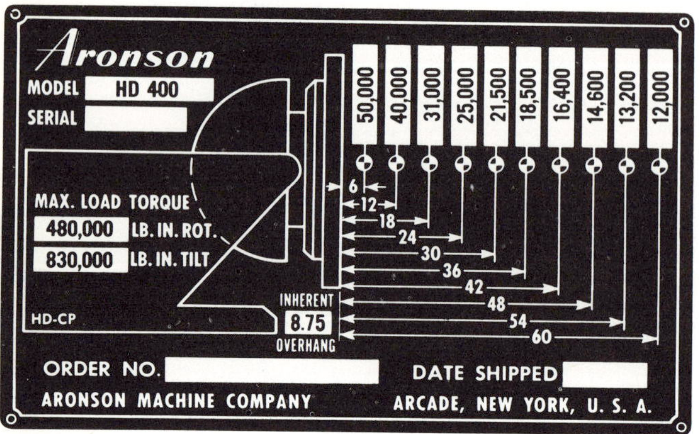


Fig. 8-30. Welding positioner load capacity levels are established by the manufacturer for safe operating limits for each type and model. (Aronson Machine Co.)

## Planishers

PLANISHERS are used to flatten and smooth long seam welds after welding to reduce finishing cost of the part. In the case of some materials, cold working by planishing will increase the mechanical properties.

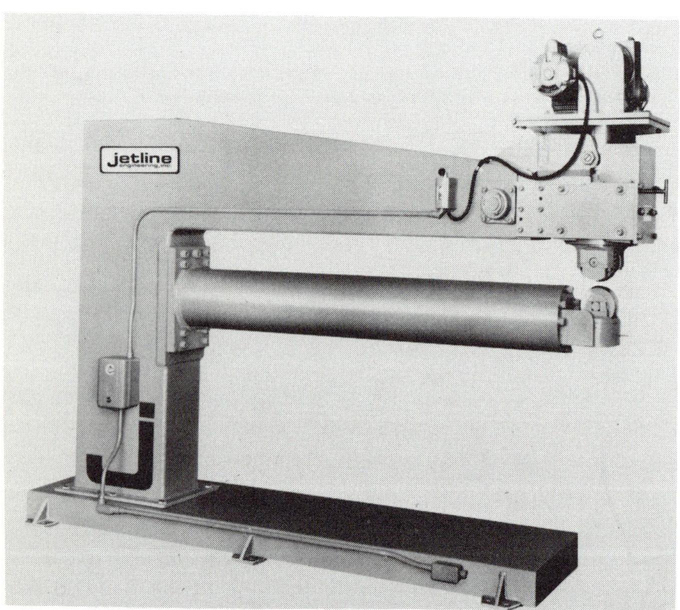


Fig. 8-31. Planishers operate on both longitudinal and circumferential welds to flatten and smooth welds. (JetLine Engineering, Inc.)

Where the part is to be formed or spun after welding, the weld cross section must be similar in size to the parent metal. The planisher flattening wheel may be specially designed for specific types of contours. Fig. 8-31 shows one type of planisher. Fig. 8-32 shows a planisher rolling operation. An "as welded" weld grain structure and a planished weld grain structure are shown in Fig. 8-33.

## Gas Atmosphere Welding Chambers

ATMOSPHERE CHAMBERS are used where the weld must be fully protected from atmospheric contamination. Reactive materials such as titanium, zirconium, tantalum, etc., readily absorb atmospheric contamination when heated during the welding operation. In some cases, tooling, trailing shields, and special torch nozzles may be used to provide the inert gas shield in the weld area. However, in many cases, this type of equipment cannot be used and gas atmosphere

Fig. 8-32. Planisher operation on a tube longseam weld. (Pathway Bellows, Inc.)

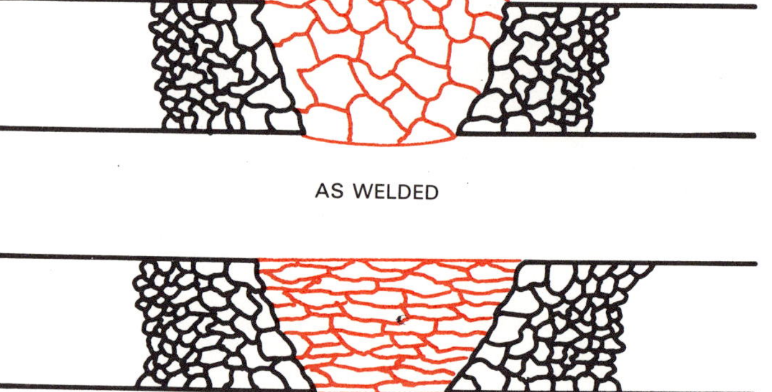


Fig. 8-33. Weld structure after welding and after planishing.

welding chambers must be used.

Atmospheric chambers may be made from a variety of materials including steel, stainless steel, and plastic. They are made in many sizes and configurations. Fig. 8-34 shows a simple chamber with work gloves installed for welding on opposite sides of the chamber. A unit of this type operates by admitting the inert gas until the atmosphere is replaced with argon. A test weld is then made on a piece of titanium scrap metal and the color of the weld is observed. An oxygen analyzer may also be used.

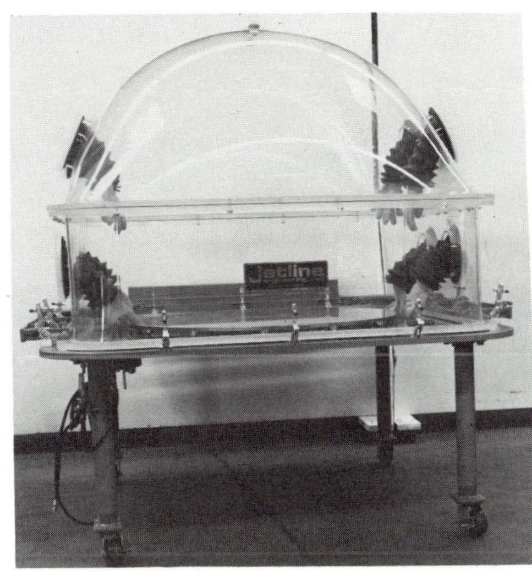


Fig. 8-34. Plastic inert gas chamber for welding titanium and many specialty metals can be adapted to various size parts by changing the lower chamber extension. (JetLine Engineering, Inc.)

Fig. 8-35. Plastic inert gas chamber with a loading port allows loading and unloading parts without affecting the atmosphere in the main chamber. A bulb tester for testing gas purity is mounted on the front of the chamber. (JetLine Engineering, Inc.)

The chamber shown in Fig. 8-35 includes a chamber for admitting or removing parts without affecting the main chamber atmosphere. Note the bulb tester to be used for oxygen testing. This may be used on either the main or loading chamber.

Chambers such as shown in Fig. 8-36 are equipped with evacuation pumps to remove the air from the chamber, then admit the inert gas into the empty chamber. These units reduce the time required to achieve a welding atmosphere. However, initial costs are quite high due to the heavier design of the chamber and the pumping equipment.

Fig. 8-36. The vacuum system mounted on this inert gas chamber reduces the time required to obtain a welding atmosphere. (JetLine Engineering, Inc.)

### Seam Trackers

SEAM TRACKERS are used in semi and automatic welding operations to align the welding torch along the weld joint. The unit operates with a sensor or probe tracking the joint. The torch then moves through the use of motorized cross slides. Fig. 8-37 shows a typical weld joint sensor and controller.

Different types of joints may be tracked by changing the type of probe. Where extensive tooling is not available and joint accuracy is required, seam trackers can be used with very good results.

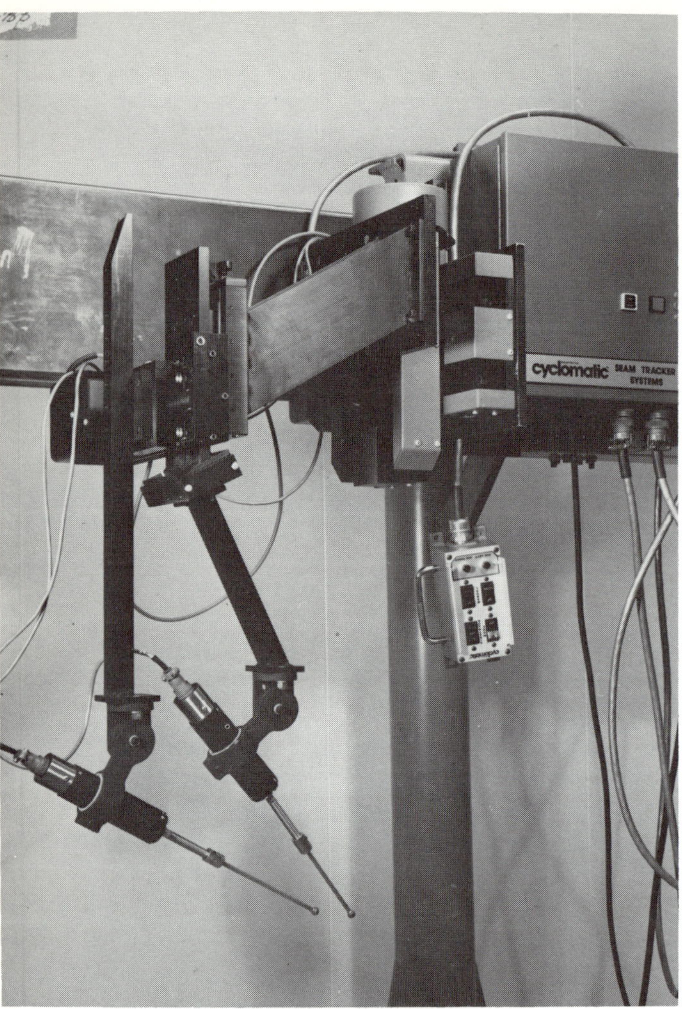


Fig. 8-37. Seam trackers allow the use of automation on components with varying contours. Combined with automatic voltage control welding heads, they are adaptable to any type of longitudinal or circumferential welding operations. (Cyclomatic Industries, Inc.)

## REVIEW QUESTIONS

1. What is "Hard Tooling?"
2. Tools that are made from clamps, angles, etc., are called ________ ________ .
3. What are the four basic types of tool designs?
4. If tools are not designed properly for expansion and shrinkage during welding, what may happen?
5. List four types of clamping pressure applications.
6. List three types of designs used in the fabrication of tool bases.
7. What materials are *not* used in the immediate weld area when using direct current for welding? Why?
8. What three materials may be used in the immediate weld area when using direct current for welding? Why?
9. Where is a shallow radius back-up groove used?
10. List the common methods used with back-up tooling for preheating the weldment.
11. Hold down bars are either a ________ or ________ type.
12. ________ bars distort more easily than ________ bars.
13. Hold down bars are made with a ________ ________ design for use on parts that are warped or distorted.
14. What are a major cause of low welds or lack of penetration along the weld seam when using back-up bars or chill bars?
15. Why are test tools important?
16. What is the function of a seamer?
17. Can manipulators be used to make longseam welds?
18. Can the argon atmosphere in a chamber be tested for purity level without an oxygen analyzer?
19. Where are seam trackers used?

# Chapter 9

# EQUIPMENT SETUP

The setup of a welding power supply and accessory equipment requires careful consideration if the welding operation is to be completed properly. Without proper setup, damage may occur to the equipment, the components or parts may be ruined, or there may be possible injury to the operator.

## WELDING PROCEDURES

In situations where the weld must be qualified by testing prior to use, weld procedures are followed. WELDING PROCEDURES list all of the specific variables and parameters obtained during the test

MANUAL GAS TUNGSTEN ARC WELDING SCHEDULE (GTAW)

PROJECT MINARC P/N 743936 CHG 00 NAME BRACKET

OPER. NO. — WELDING SPEC MINARC 43-6 PRO. QUAL. NO. —

BASE MATERIAL 6061-T6 ALUMINUM BASE METAL PREPARATION CHEMICAL Etch

TYPE OF JOINT VEE-GROOVE WELDING POSITION FLAT

FILLER MATERIAL ER-4043 SPEC. AWS DIA. 3/32"

TORCH GAS ARGON CFH 20 BACKUP GAS NONE CFH —

TOOL NO. WF NO. 631 TYPE TORCH OR HEAD LINDE HW 20

PREHEAT TEMP. 250°F INTERPASS TEMP. 250°F POST WELD HEAT TREAT NONE

TYPE CURRENT ACHF AMPERAGE 80 - 130 POWER SUPPLY A/C TRANSFORMER

VOLTAGE CONTROL N/A START METHOD HIGH FREQUENCY

WELD BEAD OVERLAP N/A BEAD WIDTH N/A CUP SIZE 8 ELECTRODE EXTENSION 1/4"-5/16"

TUNGSTEN TYPE AND CONFIGURATION:

ZIRCONIATED TUNGSTEN 3/32" DIA. /BALLED TIP.

REF: NDT INSPECTION

VISUAL AND DYE PENETRANT

| REVISIONS 00 | DATE 6-15-82 | WELDING ENGINEER W. Jones |
|---|---|---|
| | | QUALITY CONTROL |
| | | GOVERNMENT INSPECTOR |

Fig. 9-1. Welding procedure setup charts eliminate the possibility of incorrect machine settings for different jobs.

welding and must be used for the production job. A sample welding procedure for a manual welding operation is shown in Fig. 9-1.

Since all of the different variables may be recalled for each job, it is recommended that a written welding procedure or setup be established for each job. Record the variables during the main operation for future reference. Use the chart to be sure the job is properly completed.

## JOB AND EQUIPMENT CONSIDERATIONS

Prior to making first time welds or welds without established procedures, several basic considerations must be made regarding equipment. These include:

1. Power supply.
   A. Type of current to be used.
   B. Duty cycle sufficient for job.
2. Welding torch.
   A. Duty cycle sufficient for job.
3. Tungsten type and size.
   A. Pure tungsten for AC welding of aluminum and magnesium.
   B. Zirconiated tungsten for AC welding of aluminum and magnesium where tungsten spitting cannot be tolerated.
   C. One percent or two percent thoriated tungsten for all other materials using direct current straight polarity.
   D. Select tungsten size for amperage range, as shown in Fig. 9-2.
4. Tungsten preparation.
   A. Pure and zirconiated tungsten – grind radius on end to be used. Ball end to desired diameter for AC.
   B. Thoriated tungsten – grind taper on one end (do not grind painted end) for Direct Current Electrode Negative (DCSP).
5. Shielding gas. (Refer to Chapter 5)
   A. Argon – Arc easy to start. Good blanket coverage.
   B. Argon and helium mix – arc is harder to start, welds have deeper penetration. Requires higher flow rates as helium rises up from weld, and welds can be made at faster speeds.
   C. Shielding gas flow must be maintained at the torch nozzle for weld protection. If tungsten is extended beyond a 1/4 inch, use a gas lens for a more stable flow.
   D. Nozzle size should always be as large as possible to cover the weld area.

| Electrode Diameter (inches) | Current—Amperes | | Approx. Argon Gas Flow C.F.H. at 20 PSI | |
|---|---|---|---|---|
| | Pure Tungsten | Thor. Tungsten | Aluminum | Stainless Steel |
| .010 | 0–8 | 0–8 | 3–8 | 3–8 |
| .015 | 5–12 | 5–12 | 5–10 | 5–10 |
| .020 | 8–20 | 8–20 | 5–10 | 5–10 |
| .040 | 20–50 | 20–50 | 5–10 | 5–10 |
| 1/16 | 40–120 | 50–150 | 13–17 | 9–13 |
| 3/32 | 100–160 | 140–250 | 15–19 | 11–15 |
| 1/8 | 150–210 | 220–350 | 19–23 | 11–15 |
| 5/32 | 190–270 | 300–450 | 21–25 | 13–17 |
| 3/16 | 250–350 | 400–550 | 23–27 | 18–22 |
| 1/4 | 300–490 | 500–800 | 28–32 | 23–27 |

Fig. 9-2. Select tungsten size for intended job, then set amperage range on machine for the amount of current desired.

## EQUIPMENT AND WELDING MACHINE SETUP

The actual setup of the equipment and the power supply will vary considerably depending on the type and make of the equipment. Each equipment manufacturer designs equipment different from another manufacturer. However, all will have the same basic controls.

Fig. 9-3. Industrial rated (60 percent duty cycle) power supply. (Lincoln Electric Co.)

For manual all-purpose work in most shops, a machine called an industrial rated 300 ampere, 60 percent duty cycle, alternating current, direct current power supply is used. The setup of the machines may vary from one type to another.

Each manufacturer produces a booklet or instruction manual explaining in detail the operation of the machine. Prior to setting up the equipment and the machine, these instructions should be read until they are fully understood.

The major controls used in manual gas tungsten arc welding are shown on an industrial rated welding power supply in Fig. 9-3.

## EQUIPMENT AND MACHINE SETUP FOR DIRECT CURRENT ELECTRODE NEGATIVE (DCSP) GTAW

1. Connect primary power.
2. Connect argon gas supply.
   A. To torch if manual control.
   B. To machine if machine solenoid control.
3. Connect water supply (if water cooled torch).
   A. To torch if manual control.
   B. To machine if machine solenoid control.
4. Set current type switch to DCEN (DCSP).
   A. Connect torch power cable to electrode terminal.
   B. Connect ground to ground terminal.
5. Set current range switch to range required.
6. Set current control for maximum current desired. (Note: This control is 100 percent of set range).
7. Set panel/remote control.
   A. Panel setting regulates current as set on panel current control.
   B. Remote setting regulates amount of current as set on hand or foot control. Current will not exceed panel rheostat current control setting.
8. Set high frequency to start only.
   A. High frequency will only operate until arc is started.
9. Set high frequency to medium.
10. Arc stabilizer does not operate on DC.
11. Wave balancer does not operate on DC.
12. Set the soft start.
    A. Use for low starting currents on thin materials.
13. Set shielding gas post flow time if machine controlled.
    A. Post flow of gas is required to prevent electrode contamination during post weld cooling period. The time is set for the various sizes of tungstens. The gas flow should stop after the tungsten cools to a silver color.
14. A GTAW mode switch (if available).
    A. This switch actuates the machine control circuits for:
       1. Inert gas start and stop.
       2. Torch cooling water start and stop.
       3. Contactor circuit.
    B. GTAW mode switch (not available).
       1. Inert gas must be started and stopped manually.
       2. Torch cooling water must be started and stopped manually.
15. Start power supply.
16. Energize the contactor.
    A. With the mode selected to start gas and water flow:
       1. Foot control.
       2. Hand control.
       3. Panel control switch.
    B. Set argon flow rate.
17. Start arc.

## EQUIPMENT AND MACHINE SETUP FOR ALTERNATING CURRENT GTAW

1. Connect primary power.
2. Connect argon gas supply.
   A. To torch if manual control.
   B. To machine if machine solenoid control.
3. Connect water supply (if water cooled torch).
   A. To torch if manual control.
   B. To machine if machine solenoid control.
4. Set current type switch to AC.
   A. Connect torch power cable to electrode terminal.
   B. Connect ground to ground terminal.
5. Set current range switch to range required.
6. Set current control for maximum current desired. (Note: This control is 100 percent of set range).
7. Set panel/remote control.
   A. Panel setting regulates current as set on panel current control.
   B. Remote setting regulates amount of current as set on hand or foot control. Current will not exceed panel current control setting.
8. Set high frequency on continuous.
9. Set high frequency intensity to medium.
10. Set arc stabilizer to medium (if available).
    A. Adjust to remove tungsten spitting or lack of cleaning action.
11. Set wave balancer to normal.
    A. Increasing the straight polarity part of the cycle increases the penetration.
    B. Increasing the reverse polarity part of the cycle increases the cleaning action.
12. Set the soft start.
    A. Use for low starting currents on thin materials only.
13. Set shielding gas post flow time if machine controlled.
    A. Post flow of gas is required to prevent electrode contamination during post weld cooling period. The time is set for the various sizes of tungstens. The gas flow stops after tungsten

cools to a silver color.

14. GTAW mode switch (if available).
    A. This switch actuates the machine control circuits for:
        1. Inert gas start and stop.
        2. Torch cooling water start and stop.
        3. Contactor circuit.
    B. GTAW mode switch (is not available).
        1. Inert gas must be started and stopped manually.
        2. Torch cooling water must be started and stopped manually.
15. Start power supply.
16. Energize the contactor.
    A. With the mode selected to start gas and water flow:
        1. Foot control.
        2. Hand control.
        3. Panel control switch.
17. Start arc.

## EQUIPMENT AND MACHINE SETUP FOR DIRECT CURRENT ELECTRODE POSITIVE (DCRP) GTAW

The equipment for this polarity is set up the same as DCEN, however the polarity will be positive. Since all of the welding heat is generated in the electrode, a large diameter electrode must be used. A water cooled torch is required for all diameters of tungstens which may be used.

Since GTAW torches are not rated in the DCEP (DCRP) mode, the duty cycle rating may not be used. The torch temperature must be monitored by the welder. A warm torch indicates insufficient cooling and the amperage, or use time, must be lowered.

The electrode should be 2 percent thoriated to carry the high heat and the end should not be tapered. A slight radius on the end will assist in formation of a ball in this positive polarity.

## REVIEW QUESTIONS

There are no review questions for this chapter to complete. It is suggested that you obtain an operation booklet or instruction manual for an industrial rated power supply and study it thoroughly.

Each part of the machine operates in a specific manner to complete a specific task. If the machine is not operated correctly, problems may develop within the machine which may cause the machine to malfunction. This may possibly ruin the machine, or the parts or components being welded.

As an operator of the welding machines, you are responsible for its setup and operation within the duty cycle limits to do the intended job.

# Chapter 10

# MANUAL WELDING TECHNIQUES

GTAW on various weld joint designs requires different techniques depending on many factors. This chapter and Chapter 11 will cover the basic techniques used when gas tungsten arc welding steel and aluminum. These are the two most common materials used by the industry. Other materials are welded using the same basic procedures listed for steel and aluminum.

Welding skill is acquired by practice, practice, and more practice. During the practice time, watch the puddle very closely and concentrate on the weld. To assist you in concentrating on the actual weld, ask yourself questions such as:

1. Is the weld the right size?
2. Is the weld too high?
3. Is the weld too low?
4. Is this the proper travel speed?
5. Do I have the right torch angle?
6. Am I using the right wire angle?
7. Is the puddle flowing properly?
8. Are the amps correct?
9. Is the voltage correct?
10. Is the weld in the proper location?
11. Is the fillet weld leg the proper size?

Two other points to be remembered during the practice welding period are: fill the crater at the end of the weld; hold the torch over the end of the weld until it cools. This allows the post flow of shielding gas to protect the hot metal.

## GROOVE WELD DEFECTS

During the welding operation and later during the inspection of the completed weld, some common defects may be seen. Defects that are common to groove welds are listed below with references to the illustration of the groove weld defect.

1. Lack of penetration. Fig. 10-1.
2. Lack of fusion. Fig. 10-2.
3. Overlap. Fig. 10-3.
4. Undercut. Fig. 10-4.
5. Concavity. Fig. 10-5.
6. Convexity. Fig. 10-6.
7. Craters. Fig. 10-7.
8. Cracks. Fig. 10-8.
9. Porosity. Fig. 10-9.
10. Linear Porosity. Fig. 10-10.
11. Burn-through. Fig. 10-11.
12. Suck-back. Fig. 10-12.
13. Icicles. Fig. 10-13.

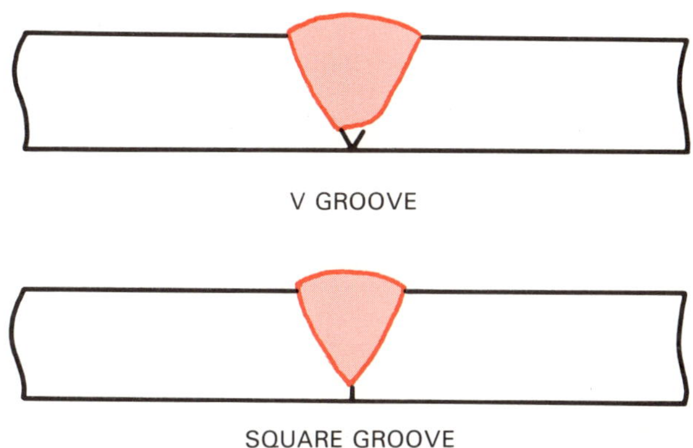


Fig. 10-1. LACK OF PENETRATION. The weld does not completely penetrate through the weld joint. This may be called INCOMPLETE PENETRATION.

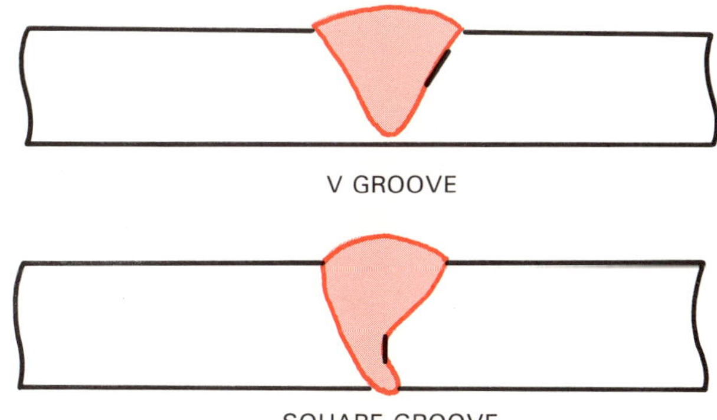


Fig. 10-2. LACK OF FUSION. This type of defect occurs more often in multiple pass welds where the weld passes do not fuse together. This may be called a COLD SHUT.

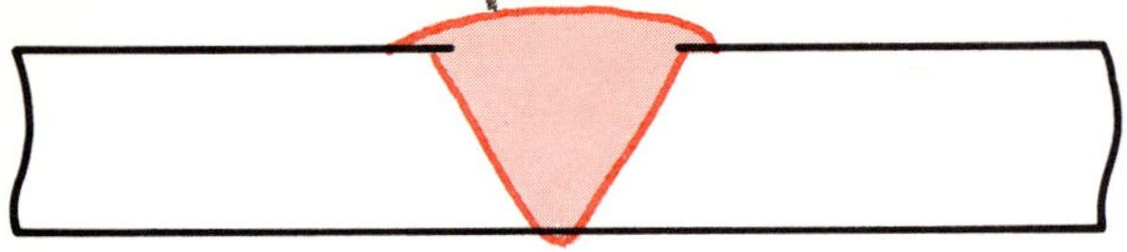

Fig. 10-3. OVERLAP. Overlapping welds usually occur on the crown beads edge where too much wire has been added.

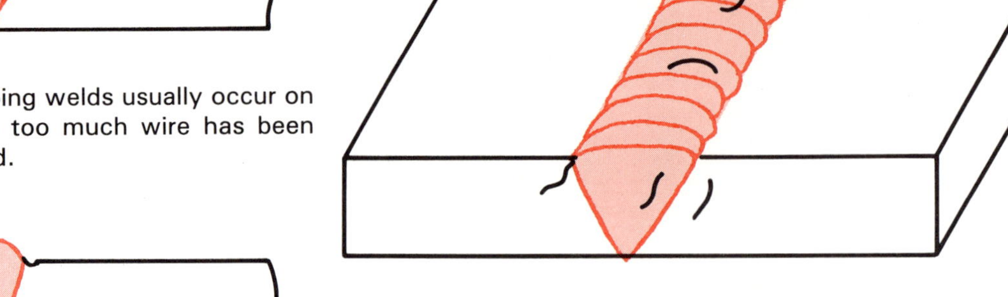

Fig. 10-8. CRACKS. Cracks may occur in any pass in welds and in the heat affected zone. They may be transverse (across) or longitudinal with the weld. They are caused by stresses imposed during the solidification and shrinkage of the molten metal.

Fig. 10-4. UNDERCUT. Undercut occurs at the edges of any pass where an insufficient amount of wire was added.

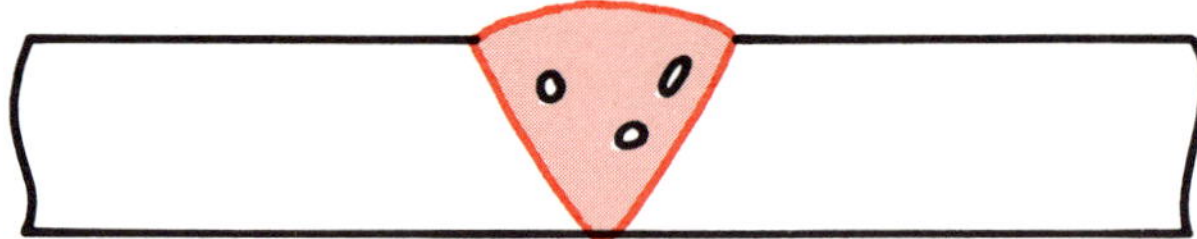

Fig. 10-9. POROSITY. A pore or multiple pores (porosity) exist where a gas bubble was formed in the molten pool and was trapped in the cooling metal. Pores may exist anywhere in the weld.

Fig. 10-5. CONCAVITY. Where a weld caves inward, it is concave. This may occur when an insufficient amount of wire was added.

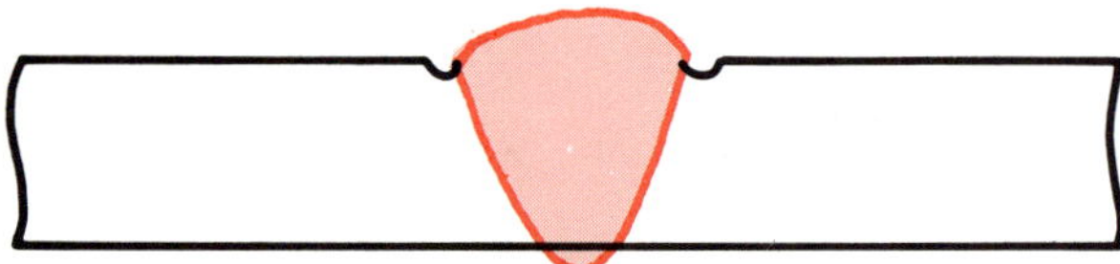

Fig. 10-6. CONVEXITY. Where a weld has too high a center section, it is convex. This happens where too much wire has been added. This weld is also undercut.

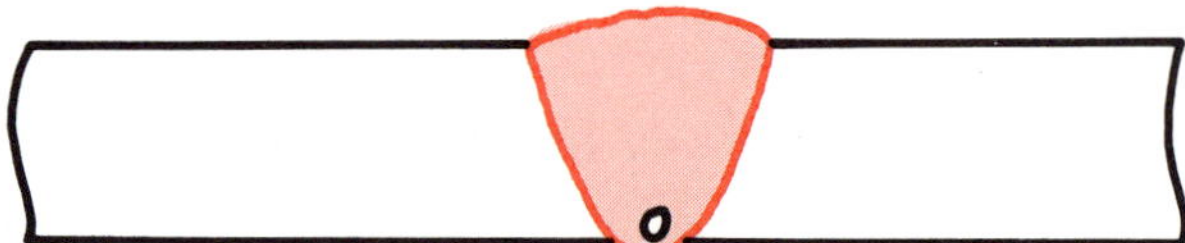

Fig. 10-10. LINEAR POROSITY. Linear (along a line) porosity is usually found in the root of a weld near the center line of the joint.

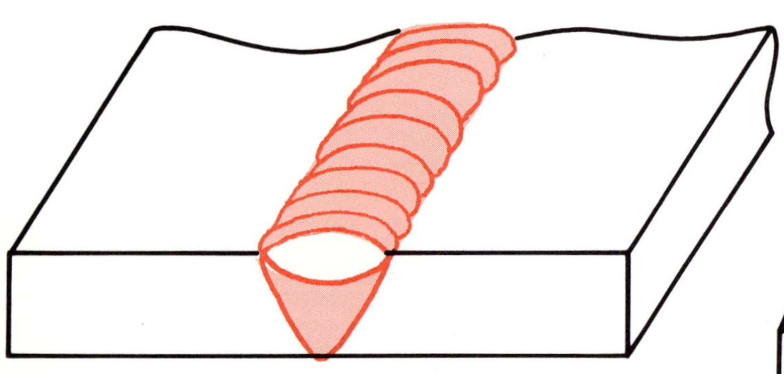

Fig. 10-7. CRATERS. Craters may exist whenever the weld ends. They may be located anywhere in the weld joint where the weld was stopped, restarted, and the crater was not properly filled. They are more often found at the end of the weld where the welder did not fill properly.

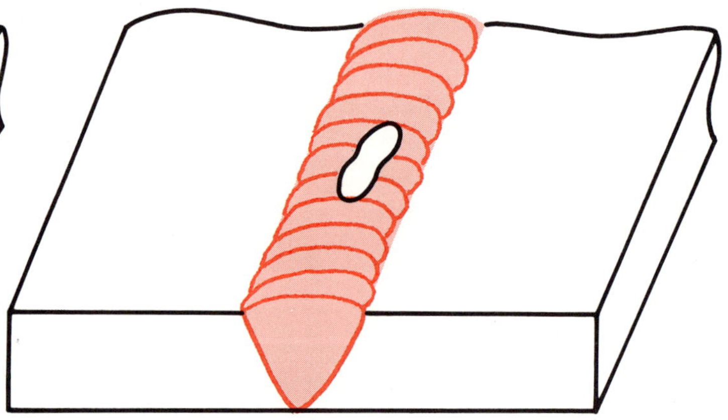

Fig. 10-11. BURN-THROUGH. Burn-through may occur in any part of the penetration weld where the molten pool is too hot.

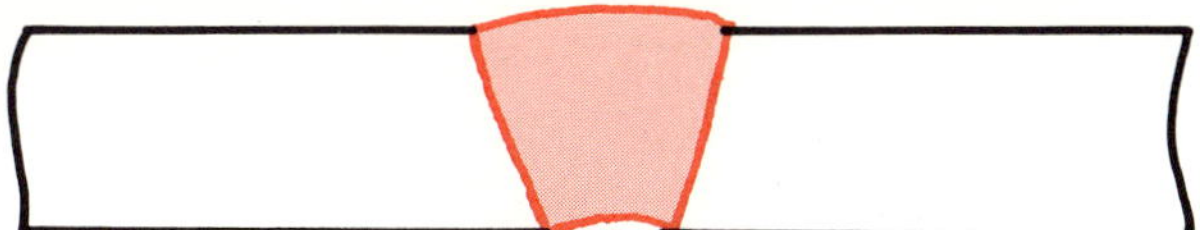

Fig. 10-12. SUCK-BACK. As the name (suck-back) implies, the molten metal is above the root of the joint instead of below the root. The molten metal actually falls during welding. However, on the cooling cycle, the metal shrinks back.

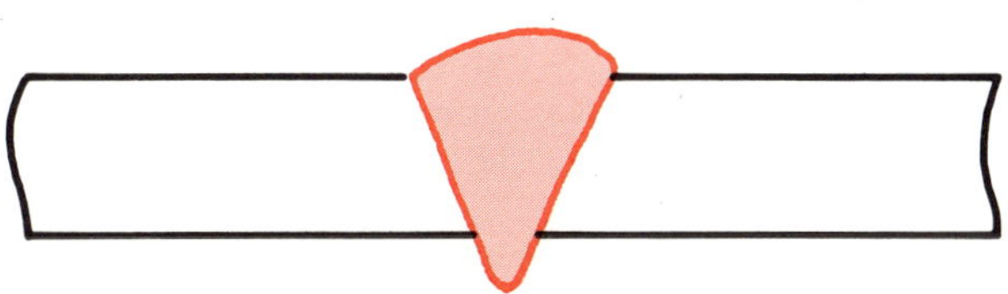

Fig. 10-13. ICICLES. Icicles are formed on the penetration side of the weld where the application of heat is uneven. If not corrected, ''burn-through'' may occur.

## FILLET WELD DEFECTS

Defects that are common to fillet welds may be noticed during the welding operation and later during the inspection of the completed weld. Defects are listed and illustrated below.

1. Lack of penetration. Fig. 10-14.
2. Lack of fusion. Fig. 10-15.
3. Overlap. Fig. 10-16.
4. Undercut. Fig. 10-17.
5. Concavity. Fig. 10-18.
6. Convexity. Fig. 10-19.
7. Craters. Fig. 10-20.
8. Cracks. Fig. 10-21.
9. Melt-through. Fig. 10-22.
10. Burn-through. Fig. 10-23.
11. Suck-back. Fig. 10-24.

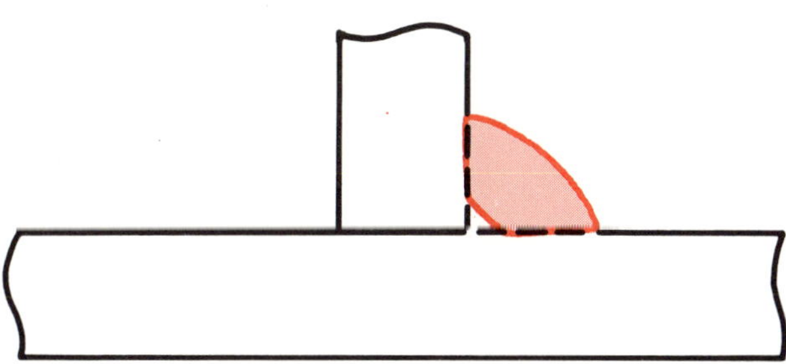

Fig. 10-14. LACK OF PENETRATION. Lack of penetration is the fault of the welder due to improper procedure or technique. This type of defect is very hard to detect by NDT (Non Destructive Testing).

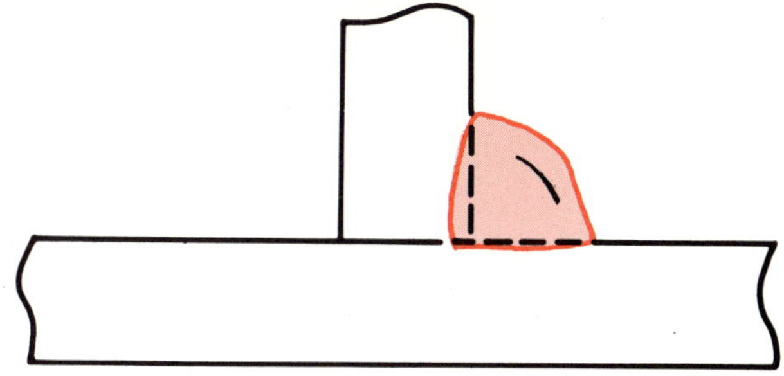

Fig. 10-15. LACK OF FUSION. This type of defect occurs more often in multiple pass welds where the weld passes do not fuse together. This may be called a COLD SHUT.

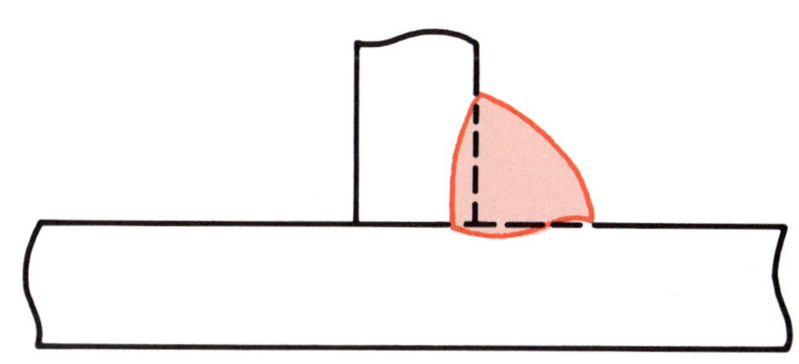

Fig. 10-16. OVERLAP. Overlaps on fillet welds usually occur on the bottom of the weld where a large amount of wire was placed into the puddle.

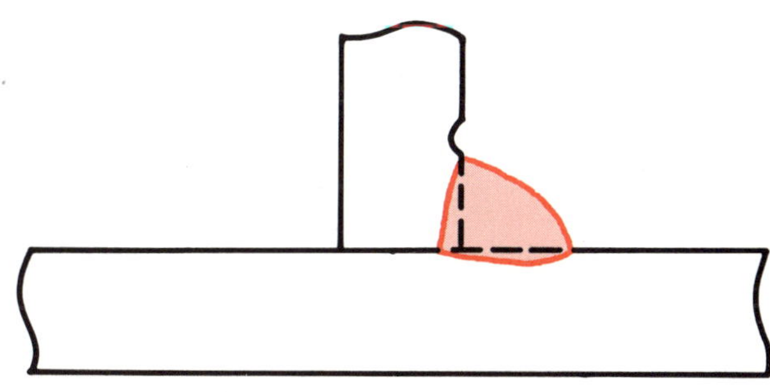

Fig. 10-17. UNDERCUT. Undercut is usually found on the upper edge of the weld. Since the welding heat rises and the metal falls, this area is very susceptible for this defect.

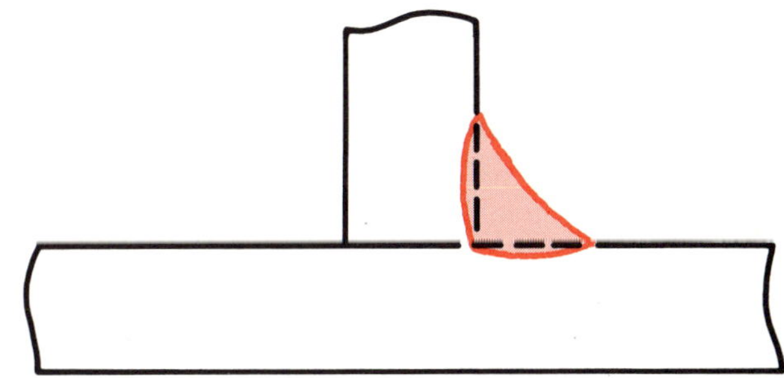

Fig. 10-18. CONCAVITY. Concave fillet welds are caused by not adding sufficient filler material to the puddle. However, be careful about adding too much or the joint will have lack of penetration.

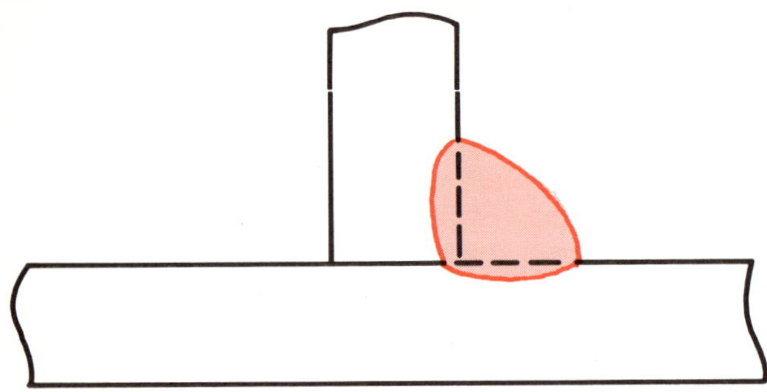

Fig. 10-19. CONVEXITY. This type of problem does not occur too often. Where it does occur, add a little less wire to the molten pool.

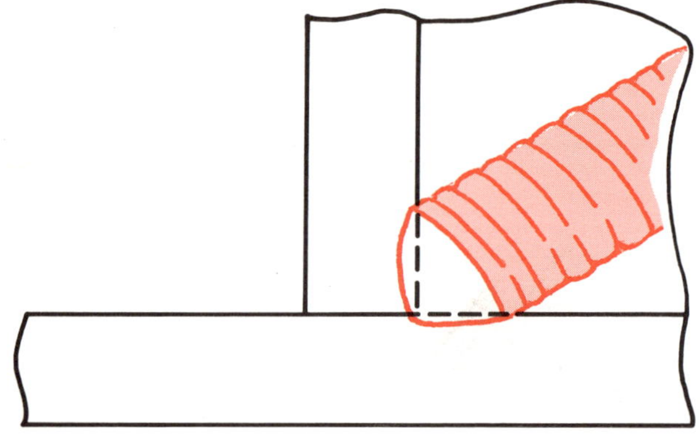

Fig. 10-20. CRATERS. Craters exist in fillet welds just as they do in groove welds. See Fig. 10-7.

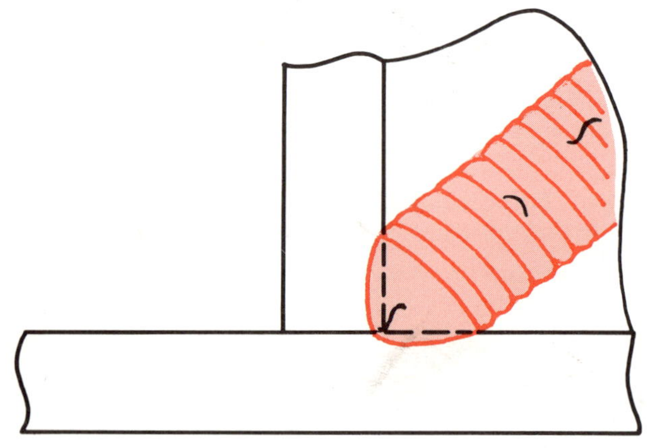

Fig. 10-21. CRACKS. Cracks exist in fillet welds just as they do in groove welds. See Fig. 10-8.

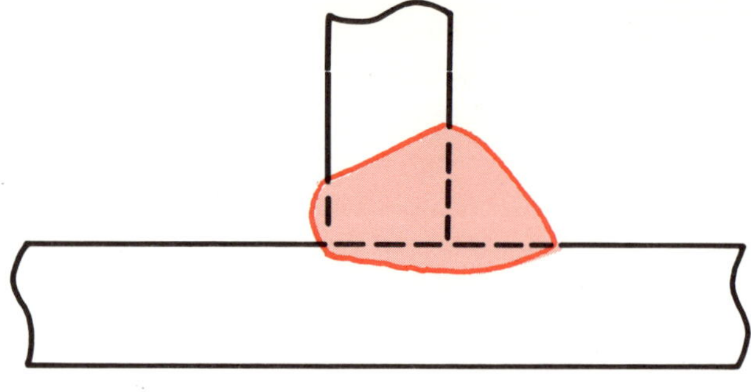

Fig. 10-22. MELT-THROUGH. Melt-through generally happens in the upper part of the fillet weld. This is caused by the flow of heat upward. It can be corrected by adding a chill bar behind the joint. Also, the torch may be angled more toward the lower part.

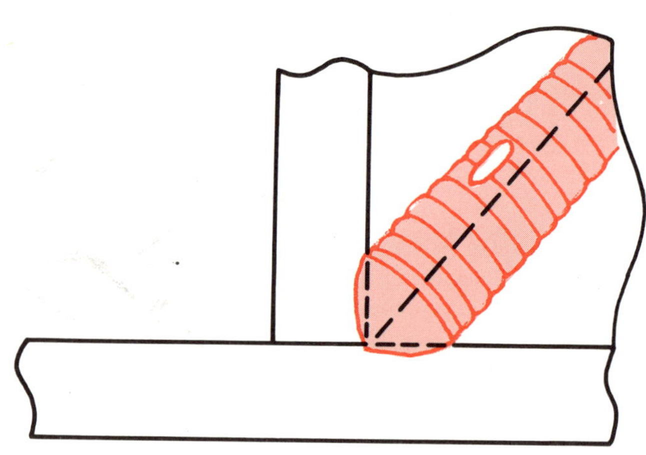

Fig. 10-23. BURN-THROUGH. Burn-through occurs when the upper leg of the part gets too hot. It may be corrected in the same manner as melt-through.

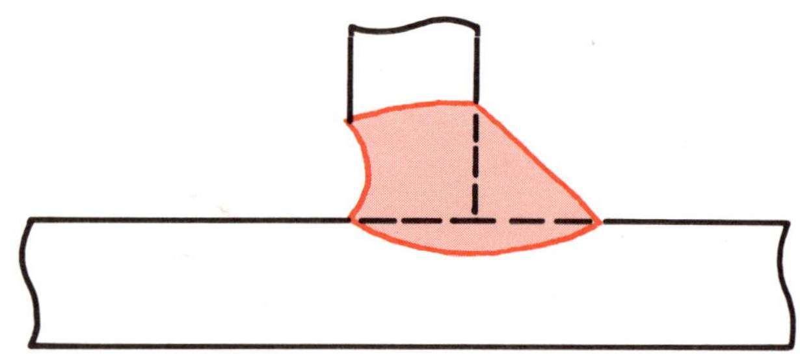

Fig. 10-24. SUCK-BACK. Suck-back occurs on either leg of the fillet due to excessive heat and improper addition of the filler material.

## PRE-WELD CLEANING

Both practice welding and production welding must be done on clean material. Welding done on improperly prepared material causes many problems during the operation and will affect final weld quality. Oxides on the material will cause porosity in the weld.

### Cleaning Cold Rolled Steel

Cold rolled steel is usually a bright metallic finish

material with a light coating of oil on the surface. Remove the oil by washing with a solvent such as alcohol or acetone. Material that has rusted should be prepared the same as hot rolled steel.

### Cleaning Hot Rolled Steel

Hot rolled steel has a blue color on the surface. This colored material is an oxide or scale that formed when the material was cooled in air after forming or rolling. The scale must be removed by grit blasting, grinding, or sanding to "bright metal" (bare metal without oxides and scale). Grit blasting, if used, must be followed by one of the other methods or wire brushed to remove imbedded grit from the surface.

Any rust that is present on the material must also be removed to "bright metal." Rust is formed by heavy oxidation. These oxides cause porosity in the weld.

### Cleaning Aluminum

All aluminum material has an oxide film on the surface. This film is created by exposing the material to air after the forming or rolling operation. This oxide film melts at approximately 3500°F while the aluminum melts at approximately 1200°F. Several methods may be used to remove this oxide.

1. Chemical etching is usually done with a commercial etch, followed by a deionized water rinse and oven drying.
2. Wire brushing with a stainless steel brush. The brush should not be the heavy wire type as it may tear the material.
3. Chipping or routing. This method of oxide removal is limited to heavier materials such as plate, castings, or forgings.
4. Abrasive pads. Abrasive pads are mineral impregnated and are used by rubbing the part. Typical pads are approximately 6 inches by 9 inches and readily form to various contours.

After the oxide film is removed, the parts should be washed with alcohol or acetone to remove any residue remaining on the material.

The oxide film will start to reform immediately after the cleaning operation ends. Therefore, if the material is not to be used immediately, store in a clean bag or paper.

### Cleaning Stainless Steel

The austenitic stainless steels are usually bright and shiny. They need only a washing with solvent, alcohol or acetone to remove oils and grease. If present, heavy oxidation must be removed prior to welding by chemical etching, grit blasting, grinding, or sanding. Grit blasting must be followed by one of the other methods or wire brushed to remove imbedded grit from the surface.

Other types of stainless steels may have a heavy oxidation on the surface caused by air cooling after forming or rolling. These materials may be cleaned using the hot rolled steel cleaning procedure.

### Cleaning Thermally Cut Edges

Joint preparations made by oxyacetylene, plasma-arc, or carbon arc gouging must be cleaned to "bright metal" before welding. These joint edges have an oxide film on the surface usually heavier than hot rolled material. Use the same procedures as the one outlined under hot rolled steel to remove this oxide. Any indentations into the side of the weld joint must be removed by grinding as they contain heavy oxides. Refer to Fig. 10-25.

Fig. 10-25. Thermal cutting gouges and oxide films must be removed to "bright metal."

### Cleaning Butted Edges

All butted edges of joints have oxides on the surface regardless of the preparation process. This oxide film must be removed to obtain high quality welds. (This is especially important on aluminum as the oxide film may prevent fusion in the root pass.) The surface can be easily cleaned by drawing a file across the edge. The operation only needs to remove a small amount of material.

## TORCH ANGLES AND BEAD PLACEMENT

The proper manipulation of the welding torch is very important in making a good weld. The torch is usually held like a pencil, as shown in Fig. 10-26. When welding in the flat position, the hand should be placed lightly on a surface, so that the hand can move across the joint evenly. Movement of the torch by the fingers alone usually results in incorrect torch angles and a poor weld.

When adding filler wire, the wire should be gripped in the fingers as shown in Fig. 10-27. The hand should

Fig. 10-26. Holding the torch in this manner allows the welder to move the torch easily to change torch angles.

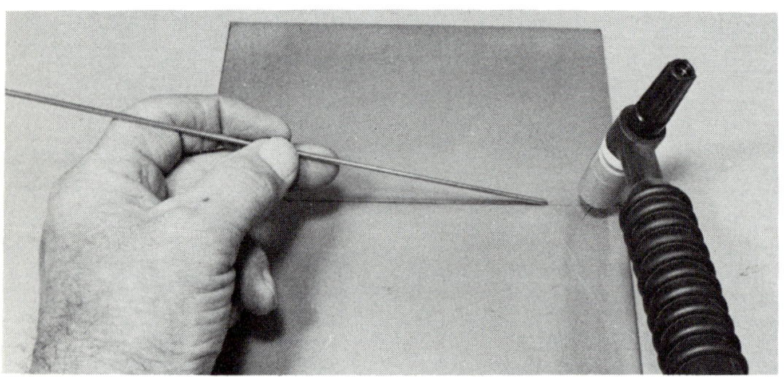

Fig. 10-27. Filler wire held in this manner can be added to the puddle as needed.

be as close as possible to the arc to hold the wire steady. The wire should move in conjunction with the torch movement. When additional wire is required, it can be moved forward through the fingers using a forward movement of the thumb. Too much extension of the filler wire from the fingers results in a wobbly wire end making addition to the puddle very uneven. Adding wire to the puddle requires steadiness and concentration to place the right amount of material at the right place, at exactly the right time. Always maintain a flat entrance angle of the wire into the puddle.

The proper position of the torch and filler wire for flat square groove butt welds is shown in Figs. 10-28 and 10-29. The proper position of the torch and filler wire for horizontal square groove butt weld is shown in Fig. 10-30. The proper position of the torch and filler wire for vertical square groove butt welds is illustrated in Fig. 10-31. The proper position of the torch and filler wire for flat V groove butt welds is demonstrated in Fig. 10-32. The proper position of the torch and filler wire for horizontal V groove butt welds is pictured in Fig. 10-33. The proper position of the torch and filler wire for vertical V groove butt welds is shown in Fig. 10-34. The proper position of the torch and filler wire for flat fillet welds is displayed in Fig. 10-35. The proper position of the torch and filler wire for horizontal fillet welds is illustrated in Fig. 10-36. The proper position of the torch and filler wire for vertical fillet welds is shown in Fig. 10-37.

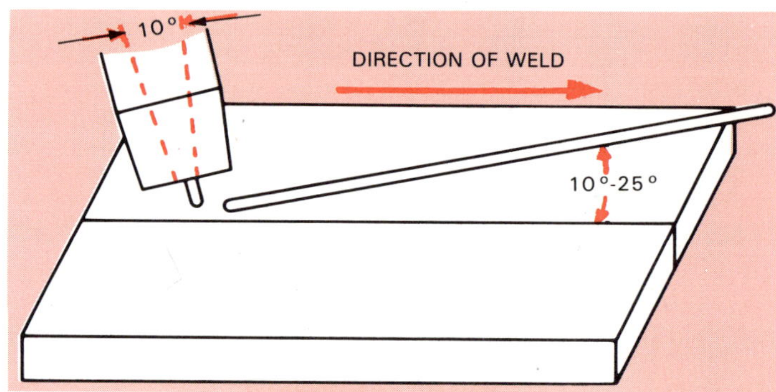


Fig. 10-28. Torch and wire positions for flat welding groove welds.

Fig. 10-29. Flat square groove butt weld joint.

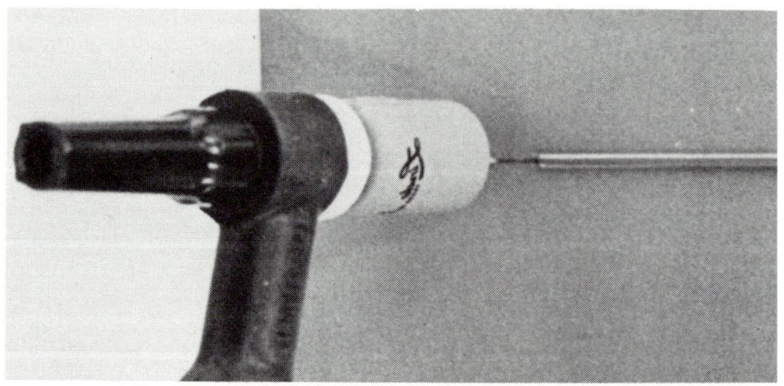

Fig. 10-30. Horizontal square groove butt weld joint.

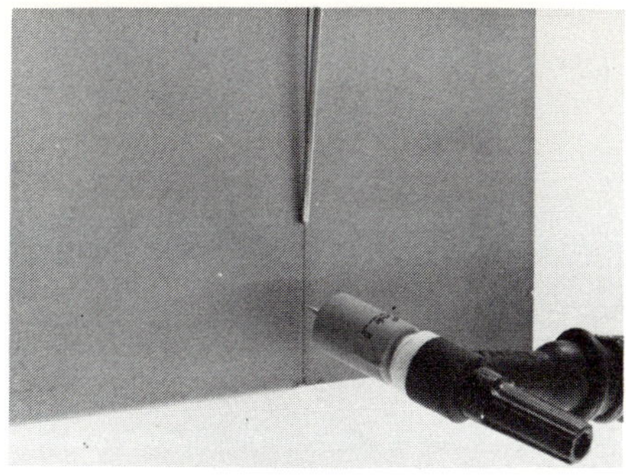

Fig. 10-31. Vertical square groove butt weld joint.

Fig. 10-35. Flat fillet weld joint.

Fig. 10-32. Flat V groove butt weld joint.

Fig. 10-36. Horizontal fillet weld joint.

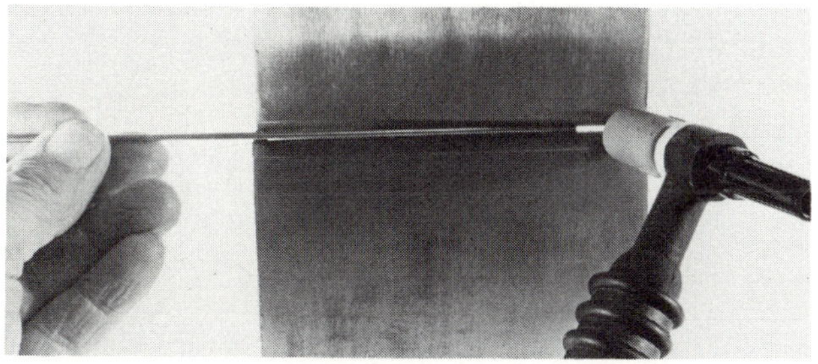

Fig. 10-33. Horizontal V groove butt weld joint.

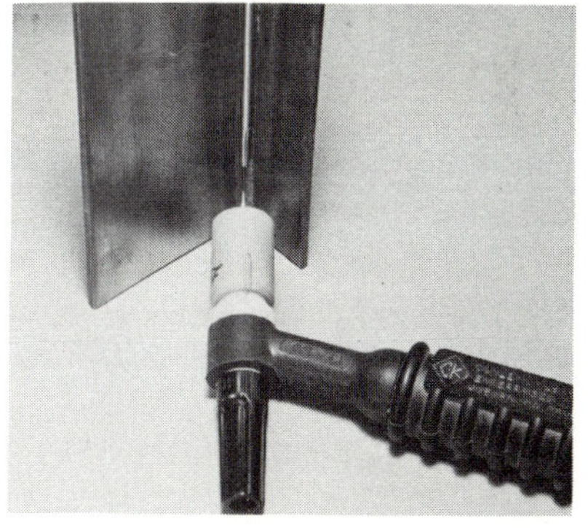

Fig. 10-37. Vertical fillet weld joint.

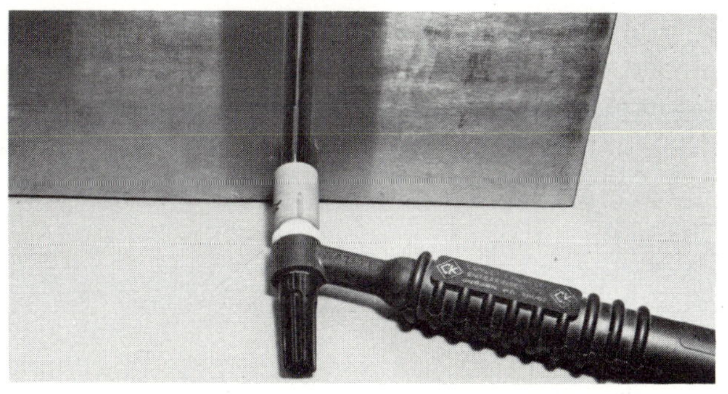

Fig. 10-34. Vertical V groove butt weld joint.

## TYPES OF BEADS/PASSES

Welds made without any side-to-side (oscillation) movement of the torch are called STRINGER BEADS or passes. Fig. 10-38 shows a stringer bead.

Welds made with side-to-side (oscillation) movement of the torch are called WASH BEADS or passes. Filler material added to the puddle should be made at the edges to prevent undercutting. Bottom part of Fig. 10-38 shows a wash bead.

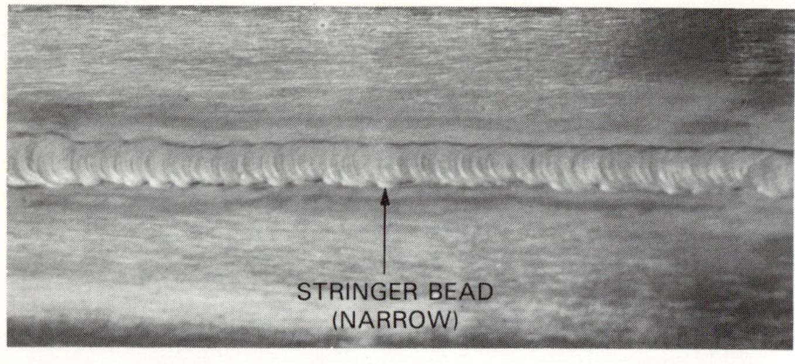


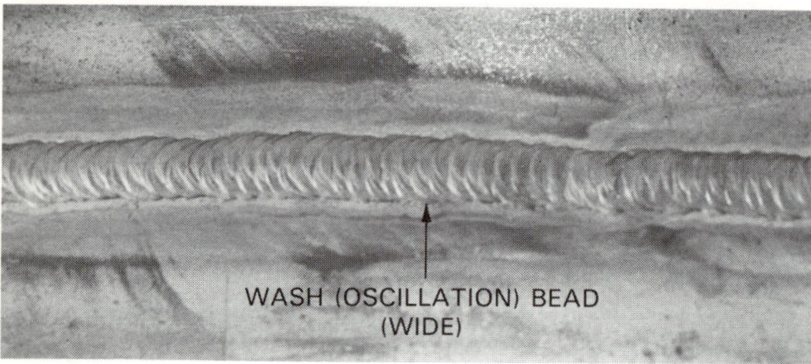


Fig. 10-38. Stringer and wash beads or passes.

## WELDING PROCEDURES

The eleven following listed procedures have been developed for gas tungsten arc welding practice and production. The material for practice can be any scrap material. The weld joint should fit properly for good results. All of the various materials should be cleaned as previously discussed before tackwelding and welding.

When welding on steel, a small white dot may form on the puddle surface. This material is silicon which separates from the base material and the filler wire. It will float on the surface as it is lighter than the metal.

Gas nozzle sizes are not specified because of the many variables involved. Always use the largest possible size. However, do not use a size that will obstruct vision of the puddle.

## WELD PROCEDURE NUMBER 10-1

WELD JOINT TYPE
Fusion Welding Sheet Metal (No Wire)

POSITION
Flat

| MATERIAL TYPE | SIZE |
|---|---|
| Cold Rolled Steel | 1/16" |

MACHINE SETUP
DCSP High Frequency Start

| SHIELDING GAS | CFH |
|---|---|
| Argon | 10-15 |

| TUNGSTEN TYPE | SIZE |
|---|---|
| 2% Thoriated | 1/16" (Tapered) |

PROCEDURE:
1. Clean material.
2. Raise part to be welded 1/8 inch above table with metal blocks.
3. Align torch to angle shown in Fig. 10-28 and lower torch to approximately 1/8 inch from top surface.
4. Start arc at low current, lower torch to approximately 1/16 inch from surface.
5. Increase amperage, and form a puddle approximately 1/8 inch to 3/16 inch diameter.
6. Move torch forward.
   A. Maintain puddle size.
   B. Maintain torch height.
   C. Maintain torch angle.
7. Stop weld at end of plate.
   A. Do not lift torch away from plate. Post flow gas will protect the hot metal during the cooling period.

Problem Areas and Corrections:
1. Uneven top weld width and depression.
   A. Variation in welding speed.
   B. As metal gets hotter increase speed or lower amperage.
2. Uneven contour.
   A. Torch angle incorrect, align torch vertical over weld.
3. Penetration uneven or lack of penetration.
   A. Insufficient amperage.
   B. Variable torch height, keep the torch the same height above the puddle.
   C. Variable travel speed, keep puddle same size.

## WELD PROCEDURE NUMBER 10-2

WELD JOINT TYPE
Fusion Welding Sheet Metal (With Wire)

POSITION
Flat

| MATERIAL TYPE | SIZE |
|---|---|
| Cold Rolled Steel | 1/16" |

| FILLER MATERIAL | SIZE |
|---|---|
| GTA Steel Welding Wire | .045"/.062" |

MACHINE SETUP
DCSP High Frequency Start

| SHIELDING GAS | CFH |
|---|---|
| Argon | 10-15 |

| TUNGSTEN TYPE | SIZE |
|---|---|
| 2% Thoriated | 1/16" (Tapered) |

PROCEDURE:
1. Clean material.
2. Tackweld plates together.
3. Raise part to be welded 1/8 inch above table with metal blocks.

4. Align torch to angle shown in Fig. 10-28 and lower to approximately 1/8 inch from top surface.
5. Hold filler wire as shown in Fig. 10-28 and move to approximately 1 inch from tungsten.
6. Start arc at low current, lower torch to approximately 1/16 inch from surface.
7. Increase amperage and form a puddle approximately 1/8 inch to 3/16 inch diameter.
8. Bring wire into the front edge of the puddle and melt enough wire to form a slight crown on the surface.
9. Draw the wire away from the puddle approximately 1/2 inch.
10. Move the torch forward approximately 1/16 inch and add wire again as in step 8.
11. Continue moving the arc across the joint, adding wire as before.
12. Stop weld at end of plate. Remember to hold torch at end until weld cools.

Problem Areas and Corrections:
1. Uneven top weld width and low crown.
   A. Variation in travel speed. Increase speed as the plates get hotter.
   B. Variation in adding wire. Add wire to form a crown.
2. Undercut.
   A. Torch not aligned vertically.
   B. Add wire to center of puddle.
3. Penetration uneven or lack of penetration.
   A. Travel speed inconsistent.
   B. Wire addition inconsistent.
   C. Insufficient amperage. Add a few amps for the added filler wire.
   D. Torch height inconsistent. Torch height variation causes changes in welding amperage.

**WELD PROCEDURE NUMBER 10-3**

WELD JOINT TYPE
Square Groove

POSITION
Flat

| MATERIAL TYPE | SIZE |
|---|---|
| Cold Rolled Steel | 1/16" |

| FILLER MATERIAL | SIZE |
|---|---|
| GTA Steel Welding Wire | .045"/.062" |

MACHINE SETUP
DCSP High Frequency Start

| SHIELDING GAS | CFH |
|---|---|
| Argon | 10-15 |

| TUNGSTEN TYPE | SIZE |
|---|---|
| 2% Thoriated | 1/16" (Tapered) |

PROCEDURE:
1. Clean material.
2. Tackweld plates together.
3. Raise part to be welded 1/8 inch above table with metal blocks.
4. Align torch to angle shown in Fig. 10-28 and lower to approximately 1/8 inch from surface.
5. Hold filler wire as shown in Fig. 10-28 and move to approximately 1 inch from tungsten.
6. Start arc at low current, lower torch to approximately 1/16 inch from surface.
7. Increase amperage and form a puddle approximately 1/8 inch to 3/16 inch diameter.
8. Bring wire into the front edge of the puddle and melt enough wire to form a slight crown on the surface.
9. Draw the wire away from the puddle approximately 1/2 inch.
10. Move the torch forward approximately 3/32 inch and add wire again as in step 8.
11. Continue moving the arc across the joint, adding wire as before.
    A. Maintain torch angle.
    B. Keep the tungsten on the joint center line.
12. Stop weld at end of plate. Remember to hold torch at end until weld cools.

Problem Areas and Correction:
1. Uneven top weld width and low crown.
   A. Variation in travel speed. Increase speed as plates get hotter.
   B. Variation in adding wire. Add wire to form a crown.
2. Undercut.
   A. Torch not aligned vertically.
   B. Add wire to center of puddle.
3. Penetration uneven or lack of penetration.
   A. Travel speed inconsistent.
   B. Wire addition inconsistent.
   C. Insufficient amperage. Add a few amps for the added filler wire.
   D. Torch height inconsistent. Torch height variation causes changes in welding amperage.
   E. Torch not maintained vertically on weld joint. Maintain torch angle on joint center line.

**WELD PROCEDURE NUMBER 10-4**

WELD JOINT TYPE
Square Groove

POSITION
Horizontal

| MATERIAL TYPE | SIZE |
|---|---|
| Cold Rolled Steel | 1/16" |

| FILLER MATERIAL | SIZE |
|---|---|
| GTA Steel Welding Wire | .045"/.062" |

MACHINE SETUP
DCSP

SHIELDING GAS: Argon — CFH: 10-15

TUNGSTEN TYPE: 2% Thoriated — SIZE: 1/16" (Tapered)

PROCEDURE:
1. Clean material.
2. Tackweld plates together.
3. Align parts to be welded with the joint in the horizontal position.
4. Align torch to angle shown in Fig. 10-30 and lower the torch to approximately 1/8 inch from surface.
5. Hold filler wire as shown in Fig. 10-30 and move to approximately 1 inch from tungsten.
6. Start arc at low current and lower torch to approximately 1/16 inch from surface.
7. Increase amperage to form a puddle approximately 1/8 inch to 3/16 inch diameter.
8. Bring wire into the upper part of the puddle and melt enough wire to form a slight crown.
9. Draw the wire away from the puddle approximately 1/2 inch.
10. Move the torch forward approximately 3/32 inch and add wire again as in step 8.
11. Continue moving the arc across the joint, adding wire as before.
    A. Maintain torch angle.
    B. Keep the tungsten at the proper angle.
12. Stop weld at end of plate. Remember to hold the torch at end until weld cools.

Problem Areas and Corrections:
1. Undercut at top of weld crown.
   A. Torch angle flat. Maintain torch angle.
   B. Add wire to top of puddle.
2. Crown sags.
   A. Add wire to top of puddle.
   B. Do not add wire in large amounts.
   C. Use a slight circular motion to hold puddle in position.
3. Penetration uneven on center line.
   A. Keep tungsten centered on joint.

**WELD PROCEDURE NUMBER 10-5**

WELD JOINT TYPE
Square Groove

POSITION
Vertical (Up-Hill)

MATERIAL TYPE: Cold Rolled Steel — SIZE: 1/16"

FILLER MATERIAL: GTA Steel Welding Wire — SIZE: .045"/.062"

MACHINE SETUP
DCSP High Frequency Start

SHIELDING GAS: Argon — CFH: 10-15

TUNGSTEN TYPE: 2% Thoriated — SIZE: 1/16" (Tapered)

PROCEDURE:
1. Clean material.
2. Tackweld plates together.
3. Align parts to be welded with the joint in the vertical position.
4. Align torch to angle shown in Fig. 10-31 and lower torch to approximately 1/8 inch from surface.
5. Hold filler wire as shown in Fig. 10-31 and move to approximately 1 inch from tungsten.
6. Start arc at low current, lower torch to approximately 1/16 inch from surface.
7. Increase amperage to form a puddle approximately 1/8 inch to 3/16 inch diameter.
8. Bring wire into the upper part of the puddle and melt enough wire to form a slight crown on the surface.
9. Draw the wire away from the puddle approximately 1/2 inch.
10. Move the torch upward approximately 3/32 inch and add wire again as in step 8.
11. Continue moving upward adding wire as before.
12. Stop weld at end of plate. Remember to hold torch at end until metal cools.

Problem Areas and Corrections:
1. Undercut.
   A. Add rod in the top and center of the puddle.
   B. Maintain torch angle.
2. High crown.
   A. Add wire in small amounts.
   B. Use smaller diameter wire.
3. Penetration uneven or lack of penetration.
   A. Travel speed inconsistent.
   B. Wire addition inconsistent.
   C. Incorrect torch angle. Maintain proper angle.
   D. Increase speed as plates will heat as the heat rises.

**WELD PROCEDURE NUMBER 10-6**

WELD JOINT TYPE
V Groove

POSITION
Flat

MATERIAL TYPE: Hot Rolled Steel — SIZE: 1/4"

FILLER MATERIAL: GTA Steel Welding Wire — SIZE: 1/16"/3/32"

MACHINE SETUP
DCSP High Frequency Start

| SHIELDING GAS | CFH |
|---|---|
| Argon | 10-15 |

| TUNGSTEN TYPE | SIZE |
|---|---|
| 2% Thoriated | 3/32" (Tapered) |

PROCEDURE:

1. Clean material and prepare as shown in Fig. 10-32.
2. Tackweld plates with 1/8 inch spacing. Mount joint approximately 1/8 inch above table with metal blocks.
3. Weld first pass with filler wire as shown in Fig. 10-32.
   A. Maintain torch on puddle with proper angle.
4. Wire brush weld to remove oxide film (all passes).
5. Realign torch and weld second pass as shown in Fig. 10-32.
   A. Add sufficient wire to have top of weld just below edge of bevel. Oscillate torch to move puddle to size desired.
6. Weld third pass with filler wire as shown in Fig. 10-32.
   A. Add sufficient wire to have approximately 1/16 inch crown.
   B. Oscillate torch to widen puddle just beyond bevel edge.

Problem Areas and Corrections:

1. Undercut.
   A. Add wire to edge of puddle on crown pass.
   B. Hold at edge when adding wire.
2. Uneven crown.
   A. Inconsistent travel speed.
   B. Inconsistent wire addition.
3. Penetration uneven.
   A. Inconsistent travel speed.
   B. Inconsistent wire addition.
   C. Tungsten not centered on puddle.

## WELD PROCEDURE NUMBER 10-7

WELD JOINT TYPE
V Groove

POSITION
Horizontal

| MATERIAL TYPE | SIZE |
|---|---|
| Hot Rolled Steel | 1/4" |

| FILLER MATERIAL | SIZE |
|---|---|
| GTAW Steel Wire | 1/16"/3/32" |

MACHINE SETUP
DCSP

| SHIELDING GAS | CFH |
|---|---|
| Argon | 10-15 |

| TUNGSTEN TYPE | SIZE |
|---|---|
| 2% Thoriated | 3/32" (Tapered) |

PROCEDURE:

1. Clean material and prepare as shown in Fig. 10-33.
2. Tackweld plates with 1/8 inch spacing. Mount joint horizontal.
3. Weld first pass with filler wire as shown in Fig. 10-33.
   A. Add filler wire to upper part of the puddle.
4. Wire brush weld to remove oxide film (all passes).
5. Realign torch and weld second pass as shown in Fig. 10-33.
   A. Do not oscillate torch.
6. Realign torch and weld third pass as shown in Fig. 10-33.
   A. Do not oscillate torch.
7. Realign torch and weld cover passes as shown in Fig. 10-33.
   A. Weld should extend over edge of bevel approximately 1/16 inch.
   B. Weld crown should not be over 1/16 inch high.

Problem Areas and Corrections:

1. Undercut.
   A. Add wire at top of puddle. Puddle will fall because of gravity.
   B. Hold torch and wire at proper angle.
   C. Do not oscillate torch. Keep puddle small.
   D. Use lower amperage range and add wire often.
   E. Allow plates to cool between passes.
2. Crown sags, uneven width, low or high crown.
   A. Determine size of each pass before starting.
   B. Determine location of each pass before starting.
   C. Maintain pass dimensions for full length of joint.
   D. Do not weld over edge of bevel until crown bead is made.
   E. Each weld on the crown should mate on the centerline of the previous weld.

## WELD PROCEDURE NUMBER 10-8

WELD JOINT TYPE
V Groove

POSITION
Vertical (Up-Hill)

| MATERIAL TYPE | SIZE |
|---|---|
| Hot Rolled Steel | 1/4" |

| FILLER MATERIAL | SIZE |
|---|---|
| GTAW Steel Wire | 1/16"/3/32" |

MACHINE SETUP
DCSP High Frequency Start

SHIELDING GAS: Argon — CFH: 10-15

TUNGSTEN TYPE: 2% Thoriated — SIZE: 3/32" (Tapered)

PROCEDURE:
1. Clean material and prepare as shown in Fig. 10-34.
2. Tackweld plates with 1/8 inch spacing. Mount joint vertical.
3. Weld first pass with filler wire as shown in Fig. 10-34.
   A. Add filler wire to the top of the puddle.
   B. Add filler wire directly on the puddle center line.
   C. Keep the tungsten on the molten pool.
4. Wire brush weld to remove oxide film (all passes).
5. Realign torch and weld second pass as shown in Fig. 10-34.
   A. Use slight oscillation for a wider bead.
   B. Add wire on the edge and always wait for a moment at this point.
6. Realign torch and weld cover pass or passes as shown in Fig. 10-34.
   A. Weld crown should extend over edge of bevel approximately 1/16 inch.
   B. Weld crown should not be over 1/16 inch high.

Problem Areas and Corrections:
1. Undercut.
   A. Add wire at top of puddle. Puddle will fall because of gravity.
   B. Hold torch and wire at proper angle.
   C. Use lower amperage and add wire often.
   D. Allow plates to cool between passes.
2. Crown uneven, crooked, uneven width, low or high crown.
   A. Determine size and position of each pass before starting.
   B. Maintain pass position for full length of joint.
   C. Do not weld over the edge of the bevel until crown bead is made. Use the edge of the bevel as a guide.
   D. Each weld on the crown should mate on the centerline of the previous weld.

## WELD PROCEDURE NUMBER 10-9

WELD JOINT TYPE: T

POSITION: Flat

MATERIAL TYPE: Cold Rolled Steel — SIZE: 1/16"/3/32"

FILLER MATERIAL: GTAW Steel Wire — SIZE: .045"/1/16"

MACHINE SETUP: DCSP High Frequency Start

SHIELDING GAS: Argon — CFH: 10-15

TUNGSTEN TYPE: 2% Thoriated — SIZE: 1/16" (Tapered)

PROCEDURE:
1. Clean materials.
2. Tackweld two plates at approximately right angles.
3. Align plates as shown in Fig. 10-35.
4. Weld using torch and wire angles as shown in Fig. 10-35.
   A. Add wire often in small amounts.
   B. Wire will flow evenly to both pieces and the crown should be flat to slightly convex.
   C. Feed the wire directly into the intersection of the joint.

Problem Areas and Corrections:
1. Concave weld crown.
   A. Insufficient wire. Add wire more often.
2. Irregular crown height or width.
   A. Improper feeding of wire. Too large a wire size will cause irregular heating of the weld puddle.
3. Suck-back or burn-through.
   A. Weld puddle is too hot. Use a smaller diameter of wire and add it often to control puddle temperature.
   B. Lower amperage.

## WELD PROCEDURE NUMBER 10-10

WELD JOINT TYPE: T

POSITION: Horizontal

MATERIAL TYPE: Cold Rolled Steel — SIZE: 1/4"

FILLER MATERIAL: GTAW Steel Wire — SIZE: 1/16"/3/32"

MACHINE SETUP: DCSP High Frequency Start

SHIELDING GAS: Argon — CFH: 10-15

TUNGSTEN TYPE: 2% Thoriated — SIZE: 3/32" (Tapered)

PROCEDURE:
1. Clean materials.
2. Tackweld two plates at approximately right angles.
3. Align plates as shown in Fig. 10-36.

4. Weld first pass using torch and wire angles as shown in Fig. 10-36.
   A. Add wire in small amounts.
   B. Feed the wire directly into the intersection of the joint for the first pass.
   C. Wire brush weld to remove oxide film (all passes).
   D. Weld second and third passes with torch and wire angles as shown in Fig. 10-36.

Problem Areas and Corrections:
1. Concave weld crown or improper contour.
   A. Insufficient wire. Add wire more often.
   B. Torch angle incorrect and welds improperly placed.
2. Leg sizes are not equal.
   A. Torch angle incorrect. Keep weld beads in proper position.
   B. Too much wire. Weld puddles too heavy.
3. Undercut on top weld.
   A. Puddle too hot. Add wire more often.
   B. Torch angle incorrect. Keep puddle even.

### WELD PROCEDURE NUMBER 10-11

WELD JOINT TYPE
T

POSITION
Vertical (Up-Hill)

| MATERIAL TYPE | SIZE |
|---|---|
| Cold Rolled Steel | 1/4" |

| FILLER MATERIAL | SIZE |
|---|---|
| GTAW Steel Wire | 1/16"/3/32" |

MACHINE SETUP
DCSP High Frequency Start

| SHIELDING GAS | CFH |
|---|---|
| Argon | 10-15 |

| TUNGSTEN TYPE | SIZE |
|---|---|
| 2% Thoriated | 3/32" (Tapered) |

PROCEDURE:
1. Clean materials.
2. Tackweld two plates at approximately right angles.
3. Align plates as shown in Fig. 10-37.
4. Weld first pass using torch and wire angles as shown in Fig. 10-37.
   A. Add wire in small amounts.
   B. Feed the wire directly into the intersection of the joint for the first pass.
   C. Wire brush to remove oxide film (all passes).
   D. Weld second and third passes with torch and wire angles as shown in Fig. 10-37.
   E. Increase speed or lower amperage as plate temperature increases.

Problem Areas and Corrections:
1. Convex weld crown.
   A. Add less wire.
   B. Increase travel speed.
2. Concave weld crown.
   A. Add more wire.
   B. Decrease travel speed.
3. Leg sizes are not equal.
   A. Torch angles incorrect. Keep welds in proper position.
   B. Keep puddle size consistent.
4. Undercut.
   A. Maintain proper torch angle.
   B. Add wire to center of puddle.

## REVIEW QUESTIONS

1. How do you develop welding skills?
2. What is a very important consideration that you must do when practicing your welding?
3. Why does welding have to be done on clean material?
4. What does the term "bright metal" define?
5. How does cold rolled steel differ from hot rolled material?
6. What is another name for rust?
7. Name two degreasers used to remove oil or grease from the material.
8. At what temperature does the oxide film on aluminum melt?
9. Name four methods which may be used to remove the oxide film from aluminum.
10. List three processes which may be used to thermally prepare weld joints for welding.
11. To prepare a groove weld joint, the abutting edge may be cleaned by drawing a __________ across the joint.
12. Why is the hand containing the filler material held close to the arc?
13. Is the wire to be added to the molten pool held at a flat or steep angle?
14. Welds made without any side-to-side movement of the torch are called __________ beads or passes.
15. Welds made with side-to-side movement of the torch are called __________ beads or passes.
16. When welding on steel a small white dot may form on the top of the molten metal. What is this material?
17. How are gas nozzles selected for use?
18. What do the letters DCEN and DCSP define?

# Chapter 11

# WELDING PROCEDURE FOR MANUAL WELDING ALUMINUM

## BASE MATERIALS

Aluminum is a non-ferrous (no-iron) material readily welded with the GTAW process. Some of the major characteristics of the material aluminum include:

1. Good thermal conductivity.
2. Good electrical conductivity.
3. Good ductility at subzero temperatures.
4. Light weight.
5. High resistance to corrosion.
6. Non-sparking.
7. Non-toxic.
8. Material does not change color when heated.

Pure aluminum melts at approximately 1200°F. The alloy materials melting range is approximately 900° to 1200°F.

The oxide film of the surface has a melting point of about 3500°F. It is this oxide film which must be removed prior to welding, or during the welding operation, if satisfactory welds are to be made. This is due to the fact that it will not melt before the base aluminum and will contaminate the weld.

### Wrought Material Identification

Aluminum and aluminum alloys are identified by the composition, thermal treatment, and work hardening characteristics. The word "wrought" identifies material made by processes other than casting and is often used in specifications and codes. The processes used in manufacture include: rolling plates and sheets, extruded material and pipe. The numbering system established by the Aluminum Association used to identify the various groups is shown in Fig. 11-1. The four digit number indicates:

1. 1st number — alloy groups.
2. 2nd number — modifications or impurity level.
3. 3rd number — aluminum alloy or aluminum purity.
4. 4th number — as above.

| Metal | Aluminum Association Number (AA No.) |
|---|---|
| Aluminum – 99.00% minimum and greater | 1xxx |
| Aluminum Alloys Grouped by Major Alloying Elements[2] | |
| Copper | 2xxx |
| Manganese | 3xxx |
| Silicon | 4xxx |
| Magnesium | 5xxx |
| Magnesium and Silicon | 6xxx |
| Zinc | 7xxx |
| Other Element | 8xxx |
| Unused Series | 9xxx |

Fig. 11-1. Wrought material identification numbers.

### Aluminum Castings

Aluminum castings are grouped by major alloying elements. Most producers of primary casting alloys use the number system shown in Fig. 11-2 to designate the alloy ingot.

### Tempers

Further identification of the material is done by identifying the condition in which the material is supplied. The Aluminum Association Temper Designation System is used for all forms of wrought and cast aluminum and aluminum alloys except ingot. The word TEMPERS is used to define materials with various mechanical properties that have been made by a series of sequences of basic treatments. The basic temper designations are as follows:

—F. As fabricated.

—O. Annealed, recrystallized (softest temper). "ANNEALED" is defined as a series of heating cycles to allow the grains of the material to reform (recrystallize) and cooling cycles to produce a material with varying degrees of

| Major Alloying Element | Number Range Assigned to Group |
|---|---|
| Silicon | 0- 99 |
| Copper | 100-199 |
| Magnesium | 200-299 |
| Silicon-Copper<br>Silicon-Magnesium<br>Silicon-Copper-Magnesium | 300-399 |
| Manganese | 400-499 |
| Nickel | 500-599 |
| Zinc | 600-699 |
| Tin | 700-799 |

Fig. 11-2. Casting material identification numbers.

softness.

–H. Strain-hardened (wrought products only). STRAIN-HARDENED identifies metal which has been strained by stretching, pulling, or forming to produce a grain structure with higher mechanical properties.

–W. Solution heat-treated. SOLUTION HEAT-TREATED identifies material which has been heated to a predetermined temperature for a suitable length of time to allow a certain element in the material to enter into a "solid solution." The alloy is then quickly cooled to hold the element in this solution.

–T. Thermally treated to produce stable tempers other than –F, –O, –H.

The –T is always followed by one or more digits. Specific sequences of basic treatments are as follows:

–T1. Naturally aged to a substantially stable condition from the as-cast condition.

–T2. Annealed (cast products only).

–T3. Solution heat-treated and then cold worked.

–T4. Solution heat-treated and naturally aged to a substantially stable condition. AGED identifies material that has been held at room temperature or a predetermined temperature for a period of time for the purpose of increasing the hardness and strength of the material.

–T5. Artificially aged from the as-cast condition.

–T6. Solution heat treated and then artificially aged.

–T7. Solution heat treated and then stabilized. STABILIZED identifies material which has been "strain-hardened," and then heated to a predetermined low temperature to slightly lower the strength and to increase the ductility.

–T8. Solution heat treated, cold worked, then artificially aged. COLD WORK identifies an operation of mechanically working material without heat.

–T9. Solution heat treated, artificially aged, then cold worked.

–T10. Artificially aged and then cold worked.

### Classification

Aluminum is classified into two categories:

1. Non-heat-treatable.
   These materials attain strength levels by addition of alloying elements such as manganese, silicon, iron and magnesium. They may also be cold worked (strain-hardened) by stretching, drawing, or swaging to increase strength levels. SWAGING identifies a process of changing the shape of the material with mechanical tools such as hammers and dies.
2. Heat-treatable.
   Initial strengths of heat-treatable alloys are produced by the addition of copper, magnesium, zinc, and silicon. Additional strengthening can be done by subjecting the material to various degrees of thermal treatment, quenching and aging. QUENCHING is the process of rapid cooling of metal from a higher temperature for the purpose of hardening.

The grouping of the non-heat treatable and heat-treatable wrought alloys is shown in Fig. 11-3.

| Non-Heat-Treatable (alloys normally cold-worked) | Heat-Treatable (alloys normally heat-treated) |
|---|---|
| 1060 | 2011 |
| 1100 | 2014 |
| 3003 | 2017 |
| 3004 | 2018 |
| 4043 | 2024 |
| 5005 | 2025 |
| 5050 | 2117 |
| 5052 | 2218 |
| 5056 | 2618 |
| 5083 | 4032 |
| 5086 | 6053 |
| 5184 | 6061 |
| 5252 | 6063 |
| 5257 | 6066 |
| 5357 | 6101 |
| 5454 | 6151 |
| 5456 | 7039 |
| 5557 | 7075 |
| 5657 | 7079 |
|  | 7178 |

Fig. 11-3. Non-heat-treatable and heat-treatable aluminum materials.

## FILLER MATERIALS FOR WELDING ALUMINUM

The filler material choice for fusion welding various aluminum alloys is important. The metal produced in the weld puddle is a combination of the parent metal and the filler material. Weld metal must have the strength, ductility, freedom from cracking, and the ability to resist corrosion required by the application.

| Base Metal | 6070 | 6061, 6063 6101, 6151 6201, 6951 | 5456 | 5454 | 5154 5254[a] | 5086 | 5083 | 5052 5652[a] | 5005 5050 | 3004 Alc. 3004 | 2219 | 2014 2024 | 1100 3003 Alc. 3003 | 1060 EC |
|---|---|---|---|---|---|---|---|---|---|---|---|---|---|---|
| 1060, EC | ER4043[h] | ER4043[h] | ER5356[c] | ER4043[e,h] | ER4043[e,h] | ER5356[c] | ER5356[c] | ER4043[i] | ER1100[c] | ER4043 | ER4145 | ER4145 | ER1100[c] | ER1100[c,d] |
| 1100, 3003 Alclad 3003 | ER4043[h] | ER4043[h] | ER5356[c] | ER4043[e,h] | ER4043[e,h] | ER5356[c] | ER5356[c] | ER4043[e,h] | ER4043[e] | ER4043[e] | ER4145 | ER4145 | ER1100[c] | |
| 2014, 2024 | ER4145 | ER4145 | | | | | | | | | ER4145[a] | ER4145[a] | | |
| 2219 | ER4043[f,h] | ER4043[f,h] | ER4043 | ER4043[h] | ER4043[h] | ER4043[i] | ER4043 | ER4043[i] | ER4043 | ER4043 | ER2319[c,f,h] | | | |
| 3004 Alclad 3004 | ER4043[e] | ER4043[b] | ER5356[e] | ER5654[b] | ER5654[b] | ER5356[e] | ER5356[e] | ER4043[e,h] | ER4043[e] | ER4043[e] | | | | |
| 5005, 5050 | ER4043[e] | ER4043[b] | ER5356[e] | ER5654[b] | ER5654[b] | ER5356[e] | ER5356[e] | ER4043[e,h] | ER4043[d] | | | | | |
| 5062, 5652[a] | ER5356[b,c] | ER5356[b,c] | ER5356[b] | ER5654[b] | ER5654[b] | ER5356[e] | ER5356[e] | ER5654[a,b,c] | | | | | | |
| 5083 | ER5356[e] | ER5356[e] | ER5183[e] | ER5356[e] | ER5356[e] | ER5356[e] | ER5183[e] | | | | | | | |
| 5086 | ER5386[e] | ER5356[e] | ER5356[e] | ER5356[b] | ER5356[b] | ER5356[e] | | | | | | | | |
| 5154, 5254[a] | ER5356[b,c] | ER5356[b,c] | ER5356[b] | ER5654[b] | ER5654[a,b] | | | | | | | | | |
| 5454 | ER5356[b,c] | ER5356[b,c] | ER5356[b] | ER5554[a,b] | | | | | | | | | | |
| 5456 | ER5356[e] | ER5356[e] | ER5556[e] | | | | | | | | | | | |
| 6061, 6063, 6101 6201, 6151, 6951 | ER4043[b,h] | ER4043[b,h] | | | | | | | | | | | | |
| 6070 | ER4043[e,h] | | | | | | | | | | | | | |

Where no filler metal is listed, base metal combination is not recommended for welding.

a. Base metal alloys 5652 and 5254 are used for hydrogen peroxide service. ER5654 filler metal is used for welding both alloys for low-temperature service (150 F and below).
b. ER5183, ER5356, ER5554, ER5556, and ER5654 may be used.
c. ER4043 may be used.
d. Filler metal with same analysis as base metal is sometimes used.
e. ER5183, ER5356, or ER5556 may be used.
f. ER4145 may be used.
g. ER2319 may be used.
h. ER4047 may be used.
i. ER1100 may be used.

Fig. 11-4. Wrought aluminum alloys—filler material selection.

The correct choice of a filler wire or alloy will eliminate or significantly reduce relatively low ductility in aluminum welds. DUCTILITY is a property of a material to deform permanently, or to exhibit plasticity without breaking while under tension (strain).

The table in Fig. 11-4 shows the various recommendations for general purpose welding. In cases where maximum strength or maximum elongation is preferred, the filler materials listed in Fig. 11-5 may be used.

Maximum weld quality can only be obtained if the filler material is clean and of high quality. Contaminates which cause weld porosity on the filler material are most often an oil or a hydrated oxide. The heat of welding releases the hydrogen from these sources which cause porosity in the weld. Aluminum filler materials are manufactured under vigorous specifications and packaged to retain cleanliness.

The filler material should be handled with the utmost care to retain it's cleanliness during storage and use in the fabrication cycle.

### Joint Preparation for Welding Aluminum

Joint edges that are prepared by the plasma arc cutting process or the carbon arc gouging process, have a heavy oxide film on the surface. This surface should be thoroughly cleaned prior to welding to prevent dross and porosity in the final weld. DROSS is oxidized metal or impurities within the weld or parent metal.

Joint edges prepared by the shearing process should be sharp without tearing or ridges. Where these conditions exist, dirt and oil may become entrapped and result in faulty welds.

### Pre-Weld Cleaning

The weld joint, the immediate weld area, and the filler material must be clean if the weld is to have satisfactory properties. The cleaning operation should be completed just before the actual welding time. An oxide film can prevent fusion between the filler metal and the base plate. An oxide film can also cause the formation of flakes of oxide. Incomplete cleaning may result in dross becoming entrapped within the weld metal. The cleaning operation may be done by:

1. Chemical cleaning.
   A. Commercial degreasers.
   B. Commercial compounds, as shown in Fig. 11-6. Surface oxide removal by a commercial cleaner is shown in Fig. 11-7.
2. Mechanical cleaning.

| Base Metal | Filler Alloys[1] | |
|---|---|---|
| | Preferred for Maximum As-welded Tensile Strength | Alternate Filler Alloys For Maximum Elongation |
| EC | 1100 | EC/1260 |
| 1100 | 1100/4043 | 1100/4043 |
| 2014 | 4145 | 4043/2319[3] |
| 2024 | 4145 | 4043/2319[3] |
| 2219 | 2319 | (4) |
| 3003 | 5183 | 1100/4043 |
| 3004 | 5554 | 5183/4043 |
| 5005 | 5183/4043 | 5183/4043 |
| 5050 | 5356 | 5183/4043 |
| 5052 | 5356/5183 | 5183/4043 |
| 5083 | 5183 | 5183 |
| 5086 | 5183 | 5183 |
| 5154 | 5356 | 5183/5356 |
| 5357 | 5554 | 5356 |
| 5454 | 5554 | 5356 |
| 5456 | 5556 | 5183 |
| 6061 | 4043/5183 | 5356[2] |
| 6063 | 4043/5183 | 5183[2] |
| 7039 | 5039 | 5183 |
| 7075 | 5183 | – |
| 7079 | 5183 | – |
| 7178 | 5183 | (4) |

The above table shows recommended choices of filler alloys for welds requiring maximum mechanical properties. For all special services of welded aluminum, inquiry should be made of your supplier.

1. Data shown are for "O" temper.
2. When making welded joints in 6061 or 6063 electrical conductor in which maximum conductivity is desired, use 4043 filler metal. However, if strength and conductivity both are required, 5356 filler may be used and the weld reinforcement increased in size to compensate for the lower conductivity of the 5356 filler metal.
3. Low ductility of weldment is not appreciably affected by filler used. Plate weldments in these base metal alloys generally have lower elongations than those of other alloys listed in this table.

Fig. 11-5. Filler materials used to obtain specific properties of completed welds.

Fig. 11-6. Commercial aluminum cleaner.

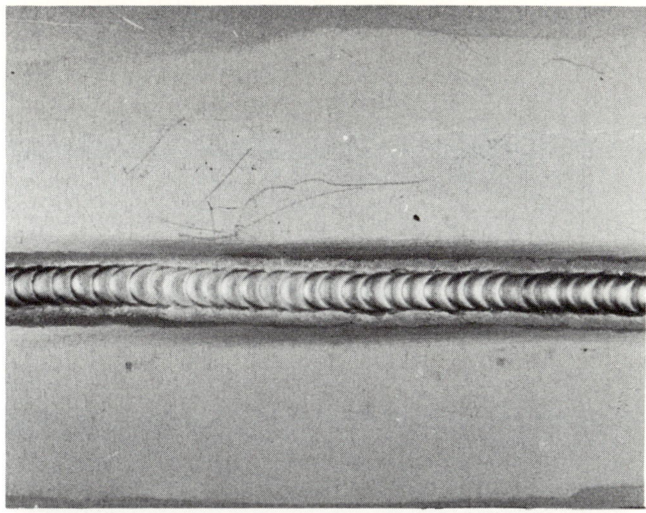

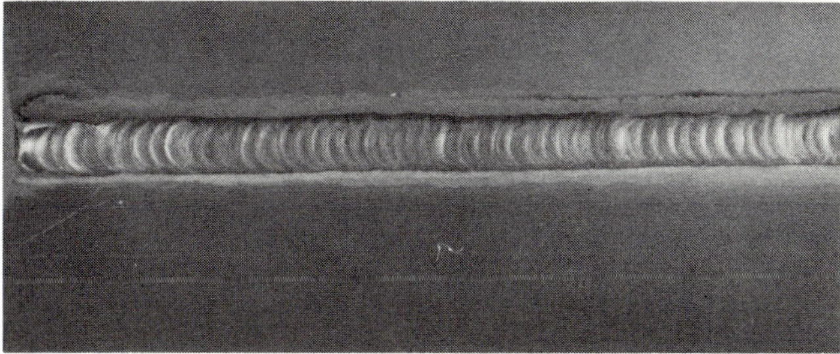

Fig. 11-7. Surface oxide removal by a commercial cleaner.

A. Filing or scraping.
B. Stainless steel wire brushes.

Immediately prior to starting to weld, the joint should be wiped with alcohol or acetone. This operation should be followed by a short drying period. A clean air blast may be used to evaporate any residual liquid. Do not attempt to weld on a wet surface.

All tooling in the weld area must also be cleaned prior to welding. Back-up bars, chill bars, clamps, etc., retain moisture, and when heated by the welding operation, release hydrogen. This part of the operation is just as important as the cleaning of the material and filler wire.

### Weld Back-Up

Two types of weld back-up are used when welding aluminum groove welds. They include:

1. INTEGRAL BACK-UP. An integral back-up is used where the joint can be designed in the manner demonstrated in Fig. 11-8. Another method is to place a back-up strip on the back side of the weld as shown in Fig. 11-9. The main problem with this design is the difficulty in achieving penetration at the root. This difficulty may be overcome if the nose (land) is not too large or a gap exists between the parts as shown in Fig. 11-10. If preheat

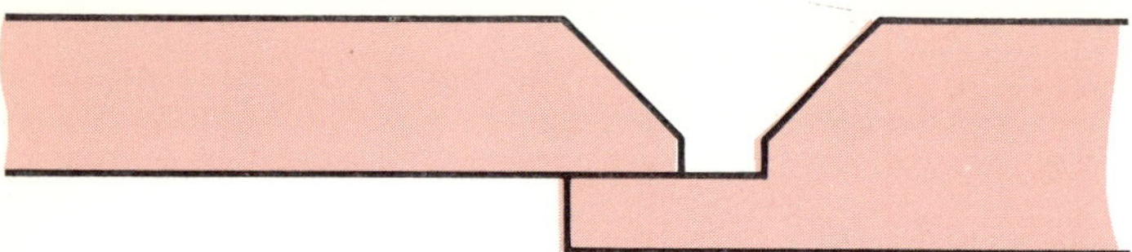

Fig. 11-8. Integral back-up design joints reduce tooling requirements.

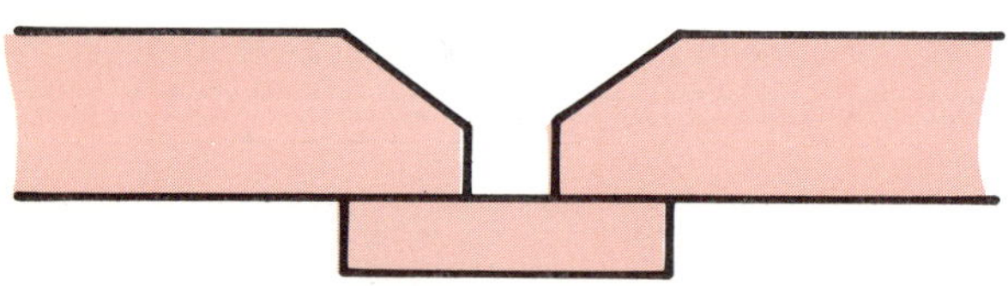

Fig. 11-9. Back-up bars are often used where tooling is difficult to install. The bar must fit tightly to the root side of the weld if a good weld is to be obtained.

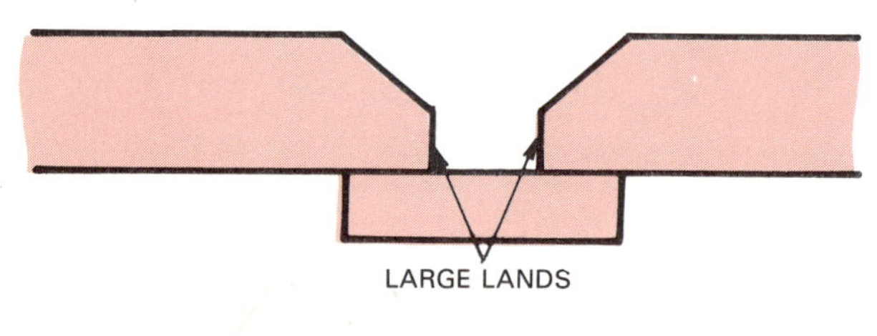


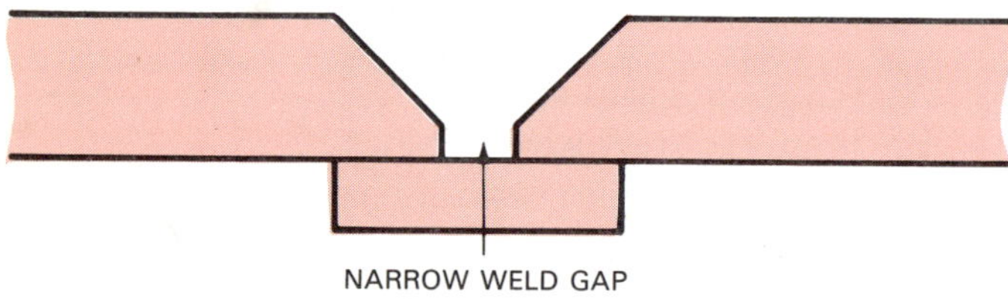


Fig. 11-10. Large lands or narrow gaps at the root of the weld obstruct penetration at the weld root. Wherever possible, use small lands or wider root spacing to assist in obtaining penetration.

is used, the welder may achieve the proper penetration.

2. REMOVABLE BACK-UP BAR. These types of bars are found in stakes and seamers, or they may be a part of the assembly tooling. The groove in the bar may be designed for forming or molding the penetration as illustrated in Fig. 11-11. Another design, as shown in Fig. 11-12, is used where argon gas is admitted through the back-up bar to prevent drop-through oxidation. (Narrow grooves cool the metal quickly which results in trapping hydrogen gas. Preheating the back-up bar allows the weld to solidify slowly. This allows the gas to rise through the melt which helps to eliminate porosity.) The argon gas type removable back-up bar is used where very high quality is required.

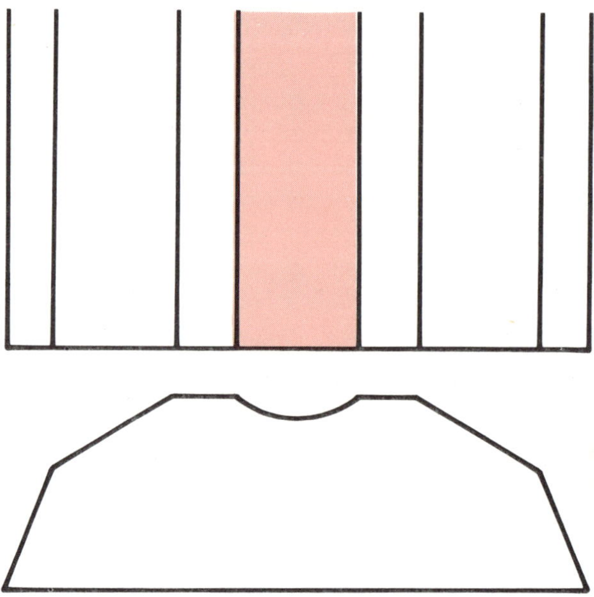

Fig. 11-11. Always use generous grooves when casting the penetration of aluminum groove welds.

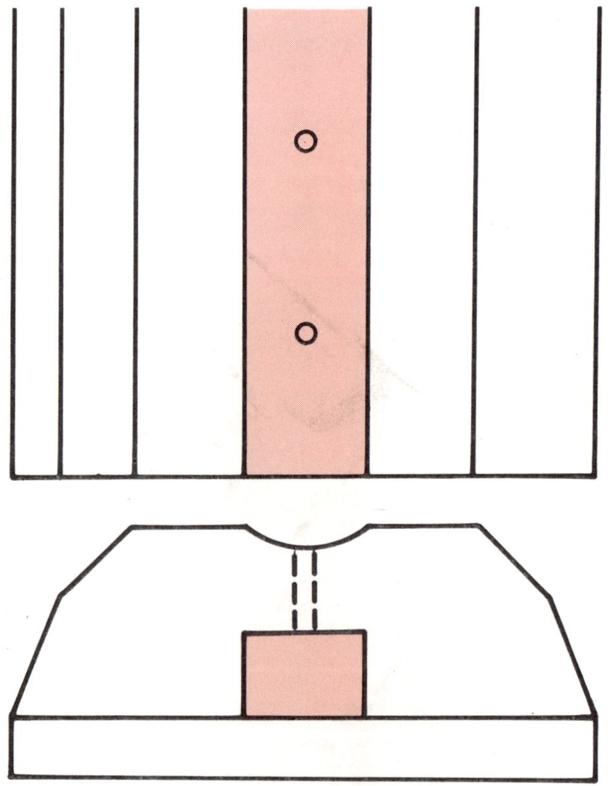

Fig. 11-12. Argon gas shields the penetration during welding and prevents the formation of oxides. This also assists in forming of the root bead and prevents oxide folds.

### Preheating Aluminum

Since aluminum spreads the heat so rapidly, it is often necessary to apply preheat to assist in obtaining penetration and proper bead contour. Preheating also allows greater travel speeds, regardless of the type of joint. If preheating is used, the entire joint area and tooling in the immediate area should be heated to the same temperature.

Temperatures of 350-400°F are normally used. Temperature indicating paints, crayons, and pellets can be used to verify the actual temperature. Refer to Fig. 11-13. Be careful that the temperature indicating material does not get into the area where the molten metal will come into contact.

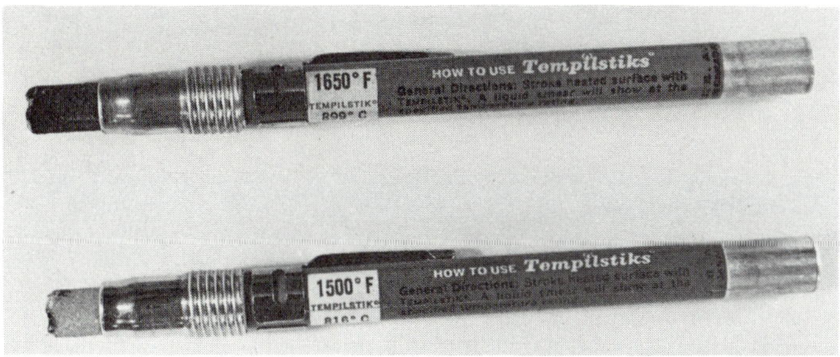


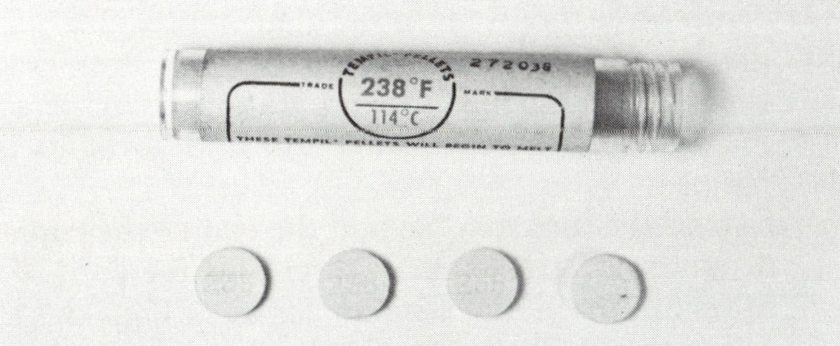


Fig. 11-13. Temperature crayons and pellets may be used to determine pre-heating temperatures prior to welding.

### Tackwelds for Aluminum

When tackwelds must be made to maintain joint alignment for welding aluminum the following considerations apply:

1. If the joint welding procedure requires preheating, the joint requires preheating before tackwelding.
2. Use the same filler material for tackwelding as will be used for the main welding procedure.
3. If the joint requires full penetration, the tackweld requires penetration.
4. Keep the tackweld as small as possible. If possible, make the tackweld smaller than the main weld.
5. If the tackweld cracks, leave it and make another close to the broken one.
6. Do not grind out tackwelds. Grinding grit in a joint makes good quality almost impossible.
7. The crater of a tackweld is the weakest point in the tackweld. Do not leave a crater in the tackweld.

### Power Supplies for Welding ACHF

1. When using a transformer or a combination AC/DC power supply not designed for GTAW, do not exceed the derated capacity of the unit. (See Chapter 3 on GTAW Equipment.)
2. If the power supply has a wave balancer control, set it on normal to start welding. Then adjust to the desired penetration or cleaning action from that point.
3. Be sure the machine is grounded properly to prevent high frequency radiation.
4. If the oxide cleaning action is lost or insufficient during welding, check the high frequency point gap. The high frequency spark is needed to maintain the start of each positive and negative cycle (straight and reverse polarity) part of the cycle. Without the reverse polarity, the cleaning action is lost. To determine if the cleaning action is adequate, make a pass with the machine operating across a test plate. Use low amperage and a high arc voltage. The cleaning action should compare favorably with Fig. 11-14.

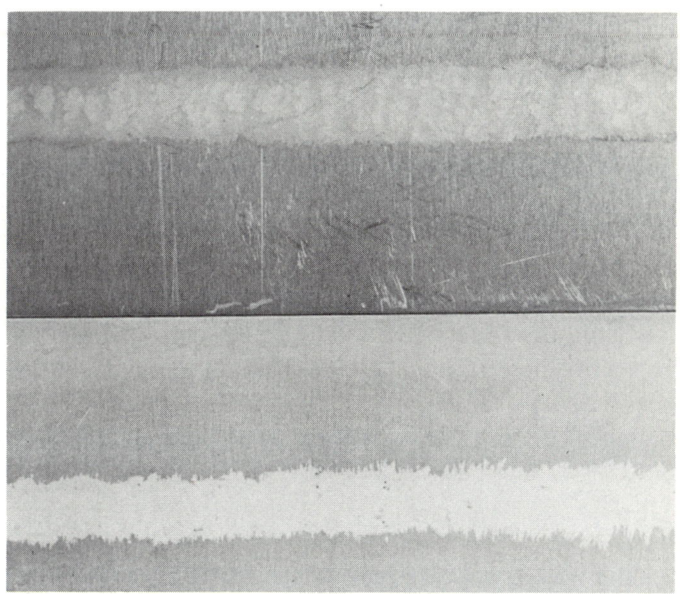

Fig. 11-14. Reverse polarity cleaning actions on "as-received" material and chemically etched material.

### Gases for Welding ACHF

Argon gas is used for thickness up to 1/8 inch thickness, a combination of 75Ar-25He will assist in

| Aluminum Thickness (inch) | Welding Position | Joint Type | Alternating Current (amperes) | Dia. of Tungsten Electrode (inch) | Argon Gas Flow | Filler Rod Diameter (inch) | Number of Passes |
|---|---|---|---|---|---|---|---|
| 1/16 | Flat<br>Horiz. & Vert.<br>Overhead | Square Butt<br>Square Butt<br>Square Butt | 70–100<br>70–100<br>60–90 | 1/16<br>1/16<br>1/16 | 20<br>20<br>25 | 3/32<br>3/32<br>3/32 | 1<br>1<br>1 |
| 1/8 | Flat<br>Horiz. & Vert.<br>Overhead | Square Butt<br>Square Butt<br>Square Butt | 125–160<br>115–150<br>115–150 | 3/32<br>3/32<br>3/32 | 20<br>20<br>25 | 1/8<br>1/8<br>1/8 | 1<br>1<br>1 |
| 1/4 | Flat<br>Horiz. & Vert.<br>Overhead | 60° Single Bevel<br>60° Single Bevel<br>100° Single Bevel | 225–275<br>200–240<br>210–260 | 5/32<br>5/32<br>5/32 | 30<br>30<br>35 | 3/16<br>3/16<br>3/16 | 2<br>2<br>2 |
| 3/8 | Flat<br>Horiz. & Vert.<br>Overhead | 60° Single Bevel<br>60° Single Bevel<br>100° Single Bevel | 325–400<br>250–320<br>275–350 | 1/4<br>3/16<br>3/16 | 35<br>35<br>40 | 1/4<br>1/4<br>1/4 | 2<br>3<br>3 |
| 1/2 | Flat<br>Horiz. & Vert.<br>Overhead | 60° Single Bevel<br>60° Single Bevel<br>100° Single Bevel | 375–450<br>250–320<br>275–340 | 1/4<br>3/16<br>3/16 | 35<br>35<br>40 | 1/4<br>1/4<br>1/4 | 3<br>3<br>4 |
| 1 | Flat | 60° Single Bevel | 500–600 | 5/16–3/8 | 35–45 | 1/4–3/8 | |

Fig. 11-15. Recommended procedures for manual alternating current welding of aluminum.

obtaining penetration on all types of aluminum joints. Using a gas mixture with larger amounts of helium may create problems with arc starting. The gas should be of high quality, with no leaks in the supply system.

### Electrodes for ACHF

Pure and zirconiated tungstens are recommended for welding with alternating current. Both have good "balling" characteristics which are required for high quality aluminum welds. Thoriated tungstens do not have this capability, and as a result will "spit" after slight usage. (Spitting is the breakdown of the tungsten. Small particles of tungsten are then separated and transferred by the arc stream into the weld puddle.)

### Techniques for ACHF Welding on Aluminum

To produce satisfactory welds on aluminum, all of the elements of the procedure must be maintained. These include:

1. Select the proper procedure from those shown in Fig. 11-15.
2. Taper large tungstens to the diameter required, as diagrammed in Fig. 11-16.
3. Blunt the tungsten on the end, start an arc on scrap material and increase amperage until a radius is formed.
4. Extend the tungsten beyond the end of the cup a distance not greater than the ID of the cup.
5. Maintain torch and wire angles as in Fig. 11-17.
6. Move the torch across the joint using the step technique, shown in Fig. 11-18. The crown of the completed weld should be similar to the weld shown in Fig. 11-19. The penetration side of the weld is shown in Fig. 11-20.
7. If the tungsten contacts the filler material in the

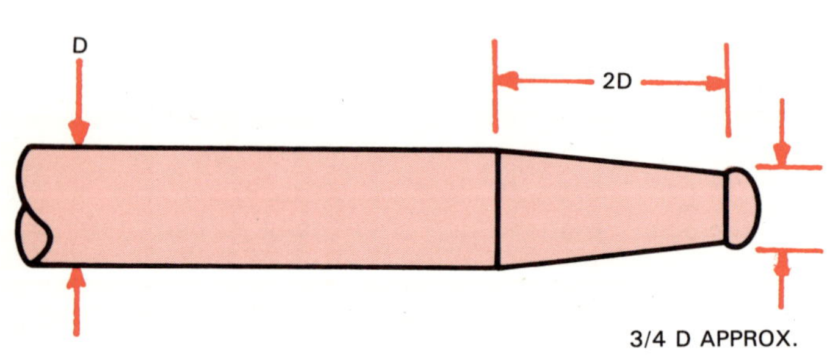


Fig. 11-16. Electrodes used for alternating current should be ground as shown. The ball or rounded tip will form as the welding current is increased.

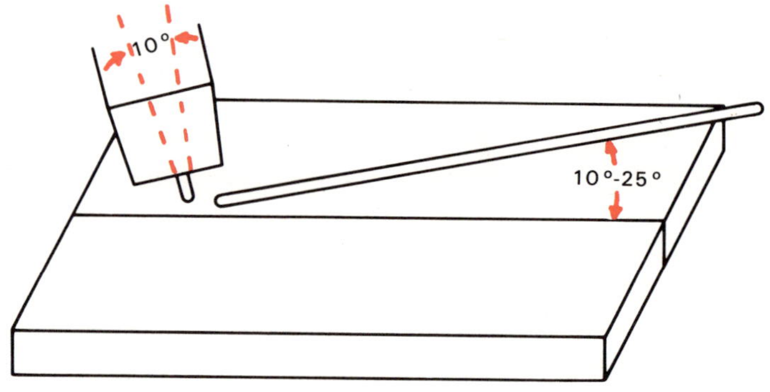


Fig. 11-17. Torch and wire positions for flat welding groove welds.

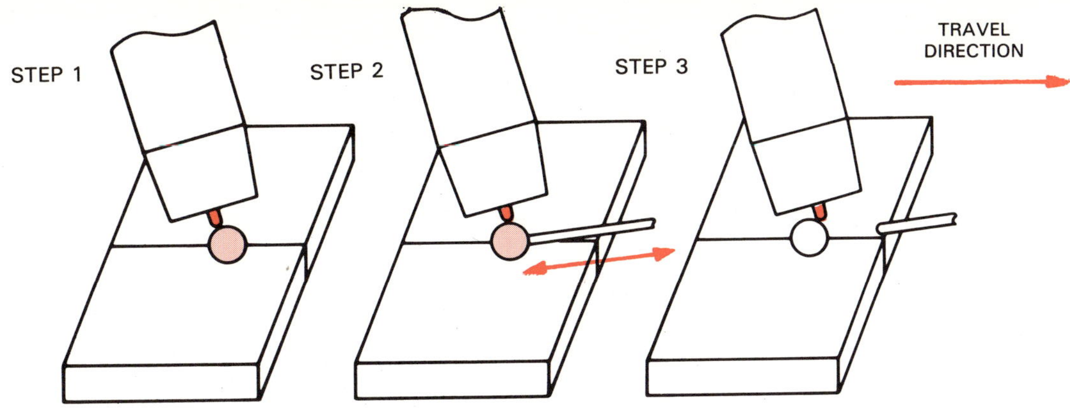


Fig. 11-18. Step welding technique for welding aluminum groove welds.

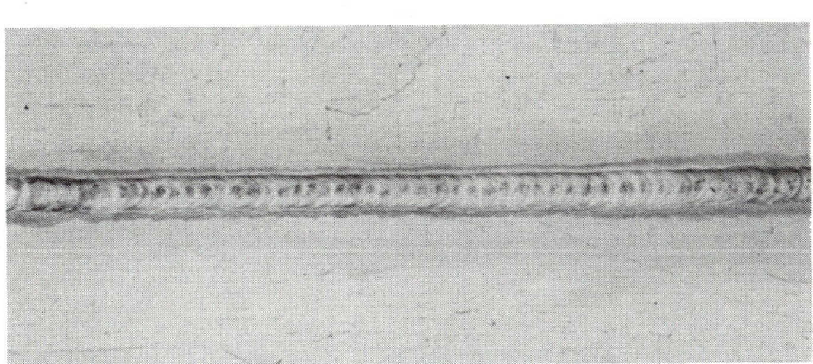

Fig. 11-19. Step welding techniques produce a definite rippled weld.

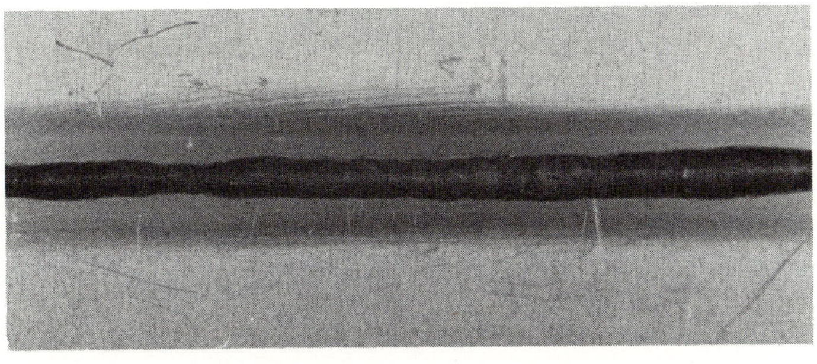

Fig. 11-20. Properly made welds have good even penetration as seen from below.

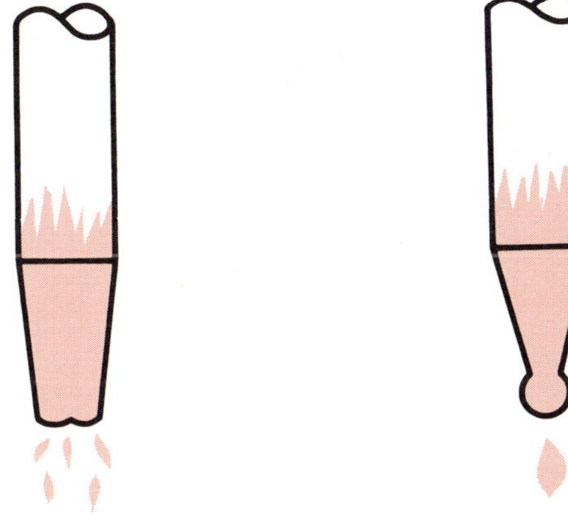

Fig. 11-21. Too large an electrode can cause "spitting" of particles across the arc. Excessive current for electrode size may cause the tip to overheat and drop off.

molten pool, stop and replace or regrind the tungsten. (If this is not done, the aluminum on the electrode will burn away and contaminate the weld.)

8. Fig. 11-21 shows the problems encountered when using too large an electrode or excessive currents.
9. Always use stainless steel brushes for removing oxides from the surface of the material or the weld.
10. Remember to fill the crater at the end of the weld.
11. Remember that aluminum is HOT SHORT. That is, aluminum has low strength at high temperatures. Therefore, it must have support near the weld area or the metal will collapse.
12. The molten pool does not change color as it changes temperature.

Fig. 11-22 explains some problem areas and correction measures which may be used when GTA welding aluminum.

## WELDING ALUMINUM WITH DCEN (DCSP)

DCEN welding is used to produce deeply penetrating welds on aluminum. The joints require a butting surface to be effective. Square groove joints, and joints with heavy lands, are preferred to prevent burn-through. Weld joint designs and procedures are described in Fig. 11-23. Since the arc does not provide any cleaning action, the joint must be very clean. If this is not done, any porosity formed in the joint will remain trapped in the weld deposit.

DCEN produces a narrow, deep weld as shown in

| BASIC CAUSE | CONTRIBUTING FACTORS | SUGGESTED CORRECTIVE MEASURES |
|---|---|---|
| Hydrogen | Dirt containing oils or other hydrocarbons. Moisture in atmosphere or on metal or a hydrated oxide film on metal. Also, moisture in gas or gas lines. Base metal may be source of entrapped hydrogen (thicker the metal, greater possibility for hydrogen). | Degrease and mechanically or chemically remove oxide from weld area. Avoid humidity, use dry metal or wipe dry. Check moisture content of gas. Check gas and water lines for leaks. Increase gas flow to compensate for increased hydrogen in thicker sections. |
| Impurities | Cleaning or other compounds, especially those containing calcium. | Use recommended cleaning compounds, keep work free of contaminants. |
| Root Penetration | Incomplete penetration when TIG welding heavy sections increases the porosity in the weld. | Preheat, use higher welding current, or redesign joint geometry. |
| Temperature | Running "too cool" tends to increase porosity, due to premature solidification of molten metal. | Maintain proper current, arc length and torch-travel speed relationship. |
| Welding Speed | Too great a welding (travel) speed may increase porosity. | Decrease welding speed and establish and maintain proper arc length and current relationship. |
| Solidification Time | Quick "freezing" of weld pool entraps any gases present, thus increasing porosity. | Establish correct welding current and speed. If work is appreciably below room temperature, use supplemental heating to reduce differential. |
| Chemical Composition of Weld Metal | Pure aluminum weld metal is more susceptible to porosity than is that containing alloying elements. | If porosity is excessive, try an alloy filler material. *Check to see that all recommended procedures for preventing porosity are being followed.* |
| Cracking | Such causes of porosity as temperature, welding time, solidification time and alloying element in parent or filler metal may also be contributing causes of cracking. Other causes may be discontinuous welds, welds that intersect, or repair welds, and cold working either before or after welding. In general, crack-sensitive alloys include those containing 0.4 to 0.6% Si, or 1.5 to 3.0% Mg, or 1.0% Cu. | Lower current and faster speeds can often help to prevent cracking. However, a change to a filler alloy which brings weld metal composition out of cracking range is recommended where possible. |

Fig. 11-22. Suggested corrective measures for possible causes of porosity and cracking in welds.

| Material Thickness (inch) | Joint Design | Current DC (amperes)(1) | Volts | Diameter of Electrode(2) (inch) | Helium Gas Flow(3) (cfh) | Travel (ipm) | Filler Rod or Wire Diameter (inch) | Number of Passes |
|---|---|---|---|---|---|---|---|---|
| 0.010 | Standing Edge | 10–15 | | 0.020 | 20–50 | | | 1 |
| 0.020 | Square Butt | 15–30 | | 0.020 | 20–50 | | .020 | 1 |
| 0.030 | Square Butt | 20–50 | | 0.020 or 0.047 | 20–50 | | .020 or .047 | 1 |
| 0.032 | Square Butt | 65–70 | 10 | 3/32 | 20–50 | 52 | None | 1 |
| 0.040 | Square Butt | 25–65 | | 3/64 | 20–50 | | .047 (3/64) | 1 |
| 0.050 | Square Butt | 35–95 | | 3/64 | 20–50 | | 3/64 | 1 |
| 0.050 | Square Butt | 70–80 | 10 | 3/32 | 20–50 | 36 | None | 1 |
| 0.060 | Square Butt | 45–120 | | 3/64 or 1/16 | 20–50 | | 3/64 or 1/16 | 1 |
| 0.070 | Square Butt | 55–145 | | 1/16 | 20–50 | | 1/16 | 1 |
| 0.080 | Square Butt | 80–175 | | 1/16 | 20–50 | | 1/16 | 1 |
| 0.090 | Square Butt | 90–185 | | 1/16 | 20–50 | | 1/16 | 1 |
| 1/8 | Square Butt | 120–220 | | 1/8 | 20–50 | | 1/8 | 1 |
| 1/8 | Square Butt | 180–200 | 12.5 | 1/8 | 20–50 | 24 | None | 1 |
| 1/4 | Square Butt | 230–340 | | 1/8 | 25–60 | | 1/8 or 3/16 | 1 |
| 1/4 | Square Butt | 220–240 | 12.5 | 1/8 | 25–60 | 22 | None | 1 |
| 1/2 | 60° V Bevel, 1/4″ Land | 300–450 | | 3/16 | 25–60 | | 1/8 or 1/4 | 1 |
| 1/2 | Square Butt | 260–300 | 13 | 5/32 | 25–60 | 20 | None | 2 |
| 3/4 | 60° V or Double-V Bevel, 3/16″ Land | 300–450 | | 3/16 | 25–60 | | 1/8 or 1/4 | 3 for single V; 2 for double-V |
| 3/4 | Square Butt | 450–470 | 9.5 | 3/16 | 40–60 | 6 | None | 2 |
| 1 | 60° V or Double-V Bevel, 3/16″ Land | 300–450 | | 3/16 | 25–60 | | 1/8 or 1/4 | 4 for single -V; 2 for double-V |
| 1 | Square Butt | 550–570 | 9.5 | 1/4 | 40–60 | 5 | None | 2 |

(1)Automatic welding is required for the higher amperages. Manual welding can be done at the lower amperages.

(2)In lighter gages of material, it is common to use larger diameter electrodes than recommended and to taper the tip.

(3)It is possible to substitute helium-argon mixtures. In automatic welding, the arc can be started in argon and the helium added to the shielding gas when welding begins. The best ratio of He-A is usually determined by experimentation. The gas flow is dependent in part on the welding speed.

Fig. 11-23. Recommended practices for DCEN welding aluminum in the flat position.

Figs. 11-24 and 11-25. Because best results on aluminum are obtained with very short arc lengths, this process is best suited for automatic welding. Arc lengths are usually about 1/16 inch for manual welding and as little as 1/64 inch for machine welding. Therefore, control of the arc length is critical. It is somewhat difficult to control arc length in manual DCEN gas tungsten arc welding.

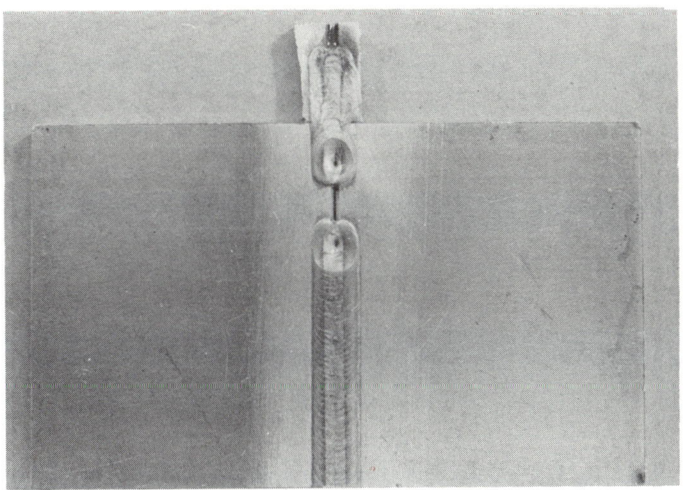

Fig. 11-24. To obtain good starts and ends in automatic welds tabs are used. These tabs also assist in eliminating cracks. The reason why the weld was stopped before the end was to determine if the size of the tackweld was sufficient to hold the parts during the welding and cooling cycle. Since the tackweld did not crack, the tackweld had sufficient quality to do the job intended. (Hold the parts in the proper alignment without failure during the welding operation).

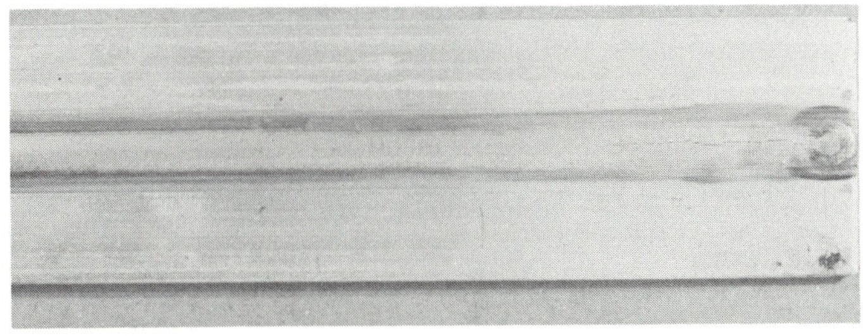

Fig. 11-25. The weld was started on the weld seam. The crown had insufficient strength to prevent cracking.

### Electrodes For Use in DCEN

All types of thoriated electrodes may be used in DCEN processes. In all cases, use an electrode that has sufficient capacity to carry the welding current used. The electrode should be tapered.

### Gases for DCEN

Helium rich mixtures (minimum 75 percent) or pure helium must be used with the DCEN process. Argon (maximum 25 percent) may be used to help stabilize the arc. Increase the gas flow since helium is lighter than argon. Refer to Fig. 11-23.

### Techniques for DCEN Welding on Aluminum

High frequency current is not needed to maintain a stable DCEN arc, as in AC. It can be used for starting the arc. The arc may also be started by scratching the electrode to the workpiece.

The high weld current will melt the base material immediately after the arc is made. Therefore, the filler material must be fed into the molten puddle immediately. Do not withdraw the wire. Move the wire forward into the puddle to supply the required filler material for the weld.

The helium gas used for shielding will leave a black sooty deposit on the top of the weld. This may be removed by wire brushing. Always remove this deposit when making multiple pass welds. Fill in the crater at the end of the weld.

There are some general considerations the welder must follow for DCEN welding on aluminum. These start with selecting the proper procedure from the chart in Fig. 11-23. Maintain the torch and wire angles as shown in Fig. 11-17.

If the tungsten contacts the filler material in the molten pool, stop and replace or regrind the tungsten. (If this is not done, the aluminum on the electrode will burn away and contaminate the weld.)

Always use stainless steel brushes for removing the oxides.

Remember to fill the crater at the end of the weld.

Aluminum is hot short. Remember that it has low strength at high temperatures. Therefore the base material must have support near the weld area or the metal will collapse.

The molten pool does not change color as it changes temperatures.

Corrective measures for GTAW on aluminum are listed in the chart shown in Fig. 11-22. Study this information.

## WELDING ALUMINUM WITH DCEP (DCRP)

DCEP welding is used to produce shallow penetration welds. Therefore, the joint thickness is usually below 1/16 inches. Designs of edge type joints or joints with very little gap tolerances are shown in Fig. 11-26.

The cleaning action of the reverse polarity is very good since the electron flow is from the workpiece to the electrode. However, good cleaning practices should be maintained as in ACHF welding.

With the use of a large electrode to carry the ma-

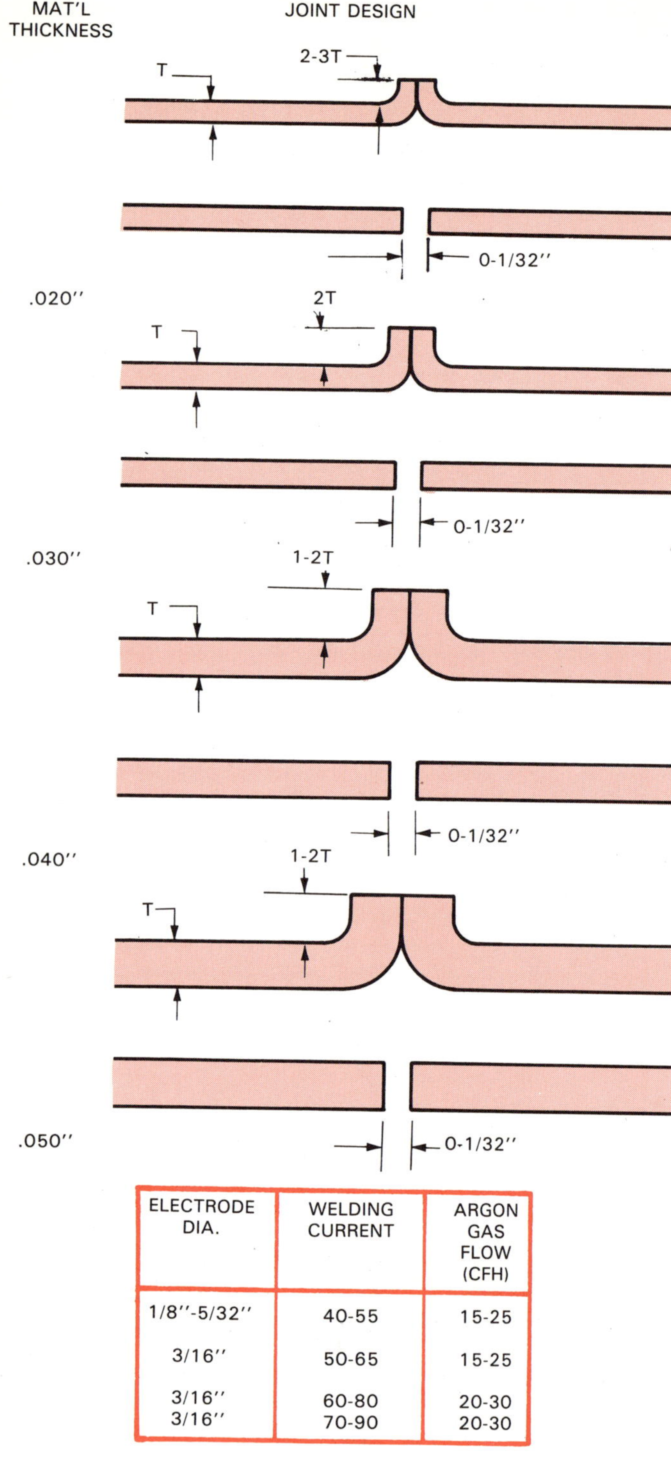


| ELECTRODE DIA. | WELDING CURRENT | ARGON GAS FLOW (CFH) |
|---|---|---|
| 1/8''-5/32'' | 40-55 | 15-25 |
| 3/16'' | 50-65 | 15-25 |
| 3/16'' | 60-80 | 20-30 |
| 3/16'' | 70-90 | 20-30 |

Fig. 11-26. Recommend procedures for DCEP welding aluminum in the flat position.

jority of the welding current, an arc gap of 3/16 inch to 1/4 inch is used.

Using this technique on thinner gauges of material requires good tooling if the joint is to stay in alignment during welding. The back-up bar must have a relief for the penetration to drop into.

### Electrodes for DCEP

Since thoriated electrodes can carry more heat for a given size, it is recommended that thoriated tungstens be used. Electrode diameters for various welding currents are shown in Fig. 11-26.

The electrode should have a slight radius on the tip. The current is applied until a "ball" is formed prior to welding. If the "ball" elongates at the required current level, use a larger tungsten.

### Gases for DCEP

Argon without helium is used in all conditions.

### Techniques for DCEP Welding on Aluminum

High frequency current is not needed to maintain a stable DCEP arc, as in AC. It can be used for starting the arc. The arc may also be started by scratching the electrode to the workpiece.

Most of the welding current is concentrated on the tungsten, so the weld operation is much slower than DCEN. Control of the arc length is not as critical as in DCEN and the weld puddle is highly visible.

Move the filler material into and out of the puddle as desired as in ACHF welding.

General considerations the welder must follow start with selecting the proper procedure from the chart in Fig. 11-26. Maintain the torch and wire angles as shown in Fig. 11-17. Move the torch across the joint using the step technique illustrated in Fig. 11-18.

If the tungsten contacts the filler material in the molten pool, stop and replace or regrind the tungsten. (If this is not done, the aluminum on the electrode will burn away and contaminate the weld.)

Always use stainless steel brushes for removing the oxides.

Remember to fill the crater at the end of the weld.

Aluminum is hot short. Remember that it has low strength at high temperatures. Therefore the base material must have support near the weld area or the metal will collapse.

The molten pool does not change color as it changes temperatures.

Corrective measures for GTAW on aluminum are listed in the chart shown in Fig. 11-22. Study this information.

## WELDING PROCEDURES

The four following listed procedures have been developed for gas tungsten arc welding practice and production. The material for practice can be any scrap aluminum material. The weld joint should fit properly for good results. All of the various materials should be cleaned as previously discussed before tackwelding and welding.

Gas nozzle sizes are not specified because of the

many variables involved. Always use the largest possible size. However, do not use a size that will obstruct vision of the puddle.

### WELD PROCEDURE NUMBER 11-1

WELD JOINT TYPE
Square Groove

POSITION
All

| MATERIAL TYPE | SIZE |
|---|---|
| Pure Aluminum | 3/32" Max. |
| **FILLER MATERIAL** | **SIZE** |
| Pure Aluminum | 3/32" |

MACHINE SETUP
ACHF

| SHIELDING GAS | CFH |
|---|---|
| Argon | 10 15 |
| **TUNGSTEN TYPE** | **SIZE** |
| Pure or Zirconia | 3/32" (Tapered And Balled) |

PROCEDURE:

The oxide film on the surface of aluminum must be removed prior to welding. This can be done by using a stainless steel brush, or a chemical etch (as described under cleaning procedures).

1. Tackweld two pieces of material at each end.
   A. Use filler wire on each tackweld.
   B. Keep joint tight (do not gap).
2. Align torch and wire and weld as shown in Fig. 11-17.
   A. The technique for welding aluminum is different from steel as the weld is made in steps. This allows good penetration and heat control over the puddle. The operational steps are shown in Fig. 11-18 and include:
      1. Form a puddle with penetration through the joint.
      2. Immediately add wire to stop the penetration from falling. Continue to add wire to form a crown.
      3. Withdraw the wire.
      4. Move forward approximately 1/8 inch and stop.
      5. Repeat steps 2, 3, and 4 until done.
      6. Increase speed as material heat increases.

Problem Areas and Corrections:

1. Incomplete penetration.
   A. Increase amperage.
   B. Wait after stop point before adding wire.
2. Penetration too heavy.
   A. Decrease amperage.
   B. Add wire immediately after stop point to cool puddle faster.
3. Uneven crown.
   A. Add or decrease amount of wire placed into the puddle.

### WELD PROCEDURE NUMBER 11-2

WELD JOINT TYPE
V Groove

POSITION
All

| MATERIAL TYPE | SIZE |
|---|---|
| Pure Aluminum | 1/4" |
| **FILLER MATERIAL** | **SIZE** |
| Pure Aluminum | 3/32" |

MACHINE SETUP
ACHF

| SHIELDING GAS | CFH |
|---|---|
| Helium-75% Argon 25% | 20-30 |
| **TUNGSTEN TYPE** | **SIZE** |
| Pure or Zirconia | 3/32"/1/8" (Tapered And Balled) |

PROCEDURE:

Prepare bevel angles by machining as shown in Fig. 11-15. Clean material as described under cleaning procedures.

1. Tackweld two pieces of material.
   A. Do not gap plates.
   B. Use filler wire on each tackweld.
2. Locate plates for position desired.
3. Preheat plates to approximately 300-400°F.
4. Weld joint using torch angles and wire positions used in Fig. 11-17.
   A. Maintain preheat temperature range for all passes to obtain even penetration and puddle control.
   B. Use the weld technique as specified in Weld Procedure Number 11-1.

Problem Areas and Corrections:

1. The major problem in welding aluminum over 1/8 inch thick is not having enough heat. Preheat is critical if the molten pool is to move correctly. Without a molten pool, penetration cannot be obtained, cold shuts will develop between passes, and the weld puddle will be very sluggish.

## WELD PROCEDURE NUMBER 11-3

| WELD JOINT TYPE | |
|---|---|
| T | |
| **POSITION** | |
| All | |
| **MATERIAL TYPE** | **SIZE** |
| Pure Aluminum | 1/16"/3/32" |
| **FILLER MATERIAL** | **SIZE** |
| Pure Aluminum | 1/16" |
| **MACHINE SETUP** | |
| ACHF | |
| **SHIELDING GAS** | **CFH** |
| Argon | 10-15 |
| **TUNGSTEN TYPE** | **SIZE** |
| Pure or Zirconia | 3/32" (Tapered And Balled) |

PROCEDURE:

Clean material as described under cleaning procedures.

1. Tackweld two plates at approximately right angles.
   A. Use filler wire on each tackweld.
   B. Keep joint tight.
2. Align plates as required.
3. Weld first pass using torch and wire angles, as for steel.
   A. Maintain torch and wire angles.
   B. To achieve penetration into the corner, the following technique must be used.
      1. Apply heat to start puddle.
      2. Apply small amount of wire.
      3. Move forward approximately 1/8 inch and stop.
      4. Add wire for the size weld desired.
      5. Remove wire from puddle.
      6. Continue on from step 3.

Problem Areas and Corrections:

1. The major problem in welding aluminum T joints is the lack of penetration into the joint intersection. The two main causes are:
   A. Insufficient heat.
      1. Increase amperage.
      2. Use smaller diameter filler wire.
      3. Do not make too large a weld.
   B. Improper torch angle.
      1. Keep the tungsten pointed to the corner.
      2. Incline the torch slightly to the bottom plate.
   C. If problem persists, check chart of recommended procedures in Fig. 11-15. Then, review all elements of 12-step procedure under Techniques for ACHF Welding on Aluminum.

## WELD PROCEDURE NUMBER 11-4

| WELD JOINT TYPE | |
|---|---|
| T | |
| **POSITION** | |
| All | |
| **MATERIAL TYPE** | **SIZE** |
| Pure Aluminum | 1/4" |
| **FILLER MATERIAL** | **SIZE** |
| Pure Aluminum | 3/32" |
| **MACHINE SETUP** | |
| ACHF | |
| **SHIELDING GAS** | **CFH** |
| Helium 75% Argon 25% | 20-30 |
| **TUNGSTEN TYPE** | **SIZE** |
| Pure or Zirconia | 3/32"/1/8" (Balled) |

PROCEDURE:

Clean materials as described under cleaning procedures.

1. Tackweld two plates of material at approximately right angles.
   A. Use filler wire on each tackweld.
   B. Keep joint tight.
2. Locate plates for position desired.
3. Preheat to approximately 300-400°F.
4. Weld first pass using torch and wire angles as required.
   A. Maintain torch and wire angles.
   B. To achieve penetration into the corner, the following techniques must be used.
      1. Apply heat to start puddle.
      2. Apply small amount of wire.
      3. Move forward approximately 1/8 inch and stop.
      4. Add wire for the size weld desired.
      5. Remove wire from puddle.
      6. Continue on from step 3.
5. Complete required passes to build weld to proper size as multipass steel fillet welds.

Problem Areas and Corrections:

1. The major problem in welding aluminum T joints is the lack of penetration into the joint intersection. The two main causes are:
   A. Insufficient heat.
      1. Increase amperage.
      2. Use smaller diameter filler wire.
      3. Do not make too large a weld.
   B. Improper torch angle.
      1. Keep the tungsten pointed to the corner.
      2. Incline the torch slightly to the bottom plate.

## REVIEW QUESTIONS

1. At what temperature does pure aluminum melt?
2. Do the aluminum alloys melt at a higher or lower temperature than the pure aluminum? What is the range?
3. Does aluminum change color when it melts?
4. Why does the oxide film have to be removed before welding?
5. Who established the aluminum identification system?
6. You have a piece of 6061-T6 aluminum. What does the "T-6" temper condition indicate?
7. Is the 5052 alloy heat treatable?
8. What is the primary welding material to be used for welding 6061-T6 aluminum to 6061-T6 aluminum? Can others be used?
9. Why must aluminum be cleaned properly?
10. What are the two types of back-up used for welding groove welds?
11. What are the functions of preheating aluminum prior to welding?
12. Why is high frequency voltage required when using alternating current for welding?
13. When welding heavier sections of aluminum manually, a gas mixture of ______ may be used to obtain deeper penetration.
14. List the two types of tungsten electrodes used for welding with alternating current.
15. Is the tungsten electrode used with alternating current pointed or blunted on the end?
16. What type of brushes are used for cleaning aluminum?
17. When a material, such as aluminum, collapses near the weld joint from heat, it is called ______ material.
18. DCEN welding of aluminum uses ______ gas for shielding.

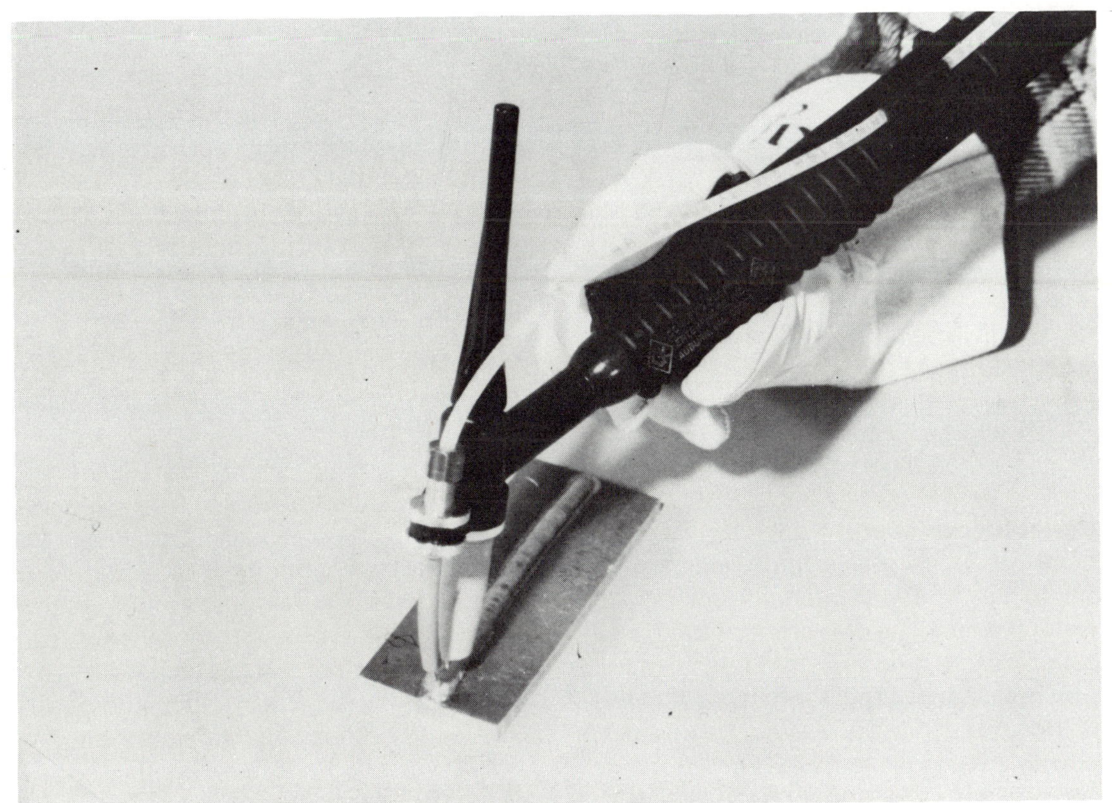

A manual weld is made on aluminum with a powered wire feed. (Conley & Kleppan Enterprises, Inc.)

# Chapter 12

# WELDING PROCEDURE FOR MANUAL WELDING MAGNESIUM

## BASE MATERIALS

Magnesium is a non-ferrous (no iron) material readily welded with the GTAW process. Some of the major characteristics of the material include:

1. Extreme lightness.
2. High strength to weight ratio.
3. Good thermal conductivity.
4. Good electrical conductivity.
5. High resistance to corrosion.
6. Non-sparking.
7. Non-toxic.
8. Material does not change color when heated.

Pure magnesium melts at approximately 1200°F. The alloy materials melt at a slightly lower temperature. The oxide film on the surface must be removed prior to welding if satisfactory welds are to be made.

Magnesium will require approximately 2/3 the amount of heat required to weld the same thickness of aluminum.

### Wrought Material Identification

The nomenclature used to designate magnesium alloys conforms to the ASTM (American Society for Testing Materials) system. The designations are based on chemical composition. The first two letters represent the two alloying elements specified in the greatest amount. These are arranged in decreasing percentages, or alphabetically if of equal percentage. The letters are followed by the respective percentages rounded off to whole numbers. These are followed by a serial letter which indicates some variation in composition.

The letters used to designate various alloying elements include:

A – Aluminum
E – Rare Earths
H – Thorium
K – Zirconium
L – Lithium
M – Manganese
Q – Silver
S – Silicon
T – Tin
Z – Zinc

Chemical compositions of wrought magnesium alloys are shown in Fig. 12-1.

### Tempers

The material can be further identified by the condition in which it is supplied. Designations used to denote tempers of magnesium mill products are:

–F as extruded.
–T5, –T51 artificially aged.
–T8, –T81 solution heat treated, strain-hardened, then artificially aged.
–0 fully annealed.
–H24, –H26 strain-hardened, then partially annealed.

## FILLER MATERIALS

The filler material choice for fusion welding various magnesium alloys is very important. The metal produced in the weld puddle is a combination of the filler material and the parent material. Weld metal must have the strength, ductility, freedom from cracking and the ability to resist corrosion required by the application.

Correct choice of the filler alloy will eliminate or significantly reduce inter-metallic compounds and low ductility in magnesium welds.

Fig. 12-2 shows the various filler materials used for joining the alloys to each other and to other combinations.

Maximum weld quality can only be obtained if the filler material is clean. Contaminates which cause weld porosity on the filler material are most often an oil or a hydrated oxide. The heat of welding releases the hydrogen from these contaminate sources which cause porosity in the weld. The filler material should be handled with the utmost care to retain it's cleanliness during storage and use in the welding cycle.

| MAGNESIUM ALLOY CHEMICAL COMPOSITION (PERCENT) | | | | | | | | | | | | | |
|---|---|---|---|---|---|---|---|---|---|---|---|---|---|
| ASTM Designation Alloy | Aluminum | Manganese, Minimum | Zinc | Zirconium | Rare Earths | Thorium | Calcium | Silicon, Maximum | Copper, Maximum | Nickel, Maximum | Iron, Maximum | Other Imp., Maximum | Magnesium |
| AZ31B | 2.5-3.5 | 0.20 | 0.7-1.3 | — | — | | 0.04 max. | 0.30 | 0.05 | 0.005 | 0.005 | 0.30 | Balance |
| AZ31C | 2.4-3.6 | 0.15 | 0.5-1.5 | — | — | — | 0.04 max. | 0.30 | 0.10 | 0.03 | — | 0.30 | Balance |
| AZ61A | 5.8-7.2 | 0.15 | 0.4-1.5 | — | — | — | — | 0.30 | 0.05 | 0.005 | 0.005 | 0.30 | Balance |
| AZ80A | 7.8-9.2 | 0.15 | 0.2-0.8 | — | — | — | — | 0.30 | 0.05 | 0.005 | 0.005 | 0.30 | Balance |
| HK31A | — | (0.15 max.) | — | 0.45-1.0 | — | 2.5-4.0 | — | — | — | — | — | 0.30 | Balance |
| HM21A | — | 0.45-1.1 | — | — | — | 1.5-2.5 | — | — | — | — | — | 0.30 | Balance |
| HM31A | — | 1.20 | — | — | — | 2.5-3.5 | — | — | — | — | — | 0.30 | Balance |
| M1A | — | 1.20 | — | — | — | — | 0.08-0.14 | 0.05 | 0.05 | 0.01 | — | 0.30 | Balance |
| ZK60A | — | — | 4.8-6.2 | 0.45 min. | — | — | — | — | — | — | — | 0.30 | Balance |

Fig. 12-1. Wrought magnesium alloy chemical requirement.

| CHART OF FILLER MATERIALS | | | | | | | | | |
|---|---|---|---|---|---|---|---|---|---|
| Alloy | AZ10A | AZ31B | AZ61A | AZ80A | HK31A | HM21A | HM31A | ZE10A | ZK60A |
| AZ10A | AZ261A* | — | — | — | — | — | — | — | — |
| AZ31B | AZ261* | AZ261* | — | — | — | — | — | — | — |
| AZ61A | AZ261* | AZ261* | AZ261* | — | — | — | — | — | — |
| AZ80A | AZ261* | AZ261* | AZ261* | AZ261* | — | — | — | — | — |
| HK31A | AZ261* | AZ261* | AZ261* | AZ261* | EZ33A | — | — | — | — |
| HM21A | AZ261* | AZ261* | AZ261* | AZ261* | EZ33A | EZ33A | — | — | — |
| HM31A | AZ261* | AZ261* | AZ261* | AZ261* | EZ33A | EZ33A | EZ33A | — | — |
| ZK60A | NR | NR | NR | NR | NR | NR | NR | NR | NR |

*Use AZ61A for best weldability of wrought alloys and greatest economy. AZ92A, AZ101A or AZ61A welding rod.
NR = Not Recommended
Dash Mark indicates a filler material of the base material combinations is the best choice.

Fig. 12-2. Filler materials used to weld magnesium and alloys. (Dow Chemical Co.)

## WELD JOINT DESIGN

Common weld joint designs for welding magnesium with alternating current and continuous high frequency are shown in Fig. 12-3. Designs for direct current straight and reverse polarity are specified in the individual welding procedures.

### Magnesium Joint Preparation

Joint edges that are prepared by the plasma arc cutting process or carbon arc gouging process have a heavy oxide film on the surface. This surface should be thoroughly cleaned prior to welding to prevent dross and porosity in the final weld.

Joint edges prepared by shearing should be sharp without tearing or ridges. Otherwise, dirt and oil may become entrapped and result in faulty welds.

### Pre-weld Cleaning of Magnesium

Magnesium is generally received from the mill with

a lightly oiled surface. The material should be degreased and cleaned with the same procedures as used with aluminum.

### Weld Back-Up

Weld back-up bars designed for the welding of aluminum work very well with magnesium. The use of an inert gas on the root side of the weld is advantageous in obtaining good penetration form in high quality welds.

### Tooling for Magnesium

Magnesium tends to distort more than aluminum. Again, the use of weld back-up bars designed for the welding of aluminum is recommended. In addition, it will be found that the use of hold down bars and chill bars in conjunction with the back-up bars will greatly reduce the distortion problem.

### Preheating

Thin structures of magnesium do not require preheat. Material over 1/4 inch may be preheated to a maximum of 300°F for increasing depth of penetration. Magnesium castings requiring welding should be heated to about 300°F to prevent shrinkage cracks.

Temperature indicating paints, crayons, and pellets as shown in Fig. 11-13 can be used to verify the actual temperature. Be careful that the temperature indicating material does not get into the area where the molten metal will contact it.

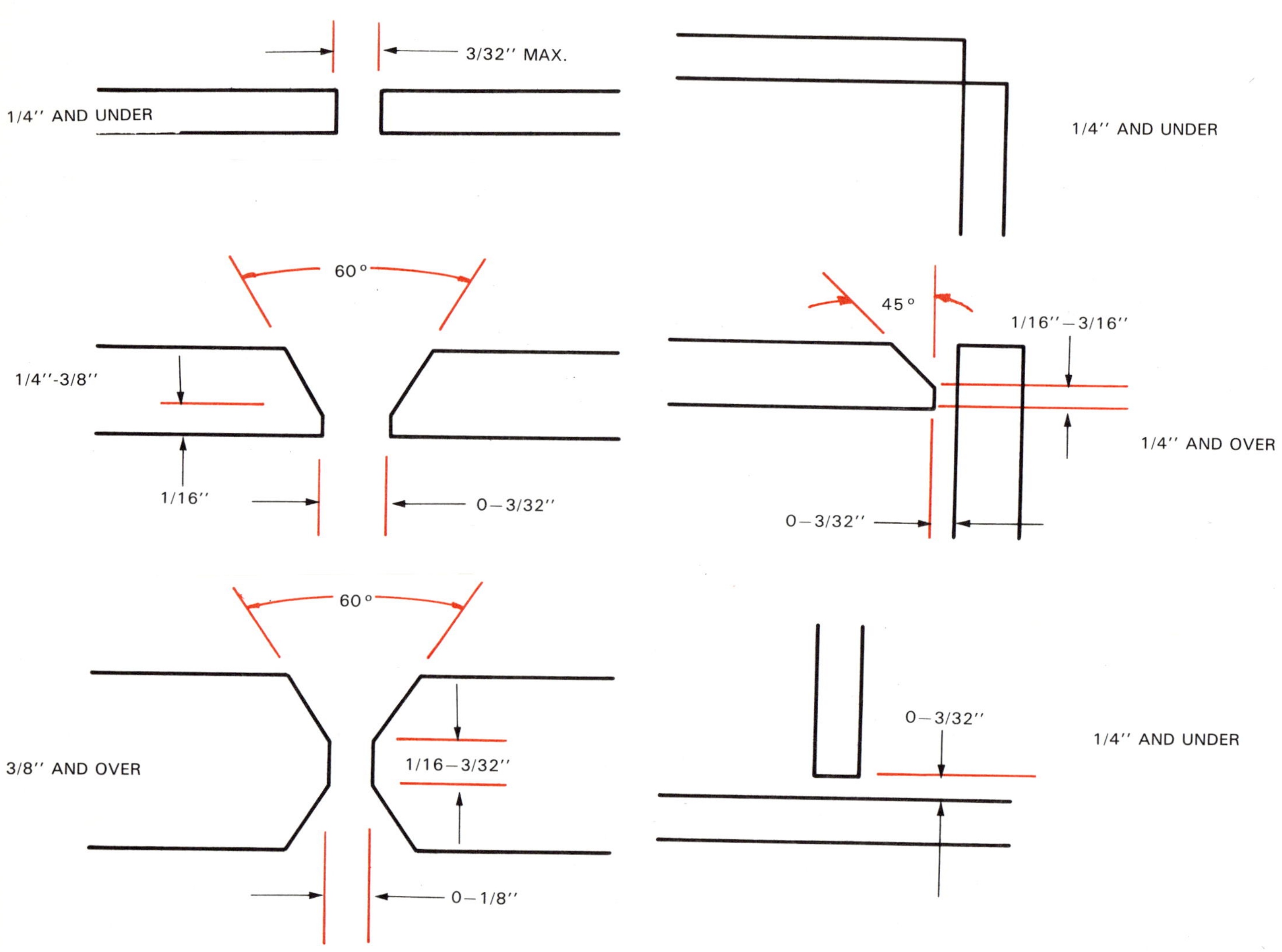


Fig. 12-3. Weld joint designs for welding magnesium with ACHF current.

### Tackwelds

When tackwelds must be made to maintain joint alignment for welding, the following considerations apply:

1. If the joint welding procedure requires preheating, the joint requires preheating before tackwelding.

2. Use the same filler material used for the main welding procedure.
3. If the joint requires full penetration, the tackweld requires penetration.
4. Keep the tackweld as small as possible. If possible, smaller than the main weld.
5. If the tackweld cracks, leave it and make another close to the broken one.
6. Do not grind out tackwelds. Grinding grit in a joint makes good quality almost impossible.
7. The crater of a tackweld is the weakest point in the tackweld. Do not leave a crater in a tackweld.

If the tackwelded assembly needs straightening, a soft mallet may be used to even the joint prior to welding. Do not get the mallet material into the open joint. Be sure joint is clean.

3. Be sure the machine is grounded properly to prevent high frequency radiation.
4. If the oxide cleaning action is lost or insufficient during welding, check the high frequency point gap. The high frequency spark is needed to maintain the start of each straight and reverse polarity part of the cycle. Without the reverse polarity, the cleaning action is lost.

### Gases for Welding Magnesium by ACHF

Argon gas is used for thicknesses up to 1/8 inch thick on all types of joints. Beyond 1/8 inch thickness, a combination of 75HE-25AR will assist in obtaining penetration on all types of joints. Using a gas mixture with larger amounts of helium creates problems with arc starting. The gas should be of high quality with no leaks in the supply system.

| MAGNESIUM WELDING PROCEDURES | | | | | | | |
|---|---|---|---|---|---|---|---|
| Material Thickness, in. | No. of Passes | Current, A[a] | Electrode Diam., in. | Weld Rod Diam., in. | Argon Flow, ft$_3$/h | Joint Design[b] | Filler Metal Consumption, lb/ft of weld |
| 0.040 | 1 | 35 | 1/16 | 3/32 | 12 | A | 0.004 |
| 0.063 | 1 | 50 | 3/32 | 3/32 | 12 | A | 0.005 |
| 0.080 | 1 | 75 | 3/32 | 3/32 | 12 | A | 0.006 |
| 0.100 | 1 | 100 | 3/32 | 3/32 | 12 | A | 0.008 |
| 0.125 | 1 | 125 | 3/32 | 1/8 | 12 | A | 0.009 |
| 0.190 | 1 | 160 | 1/8 | 1/8 | 15 | A | 0.011 |
| 0.250 | 2 | 175 | 5/32 | 1/8 | 20 | B | 0.026 |
| 0.375 | 3 | 175 | 5/32 | 5/32 | 20 | B | 0.057 |
| 0.375 | 2 | 200 | 3/16 | 1/8 | 20 | C | 0.024 |
| 0.500 | 3 | 175 | 5/32 | 5/32 | 20 | B | 0.106 |
| 0.500 | 2 | 250 | 3/16 | 1/8 | 20 | C | 0.047 |

a. Magnesium alloys containing thorium will require 20 percent higher current. The use of helium as the shield gas will reduce the weld current required by 20-30 amperes.
b. A = Square groove butt joint, zero root opening.
B = Single Vee 60° bevel butt joint, 1/16 in. root face, zero root opening.
C = Double Vee 60° bevel butt joint, 3/32 in. root face, zero root opening.

Fig. 12-4. Welding parameters used with ACHF current. (International Magnesium Assoc.)

### Power Supply Type for Welding ACHF

1. When using a transformer or a combination AC/DC power supply not designed for GTAW, do not exceed the derated capacity of the unit. (See Chapter 3).
2. If the power supply has a wave balancer control, set on normal to start welding. This adjusts the amount of current produced by negative and positive (straight and reverse) polarity cycles of alternating current. Then adjust to desired penetration or cleaning action from that point.

## ELECTRODES USED IN ACHF ON MAGNESIUM

Pure and zirconiated electrodes are recommended for welding with alternating current. Both have good "balling" characteristics which is required for high quality magnesium welds. Thoriated tungstens do not have this capability, and as a result will "spit" after slight use. (Spitting is the breakdown of the tungsten. Small particles of tungsten are then separated and transferred by the arc stream into the weld puddle.)

## ACHF TECHNIQUES IN WELDING MAGNESIUM

To produce satisfactory welds on magnesium, all of the following elements of the procedure must be maintained.

1. Select the electrode diameter from those shown in Fig. 12-4.
2. Taper the electrode as shown in Fig. 12-5.
3. Blunt the tungsten on the end. Start an arc on scrap material and increase amperage until a radius is formed as shown in Fig. 12-5.

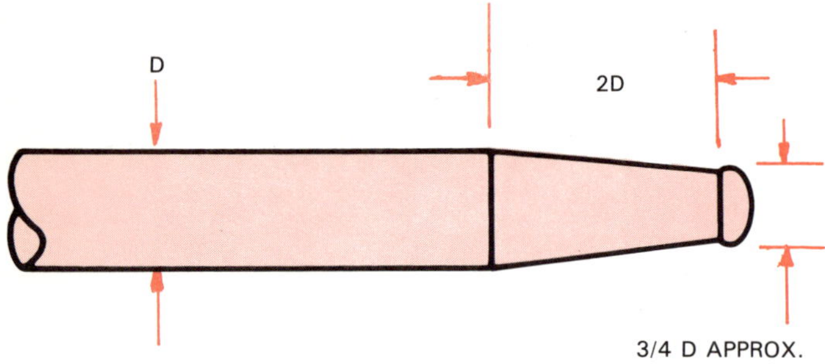


Fig. 12-5. Electrodes used for alternating current should be ground as shown. The ball or rounded tip will form as the welding current is increased.

4. Extend the tungsten beyond the end of the cup a distance not greater than the ID of the cup.
5. Maintain torch and wire angles as shown in Fig. 12-6.
6. Move the torch across the joint as shown in Fig. 12-7.
7. If tungsten contacts the filler material or the molten pool, stop and replace tungsten. If this is not done, the magnesium on the electrode will burn away and contaminate the weld.

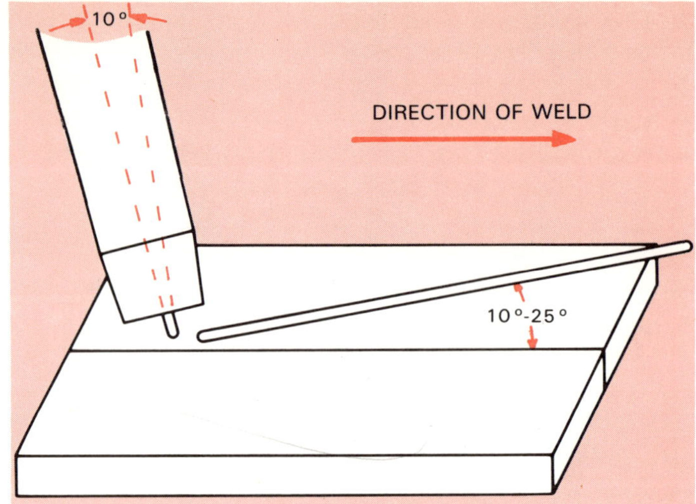


Fig. 12-6. Torch and wire positions for flat welding groove welds.

8. Always use stainless steel brushes for removing oxides from the surface of the material or the weld.
9. Remember to fill the crater at the end of the weld.
10. The molten pool does not change color as it changes temperature as does steel. The arc should be a yellowish color.
11. An orange color deposit may form on the tip of the electrode during welding. Remove the formation from the electrode by rubbing with light sandpaper between welds.

## POST WELD TREATMENT

Magnesium alloys containing more than 1.5 percent aluminum are susceptible to stress corrosion cracking. They must be stress relieved after welding. The various

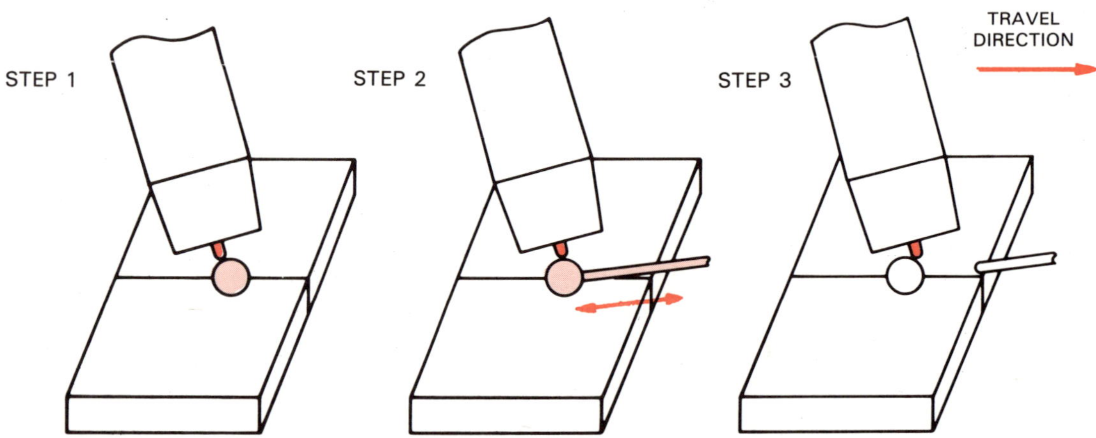


Fig. 12-7. Step welding technique for welding magnesium groove welds.

stress relief times and temperatures required are shown in Fig. 12-8.

| STRESS RELIEF | | | | | |
|---|---|---|---|---|---|
| SHEET | | | CASTINGS | | |
| Alloy | Temp., F | Time, min. | Alloy | Temp., F | Time, min. |
| AZ31B-0 | 500 | 15 | AM100A | 500 | 60 |
| AZ31B-H24 | 300 | 60 | AZ63A | 500 | 60 |
| HK31A-H24 | 550 | 30 | AZ81A | 500 | 60 |
| HM21A-T8 | 700 | 30 | AZ91C | 500 | 60 |
| HM21A-T81 | 750 | 30 | AZ92A | 500 | 60 |

Fig. 12-8. Stress relief treatment schedule.

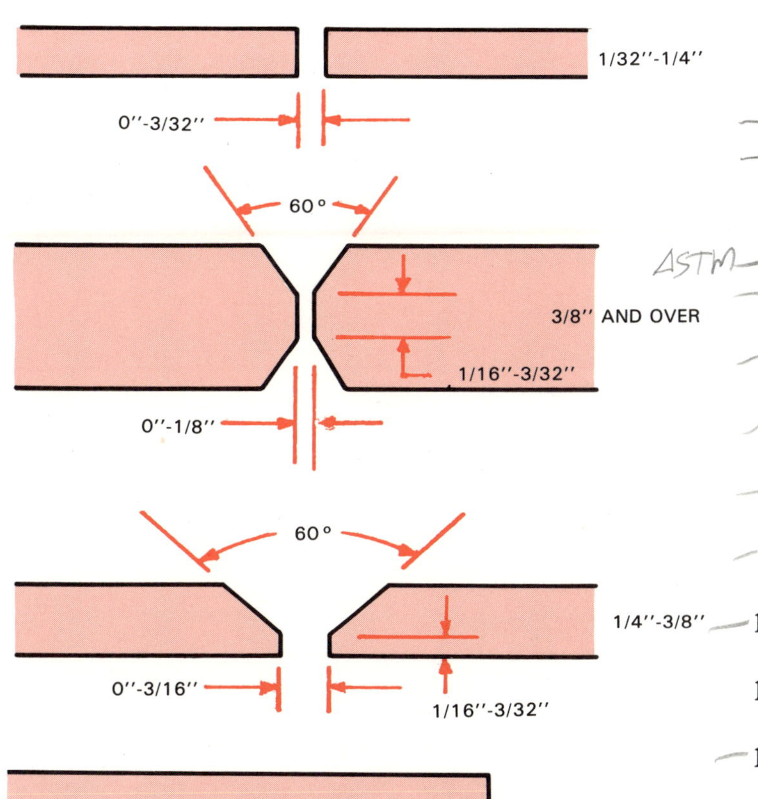


Fig. 12-9. Weld joint designs for welding magnesium with DCEN (DCSP) current.

## WELDING MAGNESIUM WITH DCEN (DCSP)

DCEN welding is used to produce deeply penetrating welds. The various types of joints which are used with this technique are shown in Fig. 12-9.

All of variables and techniques listed in Chapter 11 on DCEN apply to this process type.

Post weld treatment as specified in Fig. 12-8 also applies.

## WELDING MAGNESIUM WITH DCEP (DCRP)

DCEP welding is used to produce shallow penetration welds. The various types of joints which are used with this technique are shown in Fig. 11-26.

All of the variables and techniques listed in Chapter 11 on DCEP welding apply to this process type.

Post weld treatment as specified in Fig. 12-8 also applies.

## REVIEW QUESTIONS

1. Does magnesium contain iron?
2. At what temperature does magnesium melt?
3. Does magnesium require as much heat to melt as aluminum?
4. What system is used to identify magnesium alloys?
5. What characteristics are considered when selecting a filler material for welding the magnesium alloys?
6. What two materials most often cause porosity in magnesium welds?
7. How is magnesium generally received from the mill?
8. Can aluminum weld tooling be used for welding magnesium?
9. Which material, aluminum or magnesium, distorts the most during welding?
10. At what minimum thickness is preheat used when welding magnesium?
11. How does a temperature indicating pellet, ink or crayon operate?
12. How can a joint be straightened prior to welding magnesium?
13. What does a wave balancer do on an AC welding machine?
14. If the cleaning action is lost during ACHF welding, what may have happened in the power supply?
15. Why are pure and zirconiated tungstens used on ACHF welding?
16. During welding, the tungsten should not extend from the gas nozzle no greater than ______ ______ ______ ______.

# Chapter 13

# WELDING PROCEDURE FOR MANUAL WELDING COPPER AND COPPER ALLOYS

## BASE MATERIALS

Copper and copper alloys are a non-ferrous (no iron) group of materials. Some of the major characteristics of these materials include:

1. High thermal conductivity.
2. Good electrical conductivity.
3. Excellent corrosion resistance.
4. Good formability and ductility.
5. Some of the alloys have high hardness and good wearing properties.

Copper melts at approximately 1980°F and requires preheating with high rates of heat input during welding. The preheating is necessary because copper conducts the heat away from the weld joint very rapidly. Copper alloys may not require preheating depending on the alloy.

### Copper And Copper Alloy Identification

The Copper Development Association has assigned identification numbers to copper and copper alloys. These identification numbers are shown in Fig. 13-1. Included in this figure also are shown GTA weldability ratings.

The ratings are:

G — Good
F — Fair
NR — Not rec
E — Excellent

Note that numbers 314 through 385 contain lead which boils at welding temperatures causing porosity and cracking.

## FILLER MATERIALS FOR WELDING COPPER

Filler materials used to join copper and copper alloys are shown in Fig. 13-2. These filler materials may be obtained using the American Welding Society Specification, AWS A5.7 Copper and Copper-Alloy Welding Rods.

## WELD JOINT DESIGN AND FIT-UP

Copper weld joint designs are generally more open than steel designs. This is due to the higher thermal conductivity and the loss of preheat from the weld joint.

The type of joint design must be considered with the type of tooling used. Copper expands more than most materials. Therefore root openings must be set for welding at the preheat temperature. If groove openings are not established at preheat temperature, the joint will close during the preheat cycle due to expansion. Some common joint designs are shown in Fig. 13-3.

### Joint Preparation for GTAW of Copper

Joint edges that are prepared by the plasma arc cutting process or carbon arc gouging process have a heavy oxide film on the surface. This surface should be thoroughly cleaned prior to welding to prevent dross and porosity in the final weld.

Joint edges prepared by shearing should be sharp without tearing or ridges. Where these conditions exist, dirt, oil, and grease may become entrapped and result in faulty welds.

### Pre-Weld Cleaning

Oil, grease, etc., should be removed prior to welding with any standard degreasing material such as acetone or alcohol. If corrosion or scale is present, this must be removed prior to welding.

Commercial copper cleaners or acid baths may be used for this purpose.

### Weld Back-Up

Groove welds which will be welded from one side of the joint where 100 percent penetration is required should have a grooved copper back-up. The back-up groove will retain argon from the torch and protect the penetration area from oxidation. A typical back-up bar configuration is shown in Fig. 13-4.

| COPPER AND COPPER ALLOYS | | | |
|---|---|---|---|
| **Copper Alloy No.** | **Trade Name** | **Nominal Compo-sition, %** | **GTAW** |
| 102 | Oxygen-Free Copper (OF) | Cu, 100 | G |
| 110 | Electrolytic Tough Pitch (ETP) | Cu + 0, 100 | F |
| 113,114 | Silver-Bearing Tough Pitch (STP) | Cu + Ag<br>+0, 100 | F |
| 116<br>122 | Deoxidized, High Residual Phosphorus (DHP) | Cu + P, 100 | G |
| 210 | Gilding, 95% | Cu 95<br>Zn 5. | G |
| 220 | Commercial Bronze, 90% | Cu 90<br>Zn 10. | G |
| 226 | Jewelry Bronze, 87.5% | Cu 87.5<br>Zn 12.5 | G |
| 230 | Red Brass, 85% | Cu 85<br>Zn 15. | G |
| 240 | Low Brass, 80% | Cu 80<br>Zn 20. | G |
| 260 | Cartridge Brass, 70% | Cu 70.<br>Zn 30. | F |
| 268,270 | Yellow Brass | Cu 65<br>Zn 35. | F |
| 280 | Muntz Metal | Cu 60<br>Zn 40. | F |
| 314 | Leaded Commercial Bronze | Cu 89<br>Pb 1.9<br>Zn 9.1. | NR |
| 330 | Low-Leaded Brass Tube | Cu 66<br>Pb .5<br>Zn 33.5. | NR |
| 332 | High-Leaded Brass Tube | Cu 66<br>Pb 1.6<br>Zn 32.4. | NR |
| 335 | Low-Leaded Brass | Cu 65<br>Pb .5<br>Zn 34.5. | NR |
| 340 | Medium-Leaded Brass | Cu 65<br>Pb 1<br>Zn 34. | NR |
| 342,353 | High-Leaded Brass | Cu 64.5<br>Pb 2<br>Zn 33.5. | NR |
| 356 | Extra-High-Leaded Brass | Cu 62<br>Pb 2.5<br>Zn 35.5 | NR |
| 360 | Free-Cutting Brass | Cu 61.5<br>Pb 3<br>Zn 35.5 | NR |
| 365-368 | Leaded Muntz Metal | Cu 60<br>Pb .6<br>Zn 39.4 | NR |
| 370 | Free-Cutting Muntz Metal | Cu 60<br>Pb 1<br>Zn 39. | NR |
| 377 | Forging Brass | Cu 60<br>Pb 2<br>Zn 38. | NR |
| 385 | Architectural Bronze | Cu 57<br>Pb 3<br>Zn 40. | NR |
| 443-445 | Inhibited Admiralty | Cu 71<br>Sn 1<br>Zn 28. | G |
| 464-467 | Naval Brass | Cu 60<br>Sn .75<br>Zn 39.25. | F |
| 485 | Leaded Naval Brass | Cu 60<br>Sn .7<br>Pb 1.8<br>Zn 37.5. | NR |
| 505 | Phosphor Bronze, 1.25% E | Cu 98.7<br>Sn 1.3. | G |
| 510 | Phosphor Bronze, 5%A | Cu 95<br>Sn 5. | G |
| 521 | Phosphor Bronze, 8% C | P .2<br>Cu 92<br>Sn 8. | G |
| 524 | Phosphor Bronze, 10% D | Cu 90<br>Sn 10. | G |
| 544 | Free-Cutting Phosphor Bronze | Cu 88<br>Pb 4<br>Sn 4<br>Zn 4. | NR |
| 613,614 | Aluminum Bronze D | Cu 91<br>Al 7<br>Fe 2. | E |
| 651 | Low-Silicon Bronze B | Cu 98.5<br>Si 1.5. | E |
| 655 | High-Silicon Bronze A | Cu 97<br>Si 3. | E |
| 675 | Manganese Bronze A | Cu 58.5<br>Fe 1.4<br>Mn .1<br>Sn 1<br>Zn 39.0. | F |
| 687 | Aluminum Brass | Cu 77.5<br>Al 2<br>Zn 20.5. | F |
| 706 | Copper Nickel, 10% | Cu 88.6<br>Fe 1.4<br>Ni 10. | E |
| 715 | Copper Nickel, 30% | Cu 69.5<br>Ni 30<br>Fe .5. | F |
| 745 | Nickel Silver, 65-10 | Cu 65<br>Ni 10<br>Zn 25. | F |
| 752 | Nickel Silver, 65-18 | Cu 65<br>Ni 18<br>Zn 17. | F |
| 754 | Nickel Silver, 65-15 | Cu 65<br>Ni 15<br>Zn 20. | F |
| 757 | Nickel Silver, 65-12 | Cu 65<br>Ni 12<br>Zn 23. | F |
| 770 | Nickel Silver, 55-18 | Cu 55<br>Ni 18<br>Zn 27. | F |

Fig. 13-1. Copper and copper alloy identification.

| FILLER MATERIALS FOR COPPER | | |
|---|---|---|
| **Filler Material a** | **Common name** | **Base metal applications** |
| RCu | Copper | Coppers |
| RCuSi-A | Silicon bronze | Silicon bronzes, brasses |
| RCuSn-A | Phosphor bronze | Phosphor bronzes, brasses |
| RCuNi | Copper-nickel | Copper-nickel alloys |
| RCuA1-A2 | Aluminum bronze | Aluminum bronzes, brasses, silicon bronzes, manganese bronzes |
| RCuA1-A3 | Aluminum bronze | Aluminum bronzes |
| RCuNiA1 | — | Nickel-aluminum bronzes |
| RCuMnNiA1 | — | Manganese-nickel-aluminum bronzes |
| RBCuZn-A | Naval brass | Brasses, copper |
| RCuZn-B | Low fuming brass | Brasses, manganese bronzes |
| RCuZn-C | Low fuming brass | Brasses, manganese bronzes |

a. See AWS A5.7, Specification for Copper and Copper Alloy Bare Welding Rods and Electrodes.

Fig. 13-2. Filler material application for some common copper and copper alloys.

## Preheating Copper

Where preheating is required, an oxyacetylene torch with a neutral flame may be used. Temperature indicating crayons or liquids may be used to determine the correct temperature. Do not allow the temperature indicating material to flow into the weld joint. Heat the joint and the back-up bar (where used) thoroughly.

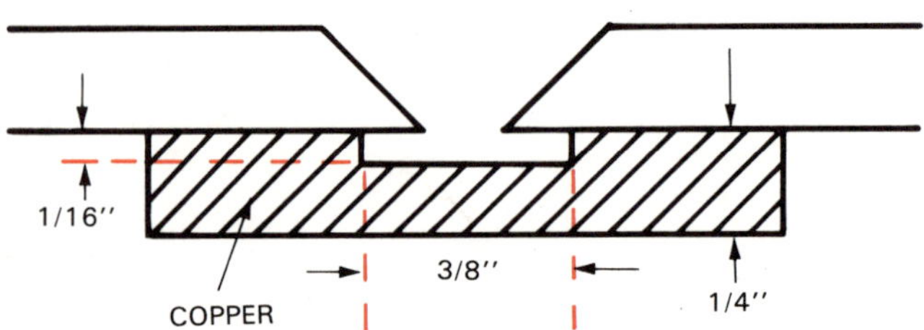


Fig. 13-4. Typical copper and copper alloy weld back-up bar design.

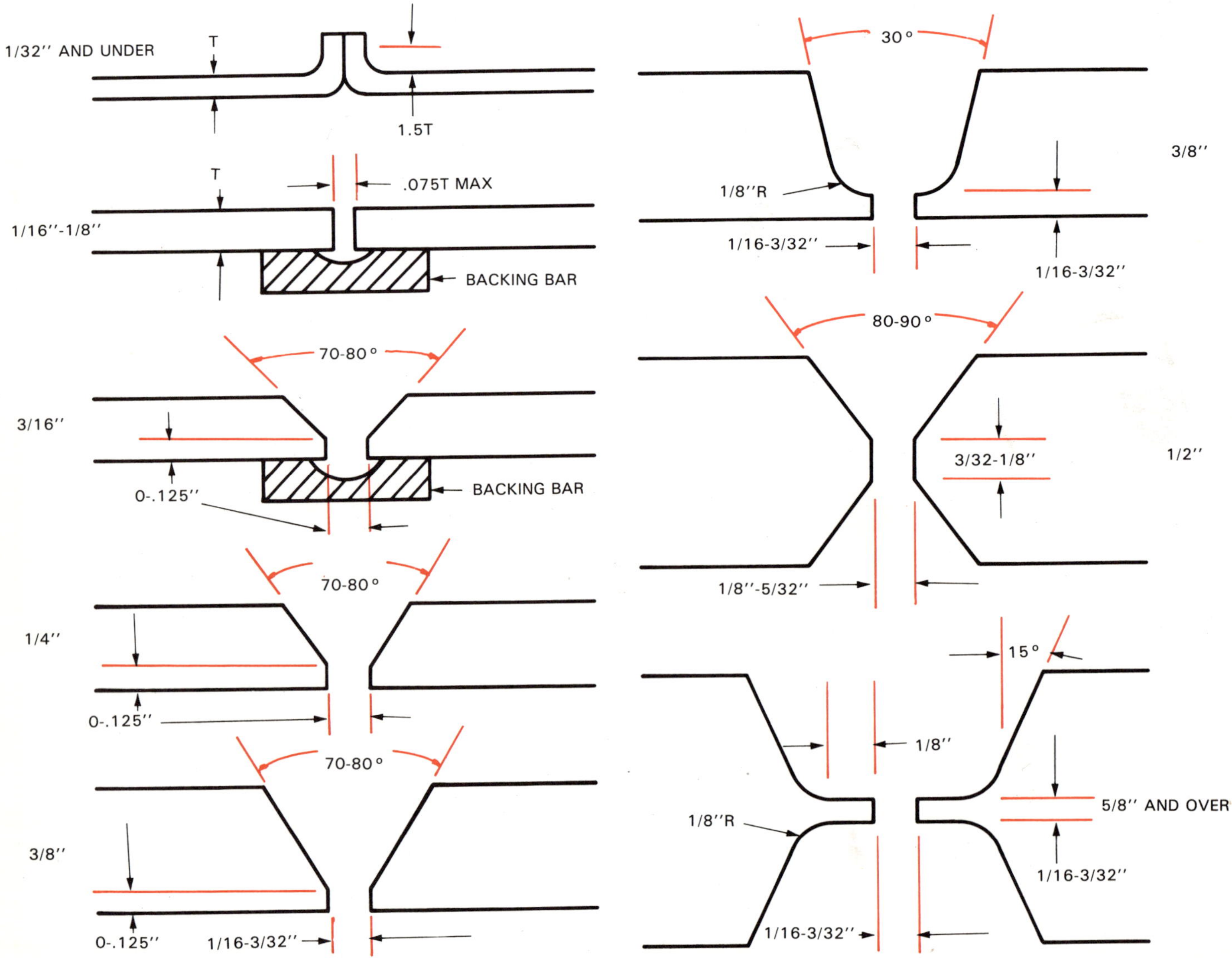

Fig. 13-3. Copper and copper alloy joint designs.

Remember that the heat will rapidly flow away from the joint. This requires close control of the preheat operation. Failure to preheat properly may cause poor fusion in the joint and incomplete penetration. Some common preheat temperatures are listed in Fig. 13-5.

| Thickness | Preheat |
|---|---|
| 1/32″ | No preheat |
| 1/32″-1/16″ | No preheat |
| 1/16″-1/8″ | 100 °F |
| 3/16″ | 100 °F |
| 1/4″ | 200 °F |
| 3/8″ | 450 °F |
| 1/2″ | 650 °F |
| 5/8″ | 750 °F |

Fig. 13-5. Recommended preheats for copper.

### Tackwelds

When tackwelds must be made to maintain joint alignment for welding, the following considerations apply:

1. If the joint welding procedure requires preheating, the joint requires preheating before tackwelding.
2. Use the same filler material used for the main welding procedure.
3. If the joint requires full penetration, the tackweld requires penetration.
4. Keep the tackweld as small as possible. If possible, make the tackweld smaller than the main weld.
5. If the tackweld cracks, leave it and make another close to the broken one.
6. Do not grind out tackwelds. If it must be removed use a rotary file. Grinding grit in a joint makes a good quality weld almost impossible.
7. The crater of a tackweld is the weakest point in the tackweld. Do not leave a crater in a tackweld.

## WELDING COPPER WITH DCEN (DCSP)

Power supply types used in welding copper or copper alloys are:

1. Motor generator or a rectifier.
2. High frequency arc start is required to prevent copper contamination of the electrode.

### Electrode Selection for DCEN

Use a 1 percent or a 2 percent thoriated electrode to weld copper. Taper the electrode to a sharp point. Always use an electrode diameter large enough to carry the required amperage.

### Gases for Welding Copper

Argon gas may be used for thicknesses up to 1/16 inch. For increasing thicknesses, helium must be added for weld puddle control and penetration. Standard gas mixes that may be used are:

1. 75 percent argon – 25 percent helium.
2. 50 percent argon – 50 percent helium.
3. 25 percent argon – 75 percent helium.

Pure helium gas creates severe arc starting problems. Helium gas is usually used in automatic operation only.

### Copper And High Copper Welding Techniques

1. Select joint design from Fig. 13-3.
2. Use RCu filler material. (Note: This filler material is deoxidized copper made specifically for welding. Do not use other types of copper for filler materials.)
3. Preheat as specified.
4. Weld with stringer beads only. Do not oscillate weld puddle. Copper is a hot short material and is prone to cracking. On multi-pass welds, the first pass must be large enough to prevent cracking.

### Copper-Zinc Alloy (Brass) Welding Techniques

1. Select joint design from Fig. 13-3.
2. Use RCuSi filler material. (Note: Copper-zinc GTA welding process. This filler material has a high zinc content and will fume during the melting of the material. This causes porosity in the weld metal.)
3. Preheat is required only to reduce current input. Do not overheat the zinc base alloys. Whitish color formations on the weld puddle indicate the puddle is too hot.
4. Maintain the arc on the filler material. The intense heat of the arc will burn the zinc alloy base material

| COPPER ALLOY (ALUMINUM BRONZE) NUMBER |
|---|
| 612 |
| 613 |
| 614 |
| 618 |
| 619 |
| 620 |
| 622 |
| 623 |
| 624 |
| 625 |
| 626 |
| 628 |
| 630 |

Fig. 13-6. Copper alloy (Aluminum Bronze) numbers which may be GTAW welded.

as explained in step 3.

Weld with stringer beads only. Do not oscillate the weld puddle. Zinc alloys are hot short and prone to cracking. On multi-pass welds, the first pass must be large enough to prevent cracks.

### Copper-Tin Alloy (Phosphor Bronze) Welding Techniques

The copper-tin alloys have a tendency to crack during welding and cooling. For this reason, GTAW is usually limited to the welding of light gauge materials and to minor repair or surfacing.

1. Use RCuSn filler material.
2. Preheat to approximately 400 degrees F.
3. Maintain inter-pass temperature during welding.
4. Use argon shielding gas.
5. Keep the weld puddle as small as possible.
6. Slowly cool to room temperature after welding.

### Copper-Aluminum Alloy (Aluminum Bronze) Welding Techniques

The copper-aluminum alloys shown in Fig. 13-6 may be welded with the GTA welding process.

1. Joint designs up to 1/8 inch thick may be square groove with approximately 1/16 inch root opening. Over 1/8 inch thick joints use a 60 degree to 70 degree V and weld from one side.
2. Alloys containing up to 10 percent aluminum may be preheated to remove residual moisture. Inter-pass temperature should not exceed 300°F. Cool after welding in still air.
3. Use argon shielding gas.
4. A commercial fluoride type flux may be applied to straight butt edges of copper-aluminum alloy sheet material. This will improve the flow and wetting of the molten puddle.
5. Filler materials used on various alloys are shown in Fig. 13-7.

| MATERIAL | FILLER ROD |
|---|---|
| Red Brasses | RCuSnA<br>RCuSnC |
| Yellow Brasses | RCuAL-A2<br>RCuSi |
| Manganese Bronze | RCuAL-A2<br>RCuSi |
| Silicon Bronze | RCuSiA |
| Phosphor Bronze | RCuSnA<br>RCuSnC |
| Aluminum Bronze | RCuAL-A2 |

Fig. 13-7. Common filler materials used to weld the brasses and bronze alloys.

### Copper-Silicon Alloy (Silicon Bronze) Welding Techniques

The silicon bronzes are the easiest to weld of all the copper alloys. They have low thermal conductivity and preheating is not required.

1. Joint design up to 1/8 inch thick may be square groove. Over 1/8 inch thick joints use a 60 degree to 70 degree V and weld from one side.
2. Inter-pass temperature should not exceed 200°F.
3. DCEN or ACHF current may be used.
4. Use argon shielding gas.
5. Filler materials used on the various alloys is RCuSi.
6. During welding, a film of silica will form on the top of the weld. Remove the film between passes by wire brushing with a stainless steel wire brush.
7. To prevent center line cracking of the weld: Allow the heat to fully penetrate material at start of weld. Add sufficient material at start of weld to provide strength at the higher temperature. Slow down the welding speed.

### Copper-Nickel Alloy Welding Techniques

The copper-nickel alloys are very weldable by the gas tungsten arc welding process. Since the thermal conductivities are about the same as steel, no preheat is required.

1. Joint design up to 1/8 inch thick may be square groove. Over 1/8 inch thickness, use a 60 degree to 70 degree V and weld from one side.
2. Argon is used for the shielding gas.
3. Filler material used on the various alloys is RCuNi.
4. Surface contaminates may cause cracking in the heat-affected zone. To prevent this condition, clean the base material thoroughly before welding.

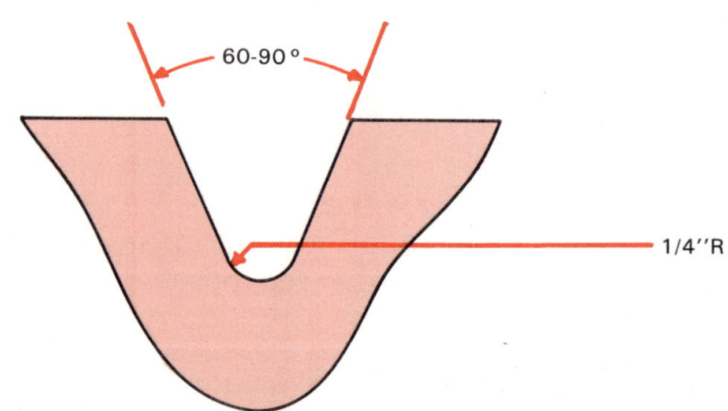


Fig. 13-8. Prepare castings for welding with a generous radius at the bottom of the groove. Taper side walls of the groove to avoid cold shuts at the edge of the weld.

5. Maintain a short arc and weld only with stringer beads.

## WELDING CASTINGS OF COPPER

The welding of castings is required for rebuilding, repairing, surfacing, and joining.

The various alloys and filler materials used are shown in Fig. 13-2.

The area to be welded must be cleaned by grinding, machining, etc., to sound metal. This removes the hard scale on the part of the casting which came into contact with the casting mold. Remove all grease, oil, dirt, and scale adjacent to the weld area to prevent contamination of the weld.

If the area to be welded extends into the casting, the groove must be tapered as shown in Fig. 13-8 to avoid cold shuts or lack of fusion.

### Brasses

1. Preheat to approximately 400°F.
2. Use 75He-25Ar for shielding gas.
3. Use DCEN (DCSP).

### Silicon Bronzes

1. Do not preheat.

| MATERIAL | TEMPERATURE |
|---|---|
| Red Brasses | 750 °F |
| Copper Alloy 443-445 | 750 °F |
| Copper Alloy 464-467 | 850 °F |
| Copper Alloy 614 | 1100 °F |
| Copper Alloy 628 | 1200 °F |
| Copper Alloy 655 | 850 °F |
| Copper Nickel Alloys | 1000 °F |

Heat slowly to temperature.
Hold for one hour minimum.

Fig. 13-9. Stress relief times and temperatures for materials susceptible to stress corrosion cracking.

2. Use argon shielding gas.
3. Inter-pass temperature not over 150°F.
4. Keep weld puddle as small as possible.
5. Mechanically clean between passes.

### Phosphor Bronzes

1. Preheat 300°F to 400°F.
2. Maintain inter-pass temperature 400°F.
3. Keep weld puddle as small as possible.
4. Slow cool after welding.

### Aluminum Bronzes

1. Preheat to approximately 300°F.
2. Inter-pass temperature should not exceed 300°F.
3. Cool in still air after welding.

### Copper-Nickel Alloys

1. Do not preheat.
2. Clean surface very well before welding.
3. Use argon gas.
4. Maintain a short arc.
5. Weld with stringer beads.

## POST WELD TREATMENT OF CASTINGS

High coppers, copper-aluminum alloys and some copper-nickel castings are susceptible to stress-corrosion cracking in some service applications. To prevent this from happening, the part should be heat treated. This operation reduces the stresses within the part. See furnace times and temperatures in Fig. 13-9.

## REVIEW QUESTIONS

1. List five major characteristics of copper and the copper alloys.
2. At what temperature does copper melt?
3. Why is copper so difficult to get hot enough to weld?
4. Copper and copper alloys are assigned identification numbers by the ________ ________
5. Why are copper alloy numbers 314 through 385 almost impossible to weld?
6. What is the major difference between steel and copper weld joint designs?
7. Why are groove openings specified at preheat temperatures rather than at room temperatures?
8. Can copper be cut with the plasma arc cutting process?
9. Sheared edges of copper must be cleaned before welding to remove ________, ________, and ________ which may be imbedded in the cut.
10. What two degreasers may be used to remove this contaminate?
11. Do 100 percent penetration welds require a back-up?
12. When using an oxyacetylene torch for preheating, what type of flame is used?
13. Failure to preheat properly may cause lack of ________ or lack of ________ in the completed weld.
14. What type of tungstens are used when welding with DCEN?
15. What type of filler material is RCu?
16. If you are welding on the copper-zinc alloys and whitish formations appear on the surface, what is wrong?
17. Why do castings require special cleaning procedures before welding?

# Chapter 14

# WELDING PROCEDURE FOR MANUAL WELDING STEEL AND STEEL ALLOYS

## BASE MATERIALS

Many types and grades of steel are included in the basic steel family. These materials are magnetic and melt at approximately 2500°F.

Carbon steels are identified as a group of steels which contain:

1. Carbon — 1.70 percent max.
2. Manganese — 1.65 percent max.
3. Silicon — 0.60 percent max.

Carbon steels which fall within this range may be further classified as:

1. Low carbon steel — up to 0.15 percent carbon.
2. Mild carbon steel — 0.15 percent to 0.29 percent carbon.
3. Medium carbon steel — 0.30 percent to 0.59 percent carbon.
4. High carbon steel — 0.60 percent to 1.70 percent carbon.

Low alloy steels contain varying amounts of carbon and a variety of alloying elements. The elements used may include chromium, molybdenum, nickel, vanadium, and manganese. These elements increase the strength and toughness of the material and in some cases, the corrosion resistance.

The alloys known as the "Quenched And Tempered Steels" are hardened and tempered for specific values prior to use in the field.

The alloys known as the chromium molybdenum steels are used in applications requiring high strength and may be used and welded in the annealed or hardened and tempered condition.

### Tool Steels

The tool steels are a combination of high carbon and alloy content. Tool steels are used for making dies, cutting bits and many other types of tools. Normal manufacturing of the tool uses annealed material. Full hardening and tempering is done prior to the use of the tool for the intended purpose.

The various types of tool steel you may see include:

1. W1, W2.
2. S1.
3. S5.
4. S7.
5. O1.
6. O6.
7. A2.
8. A4.
9. D2.
10. H11, H12, H13.
11. M1, M2, M10.

The numbers for the various tool steels are to identify the various compositions and the letters usually indicate the type of quenching required to achieve full mechanical values. For the welder this information is not that important.

### Forms And Shapes of Steel

Steel is supplied in many forms and shapes for welding. Some of the more common shapes include:

1. Hot rolled steel.
   This material may include plate and structural forms, which after rolling, are allowed to cool in air. After cooling, the oxide film on the surface is light blue in color.
2. Cold rolled steel.
   This material is final rolled to the required dimensions in the cold condition and does not have an oxide film on the surface. A light coating of oil is placed on the material to prevent rusting.
3. Castings.
   Castings are usually made in a sand mold and the surface will be rough. New castings are sand blasted which leaves a dull silver surface.
4. Forgings.
   Forgings may be made from billets, bars, round or square stock. They may be identified by the remnants of the flashing usually extending around the center of the part. The forging operation is done

while the part is hot. Therefore, a scale is formed on the surface similar to hot rolled steel.

### Classification

Steels are manufactured to specifications developed by various organizations including:

1. American Society For Testing Materials. The letter designation is ASTM and a typical material under this specification would read: ASTM A335.
2. Aeronautical Material Specification. The letter designation is AMS.
3. American Iron And Steel Institute. The letter designation is AISI.
4. Corporation Specifications are used where the material does not meet any of the above noted specifications.

## FILLER MATERIALS

Filler materials used to join the carbon, low alloy and tool steels must be selected to produce the desired properties of the weldment after welding and heat treatment if required.

Steels in the "Rimmed" or "Semi-Killed" category are prone to "bubbling" during welding, causing porosity in the weld. For this reason, a highly deoxidized filler wire should be used for all GTAW operations.

Filler materials may be selected from specifications such as:

1. AWS A5.18.
2. AWS A5.17.
3. Mil-E-23765.
4. Mil-S-6758.
5. AMS 6370.
6. QQ-W-405.

Since there are so many factors involved in wire selection (as the first paragraph states) it is impractical to list the base materials with the applicable filler material.

If the chrome-moly steel weldment is to be hardened after welding use either:

1. 1 1/4 percent Cr — 1/2 percent Mo filler material.
2. 2 1/4 percent Cr — 1 percent Mo filler material.
3. 4 to 6 percent Cr — AISI 502 filler material.

If the chrome moly steel weldment is not to be hardened after welding, the following stainless steel filler materials may be used:

1. 25 percent Cr — 20 percent Ni stainless steel filler material.
2. 25 percent Cr — 12 percent Ni stainless steel filler material.

(Do not use stainless steel filler materials for welds in service over 1000 °F. Welds subject to this temperature will have carbon migration from the steel to the stainless steel which may cause weld failure.)

When welding filler materials are required for the tool steels, refer to the manufacturer of the material for the recommended type of wire. These materials have been specially developed for specific applications and are not made to meet specifications.

### Wire Quality

Steel filler materials are supplied by the manufacturer with a:

1. Bright finish.
2. Oiled finish.
3. Copper flash finish.

All of the material is susceptible to rusting. Store the material in a dry, heated area until ready to use. Always clean the welding material with acetone or alcohol just before use. Return all unused material to the storage area when done.

## WELDING PREPARATION

### Joint Preparation

Joint edges prepared by the thermal cutting processes have a heavy oxide film on the surface, as shown in Fig. 14-1. This oxide must be completely removed prior to GTAW to prevent porosity in the weld metal.

Joint edges prepared by shearing should be sharp without tearing or ridges. Where these conditions exist, dirt or oil may become entrapped in the joint and result in faulty welds.

Fig. 14-1. The oxide scale and small gouge indentations were formed on this bevel cut by the oxyacetylene cutting process.

### Pre-weld Cleaning of Steel

The weld joint, the immediate area, and the filler material must be clean if the weld is to have satisfactory properties. The cleaning operation must remove rust and mill scale (oxide) completely from the weld area.

Several methods may be used to clean the joint. These cleaning methods include:

1. Grinding.
2. Chipping.
3. Sanding.

4. Filing.
5. Chemical etching.
6. Sand or grit blasting.

Weld joints may be sanded with a sanding wheel, Fig. 14-2, to remove oxide scale. After removal of the scale, the area should be brushed to remove oxide particles and residue from the sanding wheel. After grit blasting, the surface must be wire brushed to remove any sand particles.

Immediately prior to welding, the joint area should be washed with alcohol or acetone. This should be followed by a short drying period or a blast of clean, dry air to evaporate any residual liquid. Do not attempt to weld on a wet surface.

All tooling in the weld area must also be cleaned prior to welding. Back-up bars, chill bars, clamps, etc. retain moisture. When this tooling is heated by the welding operation it may release hydrogen. This part of the operation is just as important as the cleaning of the material and filler wire.

Fig. 14-2. The edge has been sanded to "bright metal" to prepare the part for welding. This type of sander is often called a "PG" wheel.

### Weld Back-up for Steel

Groove welds welded from one side of the joint where 100 percent penetration is required may be backed with argon gas or a solid bar. Solid bars may be copper or stainless steel.

Excluding air from the backside or penetration side of the joint will assist in the "wetting" of the penetration. A WETTED penetration will have a smooth junction and even flow of the penetration onto the parent metal. The exclusion of air will also prevent the formation of scale and oxide formation.

### Preheating Steels

Carbon steels with less than .30 percent carbon and less than 1 inch thick generally do not require preheat. An exception to this allowance is welding on highly restrained joints. These joints should be preheated to 50 to 100°F to minimize shrinkage cracks in the base metal and the weld deposit.

Low alloy steels such as the chromium-molybdenum steels will have hard heat-affected zones after welding if the preheat temperature is too low. This is caused by the rapid cooling rate of the base material and the formation of martensitic grain structures. A 200°F to 400°F preheat temperature will slow down the cooling rate and prevent the formation of a martensitic structure.

MARTENSITE is a metallurgical term that defines a type of grain structure obtained by heating and quenching. A martensitic material may be hardened by heating and quenching. Parts that are welded are in effect heated and quenched. If the carbon content of the material is sufficient (and the preheat temperature is too low) the material will harden during welding. This results in high tensile properties with very low ductility. The formation of hard "martensite" type grains increases the possibility of cracking as the weld metal cools and shrinks.

Tool steels have a very high carbon content and are prone to cracking in the heat-affected zone without sufficient preheat. Fig. 14-3 list some recommended preheats for welding tool steels.

Quenched and tempered steel require preheat and interpass temperature control to retain the original mechanical properties. The manufacturer's recommendations for these temperatures must be strictly followed.

## WELDING STEELS USING DCEN (DCSP)

Power supply type used in welding steels and steel alloys include the following two considerations:

1. Motor generator or a rectifier.
2. High frequency arc start (desirable but not mandatory).

### Electrodes Used on Steel

A 1 percent or a 2 percent thoriated electrode is recommended. Taper to a sharp point. Always use an electrode large enough to carry the required amperage.

### Gases for GTAW on Steels

Argon gas may be used for thicknesses up to 1/8 inch. For increasing thicknesses, helium may be added for puddle control and penetration. Standard gas mixes that may be used are:

1. 75 percent argon — 25 percent helium.
2. 50 percent argon — 50 percent helium.

| MATERIAL TYPE | ANNEALED BASE MATERIAL PREHEAT AND POSTHEAT TEMPERATURE | HARDENED BASE MATERIAL PREHEAT AND POSTHEAT TEMPERATURE |
|---|---|---|
| **W1, W2** | 250-450 °F | 250-450 °F |
| **S1** | 300-500 °F | 300-500 °F |
| **S5** | 300-500 °F | 300-500 °F |
| **S7** | 300-500 °F | 300-500 °F |
| **01** | 300-400 °F | 300-400 °F |
| **06** | 300-400 °F | 300-400 °F |
| **A2** | 300-500 °F | 300-400 °F |
| **A4** | 300-500 °F | 300-400 °F |
| **D2** | 700-900 °F | 700-900 °F |
| **H11, H12, H13** | 900-1200 °F | 700-1000 °F |
| **M1, M2, M10** | 950-1100 °F | 950-1050 °F |

Fig. 14-3. Preheat part to be welded thoroughly to required temperature. Do not allow the part to cool down below minimum until postheat is complete.

3. 25 percent argon — 75 percent helium.

Pure helium gas may be used when automatic welding steel or steel alloys. Gas purity must be sufficient to prevent hydrogen pick-up. Hydrogen will cause porosity and cracking.

### Techniques for Welding Steels and Steel Alloys

1. Clean the oxide scale from the weld joint and immediately weld area to "bright metal."
2. Select a filler wire for the desired properties required. When welding on porosity prone material, always use a filler wire that contains deoxidizers to prevent porosity.
3. Preheat low alloy steels and tool steels to prevent cracking in the weld and heat-affected zones.
4. Weld the porosity prone materials by maintaining the arc on the molten metal. Do not hold the arc on the parent metal. On crack sensitive materials, do not make concave welds. These welds are prone to cracking through the centerline.
5. On multi-pass welds, always remove any scale on the surface of each pass to reduce the possibility of oxide entrapment.
6. Always fill craters on crack sensitive materials.
7. Welds made on the tool steels should be very short or small stringer beads, as shown in Fig. 14-4, to reduce the expansion of the base metal and prevent cracking. Immediately after welding, the weld metal may be peened to reduce the shrinkage of the weld metal.

## POST WELD TREATMENT

Post weld treatment will depend on the type of material welded, joint restraint, and the desired mechanical values.

A stress relieving operation may be done on the carbon steels at approximately 1150°F to remove residual

Fig. 14-4. This tool steel cutter blade was repaired by welding a small overlay with a very hard material on the cutting edge. Blade can now be ground to original dimension. This operation eliminates need to scrap entire cutting blade.

stresses caused by weld shrinkage. This may be done by local or furnace heating.

The weld area should be heated to 1150°F and this temperature maintained for one hour per inch of material thickness. The part may then be air cooled.

The low alloy steels may be stress relieved in the same manner. However, the cooling period should be lengthened by covering the part with heat resistant materials.

## REVIEW QUESTIONS

1. How are carbon steels identified by element percentages?
2. List the alloying elements used to make low alloy steels.
3. In what condition can the chrome-moly steels be welded?
4. What is the normal color of hot rolled sheet or plate?
5. Castings which have been sandblasted after removal from the sand mold, have a ________ appearance.
6. Forgings are made from ________, ________, ________ or ________ stock.
7. List three specifications used for the manufacture of steel.
8. Steels that are of the rimmed or semi-killed type require the use of a ________ welding wire to prevent porosity during welding.
9. What type of limitations exist for welding chrome-moly steel with stainless steel filler material?
10. Steel filler materials are supplied to the user with one of three finishes. What are they?
11. What two degreasers may be used to clean filler material before use?
12. Why must the oxide scale be removed from the edges of thermally cut joints before use?
13. Why must a grit blasted surface be wire brushed before welding?
14. Why must tooling, back-up bars, chill bars, etc., be cleaned prior to welding?
15. What two types of back-up are used when welding 100 percent penetration steel welds?
16. Why are back-ups used for 100 percent penetration steel welds?
17. When are carbon steels with less than .30 percent carbon and less than 1 inch thick preheated?
18. What causes low alloy steels to have hard heat affected zones after welding?
19. Tool steels should be welded with very short ________ type weld beads to reduce the amount of heat input and expansion of the base material.
20. Post weld stress relief of carbon steel weldments may be done at approximately ________°F.

# Chapter 15

# GTAW PROCEDURE FOR STAINLESS STEEL

## BASE MATERIALS

The stainless steel family consists of several groups of materials. They include:

1. CHROMIUM-NICKEL MANGANESE STAINLESS STEEL, types 201 and 202 grades. These grades are austenitic, non-magnetic, and non-hardenable.
2. CHROMIUM-NICKEL STAINLESS STEEL, types 301, 302, 303, 304, 305, 308, 309, 310, 312, 314, 316, 317, 321, 347, 349 grades. These grades are austenitic, non-magnetic, and non-hardenable.
3. CHROMIUM STAINLESS STEEL types 403, 410, 414, 416, 420, 431, 440, 501, 502 grades. These grades are martensitic, magnetic, and hardenable.
4. CHROMIUM STAINLESS STEEL, types 405, 430, 446 grades. These grades are ferritic, magnetic, and non-hardenable.
5. PRECIPITATION HARDENABLE STAINLESS STEEL, types 15-5 PH, PH 15-7 MO, 17-4 PH, 17-7 PH, AM 350, AM 355, A 286. (These stainless steels are not identified by the American Iron and Steel Institute code system.)

COMMERCIALLY WROUGHT STAINLESS STEEL IDENTIFICATION (AISI)

| | COMPOSITION, PERCENT[a] | | | | | | | |
|---|---|---|---|---|---|---|---|---|
| Type | C | Mn | Si | Cr | Ni | P | S | Others |
| 201 | 0.15 | 5.5-7.5 | 1.00 | 16.0-18.0 | 3.5-5.5 | 0.06 | 0.03 | 0.25 N |
| 202 | 0.15 | 7.5-10.0 | 1.00 | 17.0-19.0 | 4.0-6.0 | 0.06 | 0.03 | 0.25 N |
| 301 | 0.15 | 2.00 | 1.00 | 16.0-18.0 | 6.0-8.0 | 0.045 | 0.03 | |
| 302 | 0.15 | 2.00 | 1.00 | 17.0-19.0 | 8.0-10.0 | 0.045 | 0.03 | |
| 302B | 0.15 | 2.00 | 2.0-3.0 | 17.0-19.0 | 8.0-10.0 | 0.045 | 0.03 | |
| 303 | 0.15 | 2.00 | 1.00 | 17.0-19.0 | 8.0-10.0 | 0.20 | 0.15 min | 0-0.6 Mo |
| 303Se | 0.15 | 2.00 | 1.00 | 17.0-19.0 | 8.0-10.0 | 0.20 | 0.06 | 0.15 Se min |
| 304 | 0.08 | 2.00 | 1.00 | 18.0-20.0 | 8.0-10.5 | 0.045 | 0.03 | |
| 304L | 0.03 | 2.00 | 1.00 | 18.0-20.0 | 8.0-12.0 | 0.045 | 0.03 | |
| 305 | 0.12 | 2.00 | 1.00 | 17.0-19.0 | 10.5-13.0 | 0.045 | 0.03 | |
| 308 | 0.08 | 2.00 | 1.00 | 19.0-21.0 | 10.0-12.0 | 0.045 | 0.03 | |
| 309 | 0.20 | 2.00 | 1.00 | 22.0-24.0 | 12.0-15.0 | 0.045 | 0.03 | |
| 309S | 0.08 | 2.00 | 1.00 | 22.0-24.0 | 12.0-15.0 | 0.045 | 0.03 | |
| 310 | 0.25 | 2.00 | 1.50 | 24.09-26.0 | 19.0-22.0 | 0.045 | 0.03 | |
| 310S | 0.08 | 2.00 | 1.50 | 24.0-26.0 | 19.0-22.0 | 0.045 | 0.03 | |
| 314 | 0.25 | 2.00 | 1.5-3.0 | 23.0-26.0 | 19.0-22.0 | 0.045 | 0.03 | |
| 316 | 0.08 | 2.00 | 1.00 | 16.0-18.0 | 10.0-14.0 | 0.045 | 0.03 | 2.0-3.0 Mo |
| 316L | 0.03 | 2.00 | 1.00 | 16.0-18.0 | 10.0-14.0 | 0.045 | 0.03 | 2.0-3.0 Mo |
| 317 | 0.08 | 2.00 | 1.00 | 18.0-20.0 | 11.0-15.0 | 0.045 | 0.03 | 3.0-4.0 Mo |
| 317L | 0.03 | 2.00 | 1.00 | 18.0-20.0 | 11.0-15.0 | 0.045 | 0.03 | 3.0-4.0 Mo |
| 321 | 0.08 | 2.00 | 1.00 | 17.0-19.0 | 9.0-12.0 | 0.045 | 0.03 | 5 × %C Ti min |
| 329 | 0.10 | 2.00 | 1.00 | 25.0-30.0 | 3.0-6.0 | 0.045 | 0.03 | 1.0-2.0 Mo |
| 330 | 0.08 | 2.00 | 0.75-1.5 | 17.0-20.0 | 34.0-37.0 | 0.04 | 0.03 | |
| 347 | 0.08 | 2.00 | 1.00 | 17.0-19.0 | 9.0-13.0 | 0.045 | 0.03 | C |
| 348 | 0.08 | 2.00 | 1.00 | 17.0-19.0 | 9.0-13.0 | 0.045 | 0.03 | 0.2 Cu[b, c] |
| 384 | 0.08 | 2.00 | 1.00 | 15.0-17.0 | 17.0-19.0 | 0.045 | 0.03 | |

a. Single values are maximum unless indicated otherwise.
b. (Cb + Ta) min — 10 × %C.
c. Ta — 0.10% max.

Fig. 15-1. Common stainless steel compositions.

## Wrought Material Identification

Commercially wrought stainless steels have been identified by the American Iron and Steel Institute (AISI). These are referred to as type numbers. Fig. 15-1 lists the type numbers and chemical analysis of commonly used stainless steels.

## Castings

Commercial stainless steel castings have been identified by the Alloy Casting Institute (ACI). These are referred to as type numbers. Fig. 15-2 lists the type numbers and chemical analysis of commonly used stainless steels.

| Alloy designation | Similar wrought type[b] | COMPOSITION, PERCENT[a] C | Si | Cr | Ni | Mo[c] | Other |
|---|---|---|---|---|---|---|---|
| **CE-30** | 312 | 0.30 | 2.00 | 26-30 | 8-11 | — | — |
| **CF-3** | 304L | 0.03 | 2.00 | 17-21 | 8-12 | — | — |
| **CF-3M** | 316L | 0.03 | 1.50 | 17-21 | 9-13 | 2.0-3.0 | — |
| **CF-8** | 304 | 0.08 | 2.00 | 18-21 | 8-11 | — | — |
| **CF-8C** | 347 | 0.08 | 2.00 | 18-21 | 9-12 | — | d |
| **CF-8M** | 316 | 0.08 | 1.50 | 18-21 | 9-12 | 2.0-3.0 | — |
| **CF-12M** | 316 | 0.12 | 1.50 | 18-21 | 9-12 | 2.0-3.0 | — |
| **CF-16F** | 303 | 0.16 | 2.00 | 18-21 | 9-12 | 1.5 | 0.20-0.35 Se |
| **CF-20** | 302 | 0.20 | 2.00 | 18-21 | 8-11 | — | — |
| **CG-8M** | 317 | 0.08 | 1.50 | 18-21 | 9-13 | 3.0-4.0 | — |
| **CH-20** | 309 | 0.20 | 2.00 | 22-26 | 12-15 | — | — |
| **CK-20** | 310 | 0.20 | 2.00 | 23-27 | 19-22 | — | — |
| **CN-7M** | — | 0.07 | 1.50 | 18-22 | 27.5-30.5 | 2.0-3.0 | 3-4 Cu |
| **HE** | — | 0.2-0.5 | 2.0 | 26-30 | 8-11 | — | — |
| **HF** | 304 | 0.2-0.4 | 2.0 | 19-23 | 9-12 | — | — |
| **HH** | 309 | 0.2-0.5 | 2.0 | 24-28 | 11-14 | — | 0.2 N |
| **HI** | — | 0.2-0.5 | 2.0 | 26-30 | 14-18 | — | — |
| **HK** | 310 | 0.2-0.6 | 2.0 | 24-28 | 18-22 | — | — |
| **HL** | — | 0.2-0.6 | 2.0 | 28-32 | 18-22 | — | — |
| **HN** | — | 0.2-0.5 | 2.0 | 19-23 | 23-27 | — | — |
| **HP** | — | 0.35-0.75 | 2.0 | 24-28 | 33-37 | — | — |
| **HT** | 330 | 0.35-0.75 | 2.5 | 15-19 | 33-37 | — | — |
| **HU** | — | 0.35-0.75 | 2.5 | 17-21 | 37-41 | — | — |

a. Single values are maximum.
Manganese — 1.50% max in CX-XX types; 2.0% max in HX types
Phosphorous — 0.04% max except for Cf-16F — 0.17% max
Sulfur — 0.04% max
b. Compositions are not exactly the same.
c. Molybdenum in HX types is 0.5% max.
d. Cb — 8 × %C (1.0% max), or Cb + Ta — 9 × %C (1.1% max)

Fig. 15-2. Casting stainless steel compositions.

| Stainless Steel | Typical Compositions, Element, Percent: C | N | Cr | Mo | Ni | Strength, lb/in.² (MPa): Tensile | Yield | Elongation, % |
|---|---|---|---|---|---|---|---|---|
| E-Brite 26-1 | 0.002 | 0.010 | 26 | 1 | — | 70,000 (480) | 50,000 (345) | 30 |
| 29-4 | 0.005 | 0.013 | 29 | 4 | — | 90,000 (620) | 75,000 (520) | 25 |
| 29-4-2 | 0.005 | 0.013 | 29 | 4 | 2 | 95,000 (650) | 85,000 (590) | 22 |
| **Carpenter** 20 Cb-3 | 0.04 | — | 20 | 2.5 | 36 | | | |

Fig. 15-3. Special stainless steel compositions.

### Special Ferritic Stainless Steels

These alloys have been developed by various companies and are specified by tradenames and/or chemical contents or composition. They include:

1. Carpenter 20 Cb-3.
2. E-Brite (26-1).
3. 29-4-2.
4. 29-4.

Fig. 15-3 lists the identification numbers and the chemical analysis for each alloy.

### Alloy Effects

The use of the various alloy elements and their effect on the base material properties is shown in Fig. 15-4.

### Austenitic Stainless Steel Properties

The austenitic stainless steel family (AISI 200 and 300 series) consists of a 18 percent chromium steel to which is added at least 8 percent nickel.

The basic grain structure is austenitic at all temperatures. Therefore, the materials are not hardenable by heat treatment. Some hardening can be obtained by cold working. This cold working does not change the basic austenitic grain structure.

The high chromium percentage in this austenitic stainless steel series imparts a good corrosion resistance to heat and acids.

Nickel is used to obtain ductility of the material from room temperatures into the cryogenic temperatures (below −250 degrees F).

The carbon content is held to a low percentage to reduce the problem of carbide precipitation and intergranular corrosion. Grades 304L and 316L, as denoted by the letter L, have a lower percentage of carbon than the standard grades. Grades 321 and 347 are stabilized with columbium and tantalum or titanium to prevent carbide precipitation.

| Element | Types of steels | Effects |
|---|---|---|
| Carbon | All types | Strongly promotes the formation of austenite. Can form a carbide with chromium that can lead to intergranular corrosion. |
| Chromium | All types | Promotes formation of ferrite. Increases resistance to oxidation and corrosion. |
| Nickel | All types | Promotes formation of austenite. Increases high temperature strength, corrosion resistance, and ductility. |
| Nitrogen | XXXN | Very strong austenite former. Like carbon, nitrogen is thirty times as effective as nickel in forming austenite. Increases strength. |
| Columbium | 347 | Primarily added to combine with carbon to reduce susceptibility to intergranular corrosion. Acts as a grain refiner. Promotes the formation of ferrite. Improves creep strength. |
| Manganese | 2XX | Promotes the stability of austenite at or near room temperature but forms ferrite at high temperatures. Inhibits hot shortness by forming MnS. |
| Molybdenum | 316, 317 | Improves strength at high temperatures. Improves corrosion resistance to reducing media. Promotes the formation of ferrite. |
| Phosphorous, selenium, or sulfur | 303, 303Se | Increases machinability, but promotes hot cracking during welding. Lowers corrosion resistance slightly. |
| Silicon | 302B | Increases resistance to scaling and promotes the formation of ferrite. Small amounts are added to all grades for deoxidizing purposes. |
| Titanium | 321 | Primarily added to combine with carbon to reduce susceptibility to intergranular corrosion. Acts as a grain refiner. Promotes the formation of ferrite. |
| Copper | CN-7M | Generally added to stainless steels to increase corrosion resistance to certain environments. Decreases susceptibility to stress-corrosion cracking and provides age hardening effects. |

Fig. 15-4. Effects of the various elements in the stainless steels.

### Martensitic Stainless Steel Properties

The martensitic stainless steels consist of a low-carbon steel to which is added 11.5 to 18 percent chromium. The lower grades of the family with approximately 12 percent chromium are used where moderate corrosion resistance is required up to approximately 1100 degrees F. The carbon content of the various grades will vary depending on the hardness requirement for the finished part. Lower carbon grades, such as 410 series, are weldable with certain precautions. Higher carbon grades with high chromium are not considered weldable materials.

With the combination of carbon and chromium the material responds to heating and air cooling for hardening.

### Precipitation Hardening Stainless Steel Properties

The precipitation hardening steels are a group of metals with chromium and nickel with alloying elements such as copper, titanium, columbium, and aluminum.

The chromium and nickel combination give the metal good corrosion and oxidation resistance. The other elements are added to promote the hardening of the material by precipitation throughout the structure during the heat treating process.

Depending on the design of the weldment, fabrication procedures, and final use of the part, various types of thermal operations may be required to achieve the desired results. All of the materials may be welded in all conditions except the A-286 type. Hot cracking in the heat-affected zone makes this A-286 type material very difficult to weld.

### Ferritic Stainless Steel Properties

This group of stainless steels contains iron-chromium-carbon and additional elements such as aluminum, columbium, molybdenum, and titanium. The added elements prevent transformation of the grain structure during heating. Therefore, they remain ferritic and non-hardenable.

The higher chromium and carbon grades are susceptible to carbide precipitation and inter-granular corrosion problems as are the austenitic stainless steels. Since these materials cannot be obtained with low carbon, they must be annealed after welding to redissolve the carbides and restore corrosion resistance.

### Special Ferritic Stainless Steel Properties

These alloys are considered ULTRA-HIGH-PURITY FERRITIC STEELS with a very low carbon and nitrogen content. They have good mechanical properties, corrosion resistance particularly to stress corrosion, cracking, and crevice corrosion due to chlorides. The lower carbon and nitrogen content reduces the possibility of carbide precipitation and inter-granular corrosion without undergoing a thermal treatment after welding.

### Carbide Precipitation of Stainless Steel

During the welding of the unstabilized austenitic stainless steels, the weld and immediate area rise into and above the 800 degree F. to 1600 degree F. range. Within this range, chromium and carbon combine and precipitate into the grain boundries as chromium-rich carbides. When this occurs, the areas next to the grain boundries have insufficient chromium to produce a protective film. This area is now sentitive to corrosion by certain materials. The sequence of events that occur are:

1. SENSITIZATION. The area is subjected to 800 degrees to 1600 degrees F. or above.
2. CARBIDE PRECIPITATION. Chromium moves out of the grains and into the grain boundry to combine with carbon that has precipitated from the solid solution to form chromium-rich carbides.
3. INTER-GRANULAR CORROSION. The corrosion takes place when the media corrodes the grains that have insufficient chromium.

### Cracking of Stainless Steel

Cracking of austenitic stainless steel welds is usually found in joints having high restraint. Cracking is caused by the lack of ductility of the weld metal during the freezing of the molten metal. This may be a problem of insufficient ferrite (crystalline form of iron) or an improper welding procedure or both.

### Distortion of Stainless Steel

The austenitic stainless steels have a very low thermal conductivity. Therefore they retain heat from welding very well. The heat remains in the weld area instead of being dispersed throughout the material. As a result, they tend to distort more than the carbon steels. For this reason, weld joints and tooling should be designed to remove welding heat from the completed weld as soon as possible. Stringer beads are favored over wash beads to reduce the amount of heat required to make the weld.

## FILLER MATERIALS FOR WELDING STAINLESS STEEL

Stainless steel filler materials are selected for use depending on the chemical analysis of the parent material. In most cases, the electrode matches or exceeds the chemical composition of the parent alloy.

The American Welding Society specification, AWS A5.9, Specification for Corrosion Resisting Chromium and Chromium Nickel Steel Bare Welding Rods lists the chemical requirements for the austenitic, martensitic and ferritic welding materials as shown in Fig. 15-5.

| AWS Classification | COMPOSITION, PERCENT[b] | | | | | | | | | | |
|---|---|---|---|---|---|---|---|---|---|---|---|
| | C | Cr | Ni | Mo | Cb plus Ta | Mn | Si | P | S | N | Cu |
| ER209[c] | 0.05 | 20.5-24.0 | 9.5-12.0 | 1.5-3.0 | — | 4.0-7.0 | 0.90 | 0.03 | 0.03 | 0.10-0.30 | 0.75 |
| ER218 | 0.10 | 16.0-18.0 | 8.0-9.0 | 0.75 | — | 7.0-9.0 | 3.5-4.5 | 0.03 | 0.03 | 0.08-0.18 | 0.75 |
| ER219 | 0.05 | 19.0-21.5 | 5.5-7.0 | 0.75 | — | 8.0-10.0 | 1.00 | 0.03 | 0.03 | 0.10-0.30 | 0.75 |
| ER240 | 0.05 | 17.0-19.0 | 4.0-6.0 | 0.75 | — | 10.5-13.5 | 1.00 | 0.03 | 0.03 | 0.10-0.20 | 0.75 |
| ER307 | 0.04-0.14 | 19.5-22.0 | 8.0-10.7 | 0.5-1.5 | — | 3.3-4.75 | 0.30-0.65 | 0.03 | 0.03 | — | 0.75 |
| ER308[d] | 0.08 | 19.5-22.0 | 9.0-11.0 | 0.75 | — | 1.0-2.5 | 0.30-0.65 | 0.03 | 0.03 | — | 0.75 |
| ER308H | 0.04-0.08 | 19.5-22.0 | 9.0-11.0 | 0.75 | — | 1.0-2.5 | 0.30-0.65 | 0.03 | 0.03 | — | 0.75 |
| ER308L[d] | 0.03 | 19.5-22.0 | 9.0-11.0 | 0.75 | — | 1.0-2.5 | 0.30-0.65 | 0.03 | 0.03 | — | 0.75 |
| ER308Mo | 0.08 | 18.0-21.0 | 9.0-12.0 | 2.0-3.0 | — | 1.0-2.5 | 0.30-0.65 | 0.03 | 0.03 | — | 0.75 |
| ER308MoL | 0.04 | 18.0-21.0 | 9.0-12.0 | 2.0-3.0 | — | 1.0-2.5 | 0.30-0.65 | 0.03 | 0.03 | — | 0.75 |
| ER309[c] | 0.12 | 23.0-25.0 | 12.0-14.0 | 0.75 | — | 1.0-2.5 | 0.30-0.65 | 0.03 | 0.03 | — | 0.75 |
| ER309L | 0.03 | 23.0-25.0 | 12.0-14.0 | 0.75 | — | 1.0-2.5 | 0.30-0.65 | 0.03 | 0.03 | — | 0.75 |
| ER310 | 0.08-0.15 | 25.0-28.0 | 20.0-22.5 | 0.75 | — | 1.0-2.5 | 0.30-0.65 | 0.03 | 0.03 | — | 0.75 |
| ER312 | 0.15 | 28.0-32.0 | 8.0-10.5 | 0.75 | — | 1.0-2.5 | 0.30-0.65 | 0.03 | 0.03 | — | 0.75 |
| ER316[d] | 0.08 | 18.0-20.0 | 11.0-14.0 | 2.0-3.0 | — | 1.0-2.5 | 0.30-0.65 | 0.03 | 0.03 | — | 0.75 |
| ER316H | 0.04-0.08 | 18.0-20.0 | 11.0-14.0 | 2.0-3.0 | — | 1.0-2.5 | 0.30-0.65 | 0.03 | 0.03 | — | 0.75 |
| ER316L[d] | 0.03 | 18.0-20.0 | 11.0-14.0 | 2.0-3.0 | — | 1.0-2.5 | 0.30-0.65 | 0.03 | 0.03 | — | 0.75 |
| ER317 | 0.08 | 18.5-20.5 | 13.0-15.0 | 3.0-4.0 | — | 1.0-2.5 | 0.30-0.65 | 0.03 | 0.03 | — | 0.75 |
| ER317L | 0.03 | 18.5-20.5 | 13.0-15.0 | 3.0-4.0 | — | 1.0-2.5 | 0.30-0.65 | 0.03 | 0.03 | — | 0.75 |
| ER318 | 0.08 | 18.0-20.0 | 11.0-14.0 | 2.0-3.0 | 8 × C min to 1.0 max | 1.0-2.5 | 0.30-0.65 | 0.03 | 0.03 | — | 0.75 |
| ER320 | 0.07 | 19.0-21.0 | 32.0-36.0 | 2.0-3.0 | 8 × C min to 1.0 max | 2.5 | 0.60 | 0.03 | 0.03 | — | 3.0-4.0 |
| ER320LR[c] | 0.025 | 19.0-21.0 | 32.0-36.0 | 2.0-3.0 | 8 × C min to 0.40 max | 1.5-2.0 | 0.15 | 0.015 | 0.020 | — | 3.0-4.0 |
| ER321[f] | 0.08 | 18.5-20.5 | 9.0-10.5 | 0.75 | — | 1.0-2.5 | 0.30-0.65 | 0.03 | 0.03 | — | 0.75 |
| ER330 | 0.18-0.25 | 15.0-17.0 | 34.0-37.0 | 0.75 | — | 1.0-2.5 | 0.30-0.65 | 0.03 | 0.03 | — | 0.75 |
| ER347[d] | 0.08 | 19.0-21.5 | 9.0-11.0 | 0.75 | 10 × C min to 1.0 max | 1.0-2.5 | 0.30-0.65 | 0.03 | 0.03 | — | 0.75 |
| ER349[g] | 0.07-0.13 | 19.0-21.5 | 8.0-9.5 | 0.35-0.65 | 1.0-1.4 | 1.0-2.5 | 0.30-0.65 | 0.03 | 0.03 | — | 0.75 |
| ER410 | 0.12 | 11.5-13.5 | 0.6 | 0.75 | — | 0.6 | 0.50 | 0.03 | 0.03 | — | 0.75 |
| ER410NiMo | 0.06 | 11.0-12.5 | 4.0-5.0 | 0.4-0.7 | — | 0.6 | 0.50 | 0.03 | 0.03 | — | 0.75 |
| ER420 | 0.25-0.40 | 12.0-14.0 | 0.6 | 0.75 | — | 0.6 | 0.50 | 0.03 | 0.03 | — | 0.75 |
| ER430 | 0.10 | 15.5-17.0 | 0.6 | 0.75 | — | 0.6 | 0.50 | 0.03 | 0.03 | — | 0.75 |
| ER630 | 0.05 | 16.0-16.75 | 4.5-5.0 | 0.75 | 0.15-0.30 | 0.25-0.75 | 0.75 | 0.04 | 0.03 | — | 3.25-4.00 |
| ER26-1 | 0.01 | 25.0-27.5 | h | 0.75-1.50 | — | 0.40 | 0.40 | 0.02 | 0.02 | 0.015 | 0.20[h] |
| ER16-8-2 | 0.10 | 14.5-16.5 | 7.5-9.5 | 1.0-2.0 | — | 1.0-2.5 | 0.30-0.65 | 0.03 | 0.03 | — | 0.75 |

a. Refer to AWS A5.9 — Specification for Corrosion Resisting Chromium and Chromium-Nickel Bare and Composite Metal Cored and Stranded Welding Electrodes and Rods.
b. Single values shown are maximum percentages except where otherwise specified.
c. Vanadium — 0.10-0.30 percent
d. These grades are available in high silicon classifications that have the same chemical composition requirements as tabulated here with the exception that the silicon content is 0.65 to 1.00 percent. These high silicon classifications are designated by the addition 'Si' to the standard classification designations in the table. The fabricator should consider carefully the use of high silicon filler metals in highly restrained fully austenitic welds. A discussion of the problem is presented in the Appendix to the specification.
e. Carbon shall be reported to the nearest 0.01 percent except for the classification E320LR for which carbon shall be reported to the nearest 0.005 percent.
f. Titanium — 9 × C min. to 1.0 max
g. Titanium — 0.10 to 0.30 percent; tungsten — 1.25 to 1.75 percent
h. Nickel, max — 0.5 minus the copper content, percent

Fig. 15-5. Austenitic, martensitic and ferritic stainless steel filler material compositions.

The filler materials shown in Fig. 15-6 list the recommended filler materials for welding the precipitation hardening materials to themselves and to other precipitation hardening materials.

The filler materials shown in Fig. 15-7 list the various filler materials which may be used to weld the common austenitic and martensitic stainless steels.

Several types of filler materials are now made with

| Material Type | 1st Choice | 2nd Choice |
|---|---|---|
| **17-4 PH or 15-5 PH** | AMS 5826 or 17-4 PH or ER 308 | ER 309 or ER 309 CB |
| **Stainless W** | AMS 5805C or A-286 or ER NiMo-3 | ER NiMo-3 or ER 309 |
| **17-7 PH** | AMS 5824A or 17-7 PH | ER 310 or ERNiCr-3 |
| **PH 15-7 Mo** | AMS 5812C (15-7 Mo) | ER 309 or ER 310 |
| **AM 350** | AMS 5774B (AM 350) | ER 308 or ER310 |
| **AM 355** | AMS 5780A (AM 355) | ER 308 or ER 309 |
| **A-286** | ER NiMo-3 | ER 309 or ER 310 |

Fig. 15-6. Precipitation hardening stainless steels may be welded together using the material listed in column 1. When welded to a dissimilar precipitation hardening material, use the filler material listed in column 2.

used as a substitute in all areas without difficulty.

Maximum weld quality can only be obtained if the filler material is clean and of high quality. Store wire in plastic containers and use only in a clean area with clean gloves. Maintain identification of each individual wire as mixed stainless wires are impossible to identify.

The flag-tag method of identification as shown in Fig. 6-11, is a safeguard that may be used to prevent mixing of welding wires or rods.

### Pre-Weld Cleaning of Stainless Steel

The weld area, just as the filler material, must be clean if the weld is to have satisfactory properties.

Oil and grease must be removed from all types of stainless steels prior to welding. Use acetone, alcohol, or any commercial degreaser.

The martensitic, ferritic and some precipitation hardening stainless materials have a hard oxide film on the surface which must be removed prior to welding. Remove this oxide film with rotary sanders to "bright metal" prior to weld, as shown in Fig. 15-8.

Heavy oxidation may be removed by grinding, sand or grit blasting. However, after blasting, rotary sand to "bright metal." Wire brushing will not sufficiently remove sand or grit particles.

All tooling in the weld area must also be cleaned prior to welding. Back-up bars, chill bars, clamps, etc., retain moisture and when heated by the welding operation, release hydrogen. This part of the operation is just as important as the cleaning of the base material and the filler wire.

| **Weld Rod** | 308 | 308L | 309 | 309Cb | 310 | 310Cb | 310Mo | 312 | 316 | 316L | 317 | 318 | 347 | 410 | 420 | 430 |
|---|---|---|---|---|---|---|---|---|---|---|---|---|---|---|---|---|
| **Base Material** | 201 | 201L | 309 | 309 | 310 | 310Cb | 310Mo | 312 | 316 | 316L | 317 | 316 | 347 | 410 | 420 | 430 |
| | 202 | 304L | 442 | 442 | 442 | 442 | | 501 | 301 | 321 | 301 | 318 | 301 | 308 | 309 | 309 |
| | 301 | 301 | 201 | 201 | 201 | | | 502 | 302 | 347 | 302 | 301 | 302 | 347 | 308 | 308 |
| | 302 | 302 | 202 | 202 | 202 | | | | 304 | | 304 | 302 | 304 | | | 308L |
| | 304 | 304 | 301 | 301 | 301 | | | | 317 | | 305 | 304 | 304L | | | 309 |
| | 305 | 305 | 302 | 302 | 302 | | | | | | | 305 | 318 | | | 309Cb |
| | 405 | 321 | 304 | 304 | 304 | | | | | | | 316 | 321 | | | 310 |
| | 409 | 347 | 305 | 305 | 305 | | | | | | | 317 | 403 | | | 310Cb |
| | 410 | | | | | | | | | | | 347 | 405 | | | 347 |
| | 430 | | | | | | | | | | | | 409 | | | |
| | | | | | | | | | | | | | 410 | | | |
| | | | | | | | | | | | | | 430 | | | |
| | | | | | | | | | | | | | 501 | | | |
| | | | | | | | | | | | | | 502 | | | |

Fig. 15-7. Filler materials listed for each column may be used to weld the materials in the same column together or in combination.

low carbon (identified with an "L" after the type number), as well as an extra low carbon (identified with "ELC" after type number). The "ELC" grade may be

### Weld Back-Up

Various types of weld back-ups are used when welding stainless steel. They include:

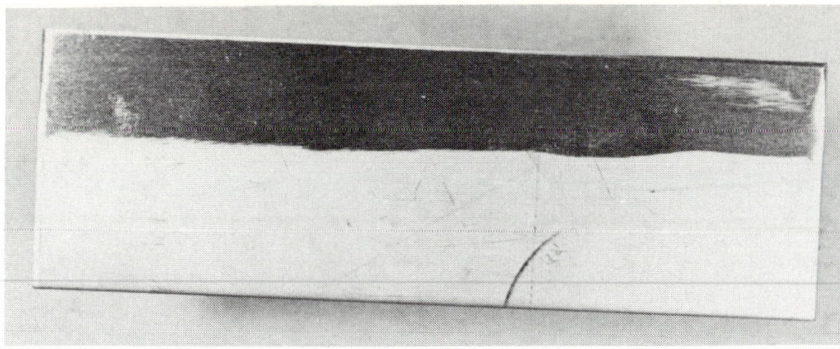

Fig. 15-8. Oxide films restrict proper metal flow and porosity within the weld. This film may be removed to "bright metal" by sanding with rotary sanders.

1. INTEGRAL. An integral back-up is used where the joint can be designed in the manner shown in Fig. 15-9. Another method is to place a backing strip on the back side of the weld, as shown in Fig. 15-10. The main problem with this design is the fit-up of the back-up bar. A loose fit allows expulsion of metal at the weld root and contamination of the penetration. These problems are shown in Fig. 15-11.

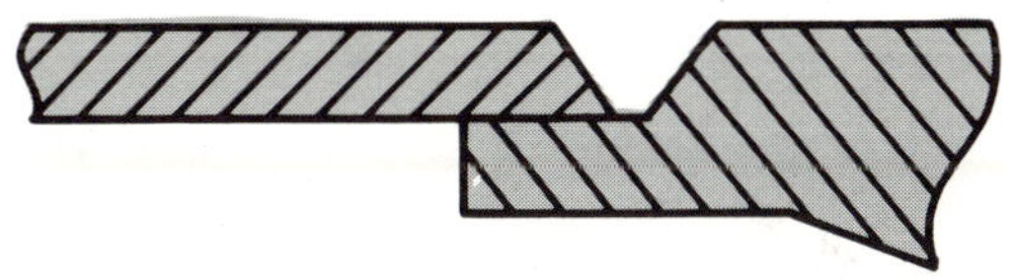

Fig. 15-9. Integral back-up bar designs are often used for pipe to fitting welds where excess stock for machining is available.

Fig. 15-10. Backing strips must fit tightly to prevent gaps between the back of the joint and the strip.

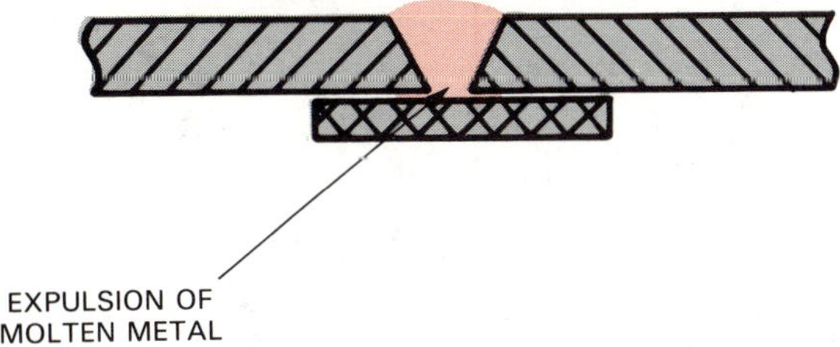


Fig 15-11. Backing strips that do not fit allow expulsion and possible contamination of the weld root.

2. REMOVABLE BACK-UP BAR. These types of back-up bars are found in stakes and seamers, or they may be a part of the assembly tooling. The groove in the bar may be designed for casting or forming of the penetration, as shown in Fig. 15-12. Another design, shown in Fig. 15-13, is used where argon gas is admitted to prevent drop-through oxidation. The argon gas type is used where very high quality is required at the back-up.

   Dimensions for the grooves of casting and argon gas back-up designs are shown in Fig. 15-14.
3. COMMERCIAL FLUX. This type of flux, Fig. 15-15, is supplied as a powder, mixed with acetone or alcohol, and applied to the part by brushing. The alcohol or acetone evaporates and the powder will remain as shown in Fig. 15-16. During welding, the powder melts and forms a barrier that prevents molten metal exposure to the air. If the flux must be removed after welding, wire brushing and hot water may be used to remove the residue.
4. NITROGEN GAS. Nitrogen gas is not an inert gas. However it may be used in some areas where it does not contact the molten metal. These areas are shown in Fig. 15-17.

## Weld Chill Bars

Weld chill bars may be used for welding all types

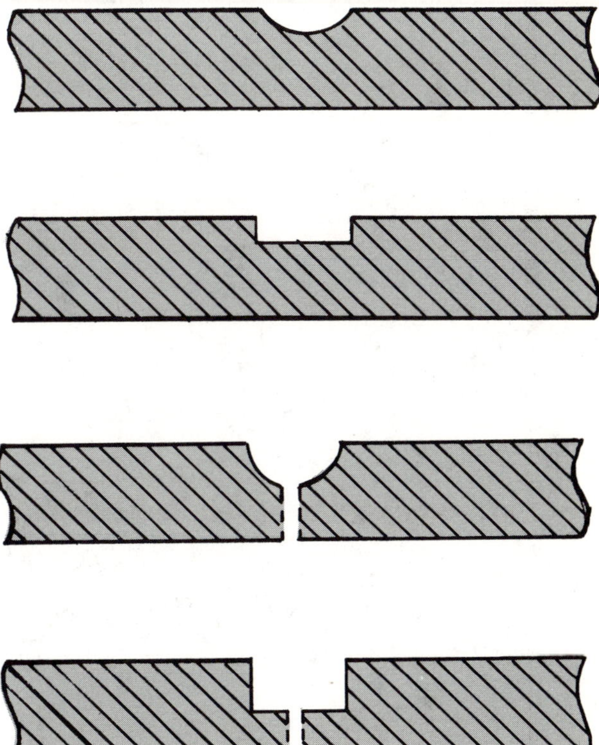

Fig. 15-12. Types of weld backing grooves for full penetration groove welds.

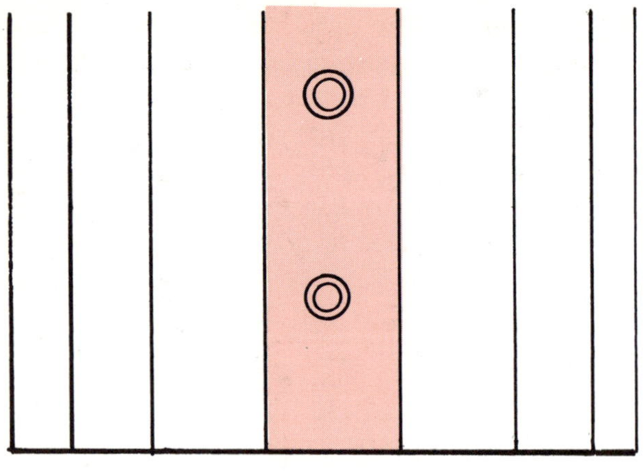

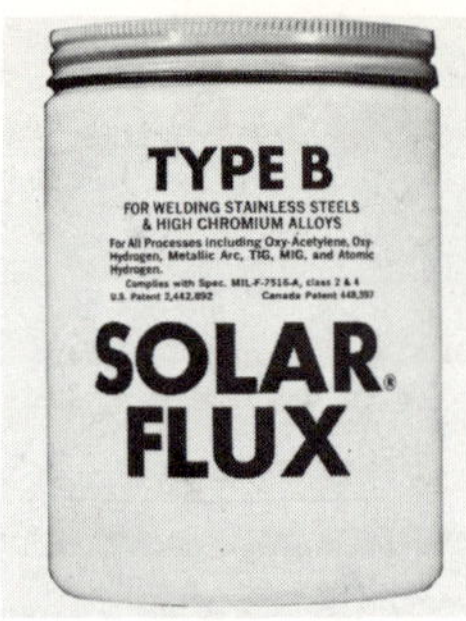


Fig. 15-15. Typical commercial flux often used when back-up bars or tooling cannot be used. (Golden Empire Corp.)

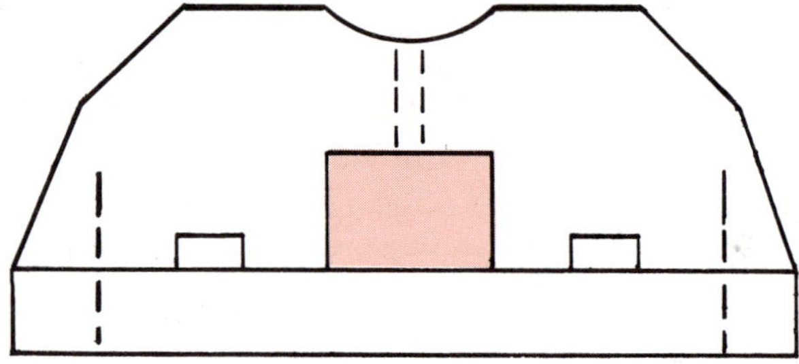

Fig. 15-13. Back-up bar design for admitting gas to the penetration side of the weld joint.

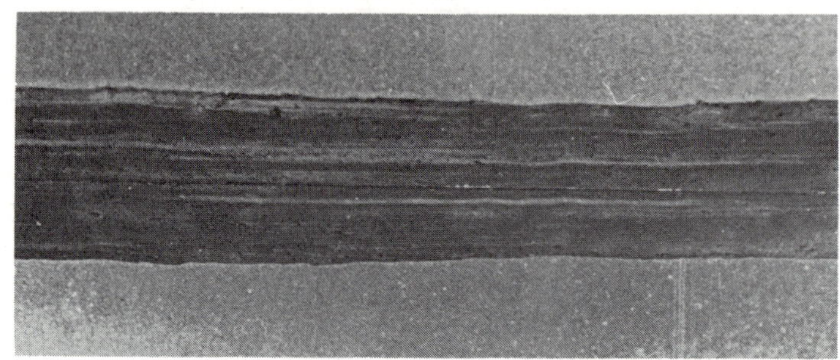

Fig. 15-16. The flux powder remains on the part after the fluid evaporates.

| METAL THICKNESS | Weld Type | Groove Dimensions For Casting Penetration | Groove Dimensions For Gas Back-Up Penetration |
|---|---|---|---|
| .005-.012 | Fusion | 040W010D | 040W125D |
| | Filler | — | |
| .013-.020 | Fusion | 063W010D | |
| | Filler | — | |
| .021-.032 | Fusion | 093W010D | 187W100D |
| | Filler | 125W020D | |
| .033-.040 | Fusion | 125W020D | |
| | Filler | 187W025D | |
| .041-.050 | Fusion | 125W020D | |
| | Filler | 187W025D | |
| .051-.062 | Fusion | 187W020D | |
| | Filler | 250W040D | |
| .063-.072 | Fusion | 187W020D | 250W100D |
| | Filler | 250W040D | |
| .073-.125 | Fusion | 250W020D | |
| | Filler | 312W040D | |
| .126-.250 | Fusion | 312W020D | 312W100D |
| | Filler | 375W050D | |
| .251-.375 | Fusion | — | |
| | Filler | — | |

Fig. 15-14. Back-up bar groove dimensions for full penetration welds.

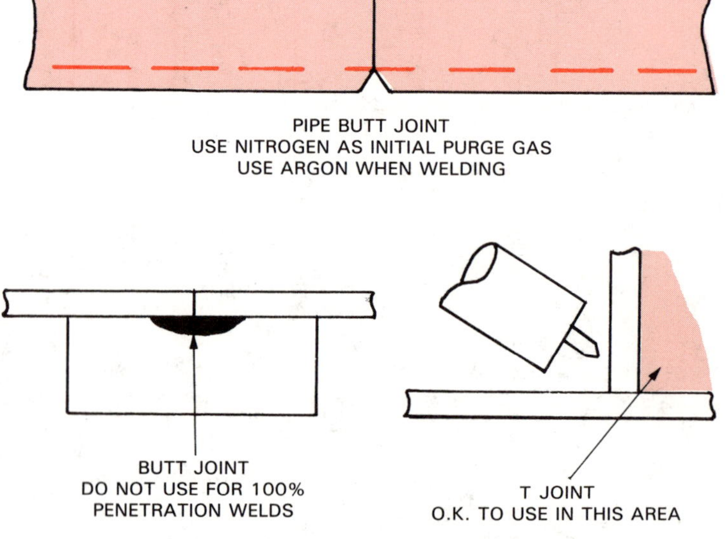


Fig. 15-17. Nitrogen used as a purge gas.

of stainless steels. For the austenitic steels, they assist in removing heat from the joint area and reduce distortion. In the martensitic steels, the bars are used to retain the preheat and slow down the cooling rate to reduce cracking. Refer to Figs. 8-20 and 8-21 showing the various designs of chill bars used on the stainless steels.

### Preheating Stainless Steel

Preheating in welding is used to slow down the cooling rate and prevent the formation of hard and brittle welds and heat affected zones.

When welding the chrome and chrome nickel stainless steels, the applications are as follows:

1. Austenitic, chromium-nickel manganese stainless steel — no preheat.
2. Austenitic, chromium-nickel stainless steel — no preheat.
3. Martensitic, chromium stainless steel (hardenable) — preheat 300 degrees F. to 500 degrees F.
4. Ferritic, chromium stainless steel (non-hardenable) — preheat approximately 300 degrees F. on welds with high restraint in grades 430, 434, 442, 446.
5. Precipitation hardenable stainless steel — no preheat.

### Tackwelds

When tackwelds must be made to maintain joint alignment for welding, the following considerations apply:

1. If the joint welding procedure requires preheating, the joint requires preheating before tackwelding.
2. Use the same filler material used for the main weld.
3. If the joint requires full penetration, the tackweld requires penetration.
4. Keep the tackweld as small as possible. If possible, smaller than the final weld.
5. If the tackweld cracks, leave it, and make another close to the broken one.
6. Do not grind out tackwelds. Grinding grit in a joint makes good quality almost impossible.
7. The crater of a tackweld is the weakest point in a tackweld. Do not leave a crater in a tackweld.
8. Always use an inert gas or flux back-up when tackwelding full penetration type joints.

## WELDING STAINLESS STEEL USING DCEN (DCSP)

The power supply type selected to weld stainless steel should be as follows:

1. Motor generator or a rectifier.
2. High frequency arc start (is desirable, but not mandatory).

### Electrode Selection for Stainless Steel

One percent or two percent thoriated electrodes, tapered to a sharp point should be selected. Always use an electrode large enough to carry the required amperage.

### Gases Used to GTAW Stainless Steel

Argon gas may be used for all thicknesses. For thicknesses over 1/8 inch, adding helium to the argon will increase the penetration.

Standard mixes that may be used are:

1. 75 percent argon — 25 percent helium.
2. 50 percent argon — 50 percent helium.
3. 25 percent argon — 75 percent helium.

Pure helium gas may be used when automatic welding the stainless steels.

## PROCEDURE FOR WELDING STAINLESS STEEL

1. Clean the oxide scale from the weld joint and immediately weld area to "bright metal" on the martensitic and ferritic materials.
2. All groove weld joints prepared by the plasma arc cutting process or carbon arc gouged must be ground as shown in Fig. 15-18 to a "bright metal" finish prior to welding.
3. Remove grease, oil, dirt, etc., with alcohol, or acetone.
4. Select a filler material listed in Fig. 15-6 or Fig. 15-7.
5. Preheat, maintain interpass temperature, and postheat on the martensitic and ferritic alloys as specified.
6. Welding procedure:
   A. Use minimum welding heat. Deep penetration into the base material is not required.
   B. Use stringer beads, do not weave. Stringer beads do not keep the metal in the 800 degrees to 1600 degrees F. range for long periods. Thus carbide precipitation is minimized.
   C. Use largest possible diameter filler wire.
   D. Use chill bars to cool weld area as fast as possible to prevent distortion on the austenitic steels.
   E. Use a gas nozzle as large as possible.
   F. When stopping weld, always maintain gas coverage from nozzle until weld cools to a

Fig. 15-18. Plasma arc cut material has a severe oxide film on the surface which must be removed to "bright metal" prior to welding.

color. Proper gas coverage from the torch will be indicated by a crown weld with a light blue, straw, or a gold color.

G. Always trim contaminated wire from used end of filler wire, as shown in Fig. 15-19.

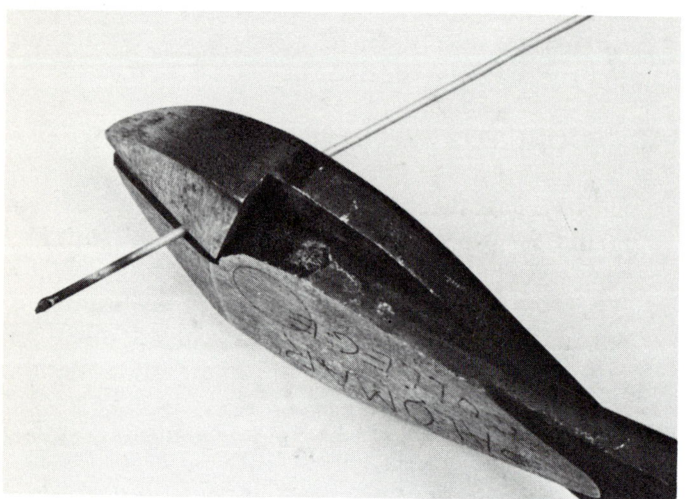

Fig. 15-19. Welding material removed from the torch shielding gas when hot is contaminated. Always remove the colored end of the wire from the rod.

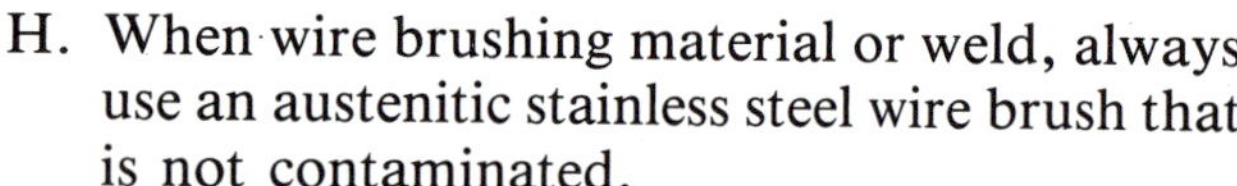

H. When wire brushing material or weld, always use an austenitic stainless steel wire brush that is not contaminated.

I. Tackwelds, root passes, and weld ends are susceptible to cracking as the weld metal cross section is thinner. To prevent cracks, make welds slightly convex and do not leave craters.

7. Welding Results

The following listed figures show stainless steel butt welds and fillet welds with the results obtained. Butt weld with chill bar tooling used to remove welding heat, Fig. 15-20. Butt weld made without chill bar tooling, Fig. 15-21. Fillet weld made without chill bar tooling, Fig. 15-22. Butt weld made without any type of shielding on the penetration, Fig. 15-23. Butt weld made with gas shielding on the penetration, Fig. 15-24. Butt weld made with a commercial flux applied to the penetration side before welding began, Fig. 15-25.

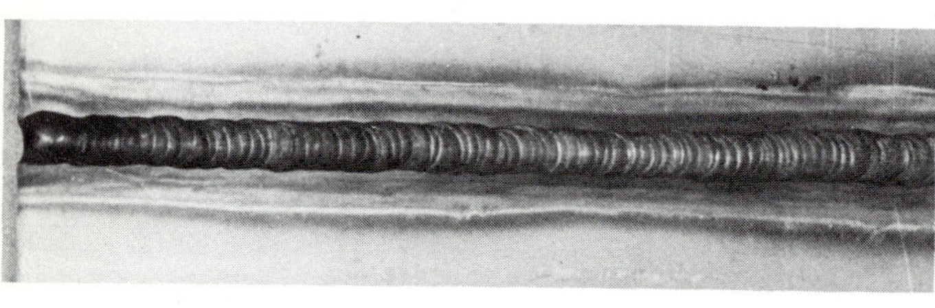

Fig. 15-20. Chill bars removed the welding heat rapidly and the weld has good crown and color and contour.

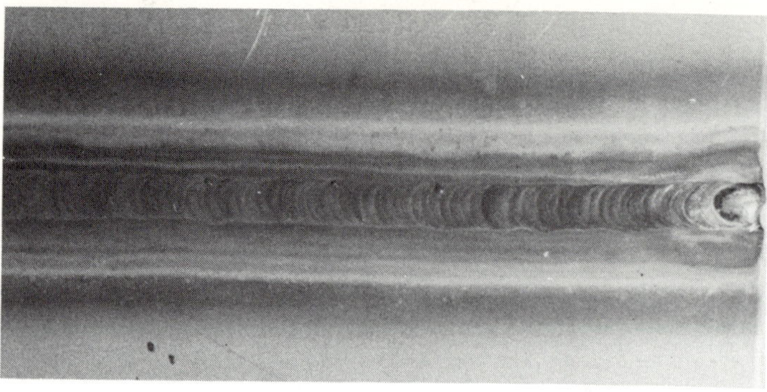

Fig. 15-21. Without chill bars, the welding heat has moved out into the plate material. The crown has no color and is flatter.

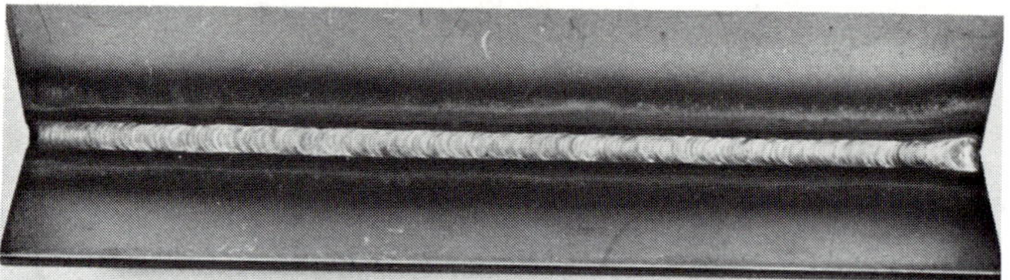

Fig. 15-22. Fillet weld made without chill bars has a wide heated zone. However, the weld has good colors due to the good gas coverage within the T design.

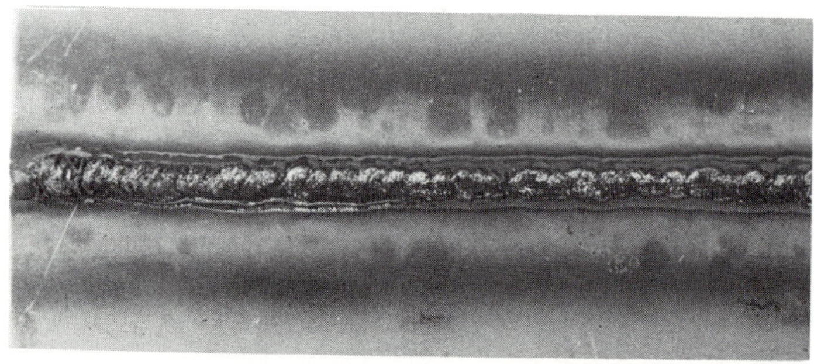

Fig. 15-23. The unprotected penetration weld side of the joint is badly oxidized from the atmosphere. Has poor bead contour and will not do the job intended. The oxidized material must be completely removed and the joint rewelded.

## POST WELD HEAT TREATMENT

During the welding of the hardenable martensitic stainless steels, some martensite will form regardless of the preheat used. If the weldment is not to be fully hardened after welding, an annealing operation is recommended. The part should be heated to 1550 degrees to 1650 degrees F., held for two hours, furnace cooled at a rate not to exceed 100 degrees F. per hour to 1100 degrees F. Then allow to cool to room temperature. Specialty type stainless steels, and the precipitation hardening alloys, should be thermally treated to the manufacturer's recommendations to achieve full mechanical values.

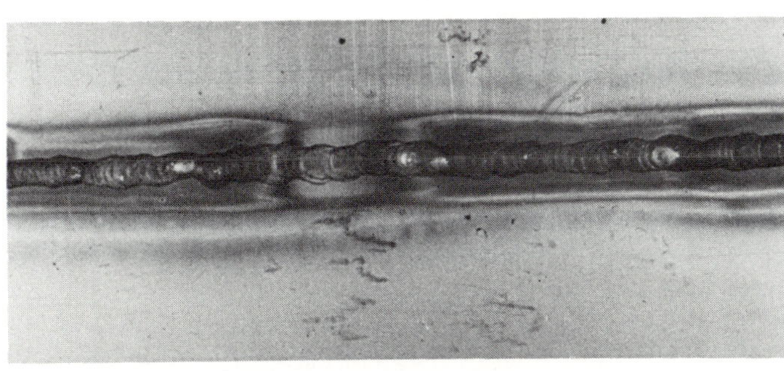

Fig. 15-24. The penetration side of this weld was protected by an inert gas (argon) and as a result has good colors, even penetration, and will do the job intended.

Fig. 15-25. A commercial flux was applied to the penetration side of the weld prior to welding. The welding heat melted the flux which then flowed over the hot metal to protect it during the cooling. The weld will not have any colors and will appear dark grey and black. This does not effect the quality of the weld. The residual flux may be removed with a hot water wash and a stainless steel brush.

## REVIEW QUESTIONS

1. List the five stainless steel families.
2. The ________ ________ ________ ________ system is used to identify the wrought materials.
3. The ________ ________ ________ system is used to identify the casting materials.
4. How are the special ferritic stainless steels identified?
5. What metal is used in the austenitic stainless steels for corrosion resistance?
6. Why is nickel used in the stainless steels?
7. Why do the austenitic stainless steels have a low carbon content?
8. What is meant by the term stabilized stainless steels?
9. What two materials are used in series 410 stainless steel? Will this steel harden?
10. What materials are added to the precipitation hardening stainless steels to promote hardening?
11. The special ferritic stainless steels are also called the ________ ________ ________ ferritic steels.
12. What does the lower carbon and nitrogen content prevent in these special ferritic stainless steels?
13. What range of temperatures are harmful to the unstabilized austenitic stainless steels?
14. Explain the sequence of events that occur when unstabilized stainless steel is heated during welding.
15. Why does stainless steel distort more than carbon steel during welding?
16. What do the letters "ELC" define?
17. How is the back-up flux powder applied to the root of the joint?
18. Can nitrogen gas be used as a backup material for stainless steel welding?
19. Which hardenable stainless steel group requires preheat prior to welding? Why?
20. Why are "stringer beads" preferred over wash beads when welding the austenitic stainless steels?
21. ________ percent or ________ percent thoriated electrodes, tapered to a sharp point, should be selected for welding stainless steel.
22. Pure ________ gas is generally used when automatic welding the stainless steel.

# Chapter 16

# GTAW PROCEDURE FOR NICKEL, NICKEL ALLOYS, AND COBALT ALLOYS

## BASE MATERIALS

There are approximately 100 different materials within the nickel, nickel alloy, and cobalt alloy group. They are made by different companies and for the most part, identified by registered tradenames. Some of the alloys are listed by the Society of Automotive Engineers Specification (SAE), Aeronautical Material Specification (AMS), The American Society for Testing Engineers Materials (ASTM), the American Society for Mechanical Engineers (ASME), and the American Welding Society (AWS). The nominal chemical composition of typical nickel alloys is shown in Fig. 16-1.

### Material Groups and Uses

1. COMMERCIAL PURE NICKEL. This material is used in the fabrication of food processing equipment, vessel liners, and piping for the food industry.
2. NICKEL-COPPER ALLOYS are used in the fabrication of processing equipment, refining equipment, and heat exchangers.
3. NICKEL-CHROMIUM ALLOYS are used in many applications involving corrosive media from cryogenic to elevated temperatures.
4. NICKEL-IRON-CHROMIUM ALLOYS have good high temperature strengths and resistance to oxidation and corrosion.
5. NICKEL-MOLYBDENUM ALLOYS have good corrosive resistance at temperatures other than elevated.
6. NICKEL-CHROMIUM-MOLYBDENUM alloys have good corrosion resistance at room temperature. They also resist oxidation at elevated temperatures.
7. PURE COBALT. This material has few uses in the fabrication field.
8. COBALT ALLOYS have good corrosion resistance and high temperature service.

Fig. 16-2 shows a typical nickel alloy material in the process of being welded.

## WELDING PROBLEM AREAS

The major problems when welding the nickel and cobalt alloys are porosity and cracking within the welds. Both of these areas can be avoided through the use of proper weld joint design, preweld cleaning techniques, and the use of good welding procedures.

The main point to remember is that these materials are designed for extreme temperatures and corrosive media. Therefore, each weld must be carefully considered as to it's placement into the service environment. Welds that are questionable at room temperature in a normal atmosphere will usually fail rapidly in service.

To prevent these types of problems you must do the following:

1. Use the proper base material for the intended service.
2. Use the proper filler material.
3. Design the weld joint, as shown in Fig. 16-3, with larger groove angles since the molten metal does not "wet" readily.
4. Wherever possible use full penetration type joints. Fillet and edge type joints promote cracking from the unwelded area.
5. Design the weld joint for minimum restraint during welding. (Allow the joint to move).
6. Design tooling to obtain maximum chilling of the weld area.
7. Use the proper cleaning techniques.
8. Use the proper welding techniques.

## FILLER MATERIALS

Filler materials for welding of nickel, nickel alloys and cobalt alloys are selected based on the chemical composition of the parent materials.

The filler material should generally match the parent material when welding the same alloys. When welding dissimilar alloys, the filler material should match the higher alloy parent material.

| Common designation[a] | Ni[b] | C | Cr | Mo | Fe | Co | Cu | Al | Ti | Cb[c] | Mn | Si | W | B | Other |
|---|---|---|---|---|---|---|---|---|---|---|---|---|---|---|---|
| **COMPOSITION, PERCENT** | | | | | | | | | | | | | | | |
| **COMMERCIALLY PURE NICKELS** | | | | | | | | | | | | | | | |
| Nickel 200 | 99.5 | 0.08 | | | 0.2 | | 0.1 | | | | 0.2 | 0.2 | | | |
| Nickel 201 | 99.5 | 0.01 | | | 0.2 | | 0.1 | | | | 0.2 | 0.2 | | | |
| Nickel 205 | 99.5 | 0.08 | | | 0.1 | | 0.08 | | 0.03 | | 0.2 | 0.08 | | | 0.05Mg |
| **SOLID SOLUTION TYPES** | | | | | | | | | | | | | | | |
| Monel 400 | 66.5 | 0.2 | | | 1.2 | | 31.5 | | | | 1 | 0.2 | | | |
| Monel 404 | 54.5 | 0.08 | | | 0.2 | | 44 | 0.03 | | | 0.05 | 0.05 | | | |
| Monel R-405 | 66.5 | 0.2 | | | 1.2 | | 31.5 | | | | 0.1 | 0.02 | | | |
| Hastelloy F | 47 | 0.05 | 22 | 6.5 | 17 | 2.5 | | | | 2 | 1.5 | 1 | 1 | | |
| Hastelloy X | 47 | 0.10 | 22 | 9 | 18 | 1.5 | | | | | 1 | 1 | 0.6 | | |
| Nichrome V | 76 | 0.1 | 20 | | 1 | | | | | | 2 | 1 | | | |
| Nichrome | 57 | 0.1 | 16 | | 25 | | | | | | 1 | 1 | | | |
| Hastelloy G | 44 | 0.1 | 22 | 6.5 | 20 | 2.5 | 2 | | | 2 | 1.5 | 1 | 1 | | |
| IN 102 | 68 | 0.06 | 15 | 3 | 7 | | | 0.4 | 0.6 | 3 | | | 3 | 0.005 | 0.03Zr, 0.02Mg |
| RA 333 | 45 | 0.05 | 25 | 3 | 18 | 3 | | | | 1 | 1.5 | 1.2 | 3 | | |
| Inconel 600 | 76 | 0.08 | 15.5 | | 8 | | 0.2 | | | | 0.5 | 0.2 | | | |
| Inconel 601 | 60.5 | 0.05 | 23 | | 14 | | | 1.4 | | | 0.5 | 0.2 | | | |
| Inconel 625 | 61 | 0.05 | 21.5 | 9 | 2.5 | | | 0.2 | 0.2 | 3.6 | 0.2 | 0.2 | | | |
| Carpenter 20Cb3 | 36 | 0.04 | 20 | 2.5 | 36 | | 3.5 | | | 0.5 | 1 | 0.5 | | | |
| Incoloy 800 | 32.5 | 0.05 | 21 | | 46 | | | 0.4 | 0.4 | | 0.8 | 0.5 | | | 0.02Mg |
| Incoloy 825 | 42 | 0.03 | 21.5 | 3 | 30 | | 2.25 | 0.1 | 0.9 | | 0.5 | 0.25 | | | |
| Hastelloy B | 61 | 0.05 | 1 | 28 | 5 | 2.5 | | | | | 1 | 1 | | | |
| Hastelloy C | 54 | 0.08 | 15.5 | 16 | 5 | 2.5 | | | | | 1 | 1 | 4 | | |
| Hastelloy D | 82 | 0.10 | | | 1 | 1.5 | 3 | | | | 1 | 9 | | | |
| Hastelloy N | 70 | 0.06 | 7 | 16.5 | 5 | | | | | | 0.8 | 0.5 | | | |
| Hastelloy W | 60 | 0.12 | 5 | 24.5 | 5.5 | 2.5 | | | | | 1 | 1 | | | |
| **PRECIPITATION HARDENABLE TYPES** | | | | | | | | | | | | | | | |
| Duranickel 301 | 96.5 | 0.15 | | | 0.3 | | 0.13 | 4.4 | 0.6 | | 0.25 | 0.5 | | | |
| Monel K-500 | 66.5 | 0.10 | | | 1 | | 29.5 | 2.7 | 0.6 | | 0.8 | 0.2 | | | |
| Waspaloy | 58 | 0.08 | 19.5 | 4 | | 13.5 | | 1.3 | 3 | | | | | 0.006 | 0.06Zr |
| Rene 41 | 55 | 0.10 | 19 | 10 | 1 | 10 | | 1.5 | 3 | | 0.05 | 0.1 | | 0.005 | |
| Nimonic 80A | 76 | 0.06 | 19.5 | | | | | 1.6 | 2.4 | | 0.3 | 0.3 | | 0.006 | 0.06Zr |
| Nimonic 90 | 59 | 0.07 | 19.5 | | | 16.5 | | 1.5 | 2.5 | | 0.3 | 0.3 | | 0.003 | 0.06Zr |
| M 252 | 55 | 0.15 | 20 | 10 | | 10 | | 1 | 2.6 | | 0.5 | 0.5 | | 0.005 | |
| Udimet 500 | 54 | 0.08 | 18 | 4 | | 18.5 | | 2.9 | 2.9 | | 0.5 | 0.5 | | 0.006 | 0.05Zr |
| Alloy 713C[d] | 74 | 0.12 | 12.5 | 4 | | | | 6 | 0.8 | 2 | | | | 0.012 | 0.10Zr |
| Inconel 718 | 52.5 | 0.04 | 19 | 3 | 18.5 | | | 0.5 | 0.9 | 5.1 | 0.2 | 0.2 | | | |
| Inconel X750 | 73 | 0.04 | 15.5 | | 7 | | | 0.7 | 2.5 | 1 | 0.5 | 0.2 | | | |
| Inconel 706 | 41.5 | 0.03 | 16 | | 40 | | | 0.2 | 1.8 | 2.9 | 0.2 | 0.2 | | | |
| Alloy 901 | 42.5 | 0.05 | 12.5 | | 36 | 6 | | 0.2 | 2.8 | | 0.1 | 0.1 | | 0.015 | |
| Ni-Span-C 901 | 42.2 | 0.03 | 5.3 | | 48.5 | | | 0.6 | 2.6 | | 0.4 | 0.5 | | | |
| IN 100[d] | 60 | 0.18 | 10 | 3 | | 15 | | 5.5 | 4.7 | | | | | 0.014 | 0.06Zr, 1.0V |
| **DISPERSION STRENGTHENED TYPES** | | | | | | | | | | | | | | | |
| TD Nickel | 98 | | | | | | | | | | | | | | 2 $ThO_2$ |
| TD Ni Cr | 78 | | 20 | | | | | | | | | | | | 2 $ThO_2$ |

a. Several of these are registered tradenames. Alloys of similar compositions may be known by other common designations or tradenames.
b. Includes small amount of cobalt if cobalt content is not specified.
c. Includes tantalum also.
d. Casting alloys.

Fig. 16-1. Nickel, nickel alloy, cobalt alloy chemical compositions.

Fig. 16-2. Nickel alloy turbine rotor vanes are being welded in place. Note the shiny weld area which indicates pre-weld cleaning. (Miller Electric Co.)

the nickel alloys are to be joined to dissimilar metals, the materials specified in Fig. 16-4 may be used.

Filler material specifications currently in use for nickel and nickel alloys include:

1. AWS A514.
2. Mil-E-21562
3. Mil-R-5031-B
4. SAE-AMS 5838, 5786, 5679, 5794, 5675, 5832, 5837.

Filler material specifications currently in use for the cobalt alloys include:

1. AMS 5789, 5796, 5801.

Maximum weld quality can only be obtained if the filler material is clean and of high quality. Store wire in plastic containers. Use only in a clean area. Handle with clean gloves. Wire that has been handled with bare hands must be cleaned before use. Use alcohol or acetone and a clean rag to remove contamination.

Maintain identification of each bare rod, as mixed rods are impossible to identify.

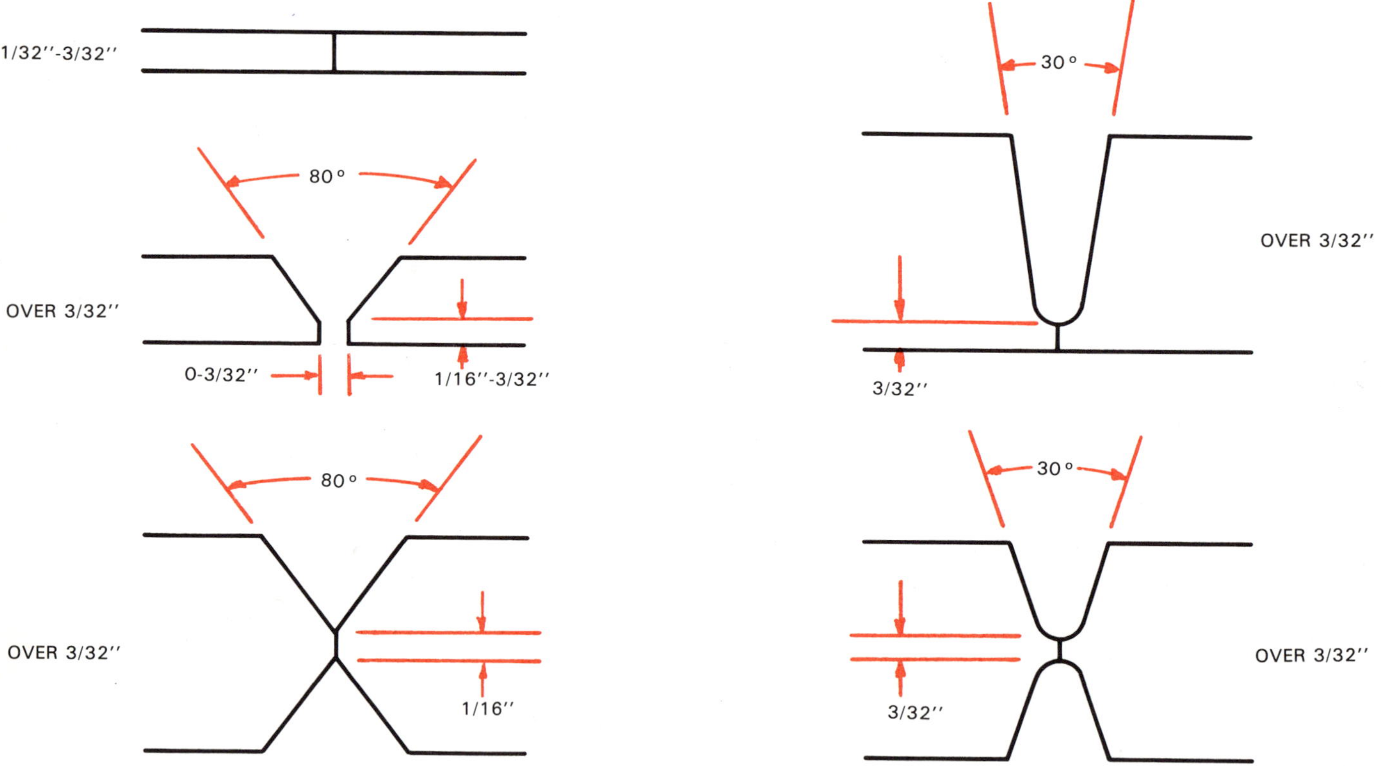


Fig. 16-3. Common weld joint designs for nickel and cobalt alloys.

Filler materials developed by individual material companies are often listed by tradenames and numbers. In some cases, they will not meet any current specification. It is suggested that before any welding is done, the manufacturer of the base material be contacted regarding the proper filler material to be used. Where

## WELD PREPARATION

Cleaning procedures for these materials must be established before any welding is done on the actual weldment. These procedures must be followed exactly or the weld may be ruined.

| NICKEL ALLOY | STAINLESS STEEL | LOW CARBON AND STAINLESS STEEL | 5-9 NICKEL STEELS | COPPER | COPPER NICKEL |
|---|---|---|---|---|---|
| **Nickel 200-201** | Inconel 82 | Inconel 82 | Inconel 82 | Monel 60 | Monel 60 |
| **Monel Alloys 400, K-500, 502** | Inconel 82 | Nickel 61 | Nickel 61 | Monel 60 | Monel 60 |
| **Inconel Alloys 600-601** | Inconel 82 | Inconel 82 | Inconel 82 | Nickel 61 | Nickel 61 |
| **Inconel Alloys 625, 706, 718, X-750** | Inconel 82 | Inconel 82 | Inconel 625 | Nickel 61 | Nickel 61 |
| **Incoloy Alloy 800** | Inconel 82 | Inconel 82 | Inconel 82 | Nickel 61 | Nickel 61 |
| **Incoloy Alloy 825** | Inconel 625 | Inconel 625 | Inconel 625 | Nickel 61 | Nickel 61 |

Fig. 16-4. Common filler materials used to join nickel alloys to dissimilar materials.

Topics and areas which are not as important for other materials are critical for these nickel and cobalt materials. Surface oxides, grease, machining coolant, lead, paint, copper, and shop dirt can lead to weld failure. Everything must be considered as a contaminate until tested and proven otherwise.

The following listed areas should be carefully considered in the preweld cleaning and setup procedure.

1. Remove all oxide film for a distance of at least 2 inches from the joint edge by sanding, grinding, routing, grit blasting, or emery cloth.
2. Thermal cut joint must have ALL heat affected zones removed by machining.
3. Any caustic material used for cleaning must be thoroughly removed, per step 1.
4. Identification marking made with pens, inks, crayons, or paints may contain lead or sulphur. These areas must be cleaned. Use a solvent, grind, or rub with emery cloth.
5. Never use sanding disc, wheels, or belts that have been used on other materials.
6. Never use brass or carbon steel brushes. Use only stainless steel brushes. Do not use stainless steel brushes that have been used on other types of materials.
7. Do not use copper, brass or bronze in any part of the weld fixture that may come into contact with the weld zone.
8. Use only clean solvents.
9. Wear clean gloves and use clean rags.
10. Do not handle parts with bare hands.
11. Do not use shop air for blowing washed parts dry. (This air may contain oil).
12. Draw file all sheared edges with clean files.
13. Wash all weld tooling that contacts the weld area with alcohol or acetone.

## Weld Back-Up

The root side of all full penetration joints on these materials must be protected from the atmosphere. Tooling must be provided to admit an inert gas for this purpose. A design for the correct back-up tooling groove is shown in Fig. 16-5. A design for incorrect back-up tooling groove is also shown in Fig. 16-5.

When welding pipe, the interior penetration side of the joint must also be provided with an inert gas. This can be done by closing off the pipe ends and purging the atmosphere from the pipe.

Consumable ring inserts of the types shown in the chapter on pipe welding are preferred over back-up rings.

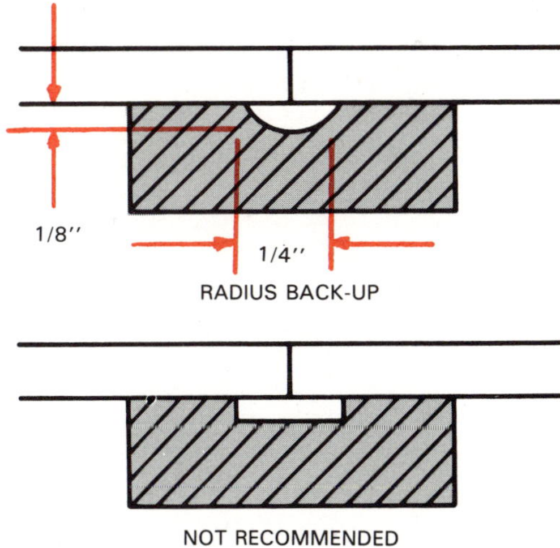


Fig. 16-5. Use radius type back-up grooves whenever possible to prevent notches on the penetration face. Casting type back-up grooves with square edges often cause notches on the penetration face.

### Tackwelds

When tackwelds must be made to maintain joint alignment for welding, the following considerations apply:

1. Use the same filler material as used for the joint weld.
2. If the joint requires full penetration, the tackweld requires penetration.
3. Keep the tackweld as small as possible. If possible, make the tackweld smaller than the final weld.
4. If the tackweld cracks, leave it and make another close to the broken one.
5. Do not grind out tackwelds. Grinding grit in a joint makes quality welds impossible.
6. The crater of a tackweld is the weakest point in a tackweld. Do not leave a crater in a tackweld.
7. Always use an inert gas back-up when tackwelding full penetration type joints.

## WELDING NICKEL, NICKEL ALLOYS, AND COBALT ALLOYS BY DCEN (DCSP)

The power supply type used to weld nickel, nickel alloys, or cobalt alloys should be as follows:

1. Motor generator or a rectifier.
2. High frequency arc start.

### Electrode For DCEN on Nickel or Cobalt

Use 1 percent or 2 percent thoriated electrodes. Taper the electrodes to a sharp point. Always use an electrode large enough to carry the required amperage.

### Gases for GTAW

Argon gas may be used for all thicknesses. For material over 1/8 inch thickness, adding helium to the argon will increase the penetration.

Standard mixes that may be used are:

1. 75 percent argon — 25 percent helium.
2. 50 percent argon — 50 percent helium.
3. 25 percent argon — 75 percent helium.
4. 95 percent argon — 5 percent hydrogen. (This mix will leave a cleaner, brighter weld crown.)

Pure helium gas may be used when automatic welding these materials.

## PROCEDURE FOR GTAW OF NICKEL AND COBALT

1. Clean the weld areas as previously described.
2. Select a filler material with the desired chemical requirements.
3. Preheat only if temperature is below 60°F.
4. Welding.
   A. Maintain the torch in as vertical a position to the weld as possible and use only enough amperage to flow the material. Deep penetration into the side walls of a joint is not required.
   B. Use the shortest possible arc length.
   C. Use stringer beads. Do not weave.
   D. Use largest possible diameter filler material.
   E. Use chill bars to cool weld area as fast as possible to prevent distortion.
   F. Use the largest possible torch gas nozzle.
   G. When stopping weld, always maintain gas coverage from torch nozzle until weld cools without oxide film. (Set the torch gas post-flow for 5 to 10 seconds longer than normal.)
   H. Always trim contaminated wire from used end of filler wire as shown in Fig. 6-16.
   I. When wire brushing material or weld always use an austenitic stainless steel wire brush.
   J. Tackwelds, root passes, and weld ends are susceptible to cracking, as the weld metal cross section is thinner. To prevent cracks, make welds convex and do not leave craters.
   K. The manufacturer of the base material should be consulted regarding post weld heat treatment.

## REVIEW QUESTIONS

1. What are some of the specifications used to identify nickel, nickel alloys, and cobalt alloys?
2. By what other means are the above alloys identified?
3. What is the main use of pure nickel?
4. What two characteristics do the cobalt alloys have?
5. What are the main problems encountered in welding nickel, nickel alloys, and cobalt alloys?
6. Does the weld metal "wet" to the base readily when welding these alloys?
7. Why are fillet and edge type joints avoided when welding these alloys?
8. Name two degreasers which may be used in the pre-weld preparation.
9. Does the oxide film on the base metal weld area need to be removed before welding?
10. How are thermal cut joints processed for welding?
11. How are identification markings made with crayons, inks, and pencils removed?
12. What type of brushes are used to brush weld joints and welds?
13. What type of gas is used besides argon to achieve deeper penetration when using manual DCSP?
14. When is preheat used?
15. Which type of weld bead is preferred when welding these alloys?

# Chapter 17

# GTAW PROCEDURE FOR TITANIUM

## BASE MATERIAL

Titanium is a metal that has many applications because of it's excellent corrosion resistance and high strength-to-weight ratios.

The material is silver colored. Pure titanium melts at approximatley 3,000 degrees F. The oxide film which forms on the surface of the material imparts a resistance to corrosion. For this reason, titanium is very useful in the chemical and processing industries.

In the pure titanium grades, oxygen is used as a strenghtening element. However, oxygen also decreases the ductility. The alloy titaniums use many different alloying elements to improve the mechanical properties into the cryogenic temperature ranges (below −250 degrees F.).

### Material Identification

Base materials are identified by the letters Ti preceding the alloy content. The alloy contents will be listed with the major alloy first, and the lesser alloys following. An example shows; Ti-6AL-4V alloy. This would indicate the material to be titanium with 6 percent aluminum and 4 percent vanadium alloy material.

### Specifications

Commercial titanium producers market titanium to commercial specifications and to the AMS (Aeronautical Material Specification). All types of titanium listed in the AMS specification are located in the 4900 number series.

ASTM specifications are used for identification of some types of titanium. These types are identified as grades. The ASTM grades are generally used for material used in the fabrication of American Society of Mechanical Engineers (ASME) weldments.

### Types of Titanium

The titanium family consists of four groups of materials and they include:

1. Commercial pure titanium.
2. Alpha alloys.
3. Alpha-beta alloys.
4. Beta alloys.

The names alpha, alpha-beta, and beta alloys apply to the type of grain structure for each metal. Since the names are often used in industry in describing the metals, the welder should be introduced to them.

Of the four groups listed, the commercially pure and the alpha alloy materials are not heat treatable for strength and ductility. Within the Alpha-beta alloys, heat treatment is restricted to specific alloys. The beta alloys are not a weldable material.

## FILLER MATERIALS FOR TITANIUM

The filler material choice for GTA welding of titanium and titanium alloys is very important. The metal produced in the weld puddle is a combination of the filler wire and the parent metal. The weld metal must have the strength, ductility, freedom from cracking, and the ability to resist corrosion (if needed) required by the weldment application.

The correct choice of a filler alloy will determine to a large degree the type of grain structure in the completed weld. It is this grain structure that determines the physical properties of the completed weld.

In many cases, a non-alloyed welding material may be used to join alloy base material. This is often done where full 100 percent joint strength is not required, or the weldment is not heat treated after welding for strength purposes. Welds in these cases will have lower tensile properties and higher ductility.

The American Welding Society has prepared welding material specifications for titanium filler materials for use in the welding of titanium. Fig. 17-1 lists the filler materials available.

Filler materials used for the welding of titanium and titanium alloys must be free from contamination and defects on the inside of the wire as well as contamination and defects on the outer surface.

Some elements and gases are trapped within the wire during manufacturing and though not detrimental in other materials, may cause failure in titanium welds.

| AWS Classification | Carbon per cent | Oxygen per cent | Hydrogen per cent | Nitrogen per cent | Aluminum per cent | Vanadium per cent | Tin per cent | Chromium per cent | Iron per cent | Molybdenum per cent | Columbium per cent | Tantalum per cent | Palladium per cent | Titanium per cent |
|---|---|---|---|---|---|---|---|---|---|---|---|---|---|---|
| RTi-1[a] | 0.03 | 0.10 | 0.005 | 0.012 | . . . | . . . | . . . | . . . | 0.10 | . . . | . . . | . . . | . . . | Remainder |
| RTi-2 | 0.05 | 0.10 | 0.008 | 0.020 | . . . | . . . | . . . | . . . | 0.20 | . . . | . . . | . . . | . . . | Remainder |
| RTi-3 | 0.05 | 0.10-0.15 | 0.008 | 0.020 | . . . | . . . | . . . | . . . | 0.20 | . . . | . . . | . . . | . . . | Remainder |
| RTi-4 | 0.05 | 0.15-0.25 | 0.008 | 0.020 | . . . | . . . | . . . | . . . | 0.30 | . . . | . . . | . . . | . . . | Remainder |
| RTi-0.2Pd | 0.05 | 0.15 | 0.008 | 0.020 | . . . | . . . | . . . | . . . | 0.25 | . . . | . . . | . . . | 0.15-0.25 | Remainder |
| RTi-3A1-2.5V | 0.05 | 0.12 | 0.008 | 0.020 | 2.5-3.5 | 2.0-3.0 | . . . | . . . | 0.25 | . . . | . . . | . . . | . . . | Remainder |
| RTiA1-2.5V-1[a] | 0.04 | 0.10 | 0.005 | 0.012 | 2.5-3.5 | 2.0-3.0 | . . . | . . . | 0.25 | . . . | . . . | . . . | . . . | Remainder |
| RTi-5A1-2.5Sn | 0.05 | 0.12 | 0.008 | 0.030 | 4.7-5.6 | . . . | 2.0-3.0 | . . . | 0.40 | . . . | . . . | . . . | . . . | Remainder |
| RTi-5A1-2.5Sn-1[a] | 0.04 | 0.10 | 0.005 | 0.012 | 4.7-5.6 | . . . | 2.0-3.0 | . . . | 0.25 | . . . | . . . | . . . | . . . | Remainder |
| RTi-6A1-2Cb-1Ta-1Mo | 0.04 | 0.10 | 0.005 | 0.012 | 5.5-6.5 | . . . | . . . | . . . | 0.15 | 0.5-1.5 | 1.5-2.5 | 0.5-1.5 | . . . | Remainder |
| RTI-6A1-4V | 0.05 | 0.15 | 0.008 | 0.020 | 5.5-6.75 | 3.5-4.5 | . . . | . . . | 0.25 | . . . | . . . | . . . | . . . | Remainder |
| RTi-6A1-4V-1[a] | 0.04 | 0.10 | 0.005 | 0.012 | 5.5-7.75 | 3.5-4.5 | . . . | . . . | 0.15 | . . . | . . . | . . . | . . . | Remainder |
| RTi-8A1-1Mo-1V | 0.05 | 0.12 | 0.008 | 0.03 | 7.35-8.35 | .75-1.25 | . . . | . . . | 0.25 | .75-1.25 | . . . | . . . | . . . | Remainder |
| RTi-13V-11Cr-3A1 | 0.05 | 0.12 | 0.008 | 0.03 | 2.5-3.5 | 12.5-14.5 | . . . | 10.0-12.0 | 0.25 | . . . | . . . | . . . | . . . | Remainder |

a — This classification of filler metal restricts the allowable interstitial content to a low level in order that the high toughness required for cryogenic applications and other special uses can be obtained in the deposited weld metal.

Note 1 — Analysis for interstitial content shall be made after the welding rod or electrode has been reduced to its final diameter.
Note 2 — Single values are maximum percentages.

Fig. 17-1. Titanium filler wire chemical compositions.

Carbon, oxygen, hydrogen, nitrogen, and iron are held to low limits in wire specifications, as they are the main cause of the lack of weld toughness.

Where high weld toughness is mandatory, the filler materials may be procurred with an extra low level of impurities.

These impurities are called *interstituals.* The welding material ordered for use with very low interstituals are called ELI materials (extra low interstituals content). The specification listed in Fig. 17-1 shows these materials with a -1 following the material type. An example of this is shown below:

ERTi-6AL-4V — Standard grade weld wire.
ERTi-6AL-4V-1 — ELI grade weld wire.

Welding material manufacturers clean titanium immediately prior to packaging. The packages are sealed to prevent the entry of dirt and contaminates. The material should be stored in these containers until required, with the packages sealed. To prevent contamination of the material, remove only the material required for immediate use and reseal the package immediately.

Handle the material required with clean gloves. Do not use bare hands. Material to be re-stored should be wiped with a clean lint free cloth that has been soaked in alcohol.

## Pre-Weld Preparation

Molten titanium has an affinity for carbon, oxygen, and hydrogen. If these contaminates are absorbed into the weld, the results may be very undesirable. Depending on the amount of contamination, porosity may form or the weld may have very low notch toughness value. (Notch toughness is the ability of a metal to resist cracking failure at a notch during stress loading.) Either of these areas may cause failure of the weld under a load far below the rated value.

These contaminates are found in many places, such as oils, grease, tools, and body salts. Therefore, it is imperative that the pre-weld cleaning must be done just prior to welding.

The cleaning operation should be complete. Tooling and the other materials must be thoroughly and completely cleaned. A sample cleaning operation may include the following steps:

1. Remove all oxide scale by acid pickling.
   A. Use a HF-$HNO_3$ solution. (HF 2 percent to 4 percent and 30 percent to 40 percent $HNO_3$.)
   B. Rinse with clean water and dry.
2. Draw file all abutting edges. (Sheared edges must be filed until smooth).
3. Wire brush entire joint and immediate area. (Brush must be stainless steel and not contaminated. After use, wash thoroughly with alcohol and store in a sealed container.)
4. Wash entire weld joint and adjacent area with a nonresidue forming solvent. Alcohol or acetone are good degreasers. After washing the weld joint, allow it to dry thoroughly. Do not blow dry with compressed air unless air is completely filtered. (An alternate to using alcohol or acetone is to alkaline wash or dip the parts. A dilute solution of sodium hydroxide or a commercial alkaline cleaner may be used. Rinse with clear water and dry.)
5. If the parts are to be installed on a fixture or tool for welding, clean the fixture with alcohol or acetone before installing parts. If the parts cannot be installed in the fixture immediately, then protect the parts by covering them with lint-free cloths or clean paper.
6. After the cleaning is done, handle the parts, fixtures, tooling, etc., with lint-free gloves and cloths. Never use the bare hands to handle cleaned titanium.

### Weld Shielding

Titanium welds must be protected from the atmosphere during welding. As previously explained, hot titanium absorbs oxygen, hydrogen, and nitrogen which contaminates the weld. The degree of contamination can be determined by the color of the contaminated weld and heat affected area.

Welds and heat affected areas that are silver in color have sufficient protection and are considered good welds.

Gold, straw, and light blue colors indicate progressive amounts of contamination. Depending on the final use of the weldment, this amount of contamination may or may not be acceptable.

Dark blue, light grey, or dark grey colors indicate excessive contamination, and the welds cannot be used for ANY purpose.

The size of the weldment, the configuration of the weldment, and the production quality will usually dictate the type of shielding needed. In all cases, the shielding gas from the torch will not be sufficient to provide shielding for all of the hot material.

To obtain a silver weld and weld area, the gas must provide coverage until the metal is cooled below approximately 500 degrees F.

Small parts may be welded in a chamber as shown in Fig. 17-2. The cleaned tools and parts are loaded into the chamber area, the top is installed and argon is fed into the chamber.

The welding can be started as soon as all of the air is removed. A test weld can be made on a scrap piece of material to test the atmosphere within the unit. A silver weld indicates a good welding atmosphere. High production jobs may have an auxiliary chamber on the side of the main chamber. The parts are passed into and out of the main chamber without damaging the main chamber atmosphere.

Welding torches should always be fitted with a gas lens device to reduce the argon gas nozzle turbulence. With this device, the gas is delivered with a blanket effect rather than a jet effect. See Fig. 4-57.

Fig. 17-2. This plexiglass dome chamber has been modified to make semiautomatic test welds. The level of gas purity may be checked using the tester shown in Fig. 5-20 after purging is complete.

Trailing shields protect the metal behind the torch as the torch is moved along the joint. Fig. 17-3 shows a shield for a circular seam weld. Fig. 17-4 shows a device for small circular welds. A test weld is shown in Fig. 17-5 using a torch trailing shield.

Back-up fixtures used to locate parts should always fit with virtually no gap between the part and the tool. If gaps are present, air may be drawn into the back-up gas system and contamination will result. Grooves

Fig. 17-3. Trailing shields protect the weld metal until it cools to a silver color.

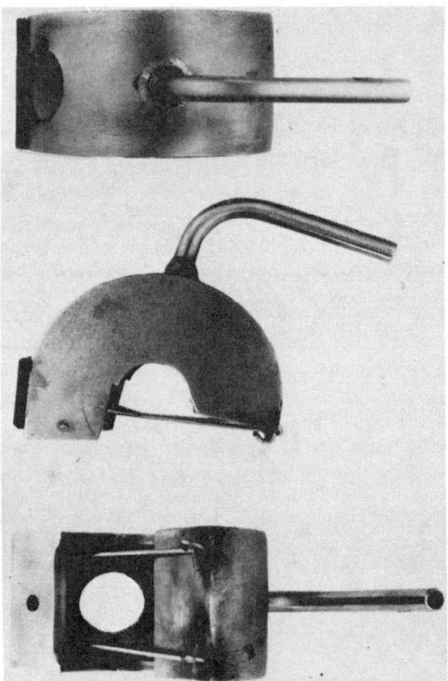

Fig. 17-4. This small manual trailing shield is very short as the travel speed is very slow.

for the back-up gas should always be as large as possible to reduce gas turbulence. Fig. 17-6 shows a hold down clamp and a backing fixture for longseam welds. The crown and penetration of a test weld made in this type of fixture is shown in Fig. 17-7.

Tubing and cylindrical parts may be blocked by tape or dams for gas shielding the backside of the weld as shown in Fig. 17-5. All of the connection fittings for the argon gas must be tight and the hoses must not leak. Any leak in the system will cause contamination of the material during welding.

Since considerable gas is used for the shielding of the metal, always have sufficient gas on hand before starting the operation. If the operation is quite long, several bottles may be interconnected for a sufficient supply.

### Tackwelds on Titanium

Tackwelds should be made to maintain joint alignment for welding. The following considerations apply:

1. The joint must be protected from contamination.
2. Use the same filler material to make the entire weld.
3. If the joint requires full penetration, the tackweld requires penetration.
4. Keep the tackweld as small as possible. If possible, make the tackweld smaller than the main weld.
5. If the tackweld cracks, leave it and then make another close to the broken tack.
6. Do not grind titanium. Use a rotary file only. Grinding grit in a joint makes good quality almost impossible.
7. The crater of a tackweld is the weakest point in a tackweld. Do not leave a crater in a tackweld.

## WELDING TITANIUM USING DCEN (DCSP)

The power supply type recommended for GTAW on titanium should be as follows:

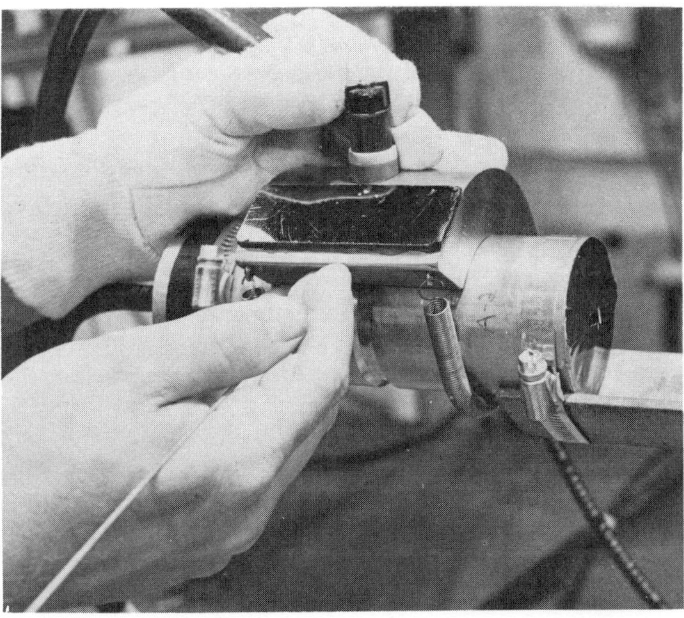

Fig. 17-5. The welder observes the puddle through the dark glass attached to the top of the shield. Note the capped end of the tube and the exit hole for the purge gas.

1. Motor generator or a rectifier.
2. High frequency arc start.

### Electrodes For Titanium

A 1 percent or 2 percent thoriated electrode tapered to a sharp point is recommended. Always use an electrode large enough to carry the required amperage.

### Gases For Use on Titanium

Argon gas is generally used for all thicknesses of titanium as the chamber, trailing shield, and back-up gas.

Helium gas may be used with argon as the torch shielding gas when automatic welding. However, trailing shield gases mix with the helium which nullifies the effect of the helium gas.

Use only gas that is known to be of the highest purity. If the gas purity is questionable, do not use until a test can be made to determine the actual gas purity.

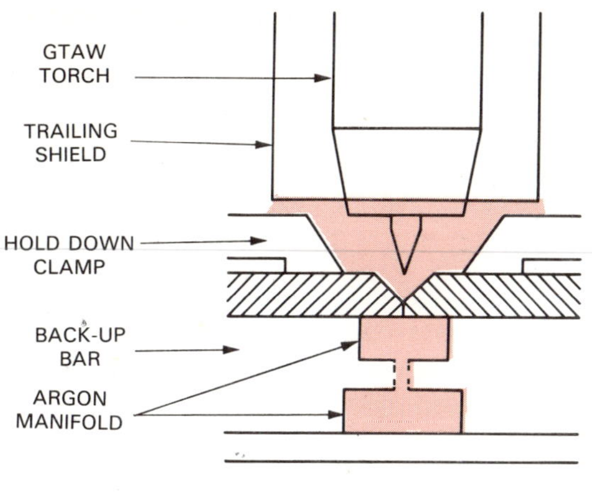


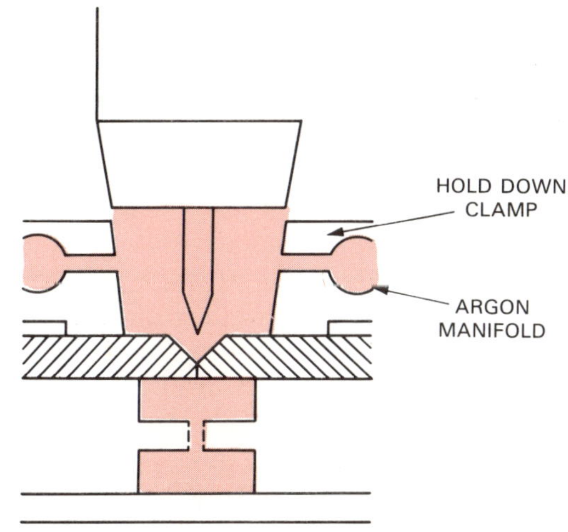


Fig. 17-6. A trailing shield is used with the design on the top. When argon is supplied from the tooling as shown on the bottom, a trailing shield is not required.

## GTAW TECHNIQUES FOR TITANIUM

1. Handle all tooling, parts, and weld wire with clean lint free gloves.
2. Check all purge gases and torch gas prior to starting.
3. Always connect ground lead with apositive clamp. Never use a spring clamp. High frequency burns on titanium can cause internal contamination and possible failure of the weldment. Fig. 17-8 shows the proper connection method of the ground lead.
4. When using a chamber for welding, always make a test weld on a scrap block to determine if the purge is adequate.

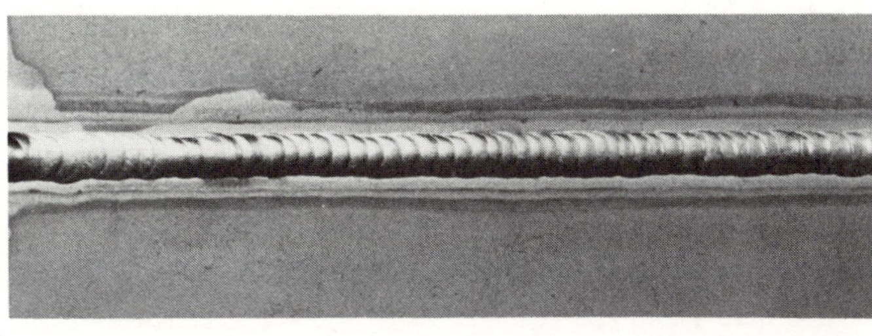

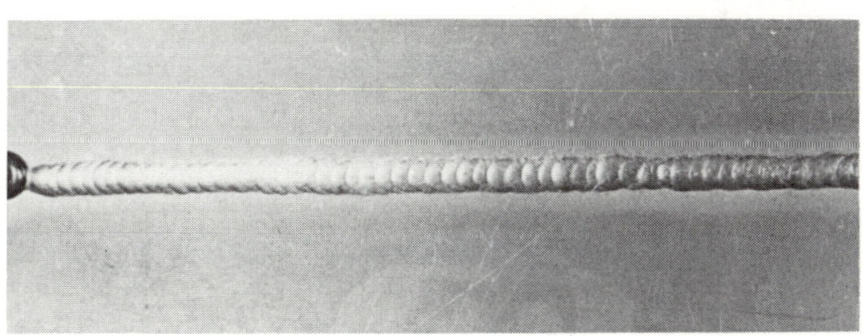

Fig. 17-7. The crown (above) and the penetration side (below) of this titanium butt weld are the same color as the base metal.

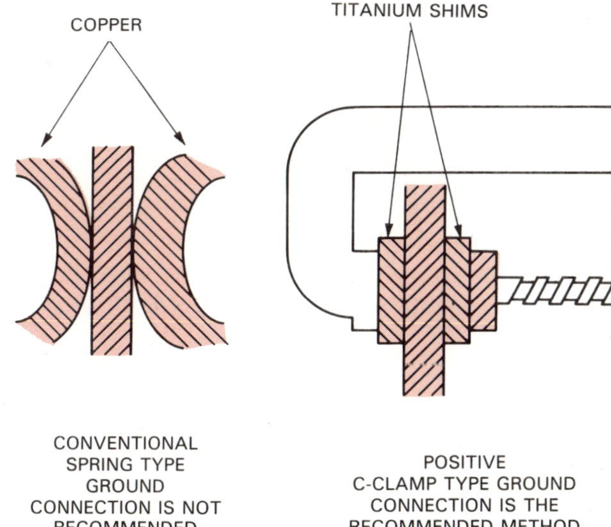


Fig. 17-8. Titanium shims are used to protect the base material from high frequency arc burns in a positive manner.

5. If the weld has a silver color the purge is good. If the weld has a gold or blue color, some contamination is present. Do not weld over this type of colored weld. If the weld is grey, the contaminated area must be completely removed by machining.
6. If the filler material is removed from the weld pool and the wire end is contaminated, it must be removed. Do not put a contaminated wire into the melt.
7. Always fill craters.

### Post Weld Heat Treatment

Post weld heat treatment shall only be done as specified by the manufacturer of the material.

## GENERAL NOTES ON GTAW OF TITANIUM

The welding of titanium is done by using proven procedures. There is no short cut method which will yield acceptable welds. Procedures that will work on many materials are not acceptable on titanium.

Some common areas that may cause contamination include:

1. Tools stored in an area where they absorb moisture.
2. Chambers opened and left open for several hours.
3. Tools, wrenches, etc., not cleaned before using in a chamber. (Always clean the back-up gas tooling to remove oil after manufacture. The oil will ruin the argon and cause severe contamination during welding.)
4. Water leaks in the water cooled welding torch.
5. Argon hoses leaking argon.
6. Argon connections loose.
7. Improper fit-up of assembly tool.
8. Argon gas supplied at too high a pressure.
9. Impure argon gas.
10. Holes in gloves in welding chamber.
11. Scratch starting arc contaminates titanium with tungsten.
12. Dirty gloves or rags.
13. Dirty alcohol or acetone.
14. Sticking tungsten into work and welding over the broken piece.

## REVIEW QUESTIONS

1. What are two merits of titanium?
2. Name two places where titanium are used.
3. What element is used to strengthen pure titanium?
4. What two specifications are used to procure titanium?
5. What are the four groups of materials in the titanium family?
6. Which group is non weldable?
7. What elements are called "interstituals?"
8. What do the letters ELI mean?
9. How are filler materials identified as ELI materials?
10. Name two chemical degreasers used to clean parts, wire, and tooling before welding.
11. What color of weld indicates good shielding without any contamination?
12. What colors of weld indicate a good weld with some contamination?
13. What colors of weld indicate the weld is badly contaminated and cannot be used under any condition?
14. Welding torches may be fitted with a special ________ ________ to reduce the gas flow turbulence from the gas nozzle.
15. Why should special precautions be taken for installation of the ground clamp to the part to be welded?
16. What operation must be done after welding wire is removed from the puddle and exposed to the air?
17. Why should back-up tooling be cleaned after manufacturing and before use in welding?

# Chapter 18

# WELDING PROCEDURE FOR MANUAL WELDING PIPE

This chapter will explain the general procedures for preparing and welding various types of pipe joints. Cleaning procedures, filler materials, pre-heating, interpass temperature, and postheating are covered in detail in the individual chapters relating to the welding of the applicable material.

## SELECTION OF DESIGN

Each type of joint has it's own merits when considering design and fabrication cost. The factors include:

1. Type of material and thickness.
2. Quality of the weld.
3. Position and accessability of the weld joint.
4. Internal flow restrictions.
5. Welder skill.
6. Production quantity.

### Socket Type Joint Design

Socket type joints are those joints which use fittings machined to accept a section of pipe. The fittings may be an elbow, bushing, reducer, T, or a special design part. Fig. 18-1 shows a typical socket type joint.

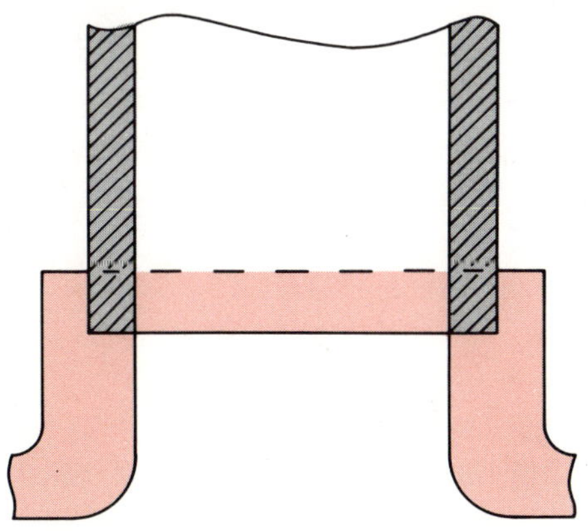

Fig. 18-1. Typical socket joint used in pipe assemblies.

### Lap Type Joint Design

The lap type of joint is used for joining ends of pipe together where a gap exists between the ends. The designs are shown in Fig. 18-2.

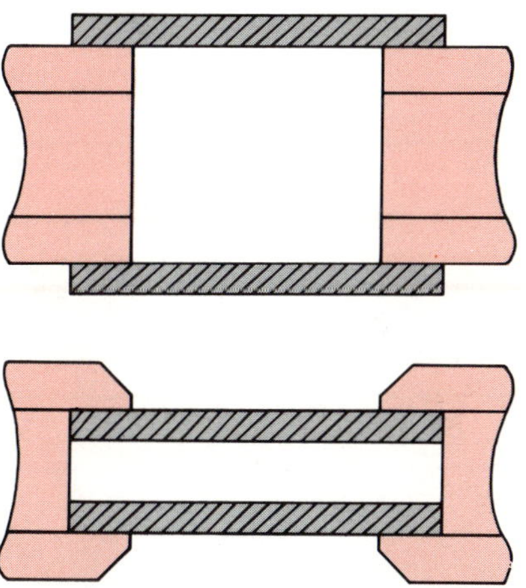

Fig. 18-2. Lap type joints are used where one pipe may be assembled inside of the other pipe.

### Laterals, T, and Y Joint Design

These types of joints are used for joining pipes with various diameters and intersections at various angles. There are many types of configurations, some of which are shown in Fig. 18-3.

### Butt Type Joint Design

Butt type joints are common throughout industry with many types of designs. These are shown in Fig. 18-4 and include:

1. Standard butt joint with land-closed joint.
2. Standard butt joint with land-open joint.
3. Special machined joint configuration.
4. Integral land joint.

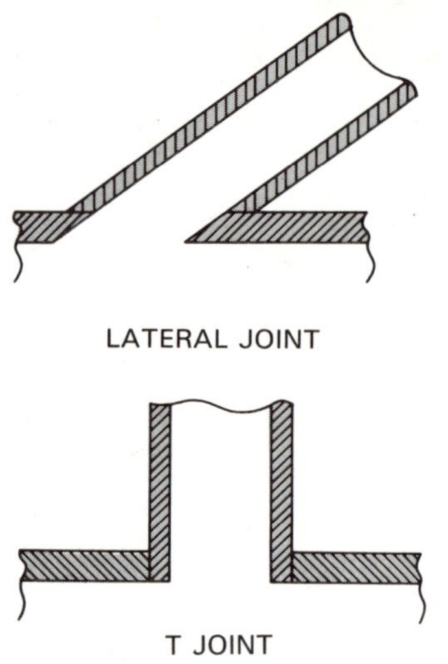


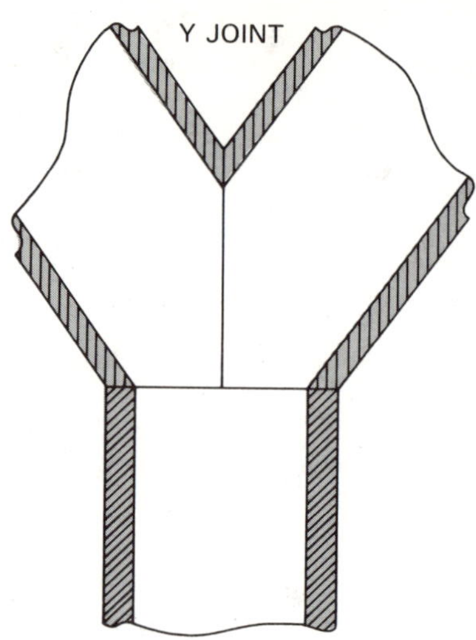


Fig. 18-3. Various types of pipe joint connections.

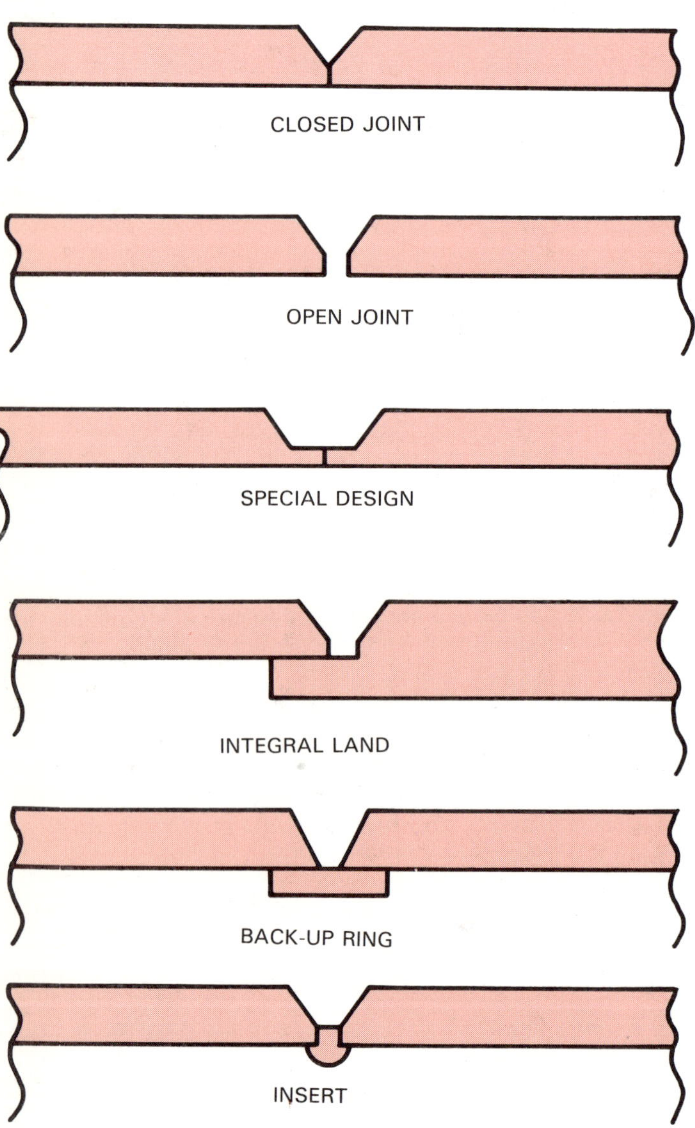


Fig. 18-4. Weld joint designs used in butt welds.

5. Back-up ring joint.
6. Consumable insert joint.

### Socket Type Joint Preparation

Socket type joints are welded with fillet welds as shown in Fig. 18-5. As there is a small amount of weld shrinkage during welding they must be prepared in a special manner. They require a preweld minimum 1/16 inch gap at the end of the pipe into the socket bore end, as shown in Fig. 18-6, or cracks may occur at the weld root.

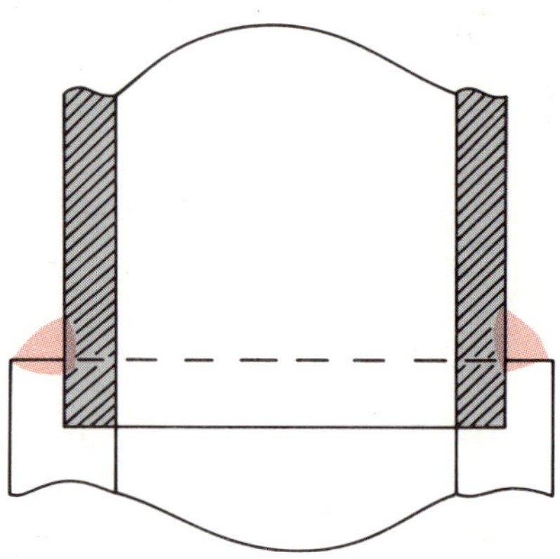

Fig. 18-5. Socket joint fillet weld.

The gap distance may be established by:

1. Inserting the pipe into the socket until it seats at the bore end. Scribe a line at the flange end, move the part out to a 1/16 inch gap.
2. Inserting a gap ring as shown in Fig. 18-7. Insert

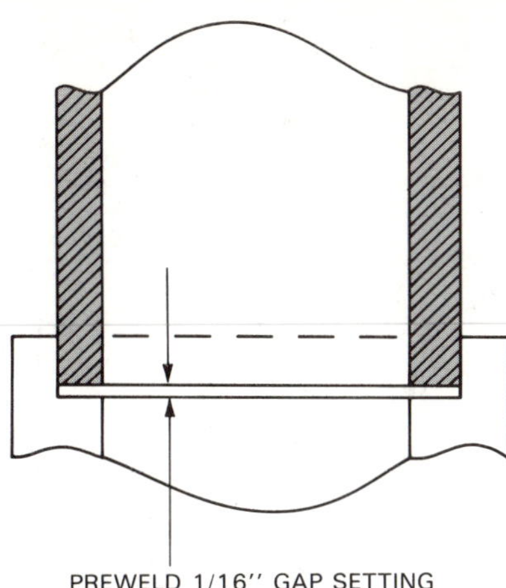


Fig. 18-6. Gap of 1/16 inch between the end of the pipe and the bore seat allows weld shrinkage to prevent cracking in the root of the weld.

Fig. 18-7. Clean the ring and pipes before assembly. Insert the ring into the recess. The end gap may be in any position. (G.A.L. Gage Co.)

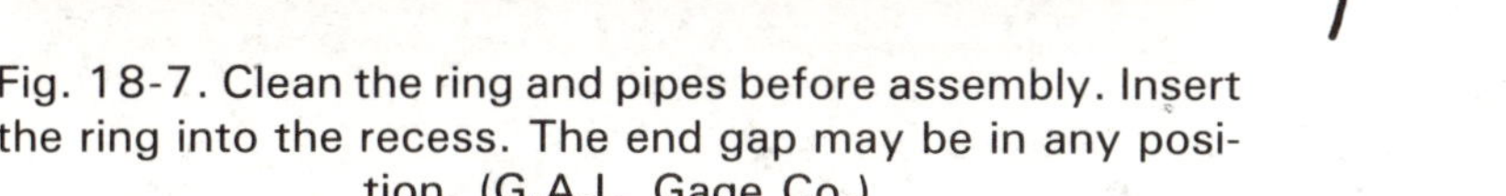

the pipe until it seats against the ring as shown in Fig. 18-8.

### Lap Type Joint Preparation

Lap type joints are welded with fillet welds. Joints of the pipe usually do not require any special preparation, as shown in Fig. 18-9.

Lap type joints with lap on the inside of the pipe may be beveled to reduce the possibility of cold laps as shown in Fig. 18-10. Cold laps are areas of the weld which have not fused with the base material.

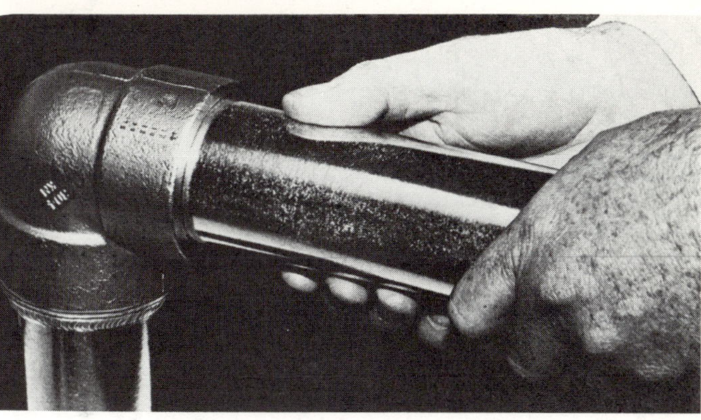

Fig. 18-8. Seat the pipe end firmly against the gap ring prior to tackwelding. (G.A.L. Gage Co.)

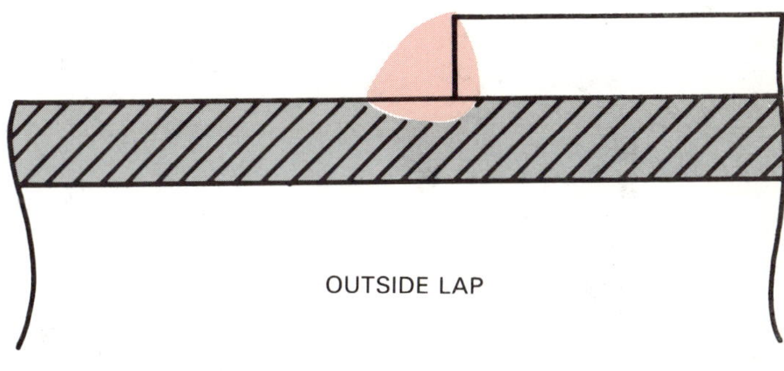


Fig. 18-9. Lap type joint with the lap on the outside of the pipe.

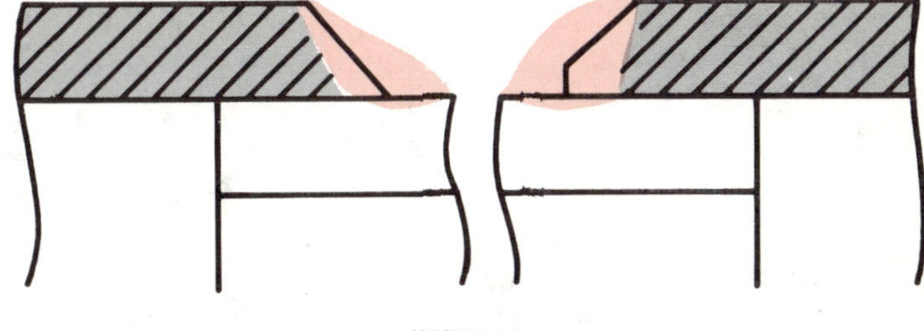


Fig. 18-10. Lap type joint with the lap on the inside of the part.

### Laterals, T, And Y Joint Preparation

These types of joints may require a butt or fillet weld or a combination of both type welds. Fillet welded joints, as shown in Fig. 18-11, usually do not require any special preparation.

Butt or combination type welds may be prepared as shown in Fig. 18-12.

### Butt Type Joint Preparation

Standard pipes and fittings are prepared at the mill

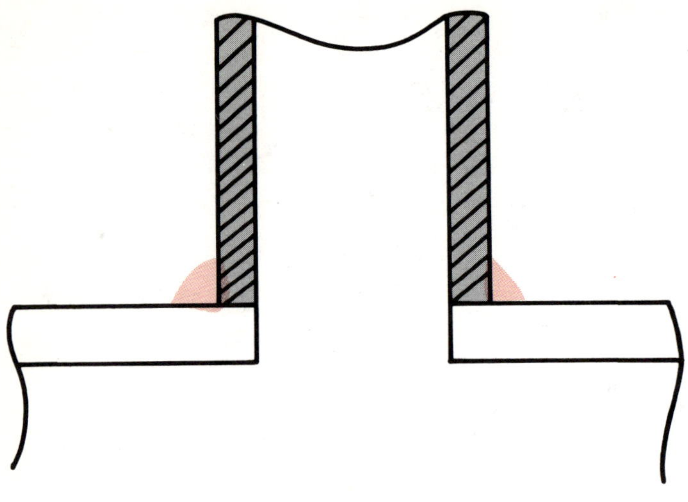

Fig. 18-11. Fillet welded T joint.

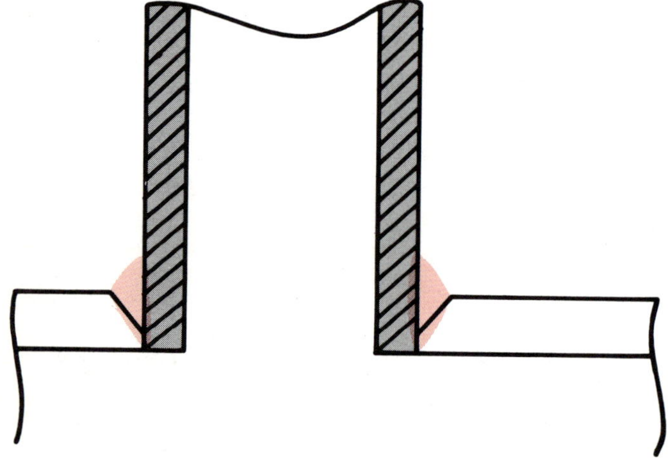

Fig. 18-12. Fillet and groove welded T joint.

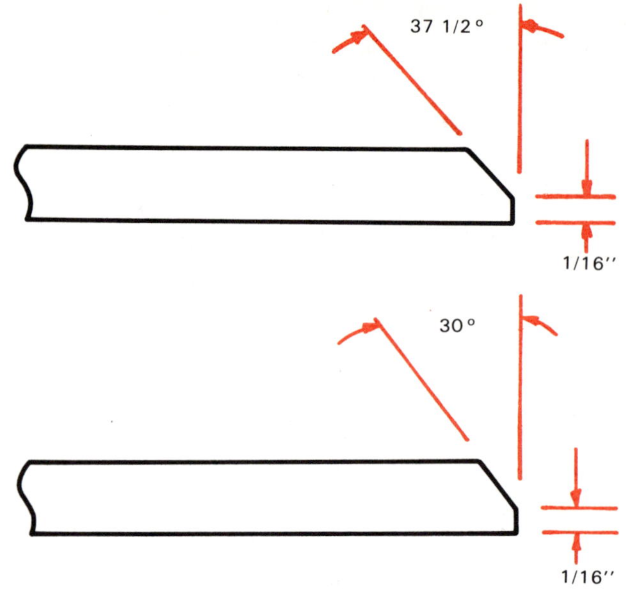


Fig. 18-13. Standard pipe weld joint preparation as prepared at the mill.

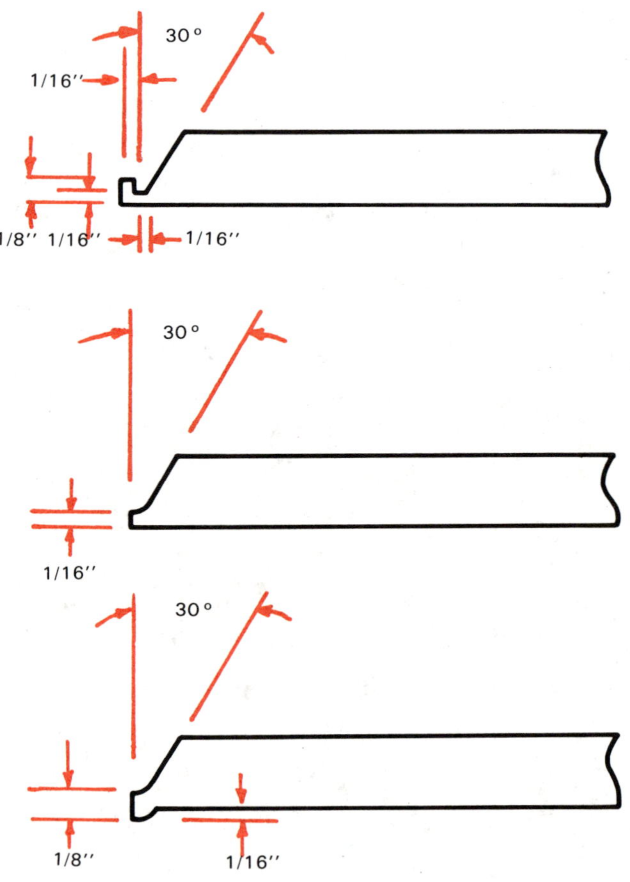


Fig. 18-14. Special purpose designed weld joints.

with a bevel and a land, as shown in Fig. 18-13. When used as an open or closed joint, no special joint preparation is required before welding.

Special machined joints are designed for a specific purpose and are more expensive to make. The designs shown in Fig. 18-14 are made from fittings, rings, or forgings where both the inside and outside dimensions may be machined. This is required if the weld joint lands are to mate evenly.

Consumable insert joints are machined to match the type of insert used. Various type joints and the inserts which may be used are shown in Fig. 18-15. Where the pipe to be prepared has considerable ovality, the pipe should be rounded prior to machining. If the ovality is not removed, the joint dimensions will not be true around the pipe face. This will cause numerous problems when welding the insert during the root pass.

Back-up ring joints require machining of the pipe inner walls to fit the back-up ring. The various types of back-up rings that may be used are shown in Fig.

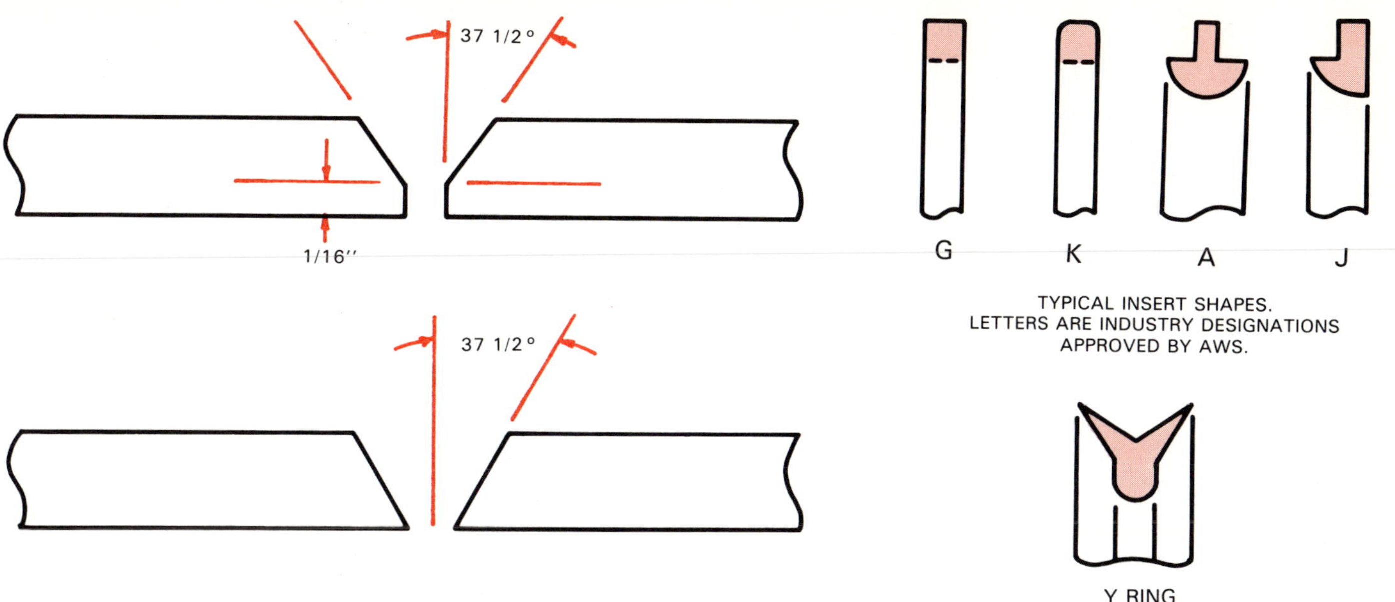


Fig. 18-15. Consumable insert types and joint preparation.

18-16. As with the other types of joints, the ovality of the pipe must be removed prior to machining. If the ovality is not removed, the pipe wall thickness may vary or the back-up ring may not fit properly during assembly.

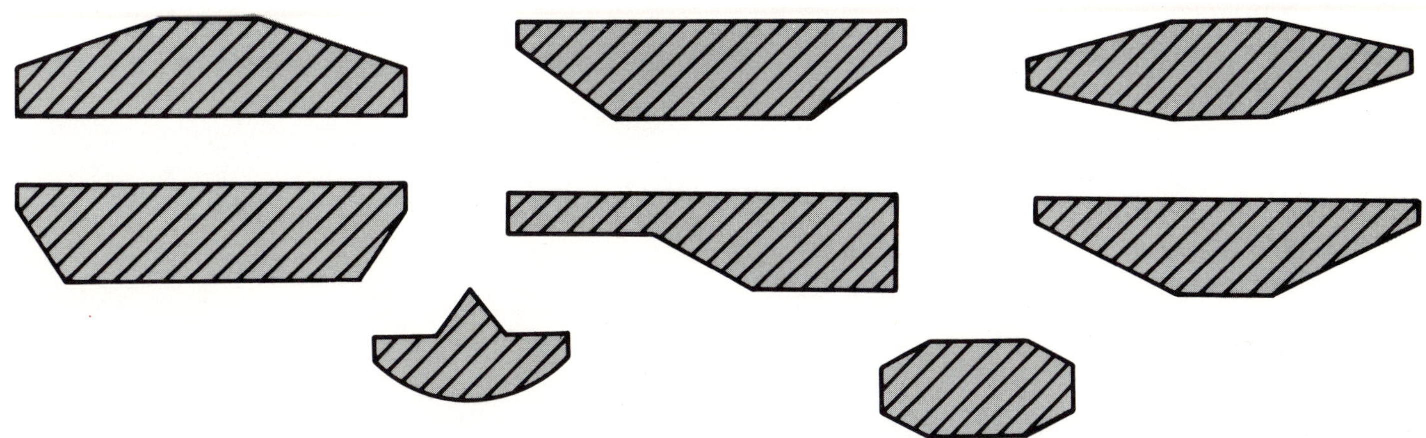

Fig. 18-16. Common commercial type back-up ring shapes.

Integral land joints require that one part of the assembly have a thicker section to make the back-up land. Some of the various designs are shown in Fig. 18-17. Ovality of the mating pipe may be a problem in this type design as in other designs, and must be considered before machining.

### Joint Preparation Protection

Joints that have been prepared by machining should be protected from damage until ready to be installed. Nicks, burrs, and dents into the machine surfaces may cause problems during assembly, welding, or both.

Steel pipe machined edges are very susceptable to rusting, especially in a heavy moisture atmosphere. To prevent rusting, you may use a light coating of oil or grease until ready for final clean-up and assembly.

## PIPE PURGING

Where pipe welds are made with the Gas Tungsten Arc Welding process, purging with an inert gas is highly recommended for the weld root and the weld area for

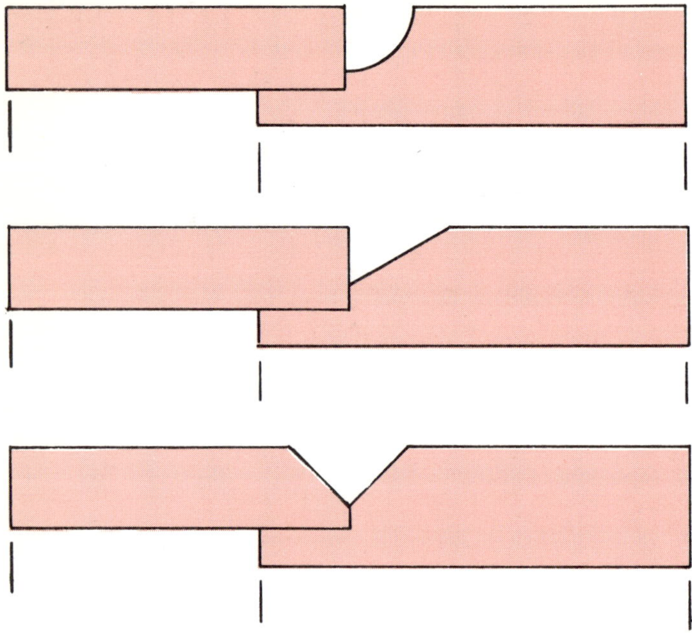

Fig. 18-17. Integral land type joint designs.

the following reasons:

1. Prevents the formation of heat scale on the ID of steel pipe. If repairs of the root pass are required later, any internal scale could be absorbed into the repair weld which may require another repair.
2. The inert atmosphere aids in the "wetting" of the penetration weld onto the inner pipe walls on butt joints that do not have back-up rings.
3. The inert atmosphere assists in the formation of the penetration weld contour on butt joints that do not have back-up rings.

The basic gases used for purging the oxygen from inside the pipe are:

1. Argon.
2. Helium.
3. Nitrogen.

The selection of the gas or gases to be used is dependent on the following criteria:

1. If the area to be purged is large enough, then nitrogen gas may be used for initial purging to remove the majority of oxygen before using argon or helium.
2. Different materials require different degrees of inertness. For example; titanium requires a very pure atmosphere to prevent contamination while steel pipe may sometimes be purged with nitrogen which may be a sufficient purge.
3. Location of the weld is an important fact to remember as helium rises and argon settles (heavier than air). For this reason, helium is a better purge for upper welds and argon is better for lower welds.
4. Weld joints using an integral back-up bar or a back-up ring may in some cases (carbon steel, stainless steels) use nitrogen as a purge gas.
5. Depending on the volume of gases required, liquidified gases are often used to provide a large volume. The use of these liquidified gases radically effects the number of cylinders required and the time lost in replacing empty cylinders.
6. Cost of the purge gases is a major factor when selecting the proper gas or gases. Nitrogen is less expensive than argon or helium. In many cases nitrogen is all that is required for proper purging. In areas where it cannot be used for the final purge, it can still be used for initial purging to considerably reduce the oxygen level.

## Purging Dams

Many types of dams or stoppers are used to restrict the purge area to a minimum. The smaller the area to purge, the lower the purging cost. In many cases, dams cannot be used and the cost rises accordingly. In each case, the design of the weldment, pipe, or joint should be considered to effect the proper purge in the minimum amount of time.

In areas where only a small number of welds are involved, the simplest method to use is shown in Fig. 18-18. The ends of the pipe are capped and sealed, purging is completed, welding is performed, and the caps removed.

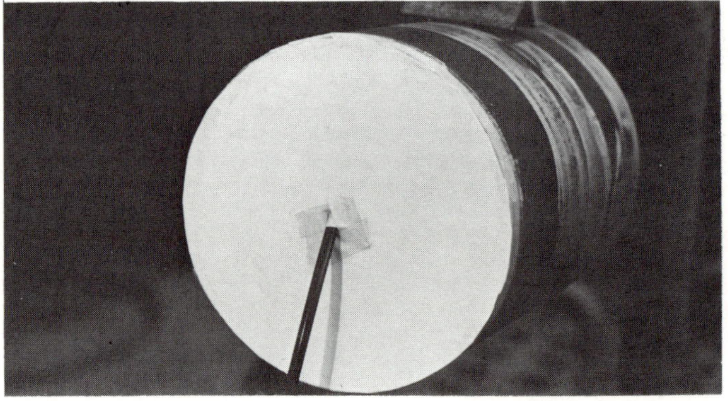

Fig. 18-18. Masking tape and cardboard are often used to make an effective seal for purging pipe.

Where several joints are to be done, various types of commercial dams may be used. These types include:

1. SOLUBLE DAMS – These dams are made of a paper which dissolves in a liquid. They are inserted into the pipe and cemented into place with a dissolvable cement. When the welding is completed, the dams are flushed away. Fig. 18-19 shows the proper installation of this type dam. This

Fig. 18-19. Special papers and cements are used where the dam is to be removed by water.

type dam cannot be used where the pipe is preheated as the heat will destroy the dam.

2. THERMALLY DISPOSABLE DAMS – These type of dams are made of heavy paper or cardboard, and inserted into the pipe ends, as shown in Fig. 18-20. They are used only where a post weld heat treatment is sufficient to ignite the material after the welding is complete. After burning, the material is flushed away.

Fig. 18-20. The paper or cardboard may be inserted in the end near the weld or anywhere in the pipe.

3. CONE PURGE DAMS – This type of a dam consists of spring loaded cones inserted into the pipe ends, as shown in Fig. 18-21 and 18-22. The cones are made of rubber or canvas, and are reusable. They are used where they can be removed after the operation is completed. This is done by pulling on the purge tube attached to the cone.
4. BLADDER DAMS – This type of dam uses small bladders or balloons placed into the pipes and expanded before purging begins. After welding, the bladders or balloons are deflated and removed from the area.

Fig. 18-21. Display of cone purge dams which may be installed near the weld joint to reduce the size of the purge area. This allows faster purge time and reduces purge gas cost. (Emmerson Hallenbeck Co., Inc.)

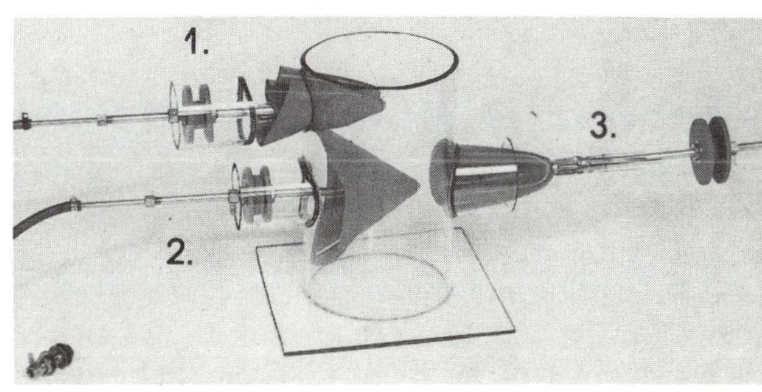


Fig. 18-22. Step 1 – Cone is inserted completely into hole. Step 2 – Cone expands to form purge area. Step 3 – Cone reverses for removal. (Emmerson Hallenbeck Co., Inc.)

## ROOT PURGING OPERATION

Purging of the weld root area requires admitting the purging gas and removing the atmosphere until the oxygen content is one percent or lower.

To achieve this low oxygen level, usually several changes in the area volume must be made. High volume, low pressure purge rates result in lower volume changes.

High flow rates work just the opposite, requiring more volume changes to reach an acceptable oxygen level. The chart in Fig. 18-23 shows typical purge inert gas flow rates and purge times in minutes for various size pipes.

When purging large volume areas, nitrogen gas is often used to reduce the initial atmosphere level. Argon is then used for the final purge. This often results in very good cost savings due to the low cost of nitrogen versus argon.

Always use a flowmeter to monitor the amount of

| Pipe Size | Purge Inert Gas Flow Rate | Purge Time Minutes | Vent Size (Inch) |
|---|---|---|---|
| 3.0 | 20 | 3 | 1/16″ |
| 4.0 | 20 | 3 | 1/16″ |
| 5.0 | 20 | 5 | 1/8″ |
| 6.0 | 20 | 6 | 1/8″ |
| 8.0 | 25 | 8 | 1/8″ |
| 10.0 | 25 | 13 | 1/8″ |
| 12.0 | 30 | 13 | 1/8″ |
| 14.0 | 30 | 16 | 1/8″ |
| 20.0 | 35 | 25 | 1/8″ |

Fig. 18-23. Typical purge times and flow rates for various pipe sizes. Rates are figured for 12 inch long pipe. Where the purge area is greater than 12 inches, increase the purge time accordingly.

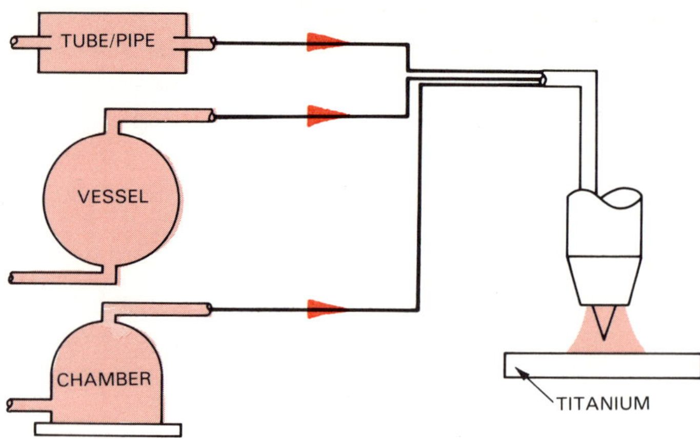


Fig. 18-24. Test to determine purity level of purge gas. Allow gas to flow over the titanium spot weld until metal has cooled to approximately 500-600 degrees F. A silver color indicates a very low oxygen level.

purge gas being delivered to the purge area. Pressure gauges do not have sufficient accuracy. Improper purging rates affect the number of gas volumes required to reach the required reduced oxygen level.

### Oxygen Level Test

To determine the oxygen level of the purge area, several methods and types of equipment may be used.

Oxygen analyzers are very reliable and offer good response time in determining oxygen level. The analyzer shown in Fig. 5-19 may be used for spot or continuous checking on single or multiple stations. Equipment of this type is often used to make purge charts of various volume areas similar to the chart shown in Fig. 18-23. After the charts are completed, volume areas can then be compared and the analyzer is not required except for very critical oxygen levels.

A simple gas purge tester is shown in Fig. 5-20. This unit uses a special light bulb with a tungsten filament and extensions for attaching the purge gas hose from the purged area. The gas is introduced into the bulb and an electrical current is used to heat the tungsten. The color of the tungsten indicates the amount of contamination. This unit does not require elaborate calibration and is considerably lower in cost than an oxygen analyzer.

Another test that can be used when no other means are available, is to connect the outlet purge gas to a GTA welding torch. Strike an arc on a piece of titanium, form a puddle, and shut off the current. Hold the torch over the puddle until it cools. A silver color of the weld indicates a very low oxygen level. Fig. 18-24 shows how the testing system is established and how the test is made.

## TACKWELDING AND WELDING

Aluminum and magnesium are welded with ACHF (Alternating Current High Frequency) type current. All other materials are welded with DCEN (DCSP) type current. Refer to the chapter covering the type of material to be welded for specific machine set-up, tungsten type, etc. High frequency arc starting and remote control of the welding current is not mandatory. However, hand controls for high frequency start and current control enables the welder to have infinite control of the weld puddle and overall weld quality.

### Fit-Up (Butt Joint Procedure)

The pipe weld joint should be fit-up if possible to eliminate:

1. Out-of-round condition. If the pipe is not machined on the outside and inside, the out-of-round may be corrected by using a tool as shown in Fig. 18-25.
2. Non parallel root faces. The faces must be filed or ground to produce an even gap to accept the insert or to make an open butt root joint.

The pipe insert (if used) should be fit-up to eliminate:

1. Uneven contact on the pipe. Bent or twisted rings must be straight and contact the joint to provide proper heat flow for penetration on the root pass.
2. End gap. Back-up rings which do not fit on the ID of the pipe must be split and fitted prior to welding. Tackweld the ring to the proper diameter before installation.

   Inserts furnished in coils must always be cut and fitted to the proper end gap before installation and tackwelding.

### Tackwelding Procedure

Prior to any tackwelding, the pipe and rings or in-

Fig. 18-25. Alignment tools are used to remove out-of-round condition and to establish groove opening. (G & W Products, Inc.)

Fig. 18-26. A dual type flowmeter instrument allows the welder to monitor both torch and purge gas at the welding station during the entire operation. (Distribution Design, Inc.)

serts, if used, must be cleaned. After cleaning, place the rings and purge dams into place and start the purge gas flow. The gas flow control unit, as shown in Fig. 18-26, may be used by the welder to monitor both the torch and purge gas flow to the weld joint.

Tackwelding may start only after the pipe is purged to the proper atmosphere. Improper purging may cause contamination of the insert or ring, and may result in incomplete fusion, melting, or wetting of the penetration onto the pipe wall.

Fig. 18-27 shows a Y ring being tackwelded into place on one end of a pipe joint using another torch as a method of purging.

Tackwelds must be feathered at each end to reduce the amount of weld cross-section before making the open joint root pass weld. This is to insure proper fusion into and out of the tackweld. If this is not done properly, incomplete fusion may result.

Do not use grinders for this purpose. Grinding wheel residue will cause porosity. High speed mill cutters or tungsten carbide ball routers as shown in Fig. 18-28 may be used for this purpose.

Fig. 18-27. Purging Y ring tackweld. (Weldring Co., Inc.)

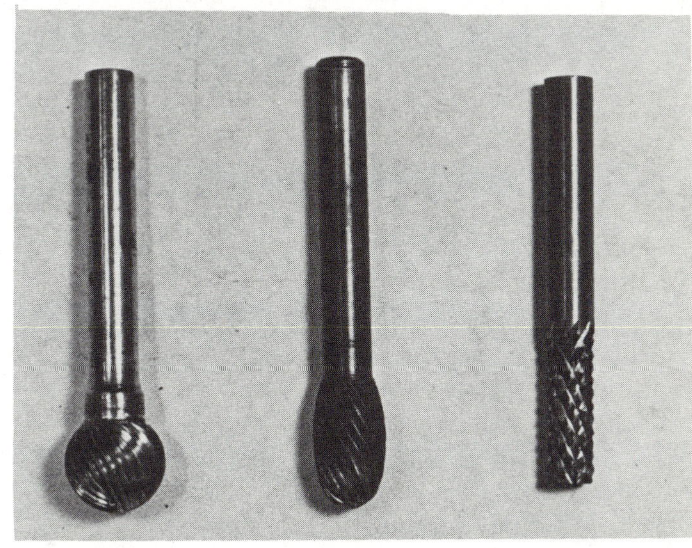

Fig. 18-28. Various types of cutters and routers used to feather tackwelds.

### Open Butt Weld Procedure

Fig. 18-29 illustrates one method which may be used to align and space a butt type of joint. Since this joint design uses no inserts, the tackwelds will shrink after welding. Always allow for this condition by using a larger spacer than the final gap desired.

The tackwelds will become part of the root pass. Therefore, care must be taken to add sufficient heat and filler material to form the under bead properly. Improperly made tackwelds cause many problems when the root pass weld is made.

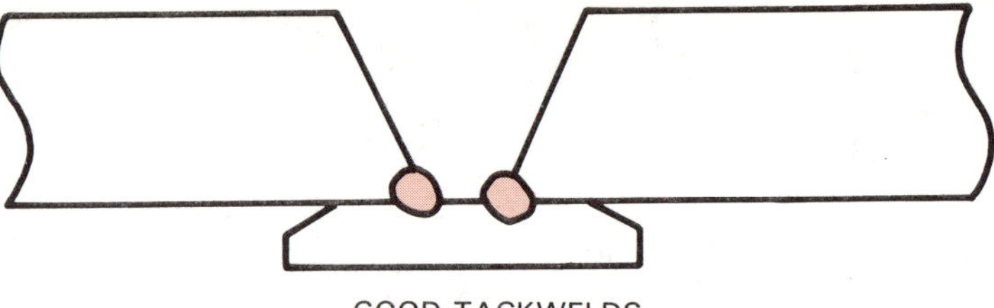


Fig. 18-30. Small fillet tackwelds adequately hold back-up rings in place.

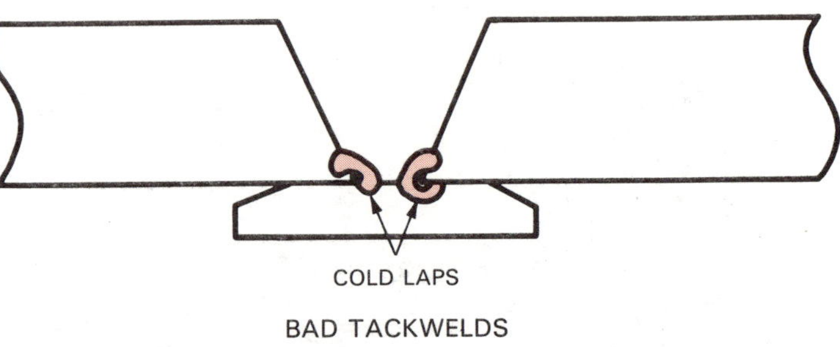


Fig. 18-31. Large fillet tackwelds may result in cold shuts between the ring and the pipe joint weld.

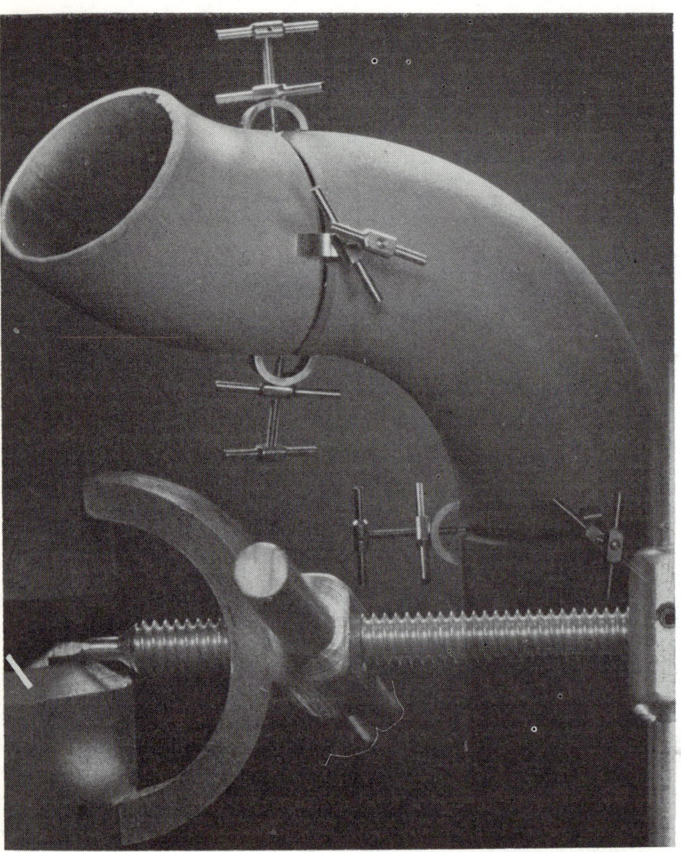

Fig. 18-29. Assembly tools for aligning pipe joints. (G & W Products, Inc.)

### Procedure for Back-Up Rings

Fig. 18-30 shows a small fillet weld securely holding the back-up ring in place. Use only enough filler wire to hold the ring in place. Penetrate into the joint to prevent cold shuts. Fig. 18-31 shows a cold shut defect which may result in a defective weld.

### Procedure Using Inserts

Fig. 18-27 shows how tackwelds are made to hold inserts in place. In most cases, filler wire is not used in tackwelding. The required metal is obtained from the insert or the pipe land. If filler wire is used, the wire composition must match the insert material.

## WELDING TECHNIQUES

Maintaining the torch and filler wire alignment during a pipe weld is extremely difficult under the best of circumstances. To assist the welder, a method called "walking the cup" is commonly used. With this method, the torch gas nozzle is rounded by grinding to form a radius and the cup is then kept in contact with the joint bevels during the welding. During welding, the torch is oscillated and at each outward swing end, the gas nozzle is "walked" forward along the joint. The welder may, by raising or lowering the tungsten from the weld, increase or decrease the amount of welding current. A dwell or hold point at the end of the oscillation also decreases the heat in the molten pool, which allows some freezing of the weld. Fig. 18-32 shows the preparation of the gas nozzle and the placement of the tungsten tip relative to the nozzle.

The placement of the torch and wire for fixed vertical up-hill welding is shown in Fig. 18-33. The placement of the torch and wire for fixed horizontal welding is shown in Fig. 18-34.

### Torch Manipulation With Pulsers

The need to manipulate the torch is greatly reduced when pulsers are used. The pulser raises and lowers the

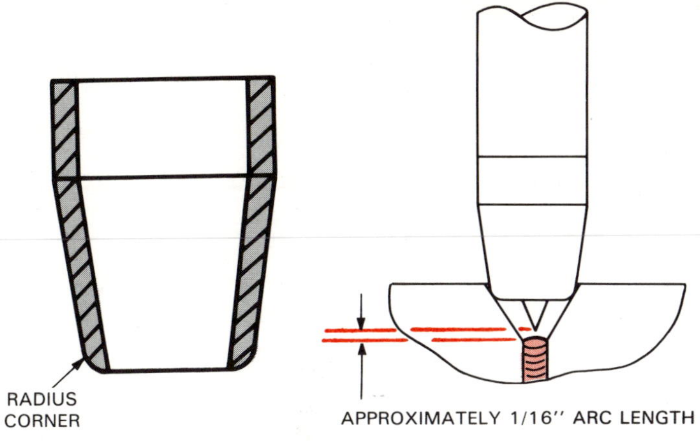


Fig. 18-32. Gas nozzle preparation and tungsten extension from the nozzle for "Walking the Cup" welding technique.

Fig. 18-33. Torch and wire positions for fixed position, vertical uphill welding.

Fig. 18-34. Torch and wire positions for fixed position, horizontal welding.

welding current automatically to preset levels which decreases the use of oscillation.

The operation of the pulser includes:

1. Current high level (molten pool forms).
2. High current level dwell time (weld wire added).
3. Low current level (molten pool freezes).
4. Low current dwell time (torch moved).

With this type of control, the weld is formed as a local spot weld. Then the torch is moved the correct distance (about one half the diameter of the spot weld) forward to form the weld again. The completed weld becomes a series of overlapping welds.

The sequencing of the pulser should be established by testing to insure adequate fusion into the root of the joint. Welder travel speed is then synchronized with the pulser operation.

Pulsers for this type of operation are available on many power supplies, or they may be placed into the system as an auxiliary item. Fig. 4-3 shows an auxiliary pulser. Where pulsers are available, they may be used to make the entire pipe joint weld regardless of the design. Pulsers generally are used to their best advantage to make root passes.

### Back-Up Ring Weld Root Passes

Welds of this type are made with the addition of filler wire. With the addition of filler wire, the cross section of the root pass is increased to prevent cracking. The wire should be always added into the front of the puddle to prevent cold shutting of the joint edges and the back-up ring. A slight oscillation of the torch may be used to assure adequate melting of the side walls into the back-up ring.

### Open Butt Weld Root Passes

The application of filler material is required in this type joint to establish the root pass with sufficient metal to form the penetration contour and prevent weld cracking. Metal thinning occurs during cooling of the weld metal and an insufficient cross section of the weld will often cause cracking.

Filler material diameter is very important in this operation. The wire diameter must be slightly larger than the opening, as shown in Fig. 18-35.

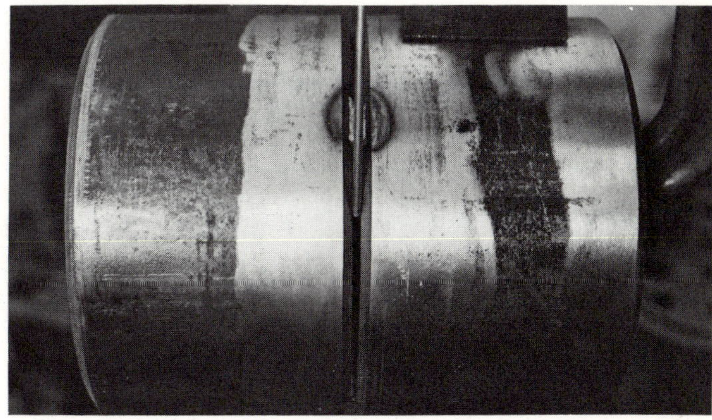

Fig. 18-35. The wire is slightly larger in diameter than the groove opening so that the wire will rest on the bottom of the V-groove walls.

Apply the wire directly in the center of the joint opening. Melt both side walls and wire together with a slight oscillation movement. Do not move the wire from the puddle. This disturbs the proper flow of metal and the cross section of the weld will vary. Fig. 18-36 shows a root pass weld with sufficient cross section, and a shrink crack caused by an undersized weld.

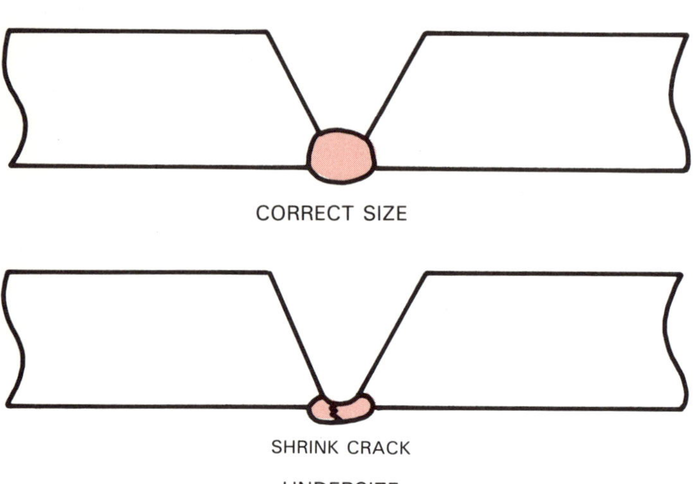


Fig. 18-36. Weld root passes with insufficient cross section may crack during or after welding.

#### Insert Ring Weld Root Passes

Weld joints with inserts rely on the melting of the insert to provide sufficient root pass cross section metal.

The addition of filler wire disturbs the heat flow with possible incorrect melting of the insert. In all cases, the insert ring must be melted completley to form the proper contour.

A slight oscillation of the torch may be used to assure adequate heating of the joint edges and proper fusion.

Fig. 18-37 shows both the proper and improper melting of an insert ring and the contour obtained.

#### Roll Welding Root Passes

This type of welding is done where the pipe may be rolled continuously or by steps to complete the operation. Fig. 18-38 shows the torch and wire alignment where the welder may continuously work on the top of the pipe.

Hold the torch at the angle shown. Always direct the tungsten to the center line of the pipe joint. Maintain the rotation speed to maintain the torch on the up-hill side of the joint center line.

#### Horizontal Welding (Pipe Vertical) Root Passes

Welding pipe in the vertical position requires very close control of the torch angle and the addition of the proper amount of filler wire at the right time. Because of gravity, the molten metal will tend to sag as shown in Fig. 18-39. Therefore, make the welds as cool as possible with the lowest possible amperage. Oscillation should be avoided. Sufficient filler metal should be added to freeze the molten pool each time wire is added. Use stringer type beads.

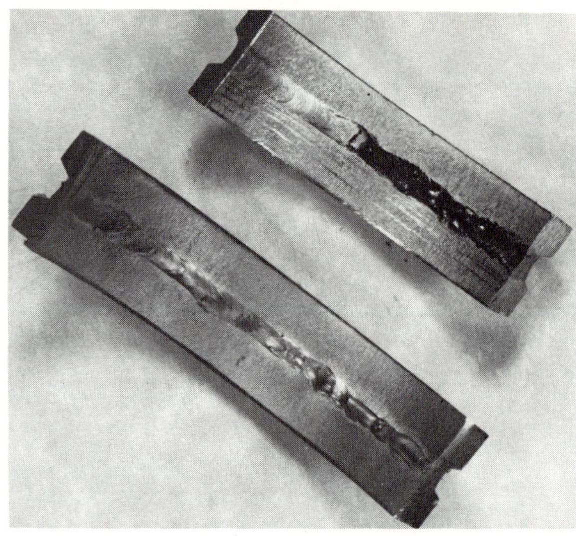

Fig. 18-37. Improper melting of the insert ring caused these defects. (Weldring Co., Inc.)

Fig. 18-38. Torch and wire alignment for welding pipe which is rolled during the operation.

#### Vertical Welding (Pipe Horizontal) Root Passes

The weld is always started near the bottom of the pipe and the weld is made vertical upward. A slight oscillation movement of the torch will insure proper melting of the side walls for complete fusion.

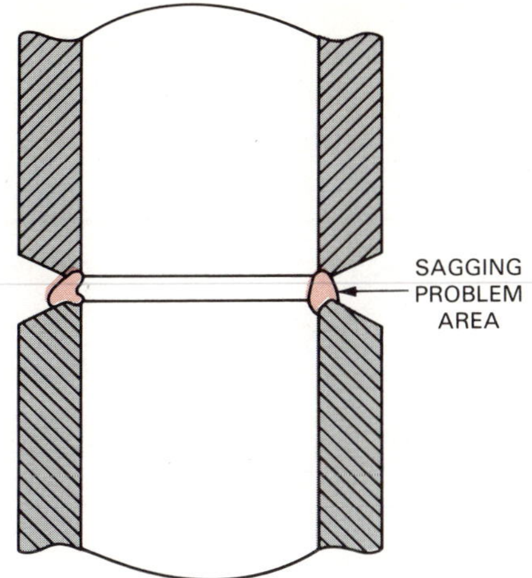


Fig. 18-39. Horizontal in-position groove weld root pass problem areas are caused by gravity sagging the weld.

The major problems encountered in this position weld with open or insert joints, is suckback on the bottom of the joint and excessive penetration at the top of the joint, as shown in Fig. 18-40. As the weld is made at the bottom, the molten pool must be chilled by adding filler metal or adjusting the torch motion to the sidewall. This allows the puddle to freeze for a moment, preventing or diminishing the suckback.

As the welder works upward, the entire joint heats which requires faster travel speed and faster application of the filler wire. The heat rising in the joint will make the puddle very fluid causing over penetration. Oscillation of the torch to the joint sidewalls will also aid in cooling of the molten pool and diminish the over penetration.

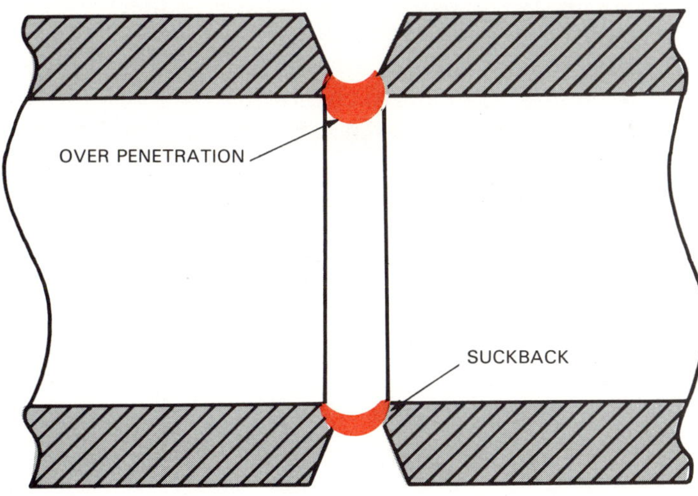


Fig. 18-40. Vertical in-position groove weld root pass problem areas.

Flat insert rings used in this type joint, may be installed offset to aid in reducing these problem area. Fig. 18-41 shows the flat ring installed with heavier section on the bottom of the joint and the lighter section at the top of the joint.

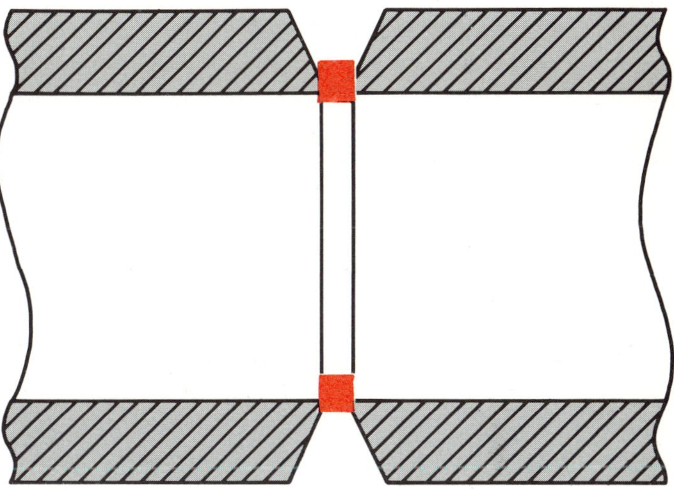

Fig. 18-41. Placement of I ring type inserts aids in correcting problems of suckback and over penetration.

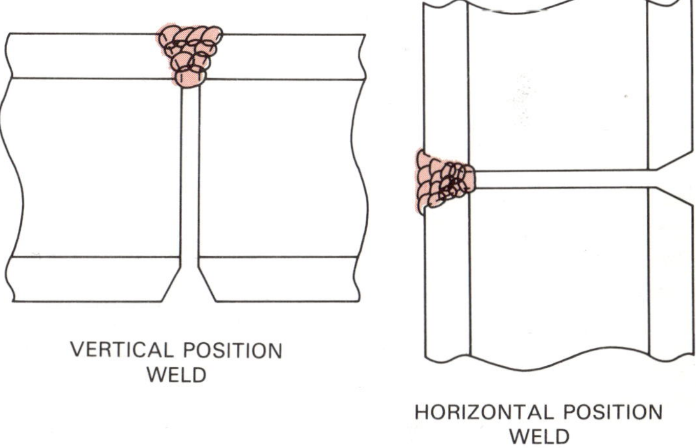


Fig. 18-42. Sequence of passes and layers for filling groove joints and making crown passes. Always keep weld layers level. When welding in the horizontal position, always start a new layer at the bottom of the joint.

## Filler and Crown Welds

Filler passes should be placed into the joint to obtain the proper fusion to the previous weld and the pipe sidewall. The addition of too much wire will cause excessive crowning and possible lack of fusion at the joint interface.

The pass placement for vertical and horizontal position welds is shown in Fig. 18-42.

Crown welds should flow evenly into the pipe walls without undercut and excessive height as shown in Figs. 18-43 and 18-44.

For specific welding procedure information, check the chapter regarding the base material to be welded.

Fig. 18-43. Multi-pass crown welds always overlap each pass one-half of the previous pass.

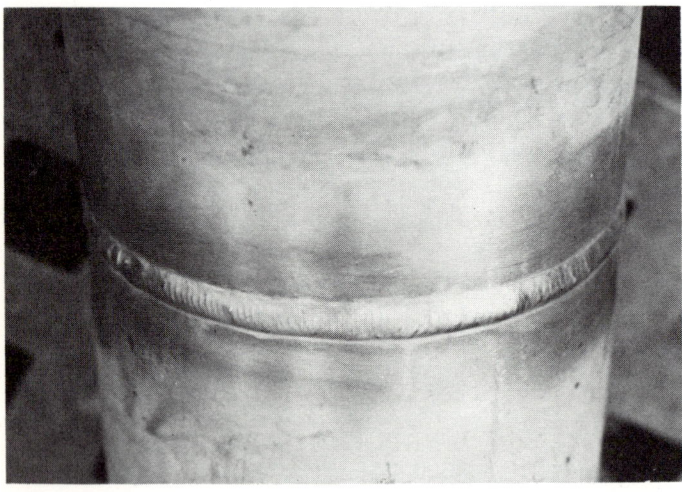

Fig. 18-44. This single pass crown weld was made by weaving back and forth across the joint. This is also called a wash bead.

## REVIEW QUESTIONS

1. What are the major factors involved in selecting pipe weld joint designs?
2. Where are socket type weld joint designs used?
3. Where would a lap type joint design be used?
4. There are 6 basic types of pipe butt joints. Name them.
5. What is the cause of weld root cracking in socket type joint welds?
6. What problem may be encountered when welding lap type joints when the lap is on the inside of the joint?
7. How is the end of standard mill pipe prepared for welding?
8. What one problem is encountered in the assembly of any pipe joint?
9. What are the benefits derived from purging butt joints where the weld extends into the interior of the pipe?
10. Which gases may be used for purging.
11. When purging large areas, which gas is the cheapest to use for the initial purging?
12. Which gas is lighter, argon or helium?
13. Which gas may be used for purging welds that have a back-up bar? Is a purge really necessary? Why?
14. Where would a liquified gas supply be used?
15. What is the basic reason for using purge dams?
16. List 4 types of purge dams.
17. What type of equipment is used for admitting purge gas?
18. Do low or high gas flow rates work best to reduce oxygen level within the purged area?
19. What is the maximum oxygen content level for purge gas before welding?
20. What operation must be done after tackwelding, and before root pass welding on open butt joints?
21. What type of equipment is used to perform the operation in question 20?
22. May pulsers be used on all types of pipe welds? Where do they perform best?
23. Is a filler material used when welding root passes with inserts?
24. What are the two main problems encountered when welding root passes on a vertical weld?
25. What is the main problem encountered in welding root and filler passes on a horizontal weld? What can be done to minimize the problem?

# Chapter 19

# WELDING PROCEDURE FOR MANUAL WELDING DISSIMILAR METALS

Welding dissimilar materials is often required to:
1. Fabricate weldments of different materials.
2. Overlay the base material to prevent corrosion, oxidation from heat, and wear.
3. Maintenance or repair of worn parts.

## MATERIAL COMBINATIONS/APPLICATIONS

Many material combinations are possible using the GTAW process to make the weld. These may include:
1. Steel alloys to steel.
2. Steel to cast iron.
3. Steel to stainless steel.
4. Steel to nickel.
5. Stainless steel to nickel.
6. Stainless steel to Inconel.
7. Copper to steel.
8. Copper nickel to steel.
9. Copper aluminum to steel.
10. Silicon bronze to steel.
11. Surfacing alloys to iron base metals.
12. Alloy metal to a non-alloy metal.

The application of the weld to join the various materials include the following:

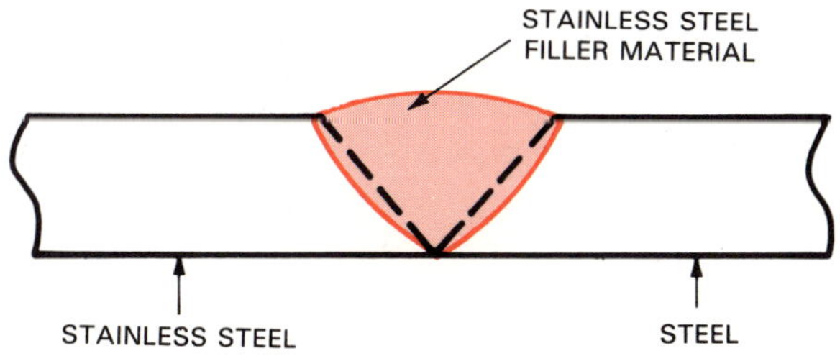


Fig. 19-1. Stainless steel filler material is used in many steel-to-stainless steel combinations where ductility is of prime importance.

1. BUTTED MATERIALS, as shown in Fig. 19-1. Depending on the alloy mix in the weld desired, one of the parent materials is used for filler material.

   Fig. 19-2 shows a dissimilar metal being used as a filler material. The filler material is compatible with both parent materials.

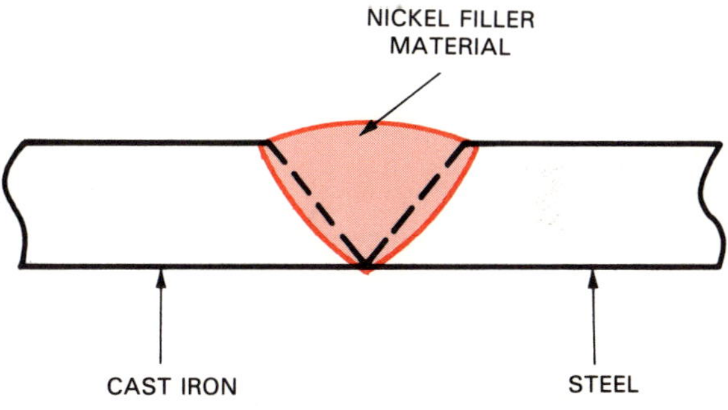


Fig. 19-2. Nickel filler material is compatible with both cast iron and steel.

2. BUTTERED MATERIALS are used to join materials that are very different. However, each material must be "buttered" with a material that is compatible with the filler material used to make the final joint. Fig. 19-3 illustrates this condition.
3. CLADDED MATERIAL as shown in Fig. 19-4 is used extensively in the manufacture of processing equipment. The "clad" is bonded to the base material at the rolling mill. The thickness of the "clad" will vary depending on the final use.

   The welding materials used must match the heavier base material and the cladding.
4. OVERLAYED MATERIALS are similar to a clad

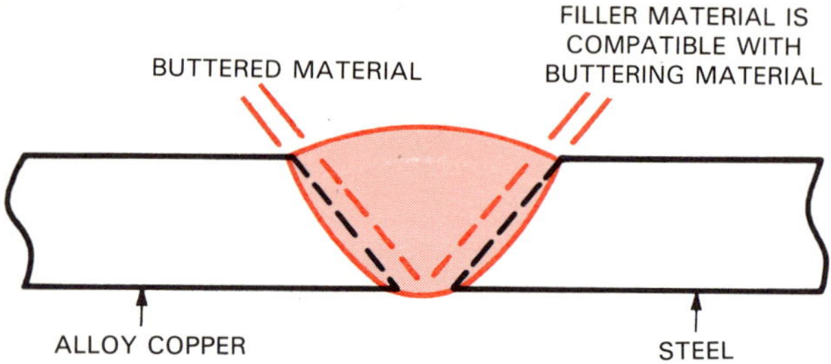


Fig. 19-3. The "buttering" material is applied to each material before the joint filler material is used.

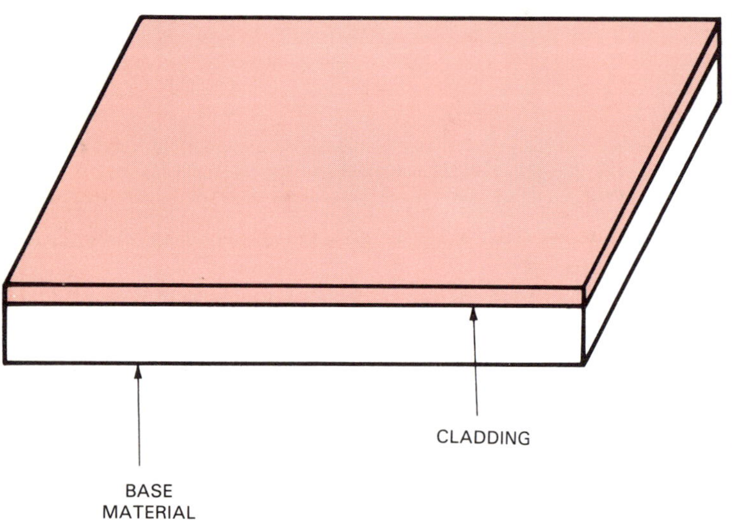


Fig. 19-4. Welds into the clad require matching filler materials. Welds into the heavier base material require filler materials for strength, ductility, and other mechanical properties.

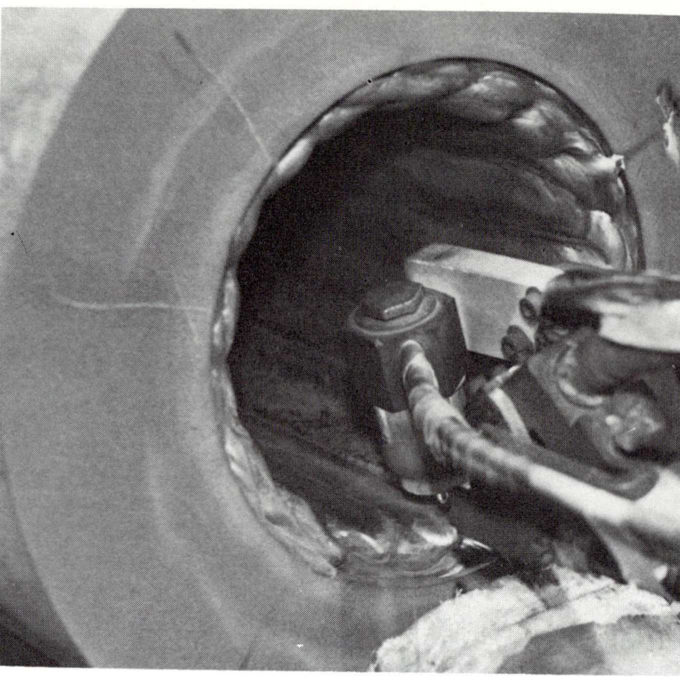

Fig. 19-6. Several layers of cladding have been applied to the bore of the tube. The special GTA welding torch is oscillated during the operation to widen the weld bead.

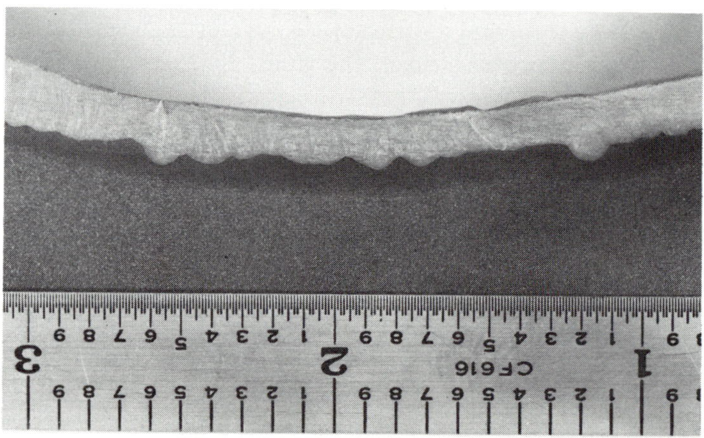

Fig. 19-7. A cross section of the part is used to measure the height of the cladding and the penetration pattern. Note the darkened heat affected zone under the cladding.

or buttered joint, however overlays are generally thicker. As the overlay thickens, the amount of dilution decreases until the deposit contains the filler material chemistry. Fig. 19-5 shows this condition. Fig. 19-6 shows a GTAW bore cladding operation. Fig. 19-7 shows a cross section of a bore cladded weld.

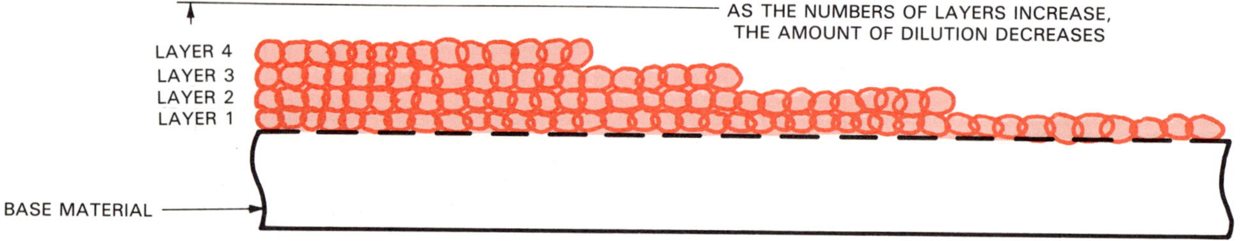


Fig. 19-5. The number of weld layers required to achieve the desired chemistry is determined by testing the final joint design preparation by chemical analysis.

### Butt Joints of Dissimilar Metals

BUTT JOINTS WITH SQUARE EDGES, as shown in Fig. 19-8, are used only where the material thickness can be welded in a single pass. This type of joint has considerable dilution between the parent materials and the filler materials.

Fig. 19-8. Square groove weld joint designs require a filler material compatible with each of the base materials due to the considerable amount of dilution.

BUTT JOINTS WITH V GROOVES welded from one side, as shown in Fig. 19-9, are used to weld thicker material with multiple passes. Dilution is high at the edges of the joint and diminishes near the center of the joint.

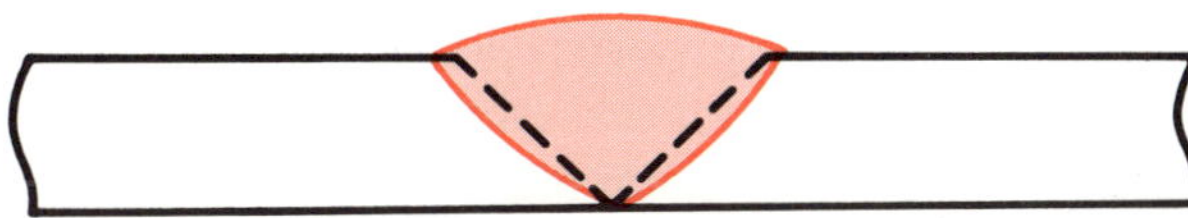

Fig. 19-9. Stringer type beads reduce penetration and dilution when welding V groove welds. Wash beads should be avoided if possible, to reduce heat input and the amount of dilution from the base materials.

BUTT JOINTS WITH DOUBLE V GROOVES welded from both sides, Fig. 19-10, are used to weld thicker material with multiple passes on each side. Distortion and dilution of the parent metal is minimized as less metal is required to fill the joint.

BUTT JOINTS WITH SINGLE OR DOUBLE BUTTERED EDGES are shown in Fig. 19-11. The buttered material is applied with sufficient height to achieve weld metal chemistry to match the filler material composition. Prior to welding the joint, the preparation desired for the final weld is prepared from the "buttered" material.

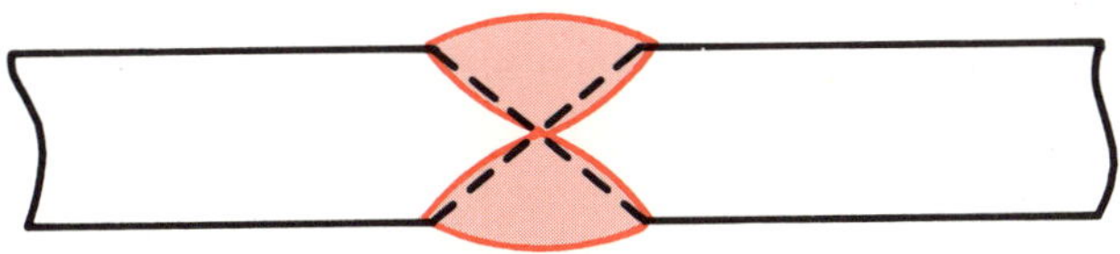

Fig. 19-10. Double V groove welds have much less dilution of the base materials since less welding is involved. Use stringer beads to further reduce dilution.

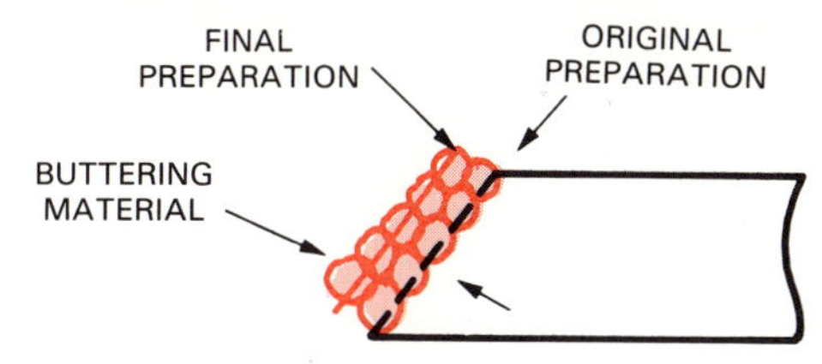


Fig. 19-11. Buttering or overlayed material must always be of sufficient height to obtain the correct material chemistry.

### Clad Material Joints

CLAD MATERIAL often requires two joint designs. One design for the base metal and one design for the cladding. The base material joint is made to standard practices. The cladding joint must be designed to allow cladding integrity.

Fig. 19-12 shows joint designs for preparing cladded materials and various weld applications.

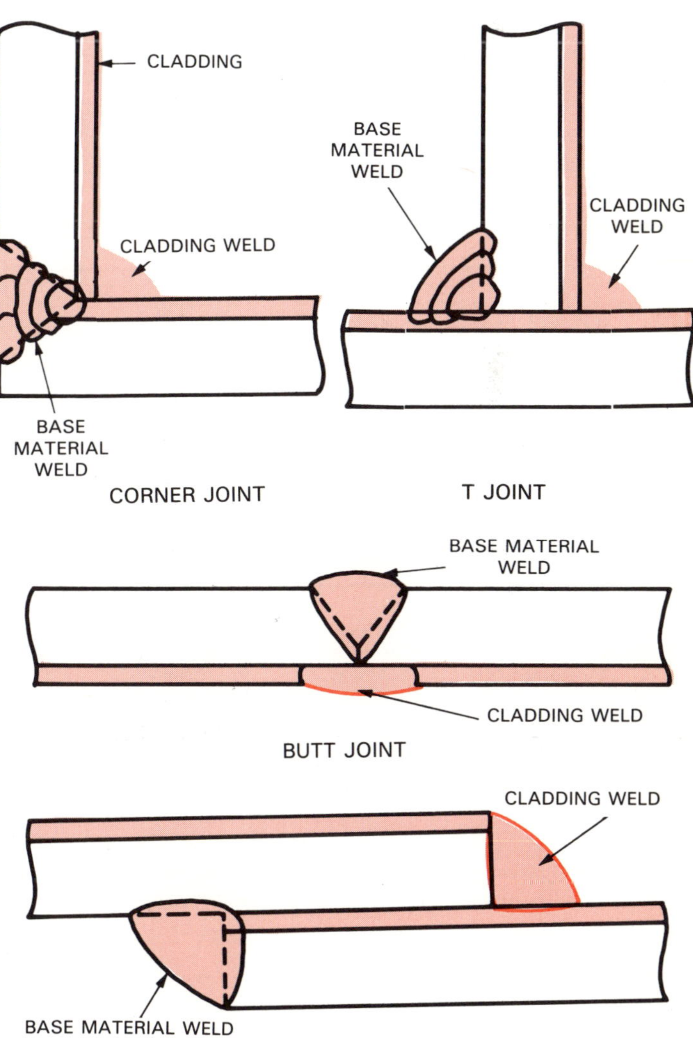


Fig. 19-12. Common joint preparations used with cladded base materials.

### Overlay Type Joints

OVERLAY TYPE JOINTS require a full weld metal chemistry at the edge of the weld. To achieve this condition, the number of layers of weld metal must be computed into the joint design, as shown in Fig. 19-13. Fig. 19-14 illustrates a grooved overlay improperly prepared which may result in improper chemistry in the final weld.

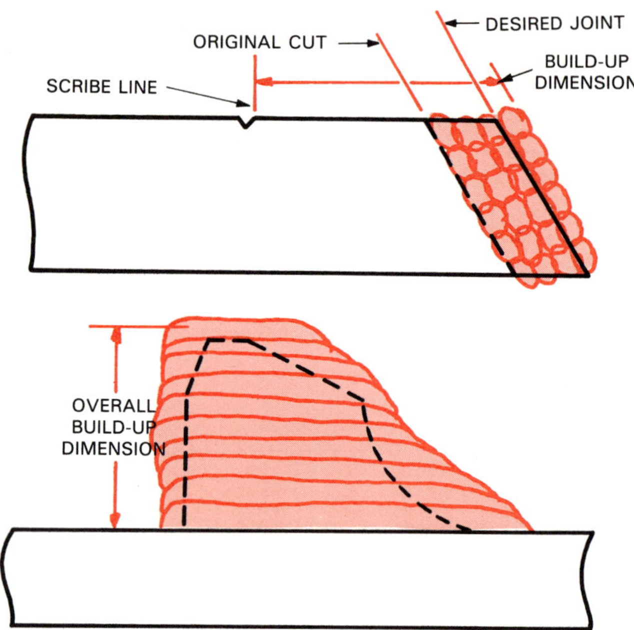


Fig. 19-13. Reference lines or points are used to assure proper overlay build-up dimensions.

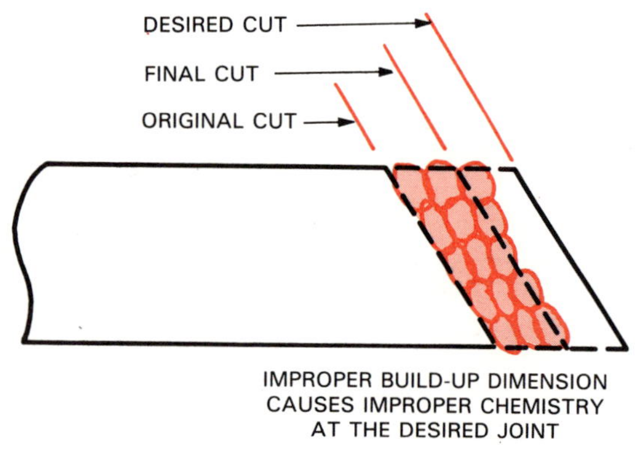


Fig. 19-14. Insufficient build-up of the overlay may result in incorrect weld metal chemistry.

## FILLER MATERIALS

The choice of filler materials for the weld joint requires analyzing the composition of the base materials, dilution percentages, and the final use of the joint. In many cases, a sufficient number of welds have been made to establish which filler materials can be used successfully. They include: Stainless steel filler metals for welding dissimilar steels, Fig. 19-15; Hardfacing and surfacing, Fig. 19-16; Filler metals for welding clad layers, Fig. 19-17; Alloys for joining clad steels, Fig. 19-18 and 19-19; Filler wires for surfacing applications, Fig. 19-20. Preheat temperatures for hardfacing are shown in Fig. 19-21.

## JOINT PREPARATION

Selecting the proper method for making the joint preparation requires consideration of various methods such as:

1. MACHINING the joint details has the advantage that closer tolerances can be held and the part is ready for welding after the machining operation. Large weldments or weldments of special design are often prepared by other means as machining cost may be prohibitive.
2. SHEARING of sheet and plate is often done as a primary operation in preparing the joint preparation. In the thinner materials, the sheared edge may be the actual joint. The sheared edge always contains a rough surface with possible entrapment of dirt or other foreign material. Further processing such as filing or grinding may be required as shown in Fig. 19-22.

   Clad materials should always be sheared with the clad surface upward. This protects the clad from scratches and nicks during the operation. The edge of the clad will have a radius formed on the corner, as shown in Fig. 19-23.

   Further processing is then required to eliminate the radius and to match the cladded areas, as shown in Fig. 19-24.
3. FLAME CUTTING of steel and steel alloys requires careful consideration of the carbon and alloy content. Higher alloy steels may harden during the operation due to the fast cooling of the metal. Improperly adjusted torches may introduce carbon onto the surface or oxidize the cut edge.

   Steels that have stainless steel or nickel cladding may be flame cut providing the clad is not more than 20 percent of the total thickness. In these cases, the heavier steel is placed on the top, as shown in Fig. 19-25. The molten steel then penetrates the alloy making the cut possible. Care must be taken to insure proper travel speeds. If the cut is lost during the operation, it is almost impossible to restart. Since the cladding is on the bottom, the cladding requires very good protection from supports to prevent scratches, nicks, etc.

   Steels with copper cladding are very difficult to cut with flame cutting. Copper conducts the heat

| Base Alloy, Type | 201, 202, 301, 302, 302B, 303[a], 304, 305, 308 | 304L | 309, 309S | 310, 310S, 314[a] | 316 | 316L | 317 | 317L | 321, 347, 348 | 330[a] | 402, 405, 410, 412, 414, 420[a] | 430, 430F, 431, 440A, 440B, 440C[a] | 446[d] | 501, 502[c,d] | 505[c,d] | Carbon Steels,[c,d] | Cr-Mo Steels[c,d] |
|---|---|---|---|---|---|---|---|---|---|---|---|---|---|---|---|---|---|
| 201, 202, 301, 302, 302B, 303[a], 304, 305, 308 | E308 | E308 | E308 | E308 | E308 | E308 | E308 | E308 | E308 | E309 | E309 | E309 | E310 | E309 | E309 | E309 | E309 |
| 304L | | E308L | E308 | E308 | E308 | E308 | E308 | E308 | E308 | E309 | E309 | E309 | E310 | E309 | E309 | E309 | E309 |
| 309, 309S | | | E309 | E309 | E309 | E309 | E309 | E309 | E309 | E309 | E309 | E309 | E310 | E309 | E309 | E309 | E309 |
| 310, 310S, 314[a] | | | | E310 | E316 | E316 | E317 | E317 | E308 | E310 | E309 | E309 | E310 | E310 | E310 | E310 | E309 |
| 316 | | | | | E316[b] | E316 | E316 | E316 | E308[b] | E309Mo | E309 | E309 | E310 | E309 | E309 | E309 | E309 |
| 316L | | | | | | E316L | E316 | E316L | E316L | E309Mo | E309 | E309 | E310 | E309 | E309 | E309 | E309 |
| 317 | | | | | | | E317 | E317 | E308[b] | E309Mo | E309 | E309 | E310 | E309 | E309 | E309 | E309 |
| 317L | | | | | | | | E317L | 308L | E309Mo | E309 | E309 | E310 | E309 | E309 | E309 | E309 |
| 321, 347, 348 | | | | | | | | | E347 | E309 | E309 | E309 | E310 | E309 | E309 | E309 | E309 |
| 330[a] | | | | | | | | | | E330 | E309 | E309 | E310 | E312 | E312 | E312 | E312 |
| 403, 405, 410, 414, 416, 420 | | | | | | | | | | | E410 | E430[e] | E410[e] | E502[e] | E505[e] | E410[e] | E410[e] |
| 430, 430F, 431, 440A, 440B, 440C | | | | | | | | | | | | E430 | E430 | E502[e] | E505[e] | E430[e] | E430[e] |
| 446 | | | | | | | | | | | | | E446 | E502[e] | E502[e] | E430[e] | E430[e] |
| 501, 502 | | | | | | | | | | | | | | E502 | E502[e] | E502[e] | E502[e] |
| 505 | | | | | | | | | | | | | | | E505 | E505[e] | E505[e] |

Notes: Grades shown are those most commonly selected for most applications; other combinations may be used. Wherever possible, recommendation is based upon the most-available and lowest-cost filler metal. Filler-metal designations are those appearing in AWS Specification, A5.9 for bare filler wire.

*a.* These alloys are sensitive to weld cracks and fissures; for this reason, E312 filler metal is a frequently recommended alternative. It is preferred especially when thick sections or highly restrained joints are required. Buttering these metals with type 312 before joining is often desirable.

*b.* E16-8-2 is preferred to lower embrittlement danger in elevated temperature service.

*c.* When joining an austenitic steel, alternate choice is to butter carbon or chromium steel with E309 and join with E308 or with filler metal similar to austenitic base metal. E307 is also commonly used for welds between austenitic stainless steel and either carbon or low-alloy steels.

*d.* ENiCrFe3 is preferred for elevated temperature service, except when sulfur compounds are present.

*e.* If austenitic weld metal is acceptable for service conditions, E309 or E310 is often employed.

Fig. 19-15. Filler materials used for joining stainless steels and dissimilar metals.

| Base Metal | Surfacing Material | CURRENT Type | Amps. | ROD Type | DEPOSIT Rc Hardness |
|---|---|---|---|---|---|
| Mild & Stainless Steels | HAYNES STELLITE Alloys | ACHF | | STELLITE #1 | 54 |
| | | ACHF | | STELLITE #6 | 39 |
| | | ACHF | | STELLITE #12 | 47 |
| | | ACHF | | STELLITE #93 | 62 |
| | | ACHF | | HASCROME | 23-43 |
| Copper | STELLITE #6 Alloy | DCSP | 180-230 for 3/16-in. material | STELLITE #6 | 42 |
| Steel, Copper & Silicon Bronze | Aluminum Bronze | DCSP | | Aluminum-Bronze Rods | 150-300 |
| Mild Steel & Cast Iron | Bronze & Copper | ACHF or DCSP | 150 for 1/2-in. material | A1-Bronze & Copper Rods | |
| Stainless Steel | Silver | ACHF | 160 for 1/2-in. material | | |
| Mild Steel | Stainless Steel | ACHF or DCSP | | | |
| Carbon & Alloy Tool Steels | Tungsten Carbide | DCSP | 300-375 | Tube of 8/15 mesh Tungsten Particles | |

Fig. 19-16. Base material and surfacing material combinations.

| Cladding Type | Filler Material |
|---|---|
| 405, 410, 410S, 429, 430 | 309,310 |
| | Inconel A, B, 182, or equiv. |
| | Inconel 82 or equiv. |
| | 430 |
| 304 | 309, 310 |
| | 309L |
| 304L | 309L |
| | 308L |
| | 309Cb |
| | 309CbL |
| 321, 347 | 309Cb, 310Cb |
| | 309Cbl |
| 316 | 347 |
| | 309Mo, 310Mo |
| | 316L |
| 316L | 309MoL |
| | 316L |
| 317 | 318 |
| | 309MoL, 310Mo, 309Mo |
| 317L | 317L |
| | 309MoL, 310Mo |
| | 317L |
| Incoloy 825 | Incoloy 65 or equiv. |
| | Incoloy 135 or equiv. |
| | Inconel 625 or equiv. |
| | Inconel 112 or equiv. |
| Inconel 600 | Inconel A, B, 182, or equiv. |
| | Inconel 82 or equiv. |
| Monel 400 | Monel |
| 70Cu-30Ni | Monel |
| 90Cu-10Ni | 70Cu-30Ni |
| Nickel | Nickel |
| Copper | Copper, Monel Nickel Inconel |

Fig. 19-17. The cladding listed in the left hand column may be welded with any of the filler materials listed.

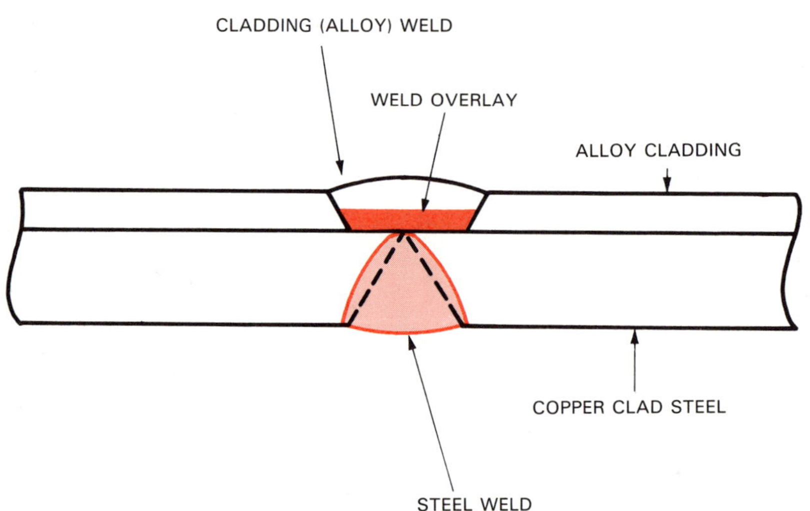


Fig. 19-18. Sequence of welding clad steels.

| CLADDING ALLOY | ALLOY FOR OVERLAY ON STEEL | ALLOY FOR WELDING CLADDING |
|---|---|---|
| Copper | RCu<br>RCuA1-A2<br>RCuSi-A<br>RNi-3 | RCu |
| Copper-Zinc | RCuA1-A2 | RCuA1-A2 |
| Copper-Tin-Zinc | RCuSn-A | RCuSn-A |
| Copper-Aluminum | RCuA1-A2 | RCuA1-A2 |
| Copper-Silicon | RCuSi-A | RCuSi-A |
| Copper-Nickel | RCuNi | RCuNi |

Fig. 19-19. Filler materials used in welding clad steels.

| Wire Designation | Brinell Hardness | Yield Strength (Thousands) PSI | Tensile Strength (Thousands) PSI | Applications |
|---|---|---|---|---|
| RCu | 50-60 | 8 | 28 | Corrosion resistant |
| RCuSn | 70-85 | | 35 | Corrosion resistant, Bearings |
| RCuSi-A | 80-100 | 25 | 50-55 | Errosion, corrosion |
| RCuNi | 60-80 | 10-20 | 50 | Bearings |
| RCuAl-A1 | 100-150 | 25-30 | 55-65 | Bearing overlay |
| RCuAl-A2 | 130-190 | 30-35 | 60-75 | Cavitation resistant |
| RCuAl-B | 140-220 | 40-45 | 90-110 | Bearing overlay |
| RCuAl-C | 180-280 | 40-45 | 90-100 | Cavitation resistant |
| RCuAl-D | 250-350 | 50-55 | 75-85 | Wear resistant |
| RCuAl-E | 290-390 | 55-70 | 70-80 | Bearings, Dies |
| NiAl Bronze | 180-200 | 55-60 | 100 | Errosion, Corrosion resistant |

Fig. 19-20. Common filler materials used for surfacing.

| Hard Surfacing → / Base Metal ↓ | | Cobalt Base Alloys | | | | | | Nickel Base Alloys | | | | | | Fe-Cr Alloys |
|---|---|---|---|---|---|---|---|---|---|---|---|---|---|---|
| | | 1 | 6 | 7 | 12 | T-400 | T-800 | 44 | 45 | 46 | 10XN | A | T-700 | Niobend |
| Carbon Steel | °F | 600 | 300 | 200 | 500 | 900 | 900 | 200 | 400 | 900 | 400 | 70 | 900 | 70 |
| .30C max. | °C | 325 | 150 | 95 | 275 | 480 | 480 | 95 | 205 | 480 | 205 | 20 | 480 | 20 |
| Carbon Steel | °F | 600 | 400 | 300 | 500 | 900 | 900 | 200 | 400 | 900 | 400 | 100 | 900 | 200 |
| .30-.50C | °C | 325 | 205 | 150 | 275 | 480 | 480 | 95 | 205 | 480 | 205 | 40 | 480 | 95 |
| Low Alloy Steels | °F | 600 | 400 | 300 | 500 | 900 | 900 | 200 | 400 | 900 | 400 | 100 | 900 | 200 |
| up to 3% total Alloys | °C | 325 | 205 | 150 | 275 | 480 | 480 | 95 | 205 | 480 | 205 | 40 | 480 | 95 |
| Medium Alloy Steels | °F | 600 | 400 | 400 | 500 | 900 | 900 | 200 | 400 | 900 | 400 | 200 | 900 | 300 |
| 3-10% total alloys | °C | 325 | 205 | 205 | 275 | 480 | 480 | 95 | 205 | 480 | 205 | 95 | 480 | 150 |
| High Alloy Steels | °F | 600 | 400 | 400 | 500 | 900 | 900 | 200 | 400 | 900 | 400 | 200 | 900 | 300 |
| Martensitic e.g. Type 410 | °C | 325 | 205 | 205 | 275 | 480 | 480 | 95 | 205 | 480 | 205 | 95 | 480 | 150 |
| High Alloy Steels | °F | 600 | 300 | 200 | 500 | 900 | 900 | 200 | 400 | 900 | 400 | 100 | 900 | 100 |
| Ferritic e.g. Type 430 | °C | 325 | 150 | 95 | 275 | 480 | 480 | 95 | 205 | 480 | 205 | 40 | 480 | 40 |
| High Alloy Steels | °F | 500 | 300 | 200 | 400 | 900 | 900 | 200 | 400 | 900 | 400 | 70 | 900 | 70 |
| Austenitic e.g. Types 304, 316 | °C | 275 | 150 | 95 | 205 | 480 | 480 | 95 | 205 | 480 | 205 | 20 | 480 | 20 |
| Nickel Alloys | °F | 500 | 300 | 200 | 400 | 900 | 900 | 200 | 400 | 900 | 400 | 70 | 900 | 70 |
| e.g. Inconel* e.g. Monel* | °C | 275 | 150 | 95 | 205 | 480 | 480 | 95 | 205 | 480 | 205 | 20 | 480 | 20 |

*Trade names of the International Nickel Co. Inc.

Fig. 19-21. Preheating of the base material reduces the amount of cracking in the surfacing alloy during the cooling period.

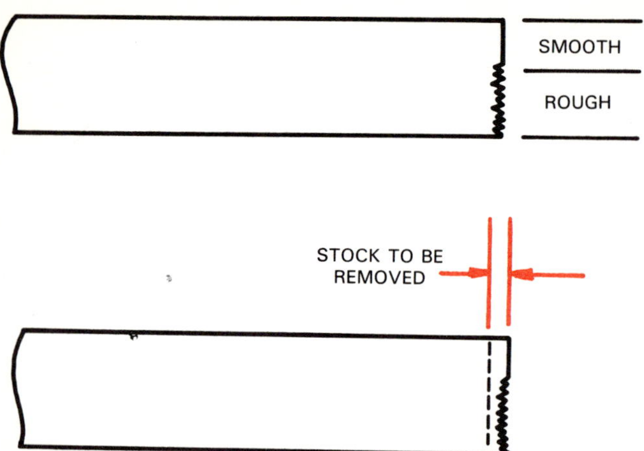


Fig. 19-22. Sufficient stock should be removed from the edge of sheared material to eliminate all rough edges.

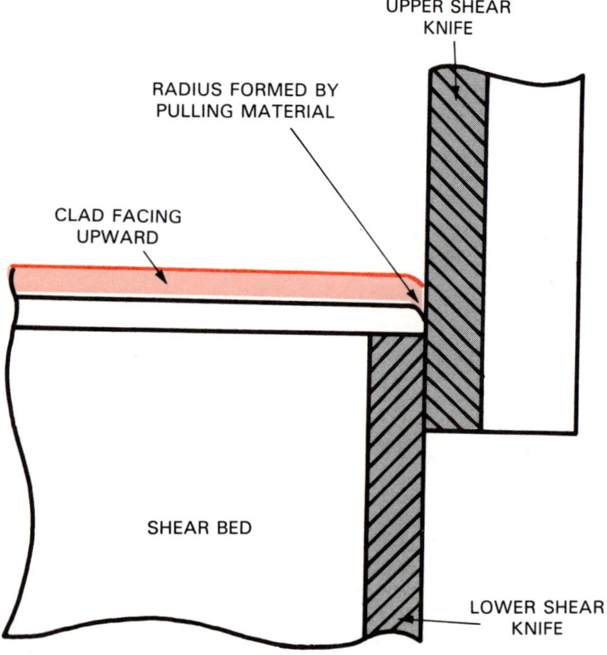


Fig. 19-23. Due to the ductility of the clad material, the metal will stretch during the shear operation. This reduces the clad thickness at the sheared edge.

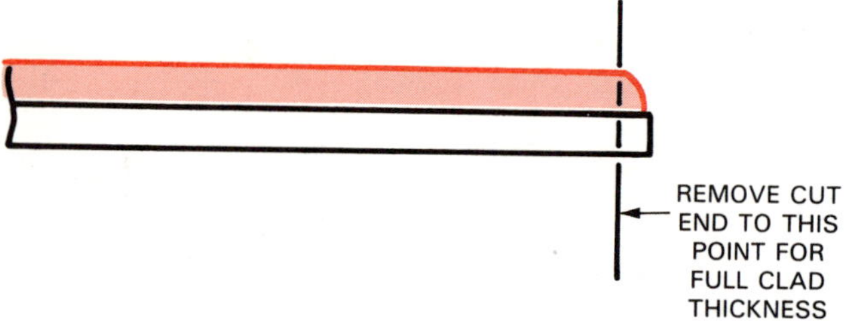


Fig. 19-24. Remove sufficient stock from the sheared edge to assure full clad thickness.

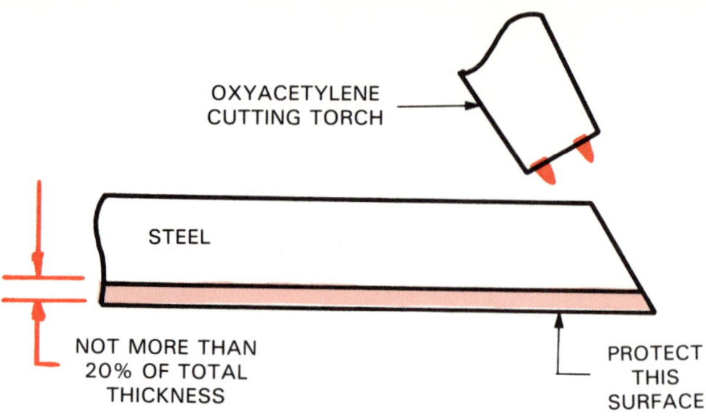


Fig. 19-25. Make a test cut on scrap clad material before making the full cut to establish the cutting parameters. (Gas pressures, tip size, travel speed.)

away from the joint very rapidly. Therefore, the copper cladding should be removed from the steel where the cut is to be made. The copper clad may be removed by grinding or a chisel, as shown in Fig. 19-26.

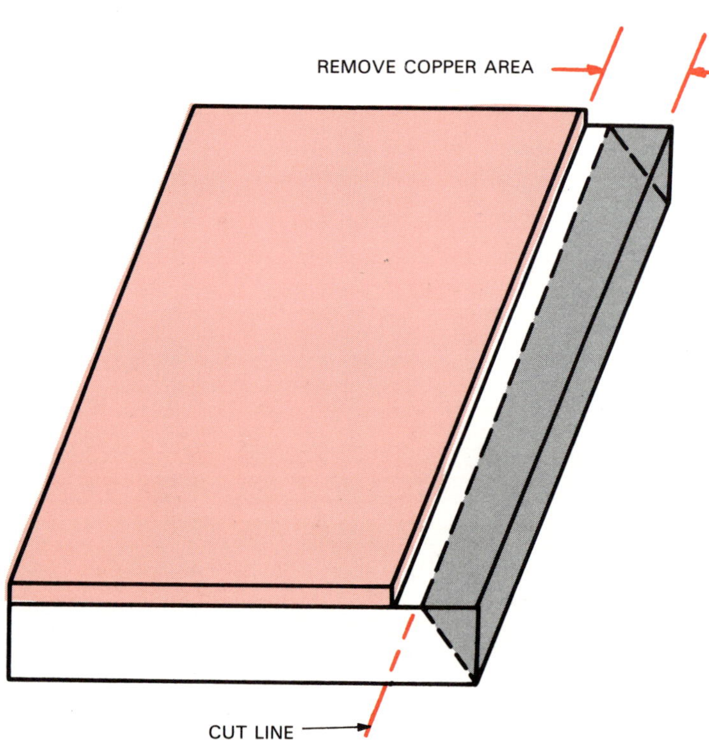


Fig. 19-26. The cladding should be removed far enough from the edge or cut line so that the copper is not melted into the burn area.

4. PLASMA ARC CUTTING may be used to cut all types of materials including the clads. Fig. 19-27 illustrates that the plasma arc cut may be made

from either side of the plate. This is an advantage in protecting the clad. The clad may face upward, thus protecting the clad from scratches, nicks, etc.

5. AIR CARBON ARC GOUGING is used for the preparation of U type joints. This type of preparation offers fast preparation time with minimum cost. Repairs, spot cladding preparation, irregular shape welds are prepared for welding by this process.

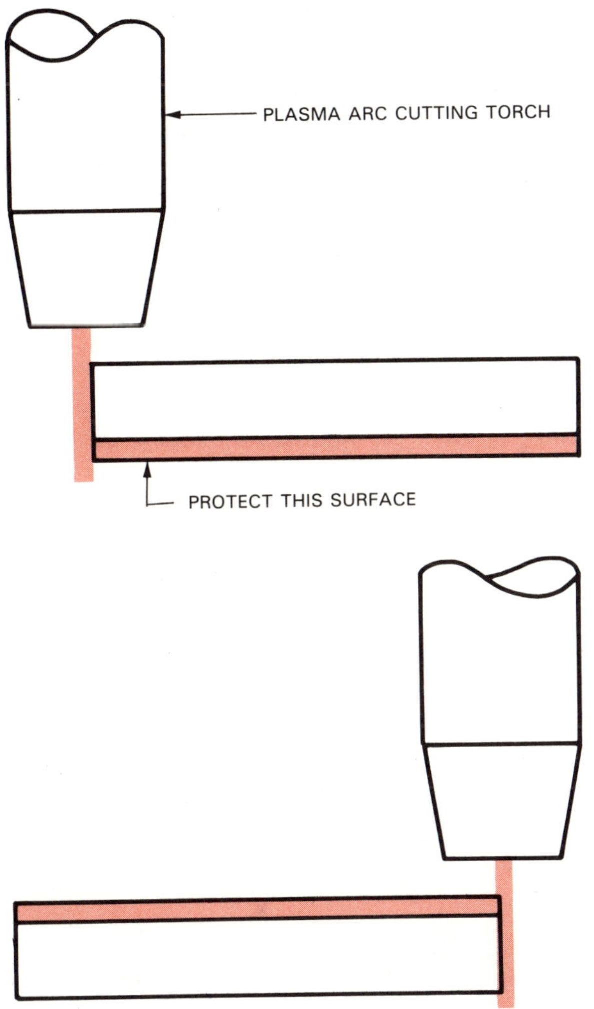


Fig. 19-27. The plasma arc cutting process will cut ferrous or non-ferrous metals. Therefore, the cut may be made from either side.

## JOINT CLEANING

Sound, defect-free welds cannot be made if the weld joint is dirty or contaminated. To insure the best possible welding conditions, the joint should be rigidly inspected before starting to weld. Slag and scale should be ground or machined from the part. Wire brushing should follow grinding to remove any particles left from the grinding wheel. Oil, grease, and pencil marks should be removed by solvent. Burrs and nicks should be removed to prevent entrapment of foreign materials in the weld joint.

## WELDING

Welding dissimilar metals requires careful consideration of the metals to be joined and the welding parameters.

1. JOINT FIT-UP. Joints that are not fit-up properly for welding can often lead to possible failure or loss of the protection capabilities. Joint and assembly tolerances should be established by test weldments when the welding parameters are made. Should changes be required in the welding parameters to correct for incorrect fit-up, they should only be allowed within a prescribed set of tolerances only. Fig. 19-28 illustrates how excessive tolerances affect the welding operation and joint integrity.

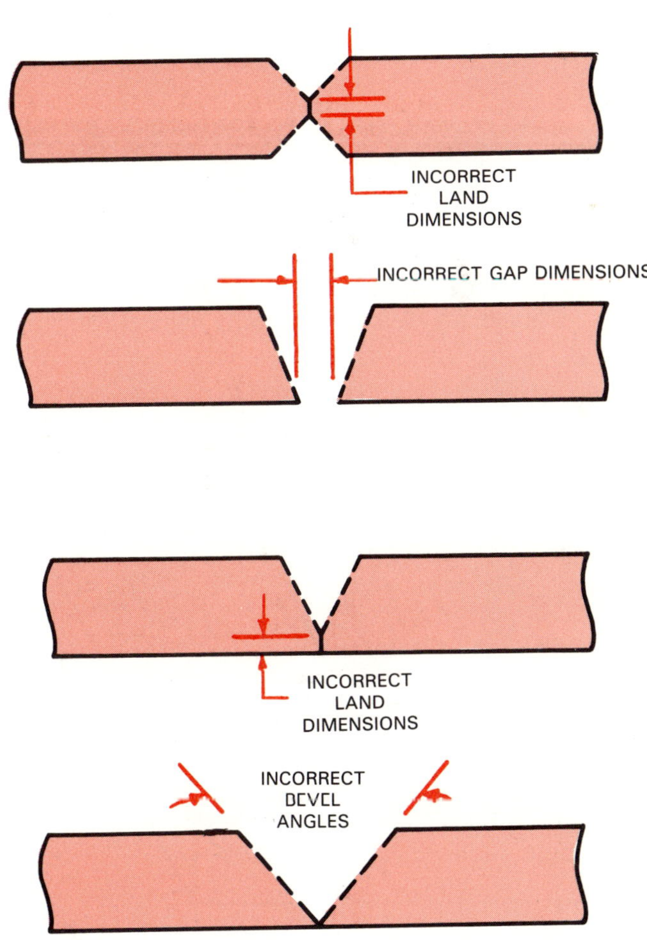


Fig. 19-28. Incorrect joint dimensions affect the alloy mix within the joint. This may affect the mechanical properties of the weld.

2. HEAT INPUT. This is dependent on amperage, voltage, travel speed, and the application of the filler material. Insufficient heat input results in lack of fusion, voids, porosity, and cold shuts within the weld joint. Too much heat can cause excessive dilution within the joint between the parent metal and the filler material. The use of pulsers to establish heating and cooling cycles will regulate amperage and reduce the overall heat input required to weld the joint.
3. WELDING TECHNIQUE. The position of the tungsten and the angle of the torch should be maintained to melt as little base material as possible. The filler material should be added to the puddle as rapidly as possible to provide a "cold puddle." Large molten pools stir the puddle and cause more dilution.

Welding done "out of position" usually generates more heat in the base material causing more penetration and dilution. Welds made in these positions should be "stringer beads" with no oscillation allowed.

Sizes of welds and sequencing of beads should be made to control dilution of one bead to another. Each weld should overlap the previous bead by one-half of the previous bead width, as shown in Fig. 19-29.

Each layer of weld should be flat without valleys or high spots, as shown in Fig. 19-30. These conditions can lead to defects within the weld if not corrected prior to starting the next layer. To correct this condition, grind all high spots even and fill or grind valleys even.

Clean each weld and visually inspect for porosity, voids, slag, silicon, etc., before applying the next bead.

Always deposit the required number of layers specified. Adding or reducing the number of layers may radically effect the dilution and composition of the final layer of the weld.

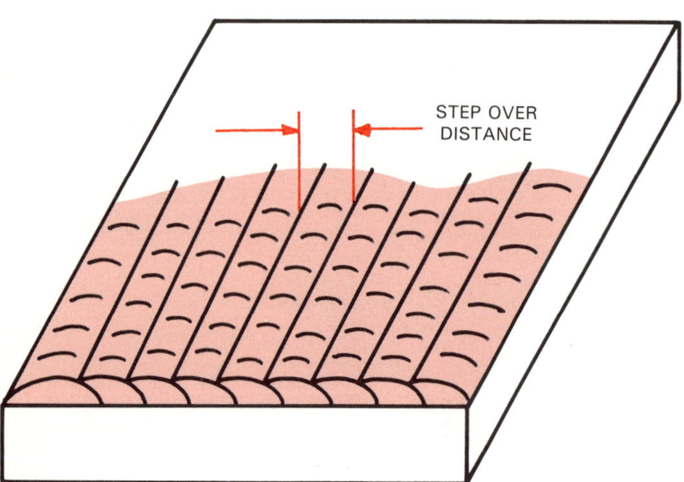


Fig. 19-29. The amount of overlap of one weld to another is called the "step-over" distance. This dimension must be established during testing and maintained in production to control the alloy mix.

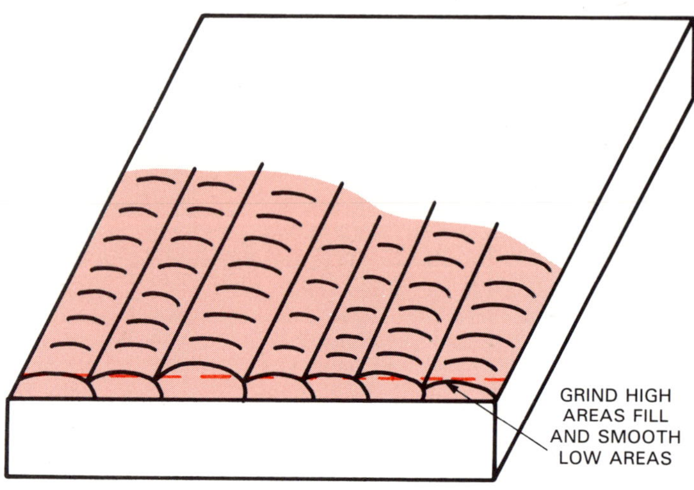


Fig. 19-30. Welds with valleys or high spots do not have the proper alloy mix. These areas may cause defects such as lack of penetration, and lack of fusion or bond, and improper dilution.

## REVIEW QUESTIONS

1. List three areas where welding of dissimilar metals are often used.
2. When another type of material is added to a joint or a build-up is made, this is called ________ or ________ .
3. Where is cladded material made? Is all cladding the same thickness?
4. Is an overlay usually thicker than a cladded surface?
5. Does a single V groove butt weld have more or less dilution than a double V groove butt weld?
6. Which type of weld in question 5 has the greatest distortion?
7. Why are clad materials usually sheared with the clad surface facing upward?
8. What happens to the clad at the sheared edge? Can this be prevented?
9. Steels with stainless or nickel cladding may be flame cut if the cladding is not more than ________ percent of the total thickness.
10. Which type of cladding metal must be thoroughly removed from steel before thermally cutting? Why?
11. Which thermal cutting process may be used to

prepare dissimilar metal joints on all types of metals?

12. Which process is used to prepare U groove type weld joints?
13. Scale and slag may be removed from the weld joint area by ____________ or ____________. This is followed by ____________ to remove particles of metal.
14. What type of equipment may be used to regulate heat input into a weld?
15. What is the major problem caused by welding dissimilar metals with a "hot puddle?"
16. Overlays are made by overlapping one weld over another. Another term used to describe this overlap is ________ ________ .
17. Why should each layer of an overlay weld be flat?

Many types of metals are welded for use in modern jet engines.
(Pratt and Whitney, Aircraft Group)

# Chapter 20

# SEMIAUTOMATIC WELDING

SEMIAUTOMATIC WELDING is a welding operation using equipment that may automatically adjust for various conditions or may be manually adjusted by the welding operator. Semiautomatic welding equipment controls may be pre-set by the operator, thus allowing the welding sequence to take place without direct operator manual direction.

Semiautomatic welding procedures are used to make individual or production type welds. The equipment may be of one or more manufacturer's types, assembled into a system and controlled by an operator. Therefore, the person who controls the semiautomatic welding operation is called a WELDING OPERATOR. It is this person's responsibility to operate the equipment in a sequence and manner to produce a satisfactory weld.

## TYPES OF WELDS

Several longseam type welds made using semiautomatic welding are shown in Fig. 20-1. The joints are usually made in the flat position. All of the joints can be made without undue difficulty, except fillet welds. Since fillet welds require penetration at the root of the joint, alignment of the joint to the tungsten tip is often difficult to obtain. Fig. 20-2 shows the effect of tungsten misalignment to the joint center line. Welding speed may be obtained by either the movement of the weldment or the movement of the welding torch, as shown in Fig. 20-3.

One type of rotational weld made with semiautomatic welding is shown in Fig. 20-4. As in longseam welding, the majority of welding is done in the flat position and fillet welds are difficult due to the joint alignment problem. Welding speed is obtained by adjusting the rotational speed of the weldment on a fixture or a positioner as shown in Fig. 20-5.

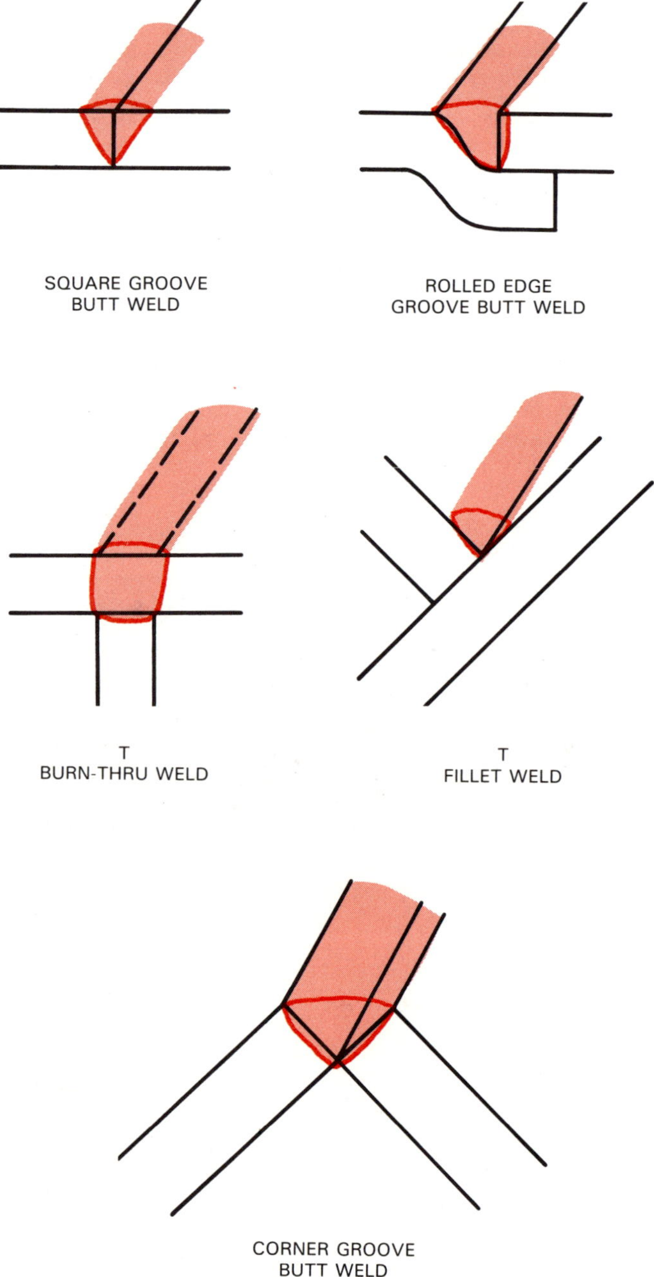


Fig. 20-1. Various types of longitudinal welds.

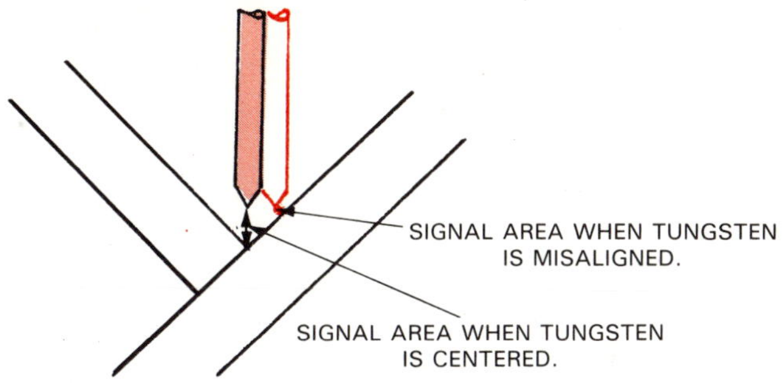


Fig. 20-2. Automatic voltage controlled welding heads will not operate properly unless joint is properly aligned.

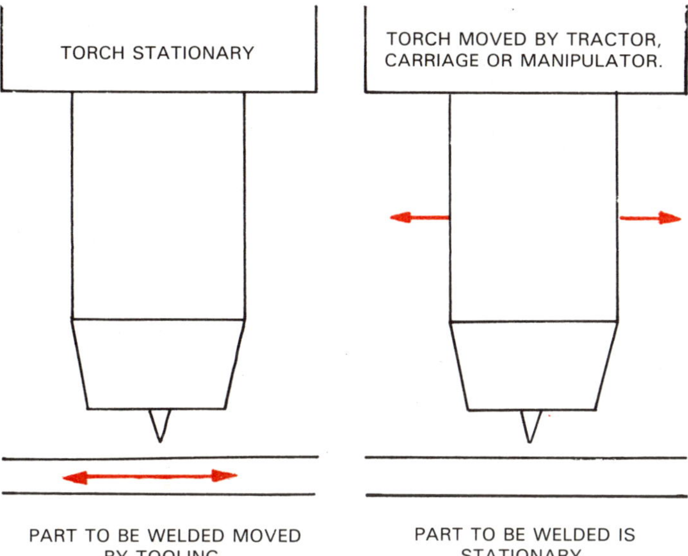


Fig. 20-3. Either the weldment or the torch or both may be moved to make the weld.

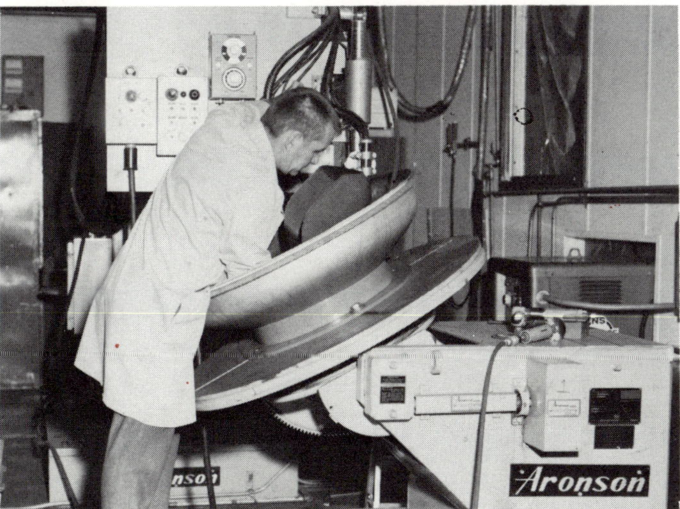


Fig. 20-4. Weldment being rotated by the welding positioner under a positioned automatic voltage controlled welding torch. (Parker-Hannifin Corp.)

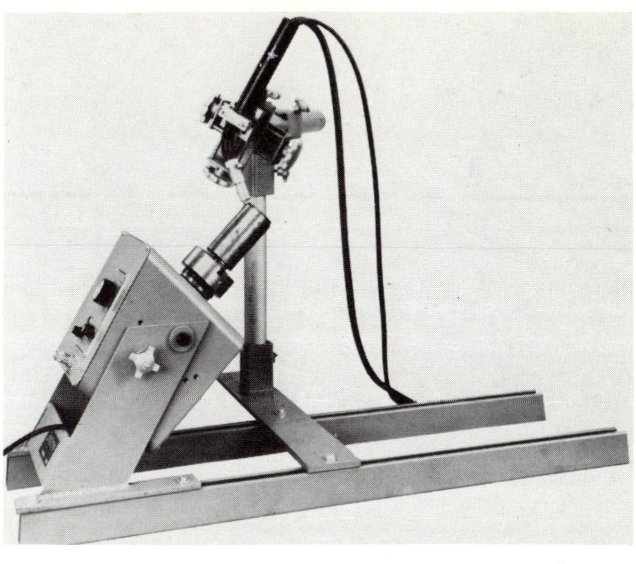

Fig. 20-5. Welding speed is controlled by the potentiometer setting on the positioner case. (Jetline Engineering Co.)

## TYPES OF MATERIALS

Virtually all materials which may be welded manually may be welded with a semiautomatic system. In addition, aluminum which is normally welded by ACHF, can be welded with DSEN (DCSP) with helium gas where a fixed arc length control is available. Thin gauge materials are often welded by semiautomatic equipment on DCEP (DCRP) where travel speeds are controlled.

## EQUIPMENT

The equipment required for semiautomatic welding either longseam or rotational may range from the most simple to very complex. The major components for the operation include the power supply, the arc voltage control, and the travel control.

Power supply considerations include:

1. Select a machine with the type of current desired.
2. The machine should have sufficient amperage for the weld.
3. The duty cycle ought to be sufficient for the length of the operation.
4. There should be the desired control of amperage available for the welder to change as required, and for starting and ending the weld operation. Remote controls, as shown in Fig. 20-6, are located at the welding site away from the power supply.

Arc voltage control considerations:

1. AUTOMATIC VOLTAGE CONTROL HEADS are often used where the gap between the part and the tungsten may vary considerably. The unit shown in Fig. 20-7 has circuits for using pulsers and alternating current.

Fig. 20-6. Remote amperage control potentiometers are mounted on this manipulator to allow for operator adjustment. Welding head has been removed in this view. (Parker-Hannifin Corp.)

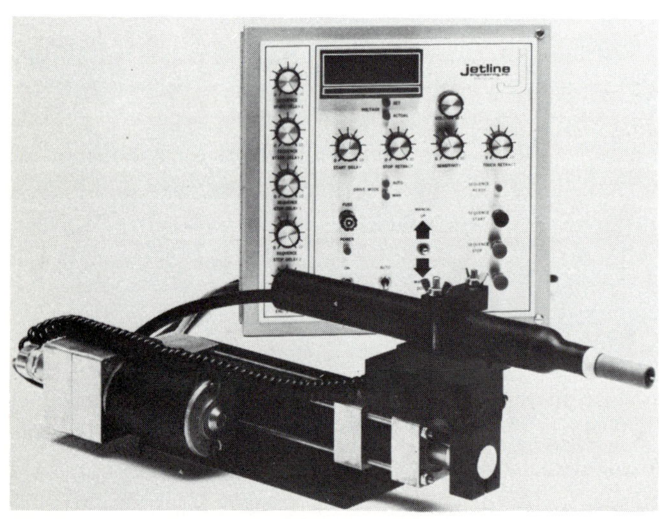


Fig. 20-7. Automatic voltage controlled welding head with sequencer for pulser lockout. (Jetline Engineering Co.)

2. MECHANICAL VOLTAGE CONTROLS used where the arc length is fixed and only requires occasional adjustment. The torch is moved up or down by adjusting screws or levers. Setting the arc gap may be done by set blocks, dial indicators, etc., as shown in Fig. 20-8. Where continuous changes may be required when welding, this control may be used in conjunction with an arc voltmeter. The voltmeter is installed into the system across the arc gap as shown in Fig. 20-9. The operator then raises or lowers the torch as required to obtain the desired voltage.

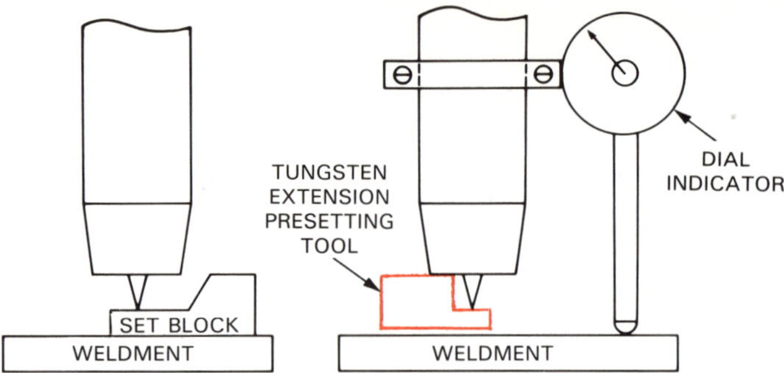


Fig. 20-8. Two methods which may be used to set tungsten to a pre-determined height from the weldment surface. The tungsten pre-setting tool must be used with the dial indicator method to locate the electrode extension only. Dial indicator readings are established before welding. Rotate indicator out of welding zone before welding.

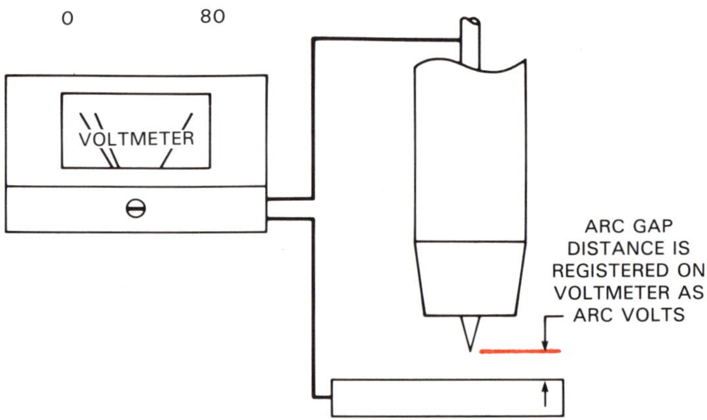


Fig. 20-9. Voltmeters may be used to determine arc gap between tungsten tip and the weldment.

## TRAVEL CONTROL

Control of welding speed is one of the major components in GTAW. If the speed is not consistent, the weld will often have areas of the weld which are not satisfactory. Two methods of drive and control systems are in common use:

1. ALTERNATING CURRENT MOTORS AND CONTROLLERS. Systems of this type include AC motors operating at a single speed with a mechanical reduction to the required welding travel speed. The controls regulate the mechanical reduction section by changing the linkage between the electrical motor and the gear box.

   Since utility power AC voltage input varies considerably, the electrical motor may not operate at a consistent RPM. This causes the weld travel speed to vary with each change in input voltage. This may effect the weld quality.

2. DIRECT CURRENT MOTORS AND CON-

TROLLERS. This type of drive system removes the mechanical reduction area by the use of a variable speed, 12 or 24 volt current electrical motor. The motors are highly efficient and easily adjustable throughout a specific range by a potentiometer located in the controller. This controller is also known as a governor.

## WELD TRAVEL

Movement of the weld puddle can be accomplished in various manners. Positioners, lathes, tractors, manipulators, carriages, and fixtures are used for longseam and rotational welding.

### Longseam Travel

Fig. 20-10 illustrates one method of moving the torch along a weld joint on a seamer fixture. In this operation, the carriage and track is a part of the machine. Since the track and the joint are parallel, tracking the joint during welding is easily accomplished.

Fig. 20-11 shows a manipulator used to make a longseam weld on a seamer.

Fig. 20-12 shows a portable tractor set-up for a longitudinal weld. The tractor drive wheels operate on a track aligned with the joint for tracking. The track also serves as a roadbed for smooth travel of the tractor.

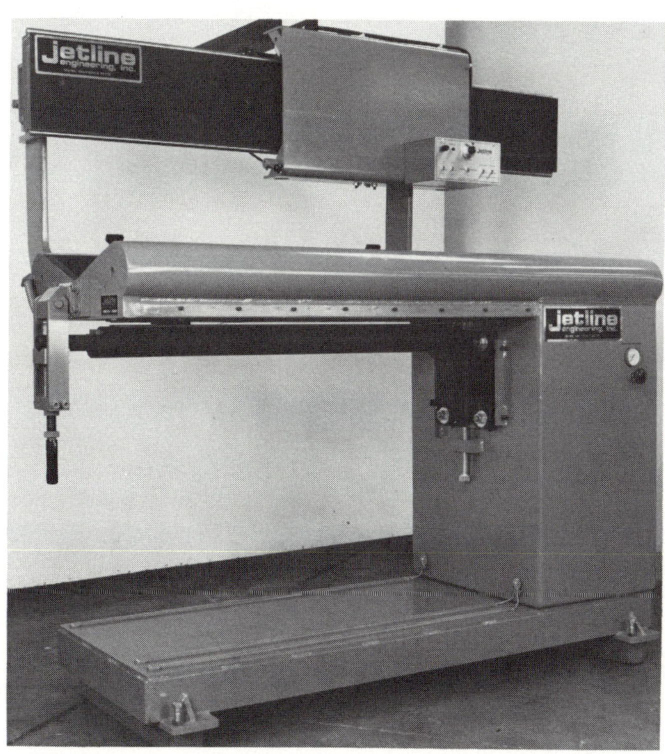


Fig. 20-10. Commercial seamer has a track and carriage that moves parallel to the weld joint. (Jetline Engineering Co.)

Fig. 20-11. The welding head and torch are mounted on the end of the manipulator. The controls are readily available to the operator. The wire feeder, cables, and hoses along with the manipulator motors are mounted out of the welding area. (Parker-Hannifin Corp.)

Fig. 20-12. Portable tractors are very useful and may be used in almost any position. (M & K Products, Inc.)

Fig. 20-13 shows a special designed system for longseam and rotational welds of various shapes. The movement of the torch or the fixture is done by combining slides and servo motors. Alignment to the shape of the part may be a combination of cams or finger sensors which direct the servo motors.

Fig. 20-13. Seam trackers are used to track weld joints of various shapes and designs. (Cyclomatic Industries, Inc.)

## Rotational Travel

Fig. 20-14 shows a lathe being used as a turning tool. The use of three jaw chucks allows rapid centering of the part and precision alignment of the torch to the weld joint. Welding speed is easily controlled through the use of the various gear box ranges.

Since welding current may arc in the lathe gears or bearings during welding, the ground connection should be located on the lathe chuck or fixture. This can be done as shown in Fig. 20-15.

Fig. 20-16 shows a manipulator aligned to track the overall length of the joint to be welded.

Fig. 20-17 shows a commercial welding positioner with a fixture mounting faceplate. Slots are provided for mounting various fixtures and tools. The center hole in the faceplate shown in Fig. 20-18 allows rapid centering of the fixture by mating with a male plug on the fixture base, as shown in Fig. 20-19.

As positioners are sold in many various sizes and rotational speeds, the welding speed at the part diameter must be computed and converted into revolutions per minute (RPM's).

Positioners are often made for other types of work rather than welding. These positioners may not have the ability to do the precision locating and turning required. Machines with both ground faceplates and machined gears reduce alignment and tracking problems.

Fig. 8-1 shows a special designed fixture for welding specific welds. The fixture is actually a lathe with a welding positioner gearbox. These types of fixtures are costly to build and are only efficient when used for welding a specific part or assembly.

Fig. 20-14. Lathe is being used as a turning tool for a weldment. Welding speed is controlled by gear box settings.

Fig. 20-15. Ground connections on rotating equipment must be accomplished to prevent arching between gears, bearings, and bushings. (Aerojet-General Corp.)

Fig. 20-16. Manipulator with welding head is set-up for rotational welding. It may also be used for longitudinal welding. (Aerojet-General Corp.)

Fig. 20-17. Commercial welding positioner has a special machined center hole for adapting tooling and slots for locking tooling to the faceplate. (Aronson Machine Co.)

The specially designed fixtures in Figs. 20-20 and 20-21 are designed for welding specific components or welds.

## ACCESSORIES

Accessories are added to a basic system to improve the quality of the weld and to assist the welding operator during the operation. The accessories include automatic voltage controls, cold wire feeder and manipulator, hot wire feeder, oscillators, pulsers, and timers and slopers.

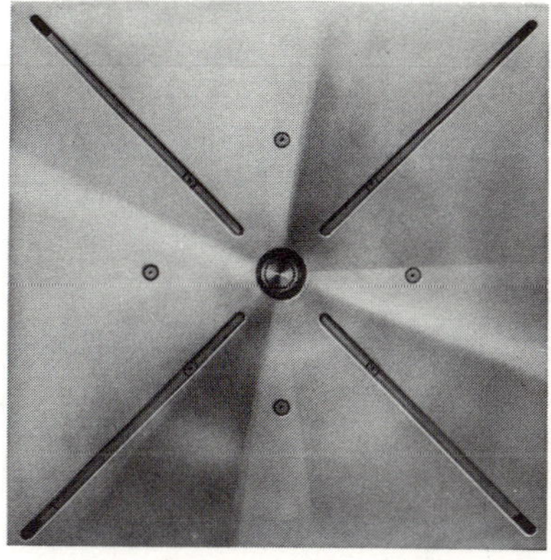

Fig. 20-18. Positioner faceplate with standard hole for locating tooling on center line. (Aronson Machine Co.)

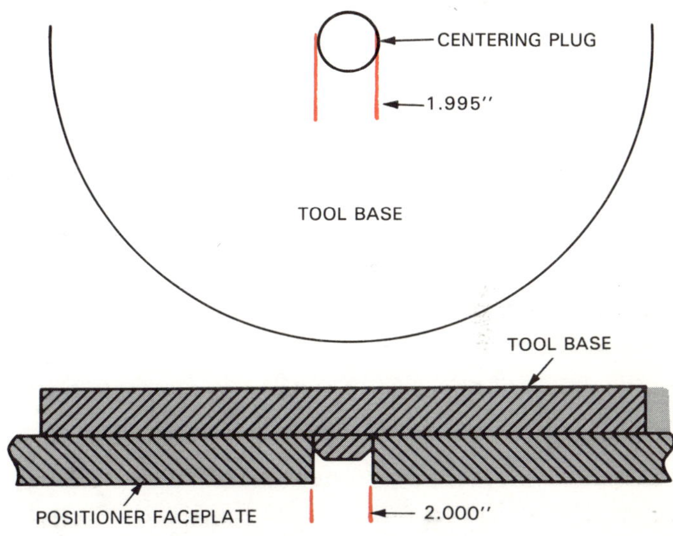


Fig. 20-19. Tool base centering plug dimensions.

### Automatic Voltage Control (AVC)

AUTOMATIC VOLTAGE CONTROL automatically regulates the distance between the workpiece and the tungsten to a preset gap. Where the arc gap may vary due to misalignment problems, these units relieve the operator of constantly making adjustments. The units are mounted with a base, torch holder, and torch as shown in Fig. 20-7. The power required is 110 volts AC, which is converted to 12 or 24 volts DC for operation of the servo motor mounted in the head.

The head operates by sending an electrical signal from the tungsten to the workpiece. Receiving the signal, the head then responds to a preset voltage and drives the head up or down as required. The unit operates very well on butt welds. However, the signal may be deflected on fillet welds as shown in Fig. 20-2 if alignment is not perfect. Fillet weld joints designed as in Fig. 20-22 remove this problem as the point of signal contact is close to the tungsten tip.

shown in Fig. 20-23. Power required for the unit is 110 volts AC which is converted to 12 or 24 volts DC for operation of the motor in the drive box.

The manipulator is used to adjust the wire fed into the puddle through the wire tip guide, shown in Fig. 4-14. The wire tip guide must be the proper size for the wire diameter used. When using hard wires, these guides need frequent replacement.

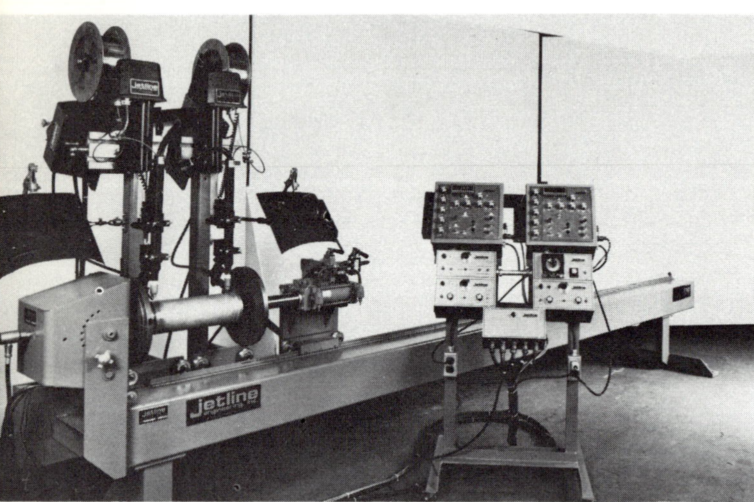


Fig. 20-20. Specially designed fixture with two welding heads. The operation can be made as single welds or double welds at the same time. (Jetline Engineering, Inc.)

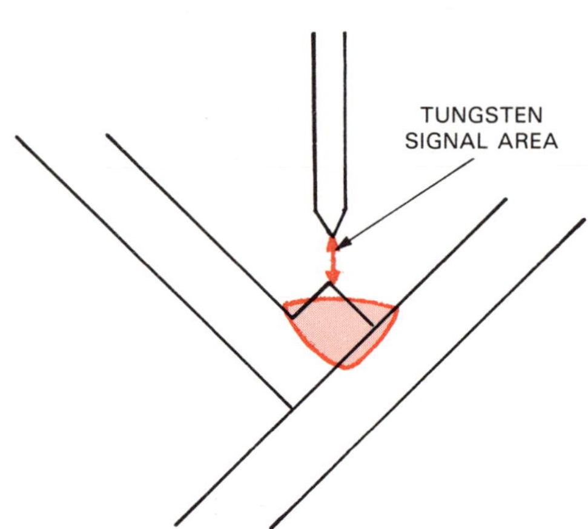


Fig. 20-22. Fillet welds designed in this manner with compatible base materials and not needing separate weld metal work well for semiautomatic welding.

Fig. 20-21. This fixture accepts three components for welding in any sequence. The welding torches may be moved along the overhead track to locate at any weld area. (Jetline Engineering, Inc.)

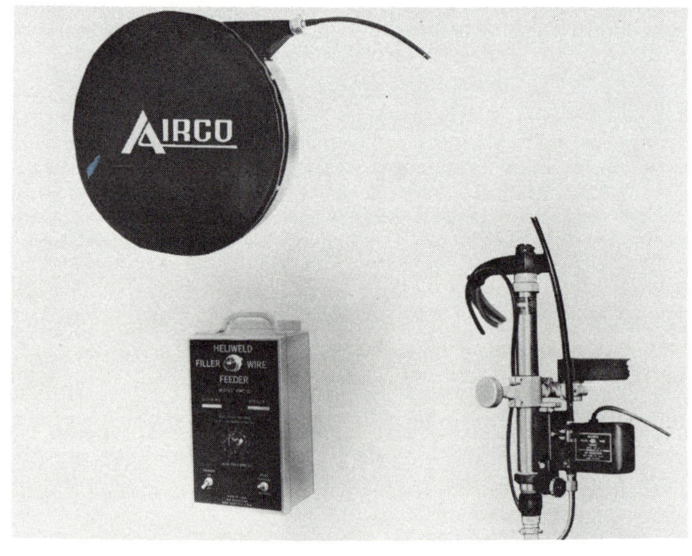


Fig. 20-23. Coldwire feeder system. (Airco.)

### Cold Wire Feeder And Manipulator

The cold wire feeder supplies welding wire into a weld joint to maintain the weld profile. The feeder consists of a mount for the weld wire supply, speed governor, DC electrical motor, gear box, wire feed drive rollers, conduit to the manipulator, and adjustment screws to move the manipulator. A complete unit is

### Hot Wire Feeder

The hot wire feeder supplies preheated welding wire into a joint to maintain the weld profile. The preheated wire has very little surface impurities, as they are burned up during the heating. Weld quality is very good since impurities are not fed into the puddle.

The system includes a mount for the wire supply,

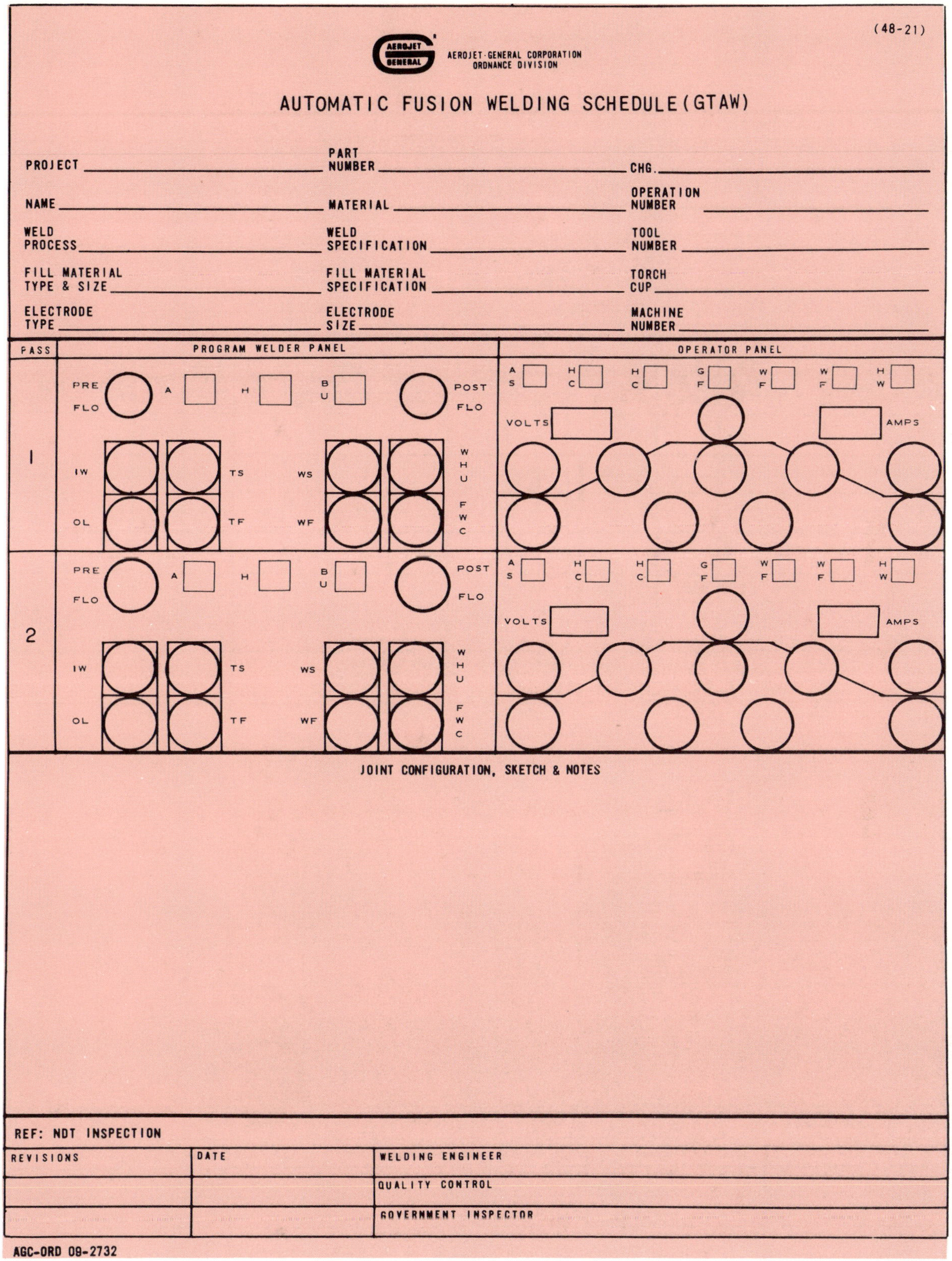

(48-21)

AEROJET GENERAL

AEROJET-GENERAL CORPORATION
ORDNANCE DIVISION

AUTOMATIC FUSION WELDING SCHEDULE (GTAW)

PROJECT ______ PART NUMBER ______ CHG. ______

NAME ______ MATERIAL ______ OPERATION NUMBER ______

WELD PROCESS ______ WELD SPECIFICATION ______ TOOL NUMBER ______

FILL MATERIAL TYPE & SIZE ______ FILL MATERIAL SPECIFICATION ______ TORCH CUP ______

ELECTRODE TYPE ______ ELECTRODE SIZE ______ MACHINE NUMBER ______

| PASS | PROGRAM WELDER PANEL | OPERATOR PANEL |
|---|---|---|
| 1 | PRE FLO, A, H, BU, POST FLO, IW, TS, WS, WHU, OL, TF, WF, FWC | AS, HC, HC, GF, WF, WF, HW, VOLTS, AMPS |
| 2 | PRE FLO, A, H, BU, POST FLO, IW, TS, WS, WHU, OL, TF, WF, FWC | AS, HC, HC, GF, WF, WF, HW, VOLTS, AMPS |

JOINT CONFIGURATION, SKETCH & NOTES

REF: NDT INSPECTION

| REVISIONS | DATE | |
|---|---|---|
| | | WELDING ENGINEER |
| | | QUALITY CONTROL |
| | | GOVERNMENT INSPECTOR |

AGC-ORD 09-2732

Fig. 20-24. GTAW schedule made for a specific type welding machine. (Aerojet-General Corp.)

(48-21)

AEROJET GENERAL

AEROJET-GENERAL CORPORATION
ORDNANCE DIVISION

AUTOMATIC FUSION WELDING SCHEDULE (GTAW)

PROJECT ______ PART NUMBER ______ CHG. ______

NAME ______ MATERIAL ______ OPERATION NUMBER ______

WELD PROCESS ______ WELD SPECIFICATION ______ TOOL NUMBER ______

FILL MATERIAL TYPE & SIZE ______ FILL MATERIAL SPECIFICATION ______ TORCH CUP ______

ELECTRODE TYPE ______ ELECTRODE SIZE ______ MACHINE NUMBER ______

MACHINE SETTINGS

| PASS NUMBER | AMPERAGE | ARC VOLTAGE | WELD SPEED IPM | WIREFEED MEASURED SPEED | SHIELDING TORCH | | SHIELDING BACKUP OR PURGE | | PRE-HEAT | INTER-HEAT | POST-HEAT | OSCILLATOR |
|---|---|---|---|---|---|---|---|---|---|---|---|---|
| | | | | | AR | HE | PURGE | TRAIL CUP | | | | |
| 1 | | | | | | | | | | | | |
| 2 | | | | | | | | | | | | |
| 3 | | | | | | | | | | | | |
| 4 | | | | | | | | | | | | |
| 5 | | | | | | | | | | | | |
| 6 | | | | | | | | | | | | |

JOINT CONFIGURATION, SKETCH & NOTES

REF: NDT INSPECTION

| REVISIONS | DATE | WELDING ENGINEER |
|---|---|---|
| | | QUALITY CONTROL |
| | | GOVERNMENT INSPECTOR |

AGC-ORD 09-2590

Fig. 20-25. GTAW schedule made for general type welding machine. (Aerojet-General Corp.)

wire feed governor, wire feed drive rollers, conduit to the hot wire heater, heater, power supply for the heater, and a wire manipulator. An assembled unit is shown in Fig. 4-22. The manipulator is used to adjust the wire into the back of the puddle. This is opposite in operation to the addition of cold wire.

### Oscillators

Oscillators are used to move the weld puddle. This will control and improve the weld profile and agitate the puddle to allow gases to rise to the surface, thus reducing the internal porosity. There are two types of oscillators:

1. MECHANICAL OSCILLATOR – This type of oscillator is mounted above the torch and wire feeder. The entire unit moves across the weld. Since the entire unit moves, the mount base must be rigid enough to support the weight and vibration during the operation. The governor control is similar to the travel governor with the addition of timers to control dwell periods at the end of each oscillation. An installed unit is shown in Fig. 4-22.
2. MAGNETIC OSCILLATOR – The magnetic type operates by deflecting the arc with an electric magnet. Therefore, the installation is quite simple and a rigid base is not required. An installed unit is shown in Fig. 4-24.

### Pulsers

Pulsers, assembled into the power supply or as a portable unit, are used to change weld current from one level to another for a set period of time. The pulsing operation develops enough heat for making the weld, then drops to a lower setting which allows the weld to solidify. A portable type pulser is shown in Fig. 4-3.

Since pulsers are set for high and low current levels, the addition of the welding wire should be also pulsed and synchronized with the amperage pulse. This requires additional equipment. It is often difficult for the operator to properly start each sequence manually. For this reason, the joint design should not require filler wire when semiautomatic welding using pulsers. Pulsing with wire and travel speed is usually restricted to full automatic welding where the sequencing can be electronically controlled.

### Timers And Slopers

Timer and sloper controls are often used to time weld current to and from one level to another level. For example, a spot weld starting at low amperage, increasing current to welding level, holding for a period of time and finally decreasing the current to zero. A sloper set-up is shown in Fig. 4-4.

This type of action is positive with accurate current levels and time periods. Operators of equipment without proper timers and slopers cannot be trained to produce welds with the level of accuracy obtained with timers and slopers.

## WELD TEST

Before the start of production welding, a WELD TEST or SAMPLE WELD should be made to determine the welding parameters required for the particular joint design. During the weld test, all of the actual parameters should be recorded for use on the production weldment. These recorded parameters are called a WELDING SCHEDULE. A sample weld schedule is shown in Figs. 20-24 and 20-25.

The weld test is also used to determine weld quality, tensile strength, and other requirements which may be placed on the weld. It can also be used as a training period for the operator and used for operator qualification. System operation is verified during weld testing, and required modifications are made prior to making the final weld test.

The weld test used to develop the weld schedule should be made on the same type of material with the same thickness and the same joint design, with the same type of tooling. In otherwords, duplicate by testing the actual joint.

### Weld Schedule For Semiautomatic GTAW

The following listed areas include the main parameters required to make a weld schedule. Others may be required to improve weld characteristics. As you study the following parameters, think about how you would complete a weld schedule similar to the schedule in Figs. 20-24 and 20-25.

### Weld Joint Design

1. Record overall thickness on butt welds.
2. Record angles and dimensions on groove welds.
3. Record wall thicknesses on fillet welds.
4. Record fit-up criteria.

Tolerances of the base material, machining and fit-up may severely effect the welding of the production part. These must be considered if the weld quality is to be the same on the weld test and the production part. If the tolerances vary considerably, then a weld test with both extremes should be made.

### Weld Tooling

Weld tooling, when used in the immediate area of the weld for heat sinks and weld back-up, has a very important effect on the welding parameters and weld quality. The test weld must be made on tooling identical to the production tooling if working parameters are desired.

1. Record back-up groove design and material type.
2. Record chill ring design and material type.
3. Record chill ring spacing.

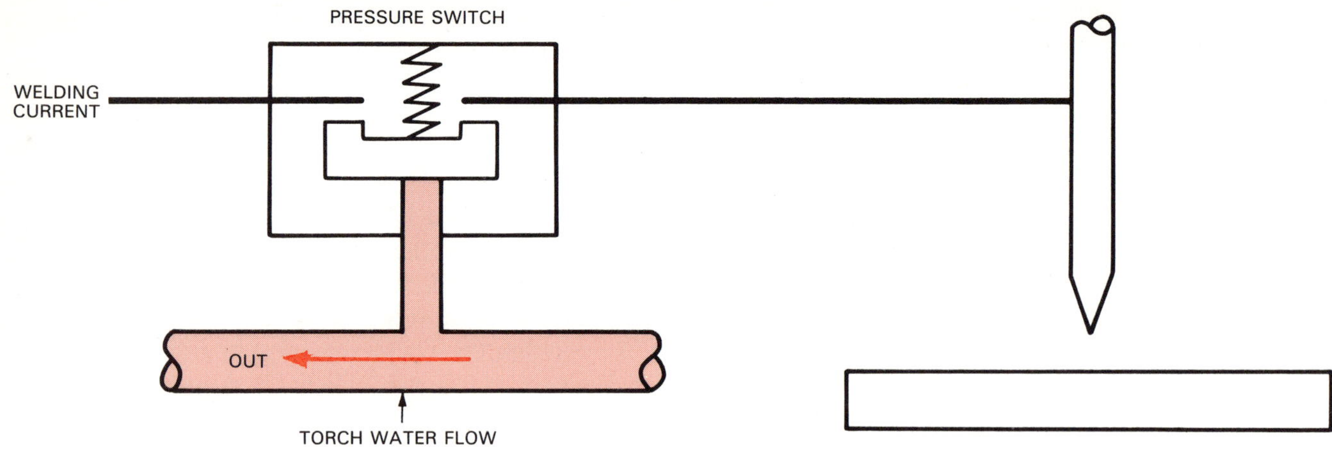


Fig. 20-26. Install pressure sensing switches on the water outlet line for proper operation.

4. Record back-up bar temperature. (If the weld is to be made at a specific temperature.)
5. Record finger bar pressure on seamers.

### Welding Current

1. Select current type and record.
2. Select amperage desired for:
   A. Initial start.
   B. Upslope.
   C. Weld current.
   D. Downslope.
3. Type of amperage control – machine or manual.
4. Type of arc start.
5. Pulser parameters (if used).
   A. Low current level.
   B. Low current time.
   C. High current level.
   D. High current time.

### High Frequency Spark

1. Off – not used.
2. Used only for starting arc with DC.
3. Used continuously with AC.

### Arc Voltage

### Automatic Voltage Control

1. Set sensitivity to midrange. Adjust during operation for smooth change in voltage corrections.
2. Set reference voltage for correct arc length during welding. For initial set-up, use 12 to 14 volts in an argon atmosphere. The same arc length with helium mixed with argon atmosphere will be several volts higher.

### Manual Voltage Control

1. Using set blocks between the tungsten tip and the workpiece, set the gap at approximately 1/16 inch.
2. Using dial indicator, set gap at approximately .062 inches. On very thin materials with low amperage, the voltage will be lower.
3. Adjust during welding to the gap desired. If a voltmeter is used, read voltmeter during welding and record.

   If a voltmeter is not used, the actual arc gap must be measured after the test weld is made. This dimension is then recorded and used for the production weld.

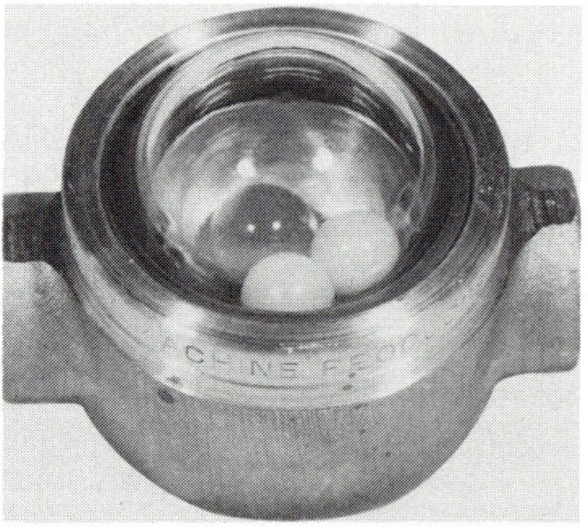

Fig. 20-27. Flow sight valves may be mounted in any position on the inlet or outlet water line. (MPC Machine Products, Inc.)

### Torch, Components, Tungsten, Gases

1. Select a torch with sufficient capacity and duty cycle for the welding operation.

   GAS COOLED (air cooled) TORCHES do not require water cooling. Where these gas cooled torches can be used, they eliminate the following from the system: Water pump, if recirculated water is used. Water filter, if city water is used. Water regulator, if city water is used. One water hose to torch. One gas hose to torch. Possibility of clogged passages in torch and hose. Possibility of water leaks in the torch head.

WATER COOLED TORCHES must have sufficient water flow to prevent overheating. To insure water flow, install a PRESSURE SWITCH in the arc start circuit, as shown in Fig. 20-26. The pressure switch will prevent the arc from starting if the pressure is too low, and will stop the arc if pressure is lost.

Another method to check water flow is the use of a VISUAL FLOW GAUGE, as shown in Fig. 20-27. This gauge should be installed where the operator can monitor the gauge as required.

2. Select the proper size and type of tungsten.
   A. Point the tungsten for DC.
   B. Ball the tungsten for AC.
3. Select the proper size collet and collet body.
   If the tungsten is to be extended from the nozzle over 1/2 inch or the area is drafty, use a gas lens. The gas lens requires a special collet and collet body.
4. Select a gas nozzle (cup).
   The gas nozzle should be the largest practical size. Small nozzles and high inert gas flow rates will draw air into the gas stream causing contamination.
5. Install the nozzle and the tungsten.
   Extend the tungsten from the nozzle approximately 1/4 inch and adjust to final extension when torch is in final position for welding.
6. Select the desired gas or gas mixture. Establish mixture percentages and flow rates. Record flow rates for all gas uses including: Torch, back-up (if used), trailing shields (if used), and purge times and rates (if used).

### Wire Feeders

1. Install wire feeder control.
2. Install conduit, manipulator, and wire feed guide tip.
3. Install filler wire of proper type and size. Check drive rollers for proper type and size. Check brake on filler wire spindle.
4. Adjust manipulator to feed the filler wire into the puddle as shown in Fig. 4-19. Adjustment screws should be centered for possible adjustment later during the welding. Adjust control to deliver amount of wire desired. When final quantity of wire is known, the wire speed should be measured and recorded in inches per minute.

### Oscillators

Setup oscillator (if used) and record the following parameters:

1. Left travel speed.
2. Left travel distance.
3. Left dwell time.
4. Right travel speed.

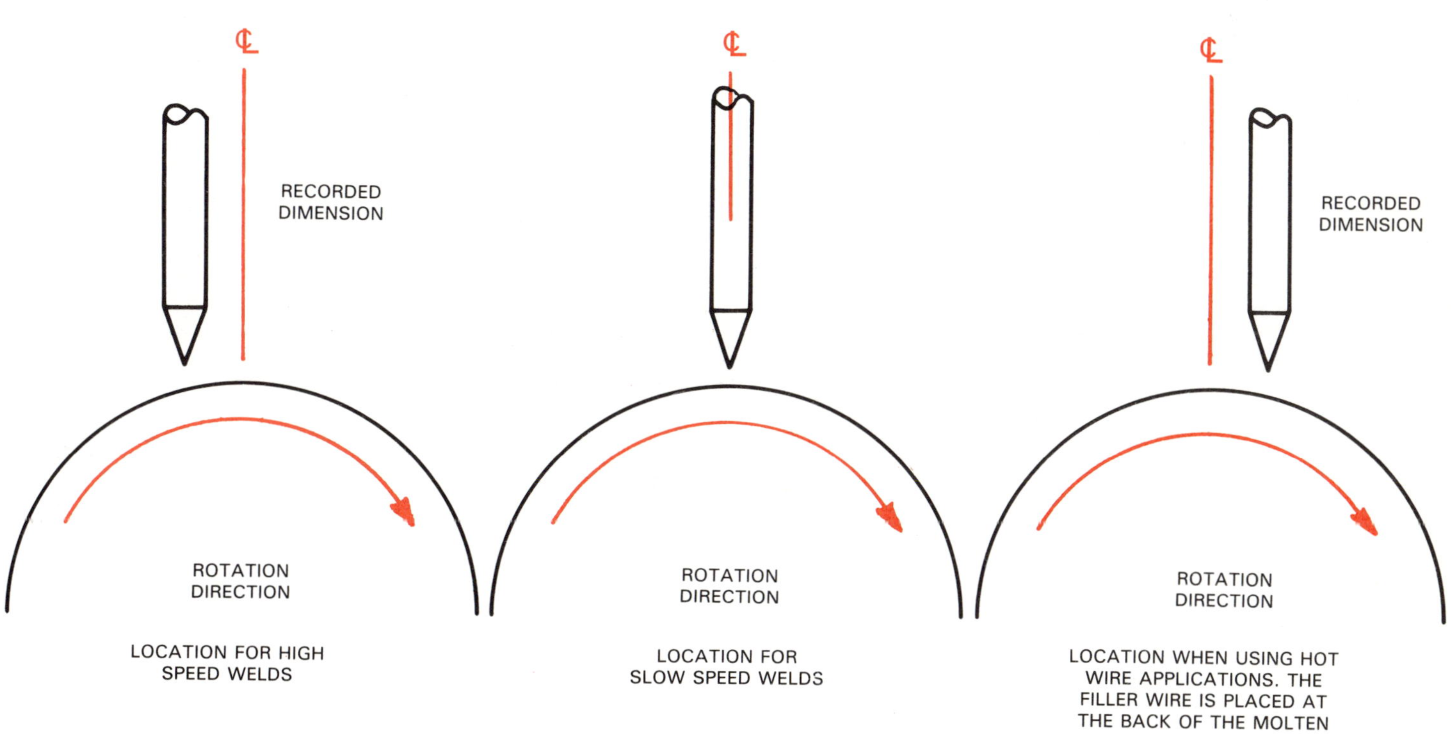


Fig. 20-28. Position of tungsten when welding on circumferences.

5. Right travel distance.
6. Right dwell time.

## TORCH AND JOINT ALIGNMENT

1. Locate the welding torch with the tungsten over the weldment center line. With the torch in position, all of the torch and adjustments (cross slides, racks, etc.) should be centered. This allows for adjustments in either direction if needed.
2. Track the tungsten completely through the entire length of the weld joint. The tungsten must track the entire weld joint very closely. Otherwise the operator will have to continuously adjust during the operation.
3. Set the travel speed of the weld. A good starting point is 6 inches per minute. Adjust as required during the weld test and record on the schedule as inches per minute.

   On circumference welds, the torch may be located on or near the part center line, as shown in Fig. 20-28. Adjust as required during the weld test and record.

## WELDING PROBLEM AREAS

The quality of the finished weld made by semiautomatic welding is dependent on several major factors. These include:

1. WELDING OPERATOR TRAINING – The operator must be thoroughly trained in the complete sequence of the operation. Since the operator must enter into the operation all of the parameters and variables at the proper time, the controls must be available when needed. "Dry runs" are used often so that the operator may become familiar with the locations of these controls.

   After the "dry runs" have been completed, several test components are actually used to test the weld schedule for reproducibility and to further train the operator prior to starting the production weld.

   Only after acceptable welds can be made on the test components should production start.
2. COMPONENT PARTS TOLERANCES – Repetitive quality welds cannot be made on parts that do not fit within the tolerances of the weld schedule. In manual welding, the welder can change many factors to correctly produce acceptable welds. However, in machine welding the operator does not have this opportunity, or capability. The operator cannot change the required parameters while welding at a fixed rate of speed.

   To prevent this type of problem, additional weld tests may be made with the component parts at the FULL TOLERANCE of the weld joint. Machine settings may then be readjusted for these conditions. This type of test also points to the possibility that additional auxiliary equipment may be required for these parts. Where these changes are not satisfactory, the tolerances must be reduced to make acceptable quality welds.
3. WELDING EQUIPMENT AND TOOLING MAINTENANCE – Worn equipment and tooling affect weld quality over a period of time. They may affect the quality to an extent that the welded parts may have to be scrapped. Old or new equipment may even malfunction due to the excessive demands placed on it due to the quantity of welds produced.

   To reduce and possibly prevent these problems, a systematic maintenance program should be established. The program should include each component of equipment. Often the maintenance checks are at the start or at the end of the day, so that production is not affected by a shut down. These maintenance checks may also indicate that major problems are coming. Major overhauls or maintenance can then be scheduled at a later period during non-production times.

## REVIEW QUESTIONS

1. Persons who operate semiautomatic welding equipment are called ________ ________.
2. Semiautomatic gas tungsten arc welding longseam welds are generally made in the ________ position.
3. What is the main problem encountered in making a fillet weld?
4. Aluminum is often welded with DCEN (DCSP) and ________ gas in the semiautomatic mode.
5. Controls for welding amperage that are located at the weld site away from the power supply are called ________ controls.
6. Do all Automatic Voltage Control units operate with all types of current and pulsers?
7. What type of instrument may be installed into the arc circuit to determine arc gap?
8. What are set blocks used for?
9. What two types of electrical motors and controllers are used to drive weld travel systems?
10. Which type of lathe chuck offers fast centering and precision alignment of the weldment?
11. Why are special grounds required on turning equipment such as lathes and positioners?
12. What is the main reason for the close hole dimension in the positioner faceplate?
13. Automatic voltage control systems, oscillators, pulsers, and wire feeders usually operate on ________ volt AC power.
14. What is the main advantage of using a hot wire

feeder?

15. What are the two types of oscillators?
16. What are oscillators used for?
17. In what type of an operation is pulsing done when using filler wire and travel speed?
18. Welding parameters are recorded on a ________ ________ .
19. List two types of equipment used to check water flow for a water cooled torch.
20. When setting up equipment for welding, in what position should slides, racks, manipulators, etc. be positioned?
21. Always track the tungsten through the ________ length of the joint before starting to weld.
22. An initial weld travel speed for semiautomatic welding is ________ inches per minute.
23. What two factors have the most affect on the quality of the semiautomatic weld?

Welder checks setup for semiautomatic welding operation on a missile motor case. (Aerojet-General Corp.)

# Chapter 21

# AUTOMATIC WELDING

AUTOMATIC WELDING is a welding operation using equipment which performs the entire welding sequence without any adjustment of the welding controls by the welding operator. The welding operator must set the parameters used during the automatic welding operation. The basic parameters to be set include amperage, voltage, wire feed, travel speed, shielding gas, and various timers and controls. Automatic welding is used with robots in many industrial settings.

Automatic welding is most often used for high production welding or where high quality repetitive welding is required. Advantages include improved quality, increased productivity, and reduced cost for most applications.

## TYPES OF EQUIPMENT

The type of equipment used for automatic welding vary considerably. Simple welds may be made using several types of standard equipment assembled into a system and operated by timers, relays or special electronic circuitry.

Welds that have changes in direction, speed of travel, pulsing modes, and other requirements, require complete electronic programming. This may be done by programmers, numerical control, or micro-processors. Examples are shown in:

1. Fig. 21-1: Programmed welder.
2. Fig. 21-2: Programmed welder.

Fig. 21-1. Each welding parameter is entered into the main program at the operator's console. The parameters were obtained from test welds that were acceptable to the welding specification limitations. (Union Carbide Corp.)

Fig. 21-2. Individual components required for the program control are assembled into a complete control unit. (Dimetrics, Inc.)

3. Fig. 21-3: Programmed welder and weld head.
4. Fig. 21-4: Computerized programmer and welding tool.
5. Fig. 4-45: Computerized programmer and welding head for automatic welding pipe joints.

Fig. 21-3. Assembled components for complete automatic welding of an aircraft duct.

### Typical Automatic Welds

The bellows assemblies shown in Fig. 21-5 are edge welds joining very thin metal discs together. The power supply must be programmed to produce all of the

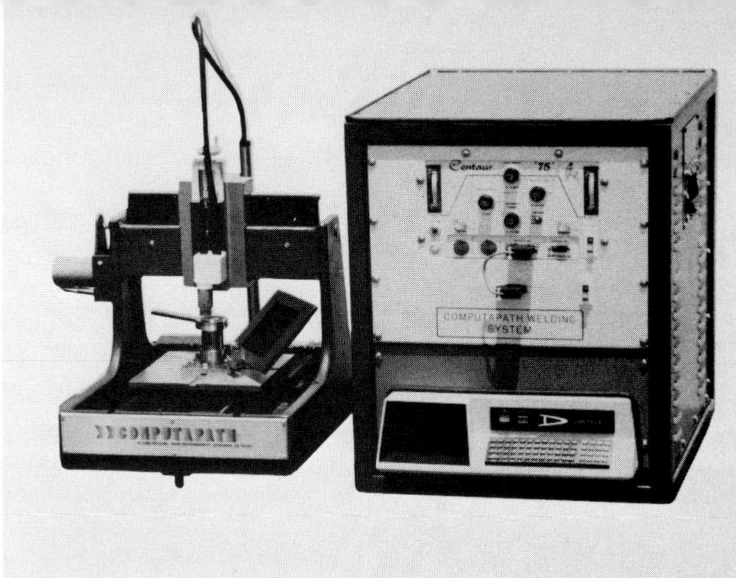


Fig. 21-4. Typical system which operates all of the welding parameters and the operation of the tooling automatically. (Dimetrics, Inc.)

Fig. 21-5. Bellows of almost all weldable materials. Edge type welds joining these discs together are rotated under a stationary arc.

needed parameters at exactly the right time to produce an acceptable weld.

The equipment shown in Figs. 4-43 and 4-44 was used to make the pipe weld shown in Fig. 21-6. A cross section of the pipe weld is shown in Fig. 21-7.

This type of weld is made in position and requires movement of the arc around the pipe. The programmer controls every parameter required to make the weld.

Fig. 21-6. The cover pass of this weld was oscillated and stepped forward at each end to provide the proper contour.

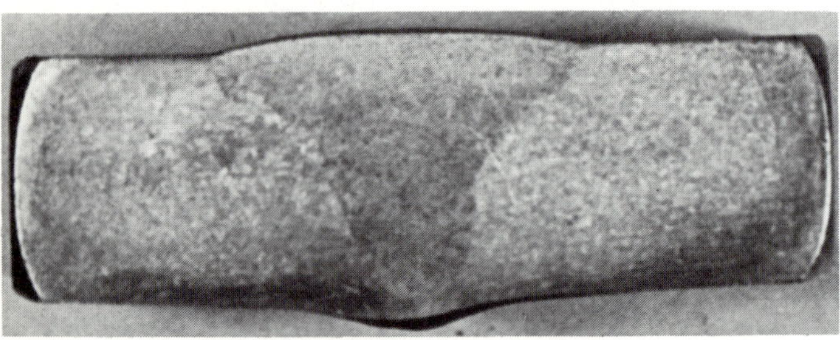

Fig. 21-7. The cross section of the weld reveals a sound weld in all aspects.

## PRE-PROGRAM TEST

For an automatic welding machine to operate properly, each parameter must be established, then entered into the programmer. In semi-automatic welding the operator can make some changes during welding. However, when using automatic programs, changes in the program may be very costly.

One of the major factors in the weld test program is the dimensions of the weld joint and the tooling. Joint tolerances and tooling have a pronounced effect on weld quality. Therefore, these tolerances must be considered in the total program.

The program established for the weld may be satisfactory for parts in tolerance. However, welds on out-of-tolerance parts may be unsatisfactory. The program should therefore include tests on component parts with minimum and maximum dimension tolerances.

### Weld Schedule Maintenance

The quality of the production weld may diminish over a period of time as the equipment and tooling is used. Therefore, periodic inspections should be scheduled to examine those areas which may deteriorate with use.

Should the machine be inoperative for a period of time, or exposed to degrading elements such as moisture or heat, the components should be checked and a sample weld test made before resuming production.

If a major change is made in the equipment such as a power supply or weld joint fixturing, a weld test will verify the existing weld schedule or that a modification is required of the existing parameters.

Since every application of automatic welding is customized for a particular set-up or industry, the welding operator must follow the parameters for the specific welding sequence. It is important that the welding operator be flexible in working with many different types and kinds of equipment. Since so many different applications exist, the welding operator must be able to read the equipment manuals and follow closely the manufacturers' directions in setting up and directing the automatic welding sequence.

Special cases and unusual requirements will continuously be requested by designers and engineers. The ability to program the parameters of these unique welding configurations will enable the business to produce the weldments in a competitive world.

# Chapter 22

# QUALITY CONTROL/INSPECTION

QUALITY CONTROL is used throughout the welding industry to monitor the quality of the items produced. All manufactured items are made to specifications. Therefore, inspections must be made during and after the manufacturing cycle, to assure that the parts meet the requirements of the specifications.

## INSPECTION AREAS

One role of the welder performing the GTAW operation is to make sure that all of the areas involved in the operation are correct. The areas of inspection, as a minimum, include:

1. Base material is as specified.
2. Joint design is as specified and within required tolerances.
3. Filler material type and size are correct.
4. The required welding equipment is available and is operating satisfactorily.
5. The tooling has been adequately tested to determine that it will support the operation properly.
6. Pre-weld inspection of the parts to determine that the parts have been cleaned properly.
7. Welder training or certification is sufficient for the weld operation.
8. Proper welding procedure is used and the welding equipment is set up properly for the operation.
9. Inspections and tests required during the welding operation are performed as specified.
10. Post weld inspections of the completed weld to ascertain that the weld will meet the visual requirements. Further nondestructive testing beyond visual inspection is usually performed by other personnel.

## TYPES OF INSPECTION

The person performing the required inspections has the responsibility of determining the acceptance or rejection of the weld to the fabrication specification. Depending on the final use of the weldment, inspections required may be simple or very complex, requiring several types of inspections.

Inspections and tests that are made on the weld that do not destroy any portion of the completed weld are called NONDESTRUCTIVE TEST (NDT).

Inspections and tests that are made on the completed weld or samples of the completed weld that destroy the weld are called DESTRUCTIVE TEST (DT).

Inspections and tests that do not fall into either of these categories may also be required to test the integrity of the welds.

## NONDESTRUCTIVE TEST

### Visual Test

VISUAL TEST (VT) is one of the most important methods of inspection and is widely used for acceptance of many welds. It is also used as a basic test prior to using other forms of inspection. This type of inspection is easy to apply, quickly done, and relatively inexpensive. The equipment used includes rulers, fillet weld gauges, squares, magnifying glasses, and reference weld samples. Some of the various tools used in weld inspection are shown in Fig. 22-1, 22-2, and 22-3. Visual inspection gives very important information with regard to the general conformity of the weldment to the specification requirements. The various visual tests may include:

A. Crown height.
B. Crown Profile.
C. Underfill.
D. Undercut.
E. Overlap.
F. Surface cracks.
G. Crater cracks.
H. Surface porosity.
I. Surface color. (Titanium welds)
J. Weld size.
K. Weld length.
L. Joint mismatch.
M. Warpage.
N. Dimensional Tolerances.
O. Root side penetration.
P. Root side profile.

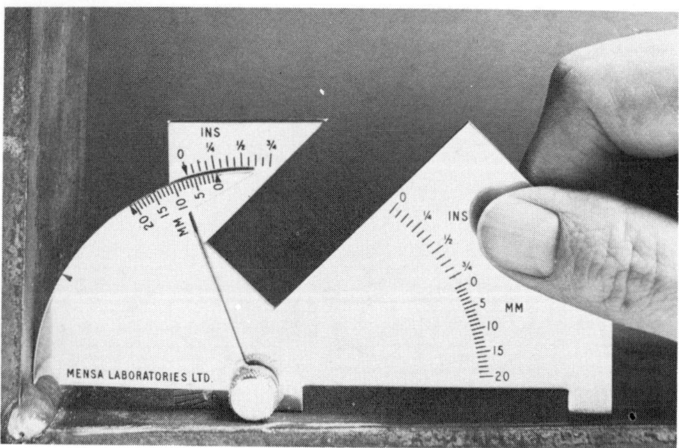


Fig. 22-1. Inspection gauge used to check a variety of dimensions including: Material thickness, bevel angle, crown height, undercut, mismatch, and fillet weld leg length and throat thickness. (British Aerospace Inc.)

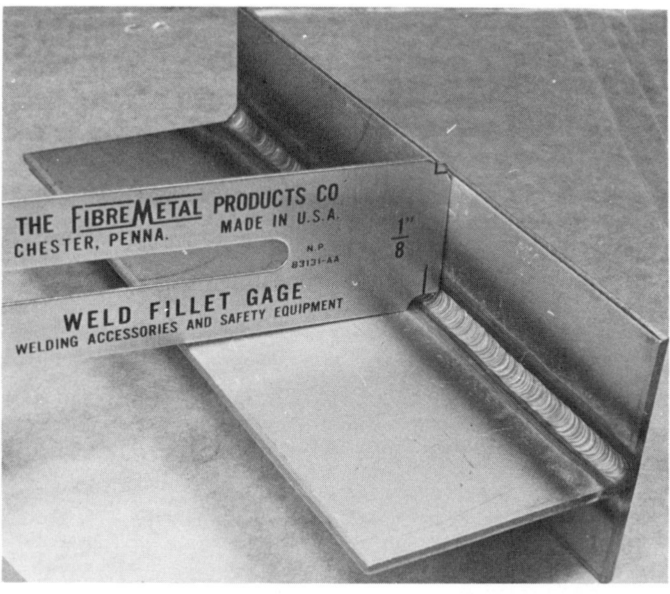


Fig. 22-2. Fillet weld size and crown gauge. (Fibre Metal)

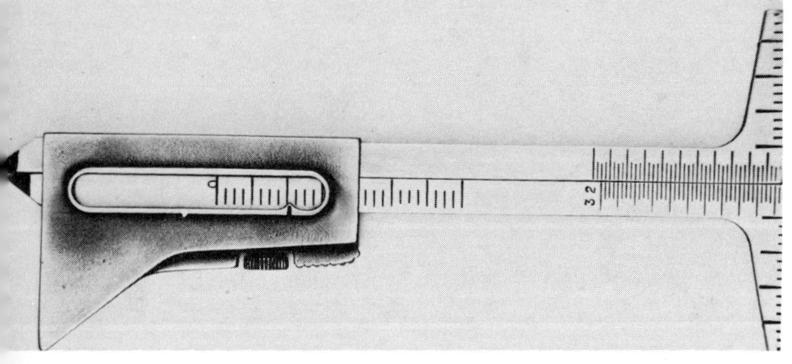

Fig. 22-3. Vernier type gauges of this design are used to check mismatch on both pipe and plate. (G.A.L. Gauge Co.)

## Penetrant Test

PENETRANT TEST (PT) is a sensitive method of detecting and locating minute discontinuities that are open to the surface of the weld. It employs a penetrating liquid (dye) which is applied over the surface of the weld. The fluid then enters the discontinuity. After a short period of time, the excess penetrant is removed from the surface. A developer is then applied to the surface and allowed to dry. The penetrant in the defect then rises to the surface by capillary action where the defect may be seen in the developer. Fig. 22-4 outlines the sequence of operations required for the inspection. This method of inspection is limited to surface examination only.

Penetrant test is particularly useful on nonmagnetic materials where magnetic particle test cannot be used. In the welding field, it is used extensively for exposing surface defects in welds on aluminum, titanium, magnesium, and austenitic stainless steel weldments. Fig. 22-5 shows how the inspection may also be used to detect leaks in all types of tanks or vessels. The two types of penetrant tests include:

DYE PENETRANT – The dye penetrant test requires only the penetrant and the developer. It may be done anywhere because it is portable. Also, it can be done in any position. The results can be detected in normal light and requires no special equipment.

FLUORESCENT PENETRANT – Fluorescent penetrant test requires an ultraviolet light (black light) to observe the test results. To utilize the black light, it may be necessary to enclose the viewing area to properly read the test results.

MAGNETIC PARTICLE TEST (MT) is a nondestructive method of detecting the presence of cracks, seams, inclusions, segregations, porosity, lack of fusion, and similar discontinuities in magnetic materials. This method will detect surface defects that are too fine to be seen with the naked eye and those defects that lie slightly below the surface.

The basic principle involved in magnetic particle testing is that when a magnetic field is established in a ferromagnetic material containing one or more defects in the path of the magnetic flux, minute poles are set up at the defects. These poles have a stronger attraction for the magnetic particles than the surrounding material.

The material is magnetized by an electric current and finely divided iron particles or iron powder are applied to the magnetized area. If the magnetic field is interrupted by a defect, the iron particles form a pattern on the surface, which is the approximate size of the defect. Fig. 22-6 shows the operation for this type of test.

Small, portable permanent magnets may be used for thin gauge materials. Heavier material will require power from transformers, generators, or rectifiers. A

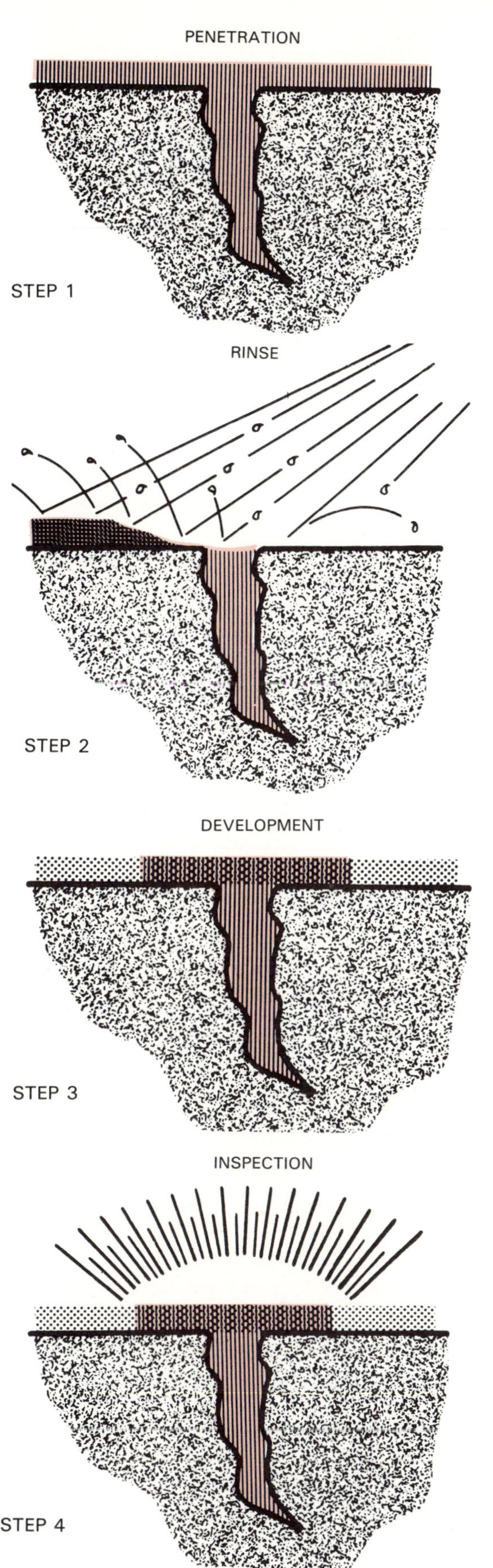


Fig. 22-4. Penetrant test sequence of operations. (Magnaflux Corp.)

Fig. 22-5. Minute leaks in tanks or vessels may be located using penetrant inspection in this manner to assure leak tight weld joints.

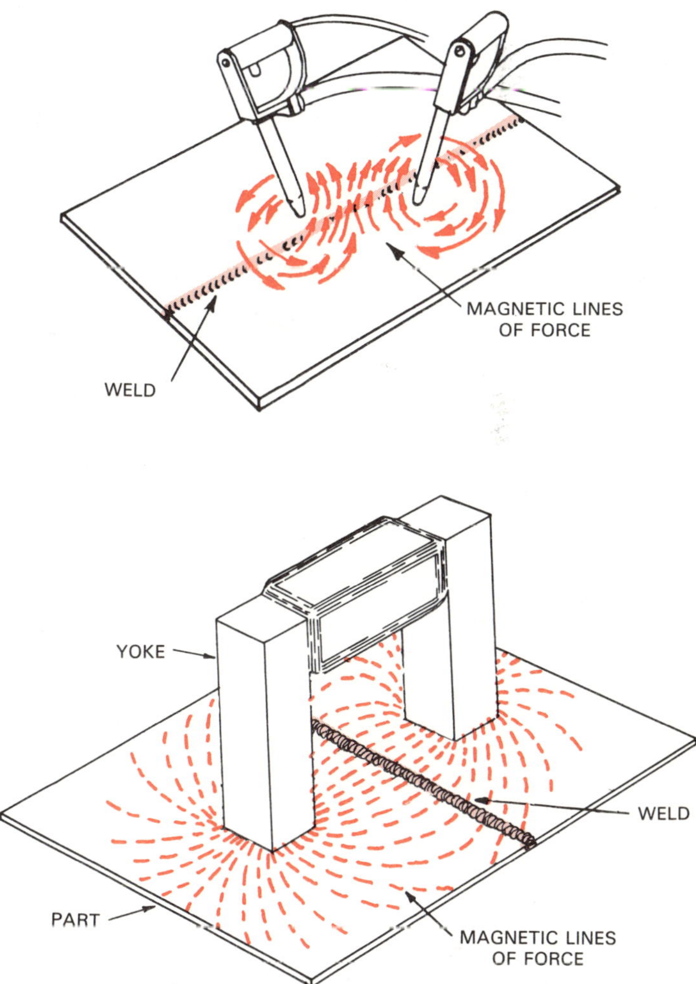


MAGNETIC PARTICLE TEST SEQUENCE OF OPERATIONS:
1. APPLY MAGNETIC FIELD USING ELECTRIC CURRENT.
2. APPLY MAGNETIC PARTICLES WHILE POWER IS ''ON''.
3. BLOW AWAY EXCESS PARTICLES.
4. INSPECT.

Fig. 22-6. Magnetic particle test operations.

typical unit is shown in Fig. 22-7.

The magnetic particle test process may be used in either a wet or dry manner depending on the type of inspection desired. The dry method uses finely divided dry particles and the wet method suspends the particles in a fluid. Depending on the material, the colors of the particles may be red or grey.

Another modification of the process is the addition of fluorescence to the particles. This requires the use of an ultrviolet light to interpret the formation of the particles at the defect. As in fluorescent penetrant testing, this may require a darkened area to properly view the test area.

Proper intrepretation of this process requires highly trained technicians. Discontinuities revealed by the test pattern may be misleading to the untrained eye and have no consequence on the actual weld quality.

Fig. 22-7. Magnetic particle units are very useful on small weldments and require only 110 volts for operation. (Magnaflux Corp.)

### Ultrasonic Test

ULTRASONIC TESTING (UT) is a nondestructive method of detecting the internal presence of cracks, inclusions, segregations, porosity, lack of fusion, and similar discontinuities in all types of metals. It may be used as the sole type of inspection or it may be used with other types of testing. The ultrasonic test is often used in conjunction with radiographic test because it will define the depth of the defect from the test surface.

The basic principal involved in ultrasonic testing is the transmittal of very high frequency sound waves through the part to be tested. The sound waves then return to sender and are visually shown on a CATHODE RAY TUBE (CRT) for interpretation.

Since sound waves of these frequencies only travel short distances in air, the test must be done with the part, the signal sender, and the receiver immersed in water, or with the transducer coupled to the workpiece. These two methods are shown in Figs. 22-8 and 22-9. Where tests are required out of perpendicular with the transducer, a wedge or angle block is placed under the transducer at the desired angle to properly scan the material, as shown in Fig. 22-10.

The UT test method has three advantages:

1. Very great penetration power. This allows the testing of thick materials.
2. Very sensitive in locating small defects in a short period of time.
3. Inspection may be done from one surface.

The major disadvantage is the proper interpretation of the results. Design of the weld, location of the defect, internal structure, and complexity of the weldment affect the interpretation of the ultrasonic signal.

In order to achieve the desired results, calibration blocks and reference weld samples are used to calibrate the equipment prior to making the test. With the proper calibration, the operator can then interpret the results to the inspection specification.

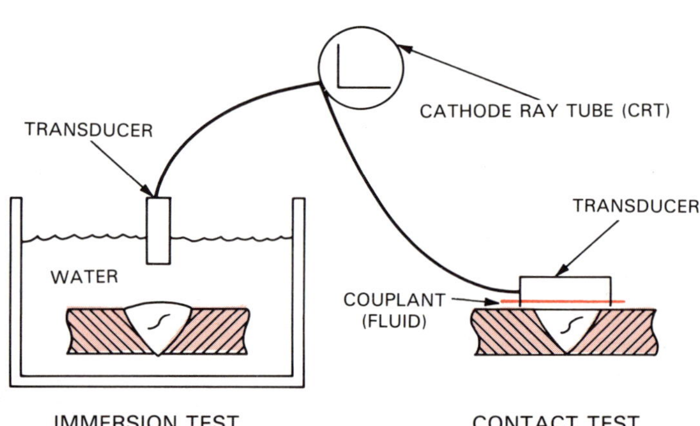


Fig. 22-8. Ultrasonic tests are made with the part and the transducer submerged in water. Where this is not practical, the transducer is coupled (connected) to the test area by a thin layer of liquid.

### Radiographic Test

RADIOGRAPHIC TEST (RT) is a nondestructive test method that shows the presence and nature of discontinuities in the interior of the welds. This test method makes use of the ability of short wave length radiations, such as x-rays or gamma rays, to penetrate material opaque to ordinary light.

X-RAYS are produced by fixed location machines where the area may be lead shielded to prevent the escape of radio activity. These locations usually have all of the support equipment such as film developing machines. The end result is a radiograph made in a minimum amount of time.

GAMMA RAYS are produced from radioactive material such as cobalt, cesium, iridum and radium. These radioactive materials must be contained in a lead shielded box and transported to the job site for in-place radiographs.

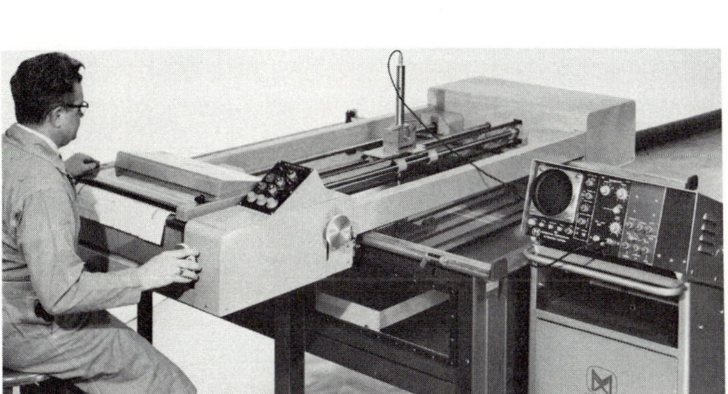

Fig. 22-9. Ultrasonic emersion test in process. (Magnaflux Corp.)

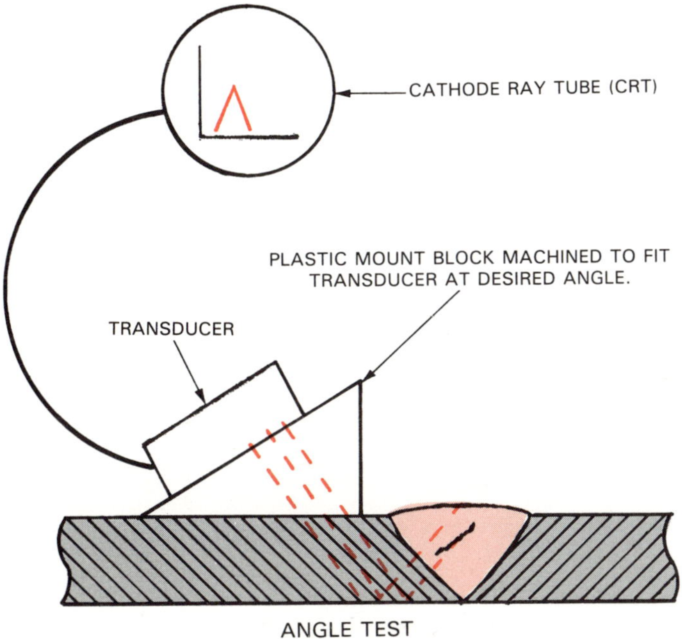


Fig. 22-10. Wedges or angle blocks are used to send and receive ultrasonic signals In areas where signal transmission may be blocked in a straight line plane.

Fig. 22-11 shows the radiograph inspection of a fusion weld. The film is developed for viewing on a special viewer.

Interpretation of the radiograph is made by a skilled technician to a specification which defines discontinuities. The film must have the sharp contrast for proper definition of the weld and any defects. (Contrast is the degree of blackness of the darker areas compared with the degree of lightness of the brighter areas.)

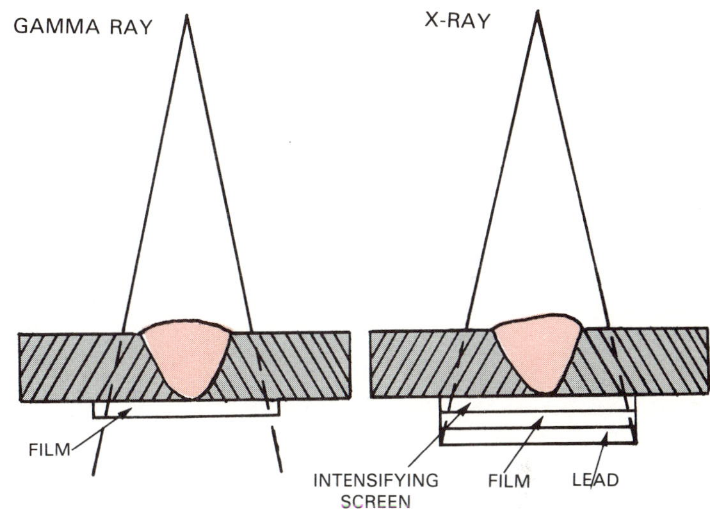


Fig. 22-11. Radiographic test operation.

To assure sharp images on the film, PENETRAMETERS are used to indicate the quality of the radiograph. Penetrameters consists of a thin shim of the base metal, usually with a thickness equal to a 2 percent of the weld thickness. One, two, or more holes with various diameters are drilled into the metal shim. Fig. 22-12 shows a sample penetrameter. The ability of the radiograph to show definite sized holes in the penetrameter establishes the radiograph quality.

Radiographs are expensive. However, they provide a permanent record of the weld quality. This is often useful for comparison with later radiographs made after the weldment has been in service.

## DESTRUCTIVE TEST

DESTRUCTIVE TEST may be done for qualification and welder certification, weld test, and on actual production welds if the weldment can be destroyed without excessive cost. Where large production runs are made, it is often less expensive to scrap a part to make the destructive quality test, rather than use more expensive nondestructive tests.

### Bend Test

BEND TESTS are used to define internal weld quality. They are made in three different forms:

1. Face bend (face of the weld tested).

## PENETRAMETERS FOR RADIOGRAPHIC TESTING

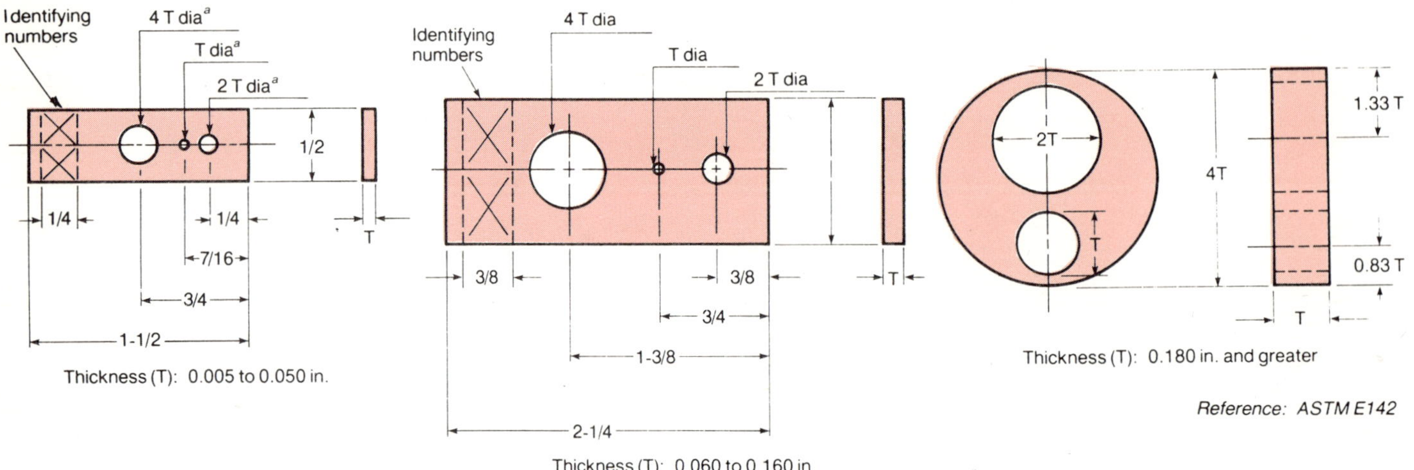


Note: Tolerances on penetrameter thickness and hole diameter shall be plus or minus 10 percent or one half of the thickness increment between penetrameter sizes, whichever is smaller.

a. Minimum diameter for 1T hole—0.010 inch
 Minimum diameter for 2T hole—0.020 inch
 Minimum diameter for 4T hole—0.040 inch

Holes true and normal to penetrameter surface; no chamfer

Fig. 22-12. Penetrameters are used to determine that the quality of the radiograph is sufficient to reveal any defects with sharpness and clarity.

2. Root bend (root of the weld tested).
3. Side bend (sides of the weld tested).

Fig. 22-13 shows the three types of bend tests. The three types of tests are made in a fixture to bend the material into a U form over a specified radius. After bending, the outer face of the U is checked for cracks and other indications, as required by the weld specification.

### Tensile Test

TENSILE TESTS are used for comparison with the parent metal mechanical values and specification requirements. A typical tensile test machine is shown in Fig. 22-14.

Tensile tests are made to define:

1. Ultimate strength of the weld. This is the point at which the weld fails under tension.
2. Yield strength of the weld. This is the point at which the weld yields or stretches under tension, and will not return to its original dimensions.
3. Elongation is the amount of stretch that occurs during the tensile test. It is measured by placing gauge marks on the sample or coupon before testing and comparing the after-break distance with the original gauge marks. The gauge mark tool is shown in Fig. 22-15.

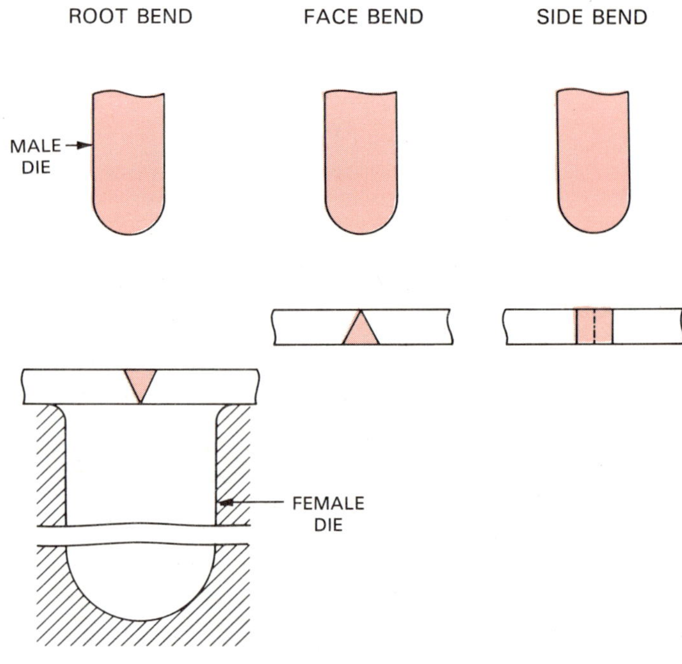


Fig. 22-13. Bend tests are made over a specified radius which is dependent on the thickness and strength of the material. The outer face of the bend may be examined by visual, penetrant, or magnetic particle test to detect for defects such as cracks, lack of fusion, and lack of penetration.

Fig. 22-14. This tensile test machine has a recorder mounted on the side to record the test operation parameters and results. (Tinius Olsen Testing Machine Co.)

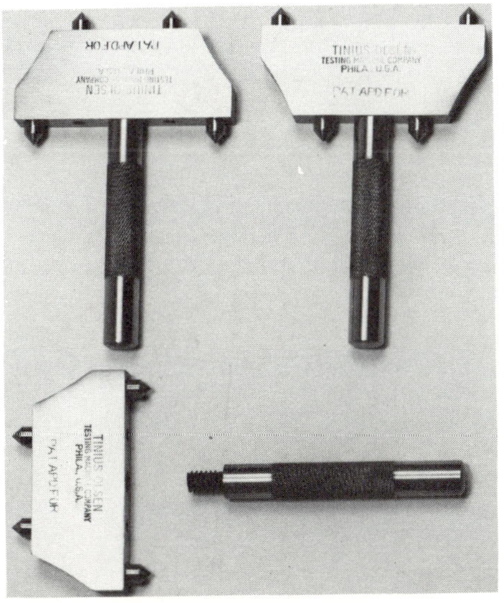

Fig. 22-15. Marking tool used for measuring 1 inch or 2 inch gauge lengths. It may also be used for marking butt weld shrinkage test. (Tinius Olsen Testing Machine Co.)

## Notch-Toughness Test

NOTCH-TOUGHNESS TESTS are used to define the ability of welds to resist cracking or crack propagation at low temperatures under loads. This test is used on welds that are placed in service in low temperature environments with pulsating or vibrating loading. The test coupons or samples are removed from the test weld. They are prepared for either Charpy or an Izod type test, shown in Fig. 22-16. The test bars are cooled to the test temperature. They are next placed into the test machine, Fig. 22-17, and then broken, Fig. 22-18. A prepared and broken test coupon is shown in Fig. 22-19. The results are measured in the energy required to make the coupon break and expressed in foot pounds. Comparisons are then made with the original material and specification requirements.

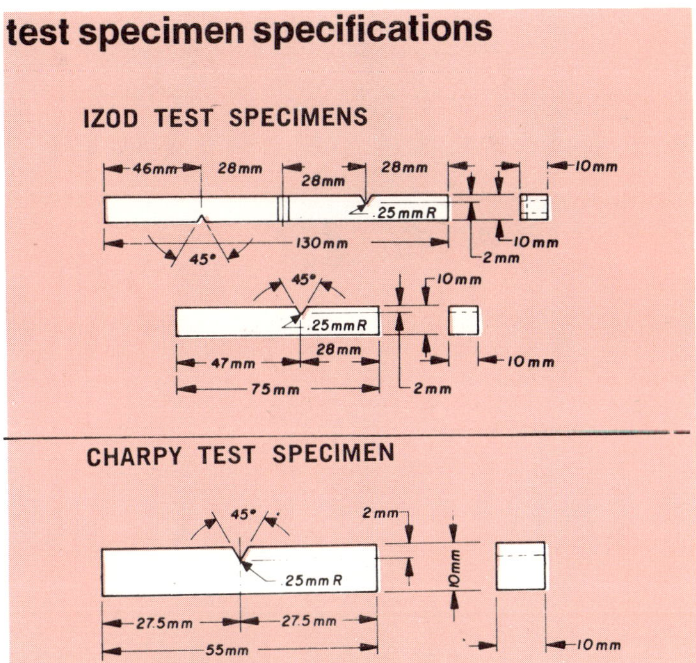


Fig. 22-16. Charpy and Izod test bar dimensions.

Fig. 22-17. Test bar is located in the holding fixture and clamped into place. (Tinius Olsen Testing Machine Co.)

Fig. 22-18. The impact test machine arm is raised into position. Releasing the arm causes the downward swing. The break of the coupon occurs at impact and the results are shown on the gauge in foot pounds. Note the tongs used to handle the very low temperature test bar. (Tinius Olsen Testing Machine Co.)

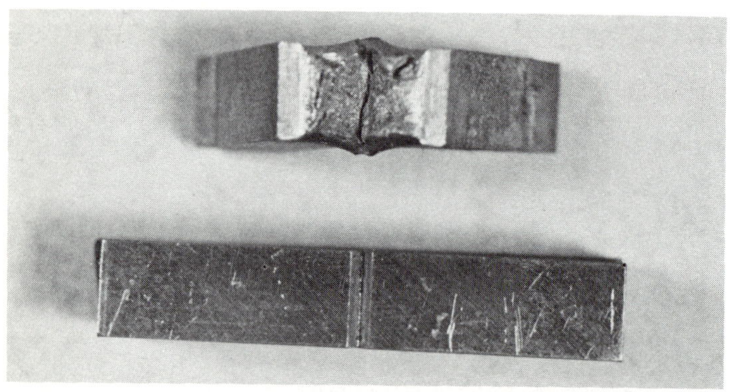

Fig. 22-19. Charpy bars before and after test.

## Cross Section Test

CROSS SECTION TESTS are used to define the internal quality and structure of the weld. The various types of tests include:

1. MACRO TESTS are polished sections of the weld for overall weld viewing with the naked eye or a magnifying glass. Increased definition of the weld layers and passes can often be obtained by etching the sample with a suitable etchant.
2. MICRO TESTS are very highly polished sections of the weld for viewing with high power microscopes. This type of test is used to determine grain size, content, and structure.
3. MICRO HARDNESS TESTS are very highly polished sections of the weld tested on special machines to determine the hardness of a very small area. The results may then be evaluated to determine the hardness variations within the grain structures and the weld zones.

## Nick-break Test

NICK-BREAK TESTS are destructive tests that are very simple to make. They are used to determine the internal quality of a weld with regard to porosity, lack of fusion, and slag.

A section of the weld to be tested is removed from the weld and prepared as shown in Fig. 22-20. The coupon is then placed into a vise as shown in Fig. 22-21 and broken with a sharp blow to the upper section. The defects can then be observed in the broken areas.

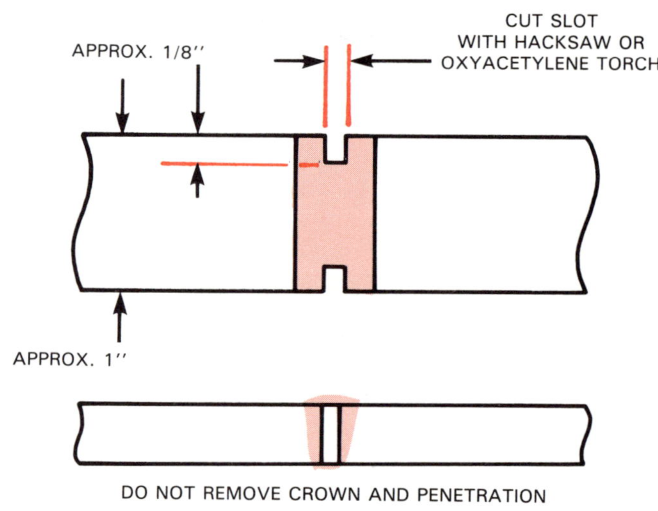


Fig. 22-20. Nick-break test dimensions.

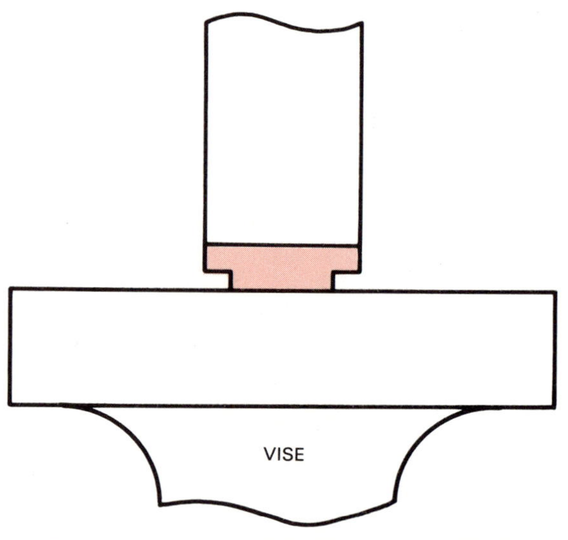


Fig. 22-21. Nick-break test set-up.

## WELD INTEGRITY INSPECTIONS

Many weldments require tests in addition to the non-destructive test requirements to verify the weld quality. These may be imposed as part of the qualification test program or they may be an additional test during or after the manufacturing cycle. The test requirements are usually a part of the fabrication specification.

### Pressure Test

PRESSURE TESTS are used for vessels, tanks, piping, and tubing. They may be an air or fluid tests. If a fluid is used, it is called a HYDROSTATIC TEST. The test program may require a number of cycles to be performed simulating the use of the part in actual service. During the actual test, the test part will expand, therefore do not restrict this growth with tools that cause undue stresses to build within the part.

Whenever conducting a pressure test, be alert to the possible failure of the unit. Plan to provide for the safety of everyone in the area.

### Hardness Test

HARDNESS TESTS are used to determine the final hardness of the weld or weldment after fabrication. Where the weldment has been heat treated (annealed, hardened, tempered, etc.) after fabrication, the hardness test will determine the effect of the operation. A hardness tester is shown in Fig. 22-22.

Hardness tests results may also be used to determine ultimate tensile strength of the material. The results of the test may be compared with standard tables. Refer to the reference section of this book. Weldment designs that include areas such as flanges and extensions may be tested by portable hardness testers as shown in Fig. 22-23.

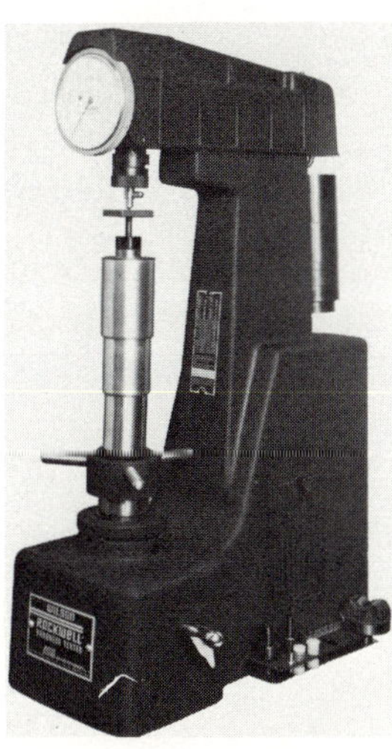

Fig. 22-22. Rockwell Hardness Testing machine for bench testing. (Page-Wilson Corp.)

Fig. 22-23. Portable Rockwell Hardness Tester for flange testing. (Page-Wilson Corp.)

Weldment designs that do not have clear areas for testing, must have test bars of the same material (welds if required) submitted with the weldment during the heat treating cycle. The bars are then tested on a stationary tensile machine as shown in Fig. 22-14, for the hardness results. A round tensile bar is shown in Fig. 22-24. A flat tensile bar is shown in Fig. 22-25.

Fig. 22-24. Round type tensile bar. The threads are machined on each end to grip the bar during the test. (Tinius Olsen Testing Machine Co.)

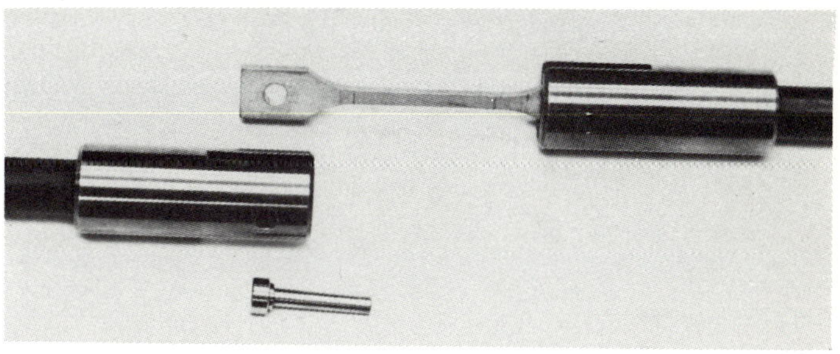

Fig. 22-25. Flat type tensile bar. The hole in the flat end is made for a pin to hold the bar and the fixture together for the test. (Tinius Olsen Testing Machine Co.)

## Corrosion Test

CORROSION TESTS are made to determine the ability of a weld to restrict corrosion by a specific material.

These tests are usually performed on test weldments simulating the weld in service. They are usually required by the specification used by the fabricator. Weld test parameters made on the qualification test weldments must be duplicated on the production part without deviation to maintain the corrosion resistance.

Fig. 22-26. Severn gauge for testing amount of ferrite in the weld. (Severn Engineering Co. and Stoody Co.)

## Ferrite Test

FERRITE TEST on completed stainless steel welds determines the amount of magnetic ferrite in an austenitic (nonmagnetic) weld. Insufficient ferrite in a weld made under high restraint is prone to cracking at red heat.

These tests are made with special designed instruments called a Severn or a Magne gauge, shown in Fig. 22-26. Limits of ferrite will depend on the use of the final weldment. These amounts are usually specified in the fabrication specification.

Weld test parameters and filler materials used on qualification test weldments must be duplicated on the production part without deviation to maintain the proper ferrite content.

## REVIEW QUESTIONS

1. What does "Quality Control" mean to the welder?
2. Should all welds be made to a specification?
3. What is a welder's role regarding inspection?
4. What do the letters NDT define?
5. What do the letters DT define?
6. With what form of inspection should every welder be very familiar.
7. List some of the tools used with the inspection in Question 6.
8. What are the limitations of penetrant inspection?
9. Where is penetrant testing generally used?
10. What is the major limitation of magnetic particle testing?
11. The material used to define reject areas in the magnetic particle test is ________ ________ .
12. What does an ultrasonic test of a weld reveal?
13. List four areas which may detract from the results of an ultrasonic test.
14. What types of tools are used to assist in the interpretation of the ultrasonic test results?
15. The two types of radiographic tests are ________ rays and ________ rays.
16. What is the purpose of a penetrameter?
17. What is the one main advantage of a radiographic test over other forms of nondestructive tests?
18. List the three types of bend tests.
19. List three values that may be obtained from the tensile test regarding mechanical values of material.
20. What are notch-toughness values?
21. List the three types of cross section tests.
22. What one very simple DT test can a welder perform to reveal the internal quality of the material?
23. What type of tool is used to determine ferrite content of a material?

# Chapter 23

# WELD REPAIR

Repair of welds that do not meet the NDT (nondestructive testing) requirements often result in unacceptable welds even though the weld repair is satisfactory. Repairing of the completed weld often buckles, distorts, or misaligns the reworked area to the point where the weld may fail when stressed to the intended load. For this reason, each intended weld repair should be thoroughly investigated and evaluated as to whether or not the defect should be removed. In many cases, the defect may be left in the weld without affecting the function of the actual weld.

Should the part require rework, a thorough welding procedure should be established which will minimize the effect of the repair on the remaining portion of the weld. This procedure must consider all of the following: the original welding parameters, the present condition of the base material, and the weld, filler material, welding sequence, inprocess inspection, tooling, and the final weld condition mechanical properties.

Incomplete consideration of any of these factors may result in further rejection of the weld repair and possible failure of the weld when placed into service.

### Dimensional Repairs

Dimensional repairs are usually required due to the insufficient addition of filler material during the welding operation. They may include:

1. Crown height too low, Fig. 23-1. Repair of this type of defect is done best with stringer beads to minimize weld shrinkage. Add only enough new filler material to build crown to requirements in height. Do not overweld.
2. Surfacing or overlay type weld height too low, Fig. 24-2. Repair of this type of defect is done best with stringer beads to minimize dilution and distortion.
3. Fillet weld size is too small, Fig. 24-3. This type of defect presents a problem as to whether a one pass weld or a two pass weld should be used. Either one will cause stresses across the throat of the weld and deformation of the bottom side of the base plate.

Regardless of the repair technique, these repairs require close control to avoid overwelding the required fillet size and adding additional stress to the original weld.

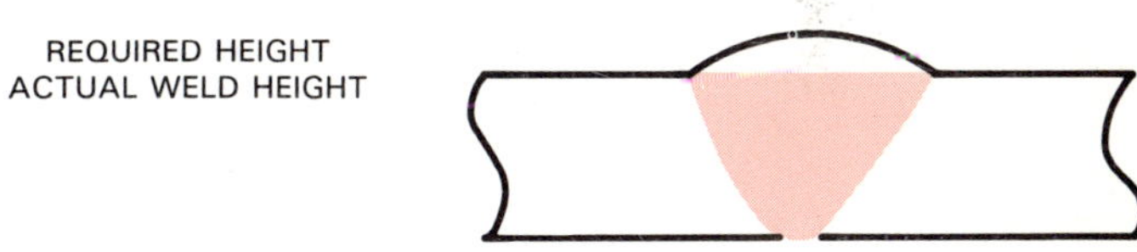


Fig. 23-1. Groove weld crown with insufficient height.

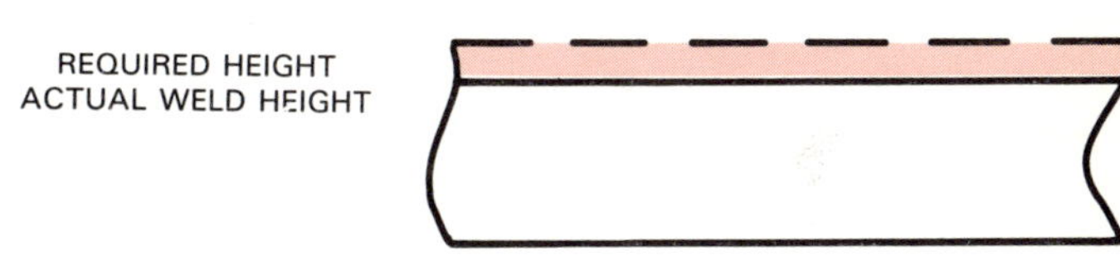


Fig. 23-2. Overlay weld deposit with insufficient height.

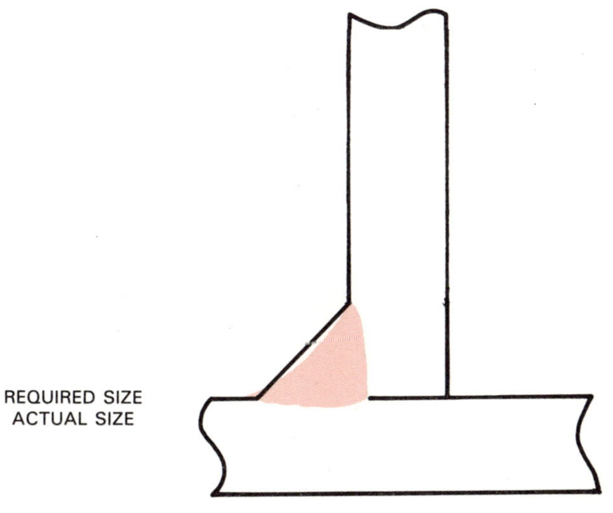


Fig. 23-3. Fillet weld leg length is satisfactory, however the weld is concave thus reducing the actual fillet weld size.

## Surface Defect Repairs

Defects which are located on the surface of the weld may also extend deeper into the weld body. For this reason, the defect should be removed. The area should be reinspected before attempting a repair. Various surface defects may include:

1. LONGITUDINAL, TRANSVERSE, OR CRATER CRACKS as shown in Fig. 23-4. Removal of the crack can be done on steel and steel alloys using small grinding wheels prepared as shown in Fig. 23-5. Remove only the required amount of metal to remove the crack.

   For all other types of metal, small rotary tungsten carbide tools, Fig. 23-6, can be used to remove the crack. Do not use grinding wheels on non-ferrous material. When repairing, add only sufficient filler material to match the adjacent weld contour.
2. UNDERCUT at the edges of the weld, Fig. 23-7. Dirt, scale and oxides may be present in the undercut area. These impurities may cause further defects if not removed prior to weld repair. They may be removed by grinding or routing as previously described. Be careful to not remove parent metal adjacent to the undercut. Since the repair will widen the original crown size, use lower currents and sufficient wire to prevent additional undercuts and underfill.
3. POROSITY or pores in the weld, Fig. 23-8. Isolated or single pores may be removed with a rotary tool for weld repair. Multiple and linear pores (aligned in a row) should be removed by grinding or machining. The weld should then be reinspected by radiographic or ultrasonic inspec-

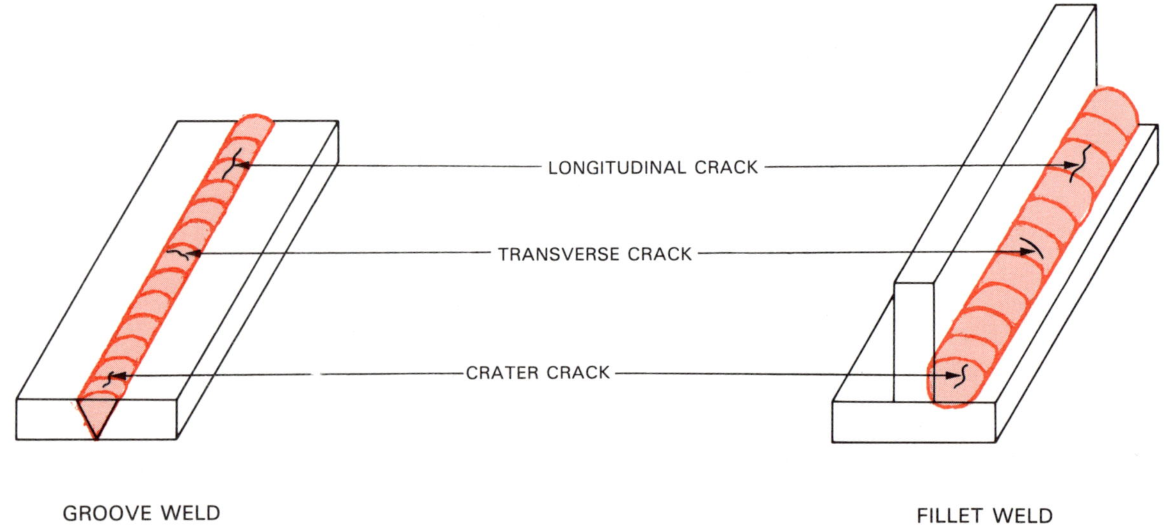


Fig. 23-4. Types of cracks which may be found on groove and fillet welds.

Fig. 23-5. Using a grinding wheel dressing stone, grind an angle on the wheel edges to the width desired.

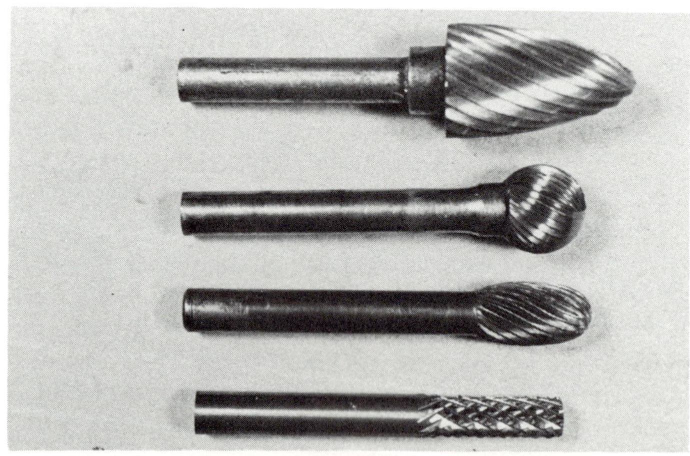

Fig. 23-6. Tungsten carbide rotary files may be obtained in many sizes and configurations.

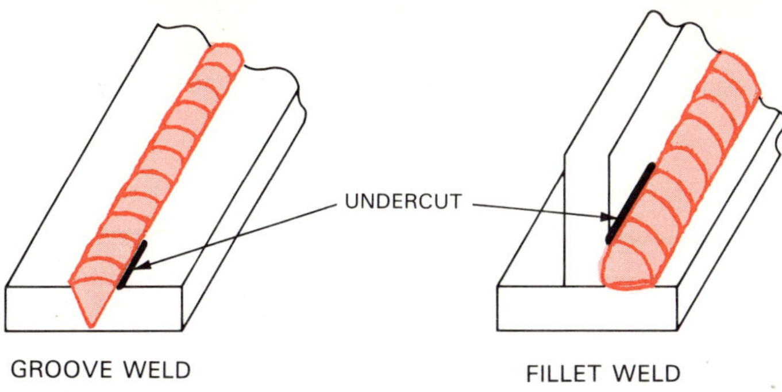


Fig. 23-7. Undercut on flat groove welds may occur on either edge of the weld. Groove and fillet welds made in the horizontal position will usually have undercut on the upper part of the weld.

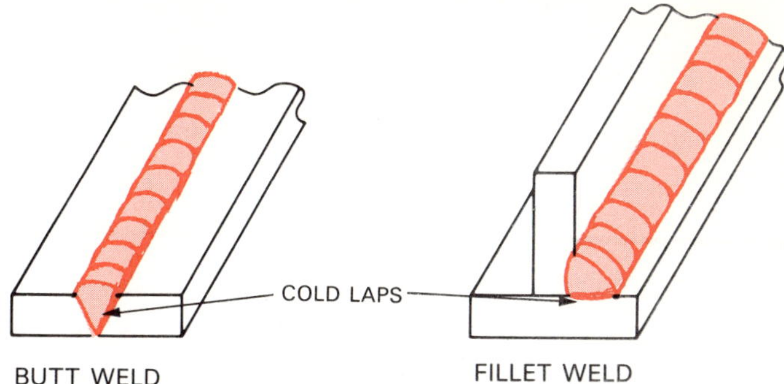


Fig. 23-9. Fillet weld cold laps are usually located on the bottom side of the weld. Butt weld cold laps may occur on either side of the weld crown.

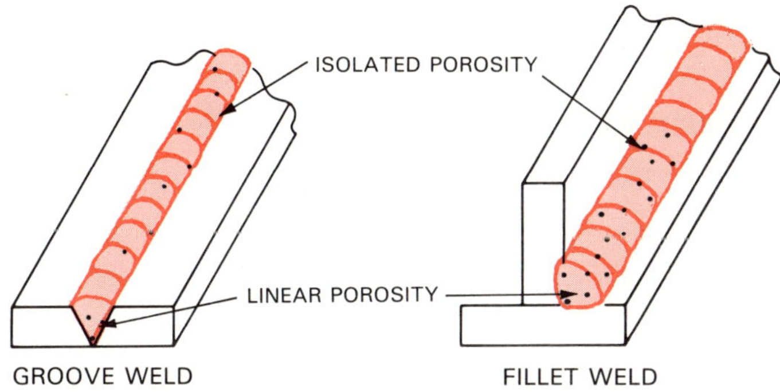


Fig. 23-8. Isolated pores may occur in any portion of the weld. Linear (aligned) pores usually are found near the bottom, along the sidewall, or at weld intersections.

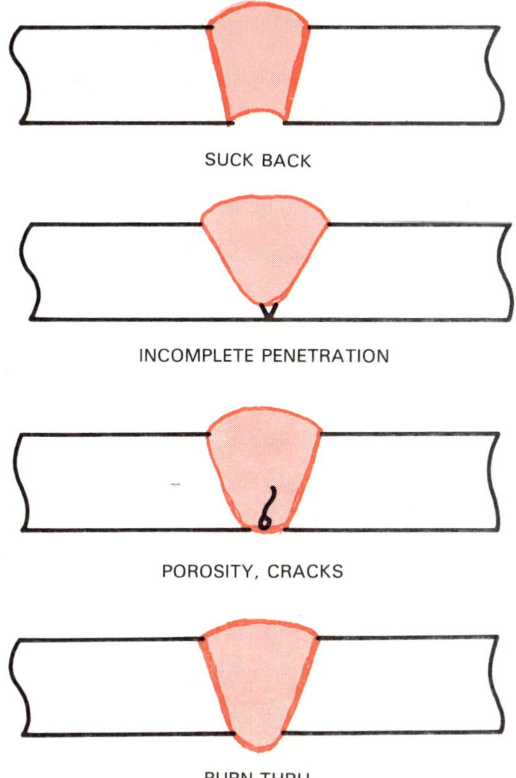


Fig. 23-10. Defects which may occur on the root side of the weld.

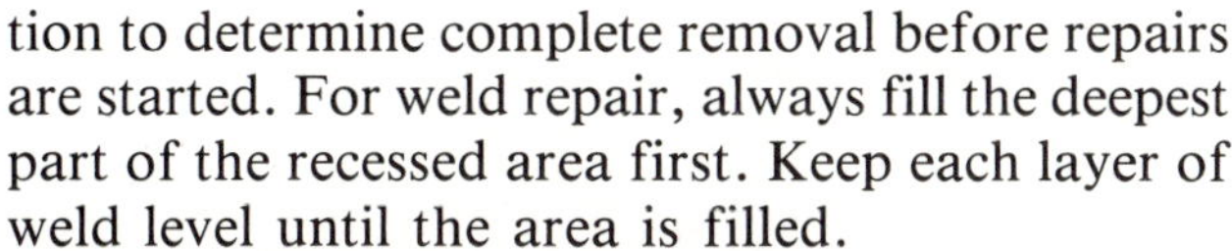

tion to determine complete removal before repairs are started. For weld repair, always fill the deepest part of the recessed area first. Keep each layer of weld level until the area is filled.

4. COLD LAPS are an area of the weld that has not fused with the base material and may occur on fillet welds or butt welds, Fig. 23-9. Since the extent of the overlap cannot be determined by NDT, the entire area must be removed by grinding or routing. Use extreme care when grinding into adjacent laps to prevent grinding into adjacent metal and creating more problems. When the overlap material has been removed, a penetrant test will determine if all of the defect is gone. Continue removing material until the penetrant test is satisfactory. If weld repair is required to satisfy crown height requirements, use low currents and sufficient wire to match the crown with adjoining material.
5. INCOMPLETE PENETRATION on the root side of butt welds, Fig. 23-10. Other types of defects may also occur with the incomplete penetration such as suck-up, cracks, porosity, burn through, etc. All of these areas must be removed by grinding or routing. To assure complete removal of all defects, penetrant test before reworking. Since oxides form in this area during welding, clean the rework area to bright metal before rewelding. Use stringer beads and add only enough wire to build a small crown.

### Internal Defect Repairs

Internal defects, Fig. 23-11, may or may not extend

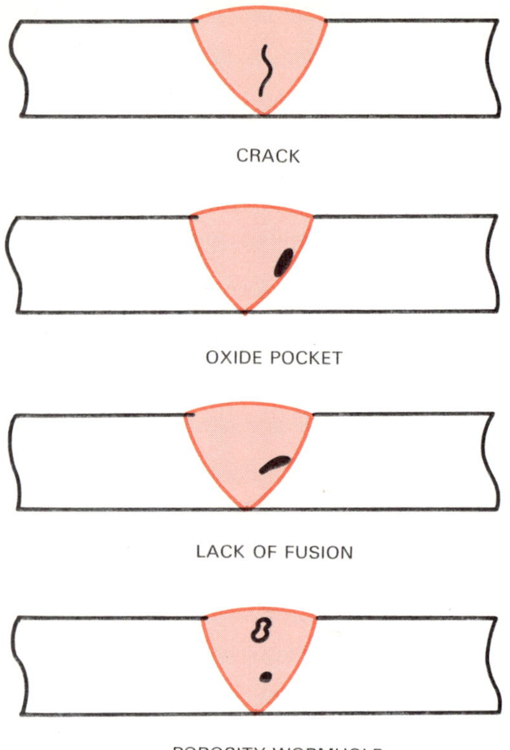


Fig. 23-11. Defects which may occur in the interior of the weld.

to the surface and may or may not be found by surface inspection. They are generally found by radiographic and ultrasonic test. Once the defect has been located by radiographic test, the next step is to mark the defect area on the weld. This can be done by marking the defect on the x-ray film. The film is aligned over the weld and a punch can be used to indent the area over the defect. The next step is to determine the depth of the defect from the top and root surface. Ultrasonic will locate the depth. Route or grind from the surface nearest to the defect.

Finding the defect in the weld by grinding or routing, requires both skill and patience. Porosity and large areas exhibiting lack of fusion are generally easy to locate and remove. Cracks and small patches of lack of fusion lines are more difficult to locate. A suggested procedure that may be followed for groove welds include:

1. If defect depth is known, remove metal to within approximately 1/16 inch from the defect. During the metal removal period, use a magnifying glass to aid in the inspection of the ground area. If the crack is in the right plane, a light blue surface may sometimes be found at the edge of the crack. This is caused by the overheating of the crack edge.
2. Penetrant test the grooved area. If no indication is seen, remove penetrant.
3. Grind approximately .010" to .015" deeper.
4. Penetrant test the grooved area again, continue with this sequence until the defect is found.
5. If the defect is still not found, x-ray to determine if defect remains.

If the crack is not found after removing metal half way through the part, it is suggested that the ground area be rewelded. Then work from the opposite surface to remove the crack.

NEVER GRIND A SLOT OR A HOLE THROUGH THE PART. Repairing a slot causes excessive distortion in the adjacent areas or shrinkage and the possibility of more defects. The removal of defects in fillet welds is extremely difficult since penetrant test cannot be used. If the penetrant is allowed to penetrate into the joint, it may cause many problems during weld repair. Assurance of defect removal on fillet welds is best done with visual or radiographic test.

### Preparing For Repair

After the defect is removed, the area should be prepared for welding by removing all rough edges on the ground area. Any oil, grease, scale, or penetrant residue may be removed with alcohol or acetone.

Do not use grit blasting in the grooved area, as particles of the material may become embedded in the material and become entrapped in the weld repair.

### Welding For Repairs

1. If possible, use stringer beads with minimum amperage for minimum shrinkage of the joint.
2. Always use a current tapering control on amperage to prevent crater cracks.
3. Clean each pass of scale and oxides.
4. Inspect each pass visually after cleaning.
5. If the weld repair is deep, have an x-ray made after 2 or 3 completed passes to confirm that new cracks have not formed. This should also be done if there is any doubt about removal of the original crack.
6. Where the root of the weld may be exposed to air,

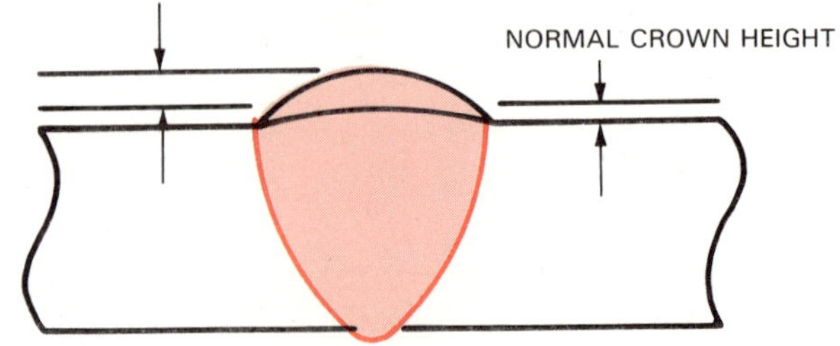


Fig. 23-12. Tempering beads are used to obtain an even structure throughout the top area of the weld. Since they add to the crown height significantly, they are usually removed after welding.

always use a backup gas.

7. Always use the original parameters for preheat, interpass temperature, and postheats where possible.
8. Do not build repair crown any higher than required. Each pass that is made stresses the base of the weld by shrinkage.
9. Where grain size must be controlled throughout the repair, the procedure shown in Fig. 23-12 may be used. The weld beads placed on the top surface of the weld will reduce the surface grain size. These beads are called TEMPERING BEADS. They are removed after welding.

### Post Weld Inspection

Weld repairs mandate that all of the NDT (non-destructive testing) required for final acceptance of the weld be completed after the weld repair is done. This means that even though several inspections were satisfactory before the weld was rejected, all of the inspections must be done over.

In areas where the weld is under high stress, it may be relevant to reinspect the entire weld to determine that other defects were not formed by the repair operation.

### Repair Review

Repairs cost money, lots of it. Repairs often detract from the appearance of the final weld. Everything possible should be done (within reason) to eliminate these costly errors.

Every repair, regardless of its severity, should be reviewed to determine the cause. Plan the possible corrective action to be taken to eliminate these areas.

Areas of review can be endless, but should include:

1. Base material.
2. Tooling.
3. Preparation for welding.
4. Inspection points of joint preparation.
5. Process application (welding variables).
6. Welder training and skill.

## REVIEW QUESTIONS

1. Why should all rejected welds be thoroughly reviewed before attempting a repair?
2. What procedures are to be considered when establishing a repair procedure?
3. What is the major cause of a dimensional repair?
4. How should dimensional repairs be made?
5. Can surface defects extend in the main body of the weld?
6. What type of tools are used to remove ferrous material defects?
7. What type of tools are used to remove non-ferrous material defects?
8. What problem may be encountered when grinding out undercut defects?
9. Single holes in the weld are called pores. What are pores in a line called?
10. What type of test is used to detect an overlap condition?
11. What type of inspection process is used to determine the depth of a defect from the surface of a groove weld?
12. What type of defects, located in the body of a weld, are very hard to locate using grinding or routing?
13. What two tools are used to assist in locating very small defects in groove welds?
14. Why should a slot or hole never be made through the part when removing a defect?
15. What material cleaning process should never be used when preparing a ground area for welding? Why?
16. What type of equipment control is most important to prevent weld crater cracks?
17. What type of beads are used to control grain size on the top portion of the weld repair?
18. Name the major review points to be used when reviewing causes of defects.

# Chapter 24

# QUALIFICATION AND CERTIFICATION

All welding is done to a specification. The specification may be only a verbal order to the welder. Or the specification may be very detailed with control over every aspect of the job. The reason for the control is to assure that the weld will perform the job intended. The amount of assurance is relative to the job requirements. For example, a weld on a steel cabinet base which may never been seen requires only very little assurance. On the opposite side, a weld on a nuclear reactor part must have very high assurance for weld quality.

To achieve this weld quality and assurance, rules are made as to procedures, methods, materials, qualification, and testing. The rules may be in the form of orders, codes, specifications or instructions, and are generally a part of the purchase order or contracts of the weldment.

### Specifications and Codes

SPECIFICATIONS and CODES are documents which detail types of materials, welding processes, preparation of joints, qualification and welding requirements, and testing. They may be specific or general in nature, and may be for one specific weld or several. Since these types of documents may be revised, it is recommended that the latest revision always be used. In cases where a specific revision is required, always use the specified revision.

As some specifications and codes are similar in content, it is strongly recommended that the referenced specification or code be studied thoroughly before using. A thorough knowledge of the requirements therein will allow the user to manufacture the weldment with a high degree of reliability.

### Procedure Specification

The procedure specification is made by the fabricator to develop the proper technique of fabrication. The PROCEDURE SPECIFICATION document details all of the requirements and instructions for welding the specific joint and includes at least:

1. Material type and condition.
2. Tooling.
3. Filler materials.
4. Joint design, weld type and position.
5. Weld process and related parameters with tolerances.
6. Final weld configuration.
7. Inspection and/or test criteria.

A procedure specification is obtained by making test welds of the various joints required by the weldment drawings. The test weld data is recorded for further use in making a weld schedule for the welder's use on the actual weldment. It is imperative that the test weld conditions parallel the actual weldment conditions. If the test and actual weldment conditions do not coincide, the welds on the production part may have substandard quality.

### Procedure Qualification

PROCEDURE QUALIFICATION results from the testing done by NDT (nondestructive testing) and DT (destructive testing) on the completed test welds. This is often called a PROCEDURE QUALIFICATION RESULT (PQR). The NDT may include one or more tests as required by the specification or code. These tests determine the quality of the weld. Acceptance or rejection of the weld quality is usually compared with standards or limits in the fabrication specification.

The final test portion of the test samples may include DT. These tests may include tensile, bend, charpy impact, macro and microhardness test to determine the mechanical values of the weld.

After the NDT and DT tests are completed and the results are satisfactory, the weld is considered qualified and ready to be used for the production weld.

Requalification is required when a major change is made in joint design, filler material, type of weld current, tooling, or any other essential variable which may affect the quality or mechanical properties of the weld. An essential variable is a major change in the operation that would affect the quality of the weld's mechanical values.

### Weld Schedule

The WELD SCHEDULE is obtained from the procedure specification data, recorded during the welding of the test samples. This WELD SCHEDULE is used by the welder to set up the joint, set up the power supply and related equipment, and make the weld.

With the many tolerances involved in material thickness, joint preparation, tooling conditions, etc., it is suggested that a tolerance be included in the welding parameters. Considerations can then be made to accommodate these tolerance areas.

### Welder/Operator Certification

WELDER OR AUTOMATIC OPERATOR CERTIFICATION is the verification of the person using the weld schedule to make an acceptable weld. The test may only allow a specific weld type, or it may include several designs with unlimited thickness. This will be dependent on the fabrication specification requirements.

The length of time a certification is effective is variable. Some certifications are for a 6 month period, or only for welding the specific part or weld. Some certification may last as long as the weld quality of the production welds are satisfactory.

Certifications may be rescinded for producing welds with poor quality or where the person does not use the process within a prescribed period of time.

Various types of welder certification tests are shown in Figs. 24-1 through 24-17.

### Specification and Testing Organizations

There is no general specification in current use because there are so many variables and requirements

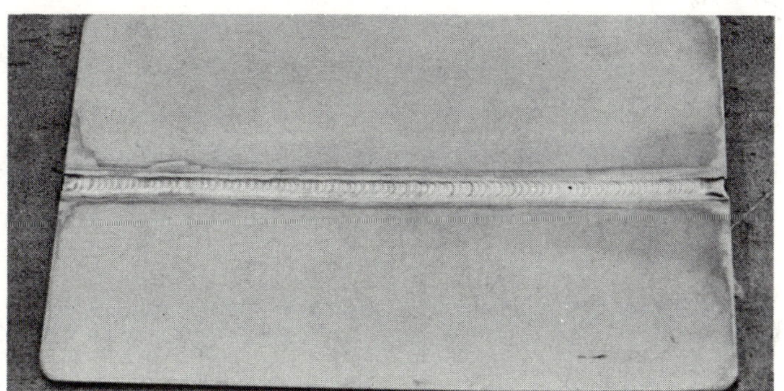

Fig. 24-1. Manual butt weld on thin gauge material. The weld is examined visually for defects.

Fig. 24-2. Manual butt weld on thin gauge material. After visual examination sections are removed and bent over a rounded mandrel.

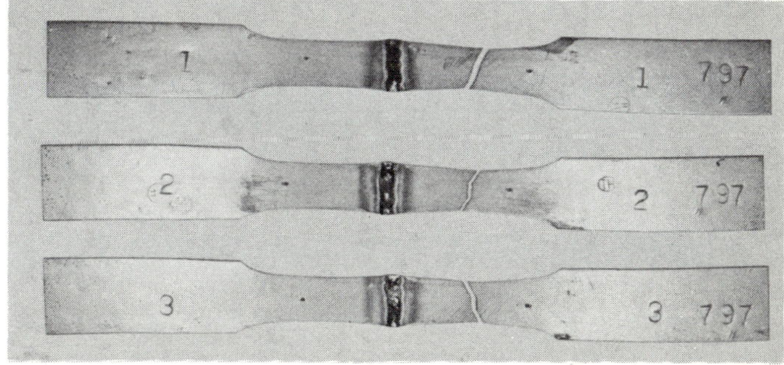


Fig. 24-3. Manual butt weld on thin gauge material has been sectioned and pieces have been machined for a reduced section tensile test.

Fig. 24-4. Reduced section tensile test pieces. These pieces have holes drilled into the ends to adapt to the test machine.

Fig. 24-5. Reduced section tensile bar removed from plate material has elongation gauge points established before tensile test. By comparing before and after test dimensions, the elongation of the joint may be computed.

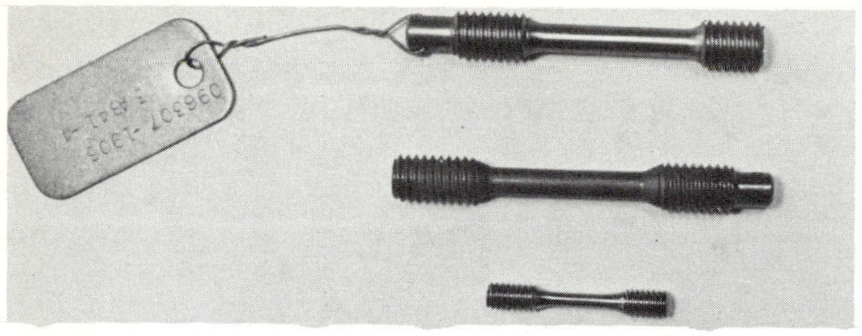

Fig. 24-6. Rounded reduced section tensile bars have threads on the ends of the bar to adapt to the test machine. These types of bars are generally used to test V-groove longitudinal weld properties.

Fig. 24-7. Guided bend tests are used to determine if the proper procedure was used to fill a V-groove joint.

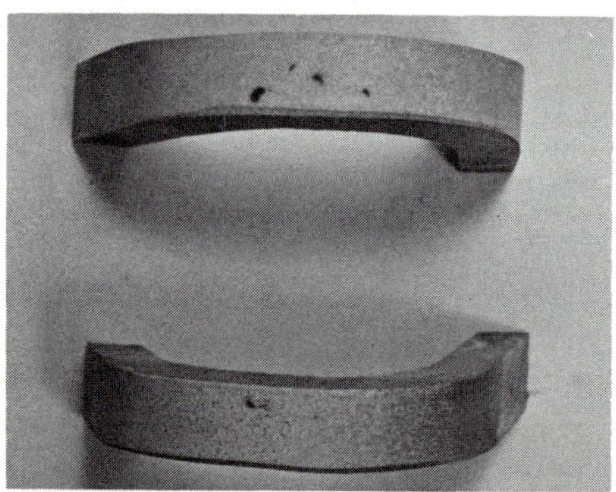

Fig. 24-8. Results of these guided bend tests revealed excessive porosity within the weld.

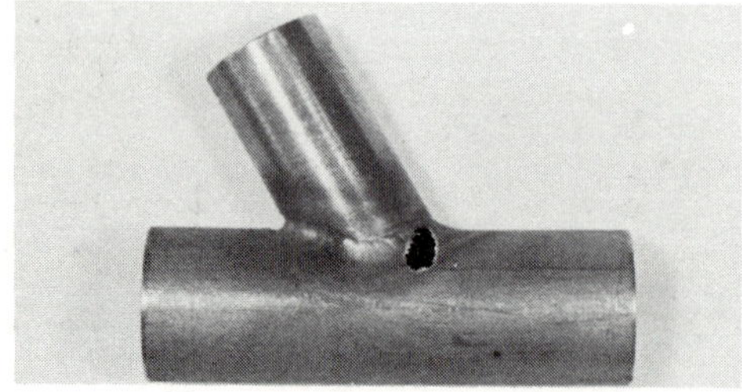

Fig. 24-9. Work quality test to determine welder proficiency. This test is visually examined and may be sectioned for internal defects.

Fig. 24-10. Work quality test to determine welder proficiency. This weld test is NDT only.

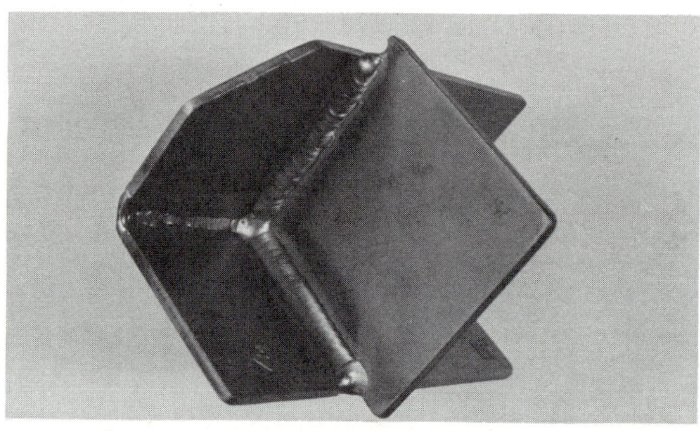

Fig. 24-11. Work quality test to determine welder proficiency. This weld test is NDT only.

Fig. 24-12. Work quality test to determine welder proficiency. This weld may use both NDT and DT.

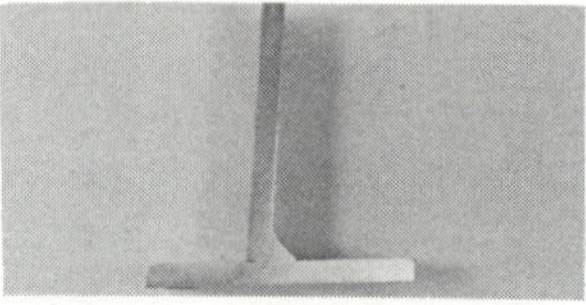

Fig. 24-13. Fillet weld test used to determine welder proficiency is usually cross sectioned and polished for visual examination. The penetration must extend to the corner of the mating edge. This test is called a macro test.

Fig. 24-14. Titanium tube plug weld test has been polished and etched with acid to reveal the weld contour. (Note the large grains revealed by the etching. This is common with titanium welds.)

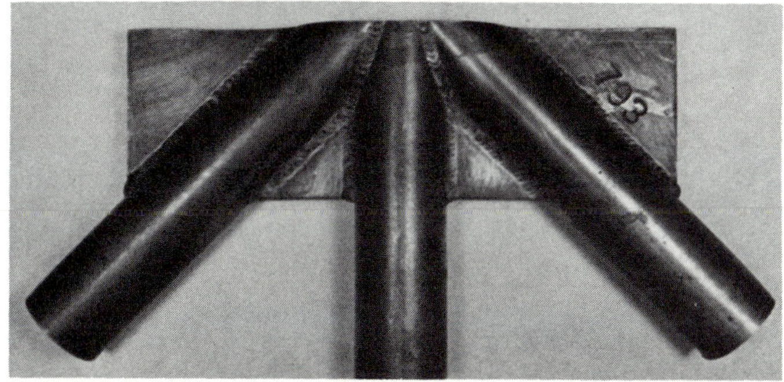

Fig. 24-15. Steel cluster was designed to test welder skill to weld typical aircraft type joints. After welding, the test was submitted to magnetic particle testing to check for cracks.

Fig. 24-16. Steel cluster test after NDT has been polished to reveal the penetration of each weld.

Fig. 24-17. Steel cross type test was designed to test welder skill in vertical position welding. NDT was completed and tensile test were made to determine joint strength.

regarding gas tungsten arc welding. Large companies that use the GTAW process usually have a company specification that states the rules for qualification, certifications, and the limits of the process within the company's structure. The ASME Boiler Code defines the limits of qualification, certification, and use of the process within Section IX of the Code. The American Welding Society (AWS) defines the limits of qualification, certification, and use of the process with the B 3.0 specification. Military organizations and contractors have for many years used the Mil-T-5021 specification for defining the limits of qualification and certification and use of the process.

The basic criteria of the specifications is that each weld type and material type shall be qualified by test procedures. Welder certification is then accomplished by the welder using qualified test procedure parameters and variables.

Qualification of the weld procedure remains in effect until an essential variable is changed. These types of changes require a new certification test and possibly a new test for the welder.

Qualification (certification) of the welder remains in effect while the welder is employed using the qualified procedure. Termination of employment automatically nullifies the welder certification.

Under the rules of these codes, welders cannot have a certification unless they are employed and certifications cannot be transferred from one employer to another.

## REVIEW QUESTIONS

1. What is the main reason for welding to a specification?
2. Where are specifications usually specified?
3. What do specifications generally specify?
4. How are specifications changed?
5. What does the Procedure Specification specify?
6. The data obtained during the Procedure Specification test is used to make ___________ .
7. What two types of tests may be made to qualify the procedure?
8. What do the letters PQR define?
9. What is an essential variable?
10. What information should the weld schedule contain?
11. What is a certified welder?
12. The types and number of tests specified for welder certification are based on the ________ ________ requirements.
13. Are all welders certified for a specific time period?
14. Can welders certifications be revoked?
15. Can a welder change jobs and still maintain his or her certification?

# Chapter 25

# ESTIMATING COST

How much will it cost? Give me a ball park figure! How long will it take? Is the welder certified? What kind of inspection is to be done on the completed weld?

These and many more questions arise when considering the cost of making a weld. The true cost of making a weld is probably never known. However, by considering the major factors, a reasonable estimated cost can be determined.

### Labor

LABOR is the area that includes all of the labor required to complete the job. This should include welding set-up time, actual welding time, and the time used after welding for checking the weld, removal of the part from the tool, etc.

When computing labor cost, the welder's efficiency or operating factor must also be considered. Actual arc hours versus clock hours for GTAW vary considerably depending on many factors. These may include:

A. Position of the weld.
B. Accessability of the weld.
C. Tungsten preparation and maintenance.
D. Welding sequence/welding equipment setup.
E. Proper gas shielding (changing bottles).
F. Maintenance of preheat, interpass, postheat.
G. Interpass cleaning.
H. Interpass weld inspection.
I. Variance in weld procedure for incorrect weld joint dimensions.
J. Addition of or disassembly of tooling during the welding sequence.
K. Application of NDT during welding sequence.
L. Welder fatigue.
M. Change of wire type or a diameter during the welding sequence.
N. Equipment duty cycle.
O. Welder training or experience for the type of weld and the equipment used.

In some welding operations, a helper is included to assist the welder to increase actual arc hours. The helper receives a lower wage than the welder, which also must be included in the actual labor cost. This way the welder can do more for what he is paid to do . . . namely weld!

Having a welder do cleaning of parts, sharpen tungstens, change gas bottles, keep records, move parts to and from the area, is not a wise use of time nor money when it can be done at a lower labor cost with a helper.

Much can be done in the area of increasing arc hours on most jobs by studying what the welder does when he is not welding. It is these areas that must be reduced to make the job more cost effective.

Consumables are the materials or parts expended during the welding operation. They may include:

A. Filler material.
B. Consumable inserts.
C. Back-up rings (welded into the joint).
D. Gases.
E. Fluxes.
F. Welding power (power is not a material or part). It is the energy consumed during the operation. This area should also include all auxiliary equipment such as:
   1. Electrical heaters.
   2. Water coolers.
   3. Wire feeders.
   4. Oscillators.
   5. Positioners and manipulators.
   6. Grinders, brushes, weld shavers.
G. Preheating, interpass and postheating oxy-fuel gas.

### Tooling and Inspection Cost

On simple shop welds, these two areas may be computed in the overhead section. Where special equipment, tooling, and inspections are required, they should receive individual attention.

Tooling cost will depend on the complexity of the weldment, the quantity of the items to be produced, and the production schedule. The total cost must also include:

A. Tool design time.
B. Tool tryout.

C. Tool modifications if required.
D. Maintenance.

Inspections that are made during the fabrication cycle are called IN-PROCESS INSPECTIONS. They are often used to verify the quality of root passes in multilayer welds. They are also used as a preliminary inspection on welds with high restraint for possible cracking. The results of these tests are not usually considered in the final NDT required for the welds.

Some specifications require the use of an official inspector for in-process and final inspection. In most cases, the in-process inspections are called "hold-points" and the part cannot be moved beyond this point until the inspection is complete. The inspector's time is then billed to the company doing the work.

### Procedure Qualification and Welder Certification

These two areas may not be required for simple tasks, or they may be included in the overhead section. For welds made to a strict specification, these two areas may have a tremendous impact on cost.

Many specifications limit changes in the procedure qualification to a few areas. Therefore, if major changes are to be made during production, new procedures must be made and qualified.

Welder training and welder qualification test must be thoroughly considered with regard to cost of the weldment. These costs are highly variable due to the ability of the welder to make acceptable test welds.

Recertification of the welder is usually required after a set period of time. In some areas, a part is removed from production and tested until destruction to verify welder certification.

Changes in the procedure qualification may nullify a welder's certification. In these cases, the welder must again certify to the new procedure.

Often a welder may not pass the certification test and will be required to have additional training. The specification may then require a double test to satisfy the requirements.

### Nondestructive Testing

These types of costs will relate to the contract or specification. All types of inspections take time and *add nothing* to the quality of the part. These may be included in the overhead section of the cost area or they may be applied as a direct cost to the job.

Where exact costs are not known for each type of inspection required, it is advisable to seek expert help in this area. Inspection labs that are experienced in these areas can relate to previous tasks and give accurate cost analysis.

When figuring cost, always remember that welds rejected by NDT must be reinspected by all NDT previous to rejection before acceptance.

### Material Processing

This cost area is considerably variable since all metals are not processed the same. One type of metal may require a chemical etch before welding, where another may require only a wire brush before welding.

Specifications may be quite explicit regarding how the material must be processed before, during, and after welding.

A method commonly used for costing of this item is to prepare a job routing sheet listing each operation as the job is processed. Then the estimated cost can be applied for the entire job.

### Repair

Estimating REPAIR COST represents a variable that is often difficult to accomplish. Welds made under very tight inspection restrictions often require special tested procedures before weld repair can be done. In areas where dimensions are held to close tolerances, repairs often ruin the part and the unit must be scrapped.

The best possible guess estimate for this area is to research previous welds similar to the estimated weld. Where repairs were required, what was done to eliminate the defects, and what was the amount of cost for the repair. Records and experience is about all that can be used in this area.

To minimize the number of repairs to the weld, it is suggested that all of the operating sequences be closely controlled. In-process inspections should be added as required to maintain weld quality in the original weld.

### Overhead

Overhead cost usually includes all items that are associated with the cost of doing business. They may include such items as:

A. Rent.
B. Telephone and office supplies.
C. Insurance.
D. Salaries or commissions of supervisors and officials.
E. Utilities.
F. Licenses, permits, donations, advertising.
G. Transportation.
H. Services: engineering, tool design, estimating.
I. Shop equipment, maintenance.
J. All other items which may be associated with the cost of doing business. This may include:
   1. Social security.
   2. Unemployment compensation.
   3. Holiday pay.
   4. Vacations.
   5. Sick pay, dental pay, etc.
   6. Incentive pay, overtime.
   7. Welder certifications, training.

## FINAL COST

To compute the final cost of the project, all of the preceding areas can be figured individually and added together.

To this figure, a profit margin is added. This figure is variable and may range from 10% to 100% or more of the computed cost. The lower percentage figure may be allowed for simple jobs. The higher figure is allowed for work that is delicate, complicated, or requires considerable development prior to or during production.

### Flat Rate Cost

Some welding facilities figure all of the cost in one rate and then consider the total hours to be expended to complete the job. This is then called the HOURLY FLAT RATE COST. Many smaller shops use this system, to which may be added material, tooling, and other special areas which may be needed for the job.

## ESTIMATING ASSIGNMENT

There are no review questions for this chapter. In place of the questions, it is suggested that a sample costing be made of an actual weldment that you would be required to make. If none are available, you may use one or more of the weldments listed in Chapter 17.

List each operation you are required to complete and then assign a cost to it. Record the actual hours spent on the task. Compare the actual cost versus the estimated cost. After the comparison is done, check the task and look for areas where time can be reduced.

On long time projects, a learning curve may also be used as this considers high cost during the start of a job and lower cost as the job progresses. As you progress in learning job costing, add a learning curve to the estimate for the actual manufacturing cycle. Review your estimates during production, then raise or lower your estimates accordingly.

## ACKNOWLEDGMENTS

The author gratefully acknowledges the assistance of many people who have contributed suggestions, ideas, photographs, and information for this book. I am deeply grateful for the graphic art work which was done by my son, Steven A. Minnick, and the manuscript typing which was done by my niece, Charlene R. Avram.

# GTAW REFERENCE SECTION



## FILTER PLATE LENSES

| WELDING PROCESS | Approximate welding range (in amps) | Federal Specification filter shade required |
|---|---|---|
| Metal-arc welding (coated electrodes)<br>Continuous covered electrode welding<br>Carbon Dioxide shield continuous covered electrode welding | 100 | 8 or 9 |
| | 100-300 | 10 or 11 |
| | Over 300 | 12 or 14 |
| Metal-arc welding (bare wire)<br>Carbon-arc welding<br>Inert-gas metal-arc welding<br>Atomic hydrogen welding | 200 | 10 or 11 |
| | Over 200 | 12 or 14 |
| Automatic carbon dioxide shield metal-arc welding (bare wire) | Over 500 | 15 or 16 |
| Inert-gas tungsten-arc welding | 15 | 8 |
| | 15-75 | 9 |
| | 75-100 | 10 |
| | 100-200 | 11 |
| | 200-250 | 12 |
| | 250-300 | 14 |

Fig. 26-1. Filter plate lenses are designed to protect welders against harmful infrared and ultraviolet rays. The lenses are heat treated and manufactured to meet or exceed ANSI Z-87.1 and Federal Specification GGG-H-211C. Heat-treated lenses are identified by the letter "H" following the shade number. (Thermacote-Welco Company)

AMERICAN WELDING SOCIETY

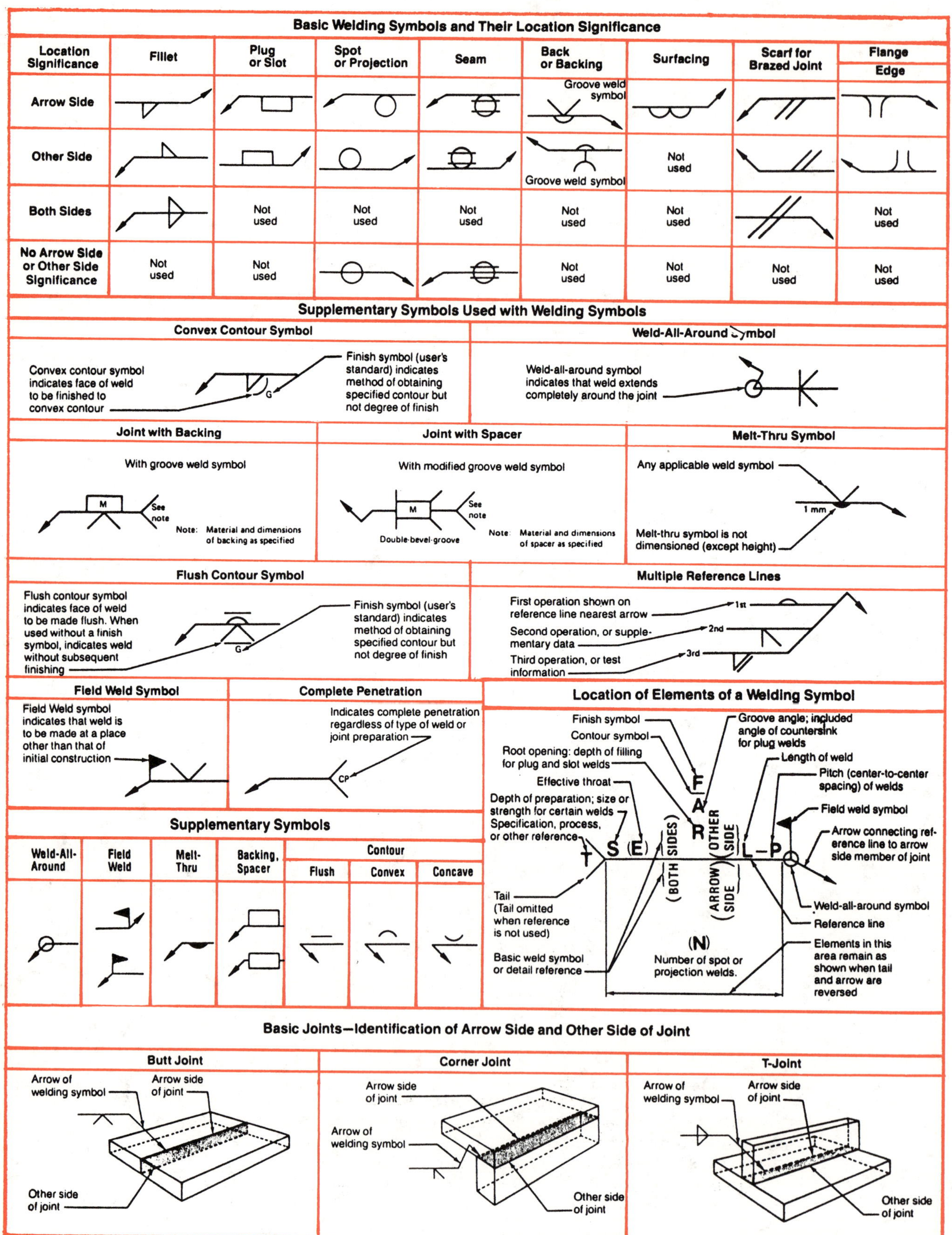


Fig. 26-2

## STANDARD WELDING SYMBOLS

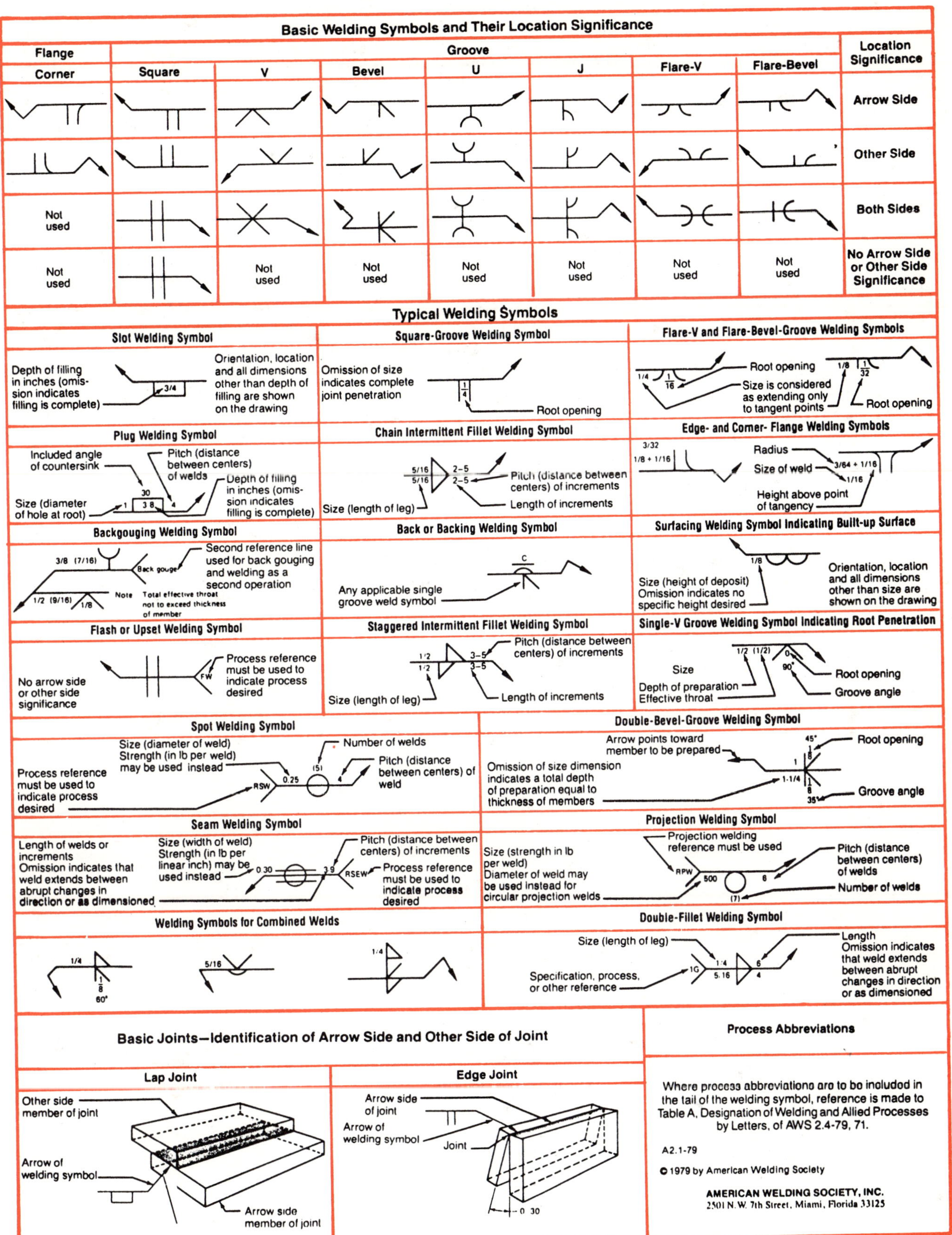


Fig. 26-3

| ETCHING REAGENTS FOR MICROSCOPIC EXAMINATION OF IRON AND STEEL | | | |
|---|---|---|---|
| **APPLICATION** | **ETCHING** | **COMPOSITION** | **REMARKS** |
| **Carbon, Low-Alloy and Medium-Alloy Steels** | 1. Nital | Nitric acid (sp gr 1.42) . 1-5 ml<br>Ethyl or Methyl alcohol . 95-99 ml | Darkens pearlite, and gives contrast between adjacent colonies; reveals ferrite boundaries; differentiates ferrite from martensite; shows case depth of nitrided steel. Etching time: 5-60 secs |
| | 2. Picral | Picric acid . . . . . . . . . . 4 g<br>Methyl alcohol . . . . . . . 100 ml | Used for annealed and quench-hardened carbon and alloy steel. Not as effective as No. 1 for revealing ferrite grain boundaries. Etching time: 5-120 secs |
| | 3. Hydrochloric and picral acids | Hydrochloric acid . . . . . 5 ml<br>Picric acid . . . . . . . . . . 1 g<br>Methyl alcohol . . . . . . . 100 ml | Reveals austenitic grain size in both quenched and quenched-and-tempered steels |
| **Alloy and Stainless Steels** | 4. Mixed acids | Nitric acid . . . . . . . . . . 10 ml<br>Hydrochloric acid . . . . . . . . 20 ml<br>Glycerol . . . . . . . . . . . . 20 ml<br>Hydrogen Peroxide . . . . 10 ml | Iron-Chromium-Nickel-and Manganese alloy steel. Etching: use fresh acid |
| | 5. Ferric Chloride Chloride | Ferric Chloride . . . . . . . 5 g<br>Hydrochloric acid . . . . . 20 ml<br>Water, distilled . . . . . . . 100 ml | Reveals structure of stainless and austenitic nickel steels |
| | 6. Marble's Reagent | Cupric Sulfate . . . . . . . . 4 g<br>Hydrochloric acid . . . . . 20 ml<br>Water, distilled . . . . . . . 20 ml | Reveals structure of various stainless steels |
| **High Speed Steels** | 7. Snyder-Graff | Hydrochloric acid . . . . . 9 ml<br>Nitric acid . 9 ml<br>Methyl alcohol . . . . . . . 100 ml | Reveals grain size of quenched and tempered high speed steels. Etching time: 15 secs to 5 min |

Fig. 26-4

| ETCHING PROCEDURES | | | | |
|---|---|---|---|---|
| **REAGENTS** | **COMPOSITION** | | **PROCEDURE** | **USES** |
| **Solutions For Aluminum** | | | | |
| Sodium Hydroxide | NaOH<br>$H_2O$ | 1 gm<br>99 ml | Swab 10 seconds | General microscopic |
| Tucker's Etch | HF<br>HCl<br>$HNO_3$<br>$H_2O$ | 15 ml<br>45 ml<br>15 ml<br>25 ml | Etch by immersion | Macroscopic |
| **Solutions For Stainless Steel** | | | | |
| Nitric & Acetic Acids | $HNO_3$<br>Acetic Acid | 30 ml<br>20 ml | Apply by swabbing | For stainless alloys and others high in nickel or cobalt. |
| Cupric sulphate | $CuSO_4$<br>HCl<br>$H_2O$ | 4 gms<br>20 ml<br>20 ml | Etch by immersion | Structure of stainless steels |
| Cupric chloride & Hydrochloric acid | $CuCl_2$<br>HCl<br>Ethyl Alcohol<br>$H_2O$ | 5 gms<br>100 ml<br>100 ml<br>100 ml | Use cold immersion or swabbing | For austenitic & ferritic steels |

Fig. 26-5

| ETCHING PROCEDURES | | | |
|---|---|---|---|
| **REAGENTS** | **COMPOSITION** | **PROCEDURE** | **USES** |
| **Solutions For Copper And Brass** | | | |
| Ammonium Hydroxide & Ammonium Persulphate | $NH_4OH$ 1 part<br>$H_2O$ 1 part<br>$(NH_4)_2S_2O_3$ (2 1/2%) 2 parts | Immersion | Polish attack of copper and some alloys. |
| Chromic Acid | Saturated aqueous solution ($CrO_3$) | Immersion or swabbing | Copper, brass, bronze, nickel silver (plain etch) |
| Ferric Chloride | $FeCl_3$ 5 parts<br>HCl 10 parts<br>$H_2O$ 100 parts | Immersion or swabbing (etch lightly) | Copper, brass, bronze, aluminum bronze. |
| **Solutions For Iron & Steel** | | | |
| **Macro Examination** | | | |
| Nitric Acid | $HNO_3$ 5 ml<br>$H_2O$ 95 ml | Immerse 30 to 60 seconds | Shows structure of welds. |
| Ammonium persulphate | $(NH_4)_2S_2O_3$ 10 gms<br>$H_2O$ 90 ml | Surface should be rubbed with cotton during etching | Brings out grain structure, recrystallization at welds. |
| Nital | $HNO_3$ 5 ml<br>Ethyl Alcohol 95 ml | Etch 5 min. followed by 1 sec. in HCl (10%) | Shows cleanness Depth of hardening carburized or decarburized surfaces, etc. |
| **Micro Examination** | | | |
| Picric acid (Picral) | Picric acid 4 gms<br>Ethyl or methyl alcohol (95%) 100 ml | Etching time a few seconds to a minute or more | For all grades of carbon steels. |

Fig. 26-6

| CHEMICAL TREATMENTS FOR REMOVAL OF OXIDE FILMS FROM ALUMINUM SURFACES | | | |
|---|---|---|---|
| **SOLUTION** | **CONCENTRATION** | **PROCEDURE** | **PURPOSE** |
| **Nitric Acid** | 50% water, 50% nitric acid, technical grade. | Immersion 15 min. Rinse in cold water, then in hot water. Dry. | Removal of thin oxide film for fusion welding. |
| **Sodium hydroxide (caustic soda) followed by** | 5% Sodium hydroxide in water. | Immersion 10-60 seconds. Rinse in cold water. | Removal of thick oxide film for all welding processes. |
| **Nitric acid** | Concentrated | Immerse for 30 seconds. Rinse in cold water, then hot water. Dry. | Removal of thick oxide film for all welding processes. |
| **Sulfuric-chromic** | $H_2SO_4$ 1 gal.<br>$CrO_3$ 45 oz.<br>Water 9 gal. | Dip for 2-3 min. Rinse in cold water, then hot water. Dry. | Removal of films and stains from heat treating, and oxide coatings. |
| **Phosphoric-chromic** | $H_3PO_3$ (75%) 3.5 gal.<br>$CrO_3$ 1.75 lbs.<br>Water 10 gals. | Dip for 5-10 min. Rinse in cold water. Rinse in hot water. Dry. | Removal of anodic coatings. |

Fig. 26-7

| TABLE OF ALLOY WIRE WEIGHTS AND MEASURES (Approx.) | | | | | | |
|---|---|---|---|---|---|---|
| Brown & Sharpe Gage No. | Decimal | Phos. Bronze Ft. per lb. | .18% Nickel Ft. per lb. | Aluminum Ft. per lb. | Copper Ft. per lb. | Brass Ft. per lb. |
| 4/0 | .4600 | 1.559 | 1.599 | 5.207 | 1.561 | 1.640 |
| 3/0 | .4096 | 1.966 | 2.016 | 6.567 | 1.968 | 2.068 |
| 2/0 | .3648 | 2.480 | 2.543 | 8.279 | 2.482 | 2.068 |
| 1/0 | .3249 | 3.127 | 3.206 | 10.44 | 3.130 | 2.608 |
| 1 | .2893 | 3.943 | 4.043 | 13.16 | 3.947 | 3.289 |
| 2 | .2576 | 4.972 | 5.098 | 16.60 | 4.977 | 4.147 |
| 3 | .2294 | 6.269 | 6.428 | 20.94 | 6.276 | 5.229 |
| 4 | .2043 | 7.906 | 8.106 | 26.40 | 7.914 | 6.594 |
| 5 | .1819 | 9.969 | 10.22 | 22.20 | 9.980 | 8.315 |
| 6 | .1620 | 12.57 | 12.89 | 41.98 | 12.58 | 10.49 |
| 7 | .1443 | 15.85 | 16.25 | 52.91 | 15.87 | 13.22 |
| 8 | .1285 | 19.99 | 20.49 | 66.73 | 20.01 | 16.67 |
| 9 | .1144 | 25.20 | 25.84 | 84.19 | 25.23 | 21.02 |
| 10 | .1019 | 31.78 | 32.59 | 106.1 | 31.82 | 26.51 |
| 11 | .09074 | 40.08 | 41.09 | 133.9 | 40.12 | 33.43 |
| 12 | .08081 | 50.53 | 51.82 | 168.8 | 50.59 | 42.15 |
| 13 | .07196 | 63.72 | 65.39 | 212.5 | 63.80 | 53.15 |
| 14 | .06408 | 80.35 | 82.34 | 268.2 | 80.44 | 66.88 |
| 15 | .05707 | 101.3 | 103.9 | 337.9 | 101.4 | 84.68 |
| 16 | .05082 | 127.8 | 131.0 | 426.9 | 127.9 | 106.6 |
| 17 | .04526 | 161.1 | 165.2 | 536.9 | 161.3 | 134.4 |
| 18 | .04030 | 203.2 | 208.3 | 678.4 | 203.4 | 169.7 |
| 19 | .03589 | 256.2 | 262.7 | 854.9 | 256.5 | 213.7 |
| 20 | .03196 | 323. | 331.2 | 1076. | 323.4 | 269.5 |
| 21 | .02846 | 407.3 | 417.7 | 1356. | 407.8 | 339.8 |
| 22 | .02535 | 513.6 | 526.7 | 1721. | 514.2 | 428.5 |
| 23 | .02257 | 647.7 | 664.1 | 2157. | 648.4 | 540.2 |
| 24 | .02010 | 816.7 | 837.4 | 2727. | 817.7 | 681.3 |
| 25 | .01790 | 1030. | 1056. | 3439. | 1031. | 859. |
| 26 | .01594 | 1299. | 1332. | 4358. | 1300. | 1083. |
| 27 | .0142 | 1638. | 1679. | 5464. | 1639. | 1366. |
| 28 | .01264 | 2065. | 2117. | 6940. | 2067. | 1723. |

Fig. 26-8

| INCHES PER POUND OF WIRE | | | | | | | | | | | | |
|---|---|---|---|---|---|---|---|---|---|---|---|---|
| Wire Diameter | | | | | | | | | | | | |
| Decimal | Fraction | Mag. | Alum. | Alum. Bronze (10)% | Stain-less Steel | Mild Steel | Stain-less Steel | Si. Bronze | Copper Nickel | Nickel | De-ox. Copper | Ti |
| .020 | | 50500 | 32400 | 11600 | 11350 | 11100 | 10950 | 10300 | 9950 | 9900 | 9800 | 19524 |
| .025 | | 34700 | 22300 | 7960 | 7820 | 7680 | 7550 | 7100 | 6850 | 6820 | 6750 | 12492 |
| .030 | | 22400 | 14420 | 5150 | 5050 | 4960 | 4880 | 4600 | 4430 | 4400 | 4360 | 8776 |
| .035 | | 16500 | 10600 | 3780 | 3720 | 3650 | 3590 | 3380 | 3260 | 3240 | 3200 | 6372 |
| .040 | | 12600 | 8120 | 2900 | 2840 | 2790 | 2750 | 2580 | 2490 | 2480 | 2450 | 4884 |
| .045 | 3/64 | 9990 | 6410 | 2290 | 2240 | 2210 | 2210 | 2040 | 1970 | 1960 | 1940 | 3852 |
| .062 | 1/16 | 5270 | 3382 | 1220 | 1180 | 1160 | 1140 | 1070 | 1040 | 1030 | 1020 | 2028 |
| .078 | 5/64 | 3300 | 2120 | 756 | 742 | 730 | 718 | 675 | 650 | 647 | 640 | |
| .093 | 3/32 | 2350 | 1510 | 538 | 528 | 519 | 510 | 480 | 462 | 460 | 455 | 964.8 |
| .125 | 1/8 | 1280 | 825 | 295 | 289 | 284 | 279 | 263 | 253 | 252 | 249 | 499.92 |

Fig. 26-9

## HARDNESS CONVERSION TABLE

| BRINELL | | | ROCKWELL | | | |
|---|---|---|---|---|---|---|
| **Dia. in mm, 3000 kg. load 10 mm ball** | **Hardness No.** | **Vickers or Firth Hardness No.** | **C 150 kg. load 120° Diamond Cone** | **B 100 kg. load 1/16 in. dia. ball** | **Scleroscope No.** | **Tensile Strength 1000 psi** |
| 2.05 | 898 | | | | | 440 |
| 2.10 | 857 | | | | | 420 |
| 2.15 | 817 | | | | | 401 |
| 2.20 | 780 | 1150 | 70 | | 106 | 384 |
| 2.25 | 745 | 1050 | 68 | | 100 | 368 |
| 2.30 | 712 | 960 | 66 | | 95 | 352 |
| 2.35 | 682 | 885 | 64 | | 91 | 337 |
| 2.40 | 653 | 820 | 62 | | 87 | 324 |
| 2.45 | 627 | 765 | 60 | | 84 | 311 |
| 2.50 | 601 | 717 | 58 | | 81 | 298 |
| 2.55 | 578 | 675 | 57 | | 78 | 287 |
| 2.60 | 555 | 633 | 55 | 120 | 75 | 276 |
| 2.65 | 534 | 598 | 53 | 119 | 72 | 266 |
| 2.70 | 514 | 567 | 52 | 119 | 70 | 256 |
| 2.75 | 495 | 540 | 50 | 117 | 67 | 247 |
| 2.80 | 477 | 515 | 49 | 117 | 65 | 238 |
| 2.85 | 461 | 494 | 47 | 116 | 63 | 229 |
| 2.90 | 444 | 472 | 46 | 115 | 61 | 220 |
| 2.95 | 429 | 454 | 45 | 115 | 59 | 212 |
| 3.00 | 415 | 437 | 44 | 114 | 57 | 204 |
| 3.05 | 401 | 420 | 42 | 113 | 55 | 196 |
| 3.10 | 388 | 404 | 41 | 112 | 54 | 189 |
| 3.15 | 375 | 389 | 40 | 112 | 52 | 182 |
| 3.20 | 363 | 375 | 38 | 110 | 51 | 176 |
| 3.25 | 352 | 363 | 37 | 110 | 49 | 170 |
| 3.30 | 341 | 350 | 36 | 109 | 48 | 165 |
| 3.35 | 331 | 339 | 35 | 109 | 46 | 160 |
| 3.40 | 321 | 327 | 34 | 108 | 45 | 155 |
| 3.45 | 311 | 316 | 33 | 108 | 44 | 150 |
| 3.50 | 302 | 305 | 32 | 107 | 43 | 146 |
| 3.55 | 293 | 296 | 31 | 106 | 42 | 142 |
| 3.60 | 285 | 287 | 30 | 105 | 40 | 138 |
| 3.65 | 277 | 279 | 29 | 104 | 39 | 134 |
| 3.70 | 269 | 270 | 28 | 104 | 38 | 131 |
| 3.75 | 262 | 263 | 26 | 103 | 37 | 128 |
| 3.80 | 255 | 256 | 25 | 102 | 37 | 125 |
| 3.85 | 248 | 248 | 24 | 102 | 36 | 122 |
| 3.90 | 241 | 241 | 23 | 100 | 35 | 119 |
| 3.95 | 235 | 235 | 22 | 99 | 34 | 116 |
| 4.00 | 229 | 229 | 21 | 98 | 33 | 113 |
| 4.05 | 223 | 223 | 20 | 97 | 32 | 110 |
| 4.10 | 217 | 217 | 18 | 96 | 31 | 107 |
| 4.15 | 212 | 212 | 17 | 96 | 31 | 104 |
| 4.20 | 207 | 207 | 16 | 95 | 30 | 101 |
| 4.25 | 202 | 202 | 15 | 94 | 30 | 99 |
| 4.30 | 197 | 197 | 13 | 93 | 29 | 97 |
| 4.35 | 192 | 192 | 12 | 92 | 28 | 95 |
| 4.40 | 187 | 187 | 10 | 91 | 28 | 93 |
| 4.45 | 183 | 183 | 9 | 90 | 27 | 91 |
| 4.50 | 179 | 179 | 8 | 89 | 27 | 89 |
| 4.55 | 174 | 174 | 7 | 88 | 26 | 87 |
| 4.60 | 170 | 170 | 6 | 87 | 26 | 85 |
| 4.65 | 166 | 166 | 4 | 86 | 25 | 83 |
| 4.70 | 163 | 163 | 3 | 85 | 25 | 82 |
| 4.75 | 159 | 159 | 2 | 84 | 24 | 80 |
| 4.80 | 156 | 156 | 1 | 83 | 24 | 78 |
| 4.85 | 153 | 153 | | 82 | 23 | 76 |
| 4.90 | 149 | 149 | | 81 | 23 | 75 |
| 4.95 | 146 | 146 | | 80 | 22 | 74 |
| 5.00 | 143 | 143 | | 79 | 22 | 72 |
| 5.05 | 140 | 140 | | 78 | 21 | 71 |
| 5.10 | 137 | 137 | | 77 | 21 | 70 |
| 5.15 | 134 | 134 | | 76 | 21 | 68 |
| 5.20 | 131 | 131 | | 74 | 20 | 66 |
| 5.25 | 128 | 128 | | 73 | 20 | 65 |
| 5.30 | 126 | 126 | | 72 | | 64 |
| 5.35 | 124 | 124 | | 71 | | 63 |
| 5.40 | 121 | 121 | | 70 | | 62 |

*Continued page 254*

| HARDNESS CONVERSION TABLE | | | | | | |
|---|---|---|---|---|---|---|
| **BRINELL** | | | **ROCKWELL** | | | |
| **Dia. in mm, 3000 kg. load 10 mm ball** | **Hardness No.** | **Vickers or Firth Hardness No.** | **C 150 kg. load 120° Diamond Cone** | **B 100 kg. load 1/16 in. dia. ball** | **Scleroscope No.** | **Tensile Strength 1000 psi** |
| 5.45 | 118 | 118 | | 69 | | 61 |
| 5.50 | 116 | 116 | | 68 | | 60 |
| 5.55 | 114 | 114 | | 67 | | 59 |
| 5.60 | 112 | 112 | | 66 | | 58 |
| 5.65 | 109 | 109 | | 65 | | 56 |
| 5.70 | 107 | 107 | | 64 | | 56 |
| 5.75 | 105 | 105 | | 62 | | 54 |
| 5.80 | 103 | 103 | | 61 | | 53 |
| 5.85 | 101 | 101 | | 60 | | 52 |
| 5.90 | 99 | 99 | | 59 | | 51 |
| 5.95 | 97 | 97 | | 57 | | 50 |
| 6.00 | 95 | 95 | | 56 | | 49 |

Fig. 26-10

| GTAW AMPERAGE RANGES & ARGON GAS FLOW | | | | | | | | | |
|---|---|---|---|---|---|---|---|---|---|
| **Electrode Dia. (inches)** | **Cup Size** | **Welding Current (Amperage)** | | | | **Argon Flow (CFH) Ferrous Metals** | | **Argon Flow (CFH) Aluminum** | |
| | | **AC Pure** | **AC Thoriated** | **DCSP Pure** | **DCSP Thoriated** | **Standard Collet Body** | **Gas Lens Collet Body** | **Standard Collet Body** | **Gas Lens Collet Body** |
| .020 | 4-5 | 5-15 | 5-20 | 5-15 | 5-20 | 5-8 | 5-8 | 5-8 | 5-8 |
| .040 | 4-5 | 10-60 | 15-80 | 15-70 | 20-80 | 5-10 | 5-8 | 5-12 | 5-10 |
| 1/16 | 4-5-6 | 50-100 | 70-150 | 70-130 | 80-150 | 7-12 | 5-10 | 8-15 | 7-12 |
| 3/32 | 6-7-8 | 100-160 | 140-235 | 150-220 | 150-250 | 10-15 | 8-10 | 10-20 | 10-15 |
| 1/8 | 7-8-10 | 150-210 | 220-325 | 220-330 | 240-350 | 10-18 | 8-12 | 12-25 | 10-20 |
| 5/32 | 8-10 | 200-275 | 300-425 | 375-475 | 400-500 | 15-25 | 10-15 | 15-30 | 12-25 |
| 3/16 | 8-10 | 250-350 | 400-525 | 475-800 | 475-800 | 20-35 | 12-25 | 25-40 | 15-30 |
| 1/4 | 10 | 325-700 | 500-700 | 750-1000 | 700-1100 | 25-50 | 20-35 | 30-55 | 25-45 |

Fig. 26-11

| WELD METAL REQUIREMENTS FOR FILLET WELDS | | | | |
|---|---|---|---|---|
| **Size of Fillet** | **45 Deg. Fillets** | | **30-60 Deg. Fillets** | |
| **Inches** | **Lbs. of Metal Per Foot** | **Lbs. of Rod Per Foot** | **Lbs. of Metal Per Foot** | **Lbs. of Rod Per Foot** |
| 1/8 | .027 | .039 | .054 | .078 |
| 3/16 | .063 | .090 | .126 | .180 |
| 1/4 | .106 | .151 | .212 | .302 |
| 5/16 | .166 | .237 | .332 | .474 |
| 3/8 | .239 | .342 | .478 | .684 |
| 7/16 | .325 | .465 | .650 | .930 |
| 1/2 | .425 | .607 | .850 | 1.214 |
| 5/8 | .663 | .948 | 1.226 | 1.896 |
| 3/4 | .955 | 1.364 | 1.910 | 2.728 |
| 7/8 | 1.300 | 1.857 | 2.600 | 3.714 |
| 1 | 1.698 | 2.425 | 3.396 | 4.850 |

Fig. 26-12

# Liquid to gas equivalents

| GALLONS OF LIQUID at sea level pressure | Approximate Number Of 244 SCF Oxygen Cylinders | STANDARD CUBIC FEET of GAS at 70° F, sea level pressure | | |
|---|---|---|---|---|
| | | Oxygen | Nitrogen | Argon |
| 1 | ½ | 115.6 | 92.9 | 113.2 |
| 10 | 5 | 1,156 | 929 | 1,132 |
| 20 | 9 | 2,312 | 1,858 | 2,264 |
| 30 | 14 | 3,468 | 2,787 | 3,396 |
| 40 | 19 | 4,624 | 3,716 | 4,528 |
| 50 | 24 | 5,780 | 4,645 | 5,660 |
| 60 | 28 | 6,936 | 5,574 | 6,792 |
| 70 | 33 | 8,092 | 6,503 | 7,924 |
| 80 | 38 | 9,248 | 7,432 | 9,056 |
| 90 | 43 | 10,404 | 8,361 | 10,188 |
| 100 | 47 | 11,560 | 9,290 | 11,320 |
| 150 | 71 | 17,340 | 13,935 | 16,980 |
| 200 | 95 | 23,120 | 18,580 | 22,640 |
| 300 | 142 | 34,680 | 27,870 | 33,960 |
| 400 | 190 | 46,240 | 37,160 | 45,280 |
| 500 | 237 | 57,800 | 46,450 | 56,600 |
| 600 | 284 | 69,360 | 55,740 | 67,920 |
| 700 | 332 | 80,920 | 65,030 | 79,240 |
| 800 | 379 | 92,480 | 74,320 | 90,560 |
| 900 | 426 | 104,040 | 83,610 | 101,880 |
| 1000 | 474 | 115,600 | 92,900 | 113,200 |
| 1200 | 569 | 138,720 | 111,400 | 135,840 |
| 1500 | 711 | 173,400 | 139,350 | 169,800 |
| 2000 | 948 | 231,200 | 185,800 | 226,400 |
| 2500 | 1184 | 289,000 | 232,250 | 283,000 |
| 3000 | 1421 | 346,800 | 278,700 | 339,600 |
| 5000 | 2369 | 578,000 | 464,500 | 566,000 |

Chart courtesy of Ronan & Kunzl, Inc.

The difference in volume of oxygen, for example, as a liquid and as a gas at normal temperature and pressure (70 F and 14.7 lb per sq in. absolute) is 862 to 1.

## Boiling points

| | C | F | K | R |
|---|---|---|---|---|
| Oxygen | –183.0 | –297.4 | 90.1 | 162.2 |
| Nitrogen | –195.8 | –320.4 | 77.3 | 139.1 |
| Argon | –185.7 | –302.3 | 87.4 | 157.3 |

C = 5/9 (F – 32)  F = 9/5 (C) + 32  0°K = –273.19 C

***Properties of elements and metal compositions.***

| Elements | Symbol | Density (specific gravity) | Weight per cu. ft. | Specific heat | Melting point °C. | Melting point °F. |
|---|---|---|---|---|---|---|
| Aluminum | Al | 2.7 | 166.7 | 0.212 | 658.7 | 1217.7 |
| Antimony | Sb | 6.69 | 418.3 | 0.049 | 630 | 1166 |
| Armco iron | ... | 7.9 | 490.0 | 0.115 | 1535 | 2795 |
| Carbon | C | 2.34 | 219.1 | 0.113 | 3600 | 6512 |
| Chromium | Cr | 6.92 | 431.9 | 0.104 | 1615 | 3034 |
| Columbium | Cb | 7.06 | 452.54 | ... | 1700 | 3124 |
| Copper | Cu | 8.89 | 555.6 | 0.092 | 1083 | 1981.4 |
| Gold | Au | 19.33 | 1205.0 | 0.032 | 1063 | 1946 |
| Hydrogen | H | 0.070* | 0.00533 | ... | −259 | −434.2 |
| Iridium | Ir | 22.42 | 1400.0 | 0.032 | 2300 | 4172 |
| Iron | Fe | 7.865 | 490.9 | 0.115 | 1530 | 2786 |
| Lead | Pb | 11.37 | 708.5 | 0.030 | 327 | 621 |
| Manganese | Mn | 7.4 | 463.2 | 0.111 | 1260 | 2300 |
| Mercury | Hg | 13.55 | 848.84 | 0.033 | −38.7 | −37.6 |
| Nickel | Ni | 8.80 | 555.6 | 0.109 | 1452 | 2645.6 |
| Nitrogen | N | 0.97* | 0.063 | ... | −210 | −346 |
| Oxygen | O | 1.10* | 0.0866 | ... | −218 | −360 |
| Phosphorus | P | 1.83 | 146.1 | 0.19 | 44 | 111.2 |
| Platinum | Pt | 21.45 | 1336.0 | 0.032 | 1755 | 3191 |
| Potassium | K | 0.87 | 54.3 | 0.170 | 62.3 | 144.1 |
| Silicon | Si | 2.49 | 131.1 | 0.175 | 1420 | 2588 |
| Silver | Ag | 10.5 | 655.5 | 0.055 | 960.5 | 1761 |
| Sodium | Na | 0.971 | 60.6 | 0.253 | 97.5 | 207.5 |
| Sulfur | S | 1.95 | 128.0 | 0.173 | 119.2 | 246 |
| Tin | Sn | 7.30 | 455.7 | 0.054 | 231.9 | 449.5 |
| Titanium | Ti | 5.3 | 218.5 | 0.110 | 1795 | 3263 |
| Tungsten | W | 17.5 | 1186.0 | 0.034 | 3000 | 5432 |
| Uranium | U | 18.7 | 1167.0 | 0.028 | | |
| Vanadium | V | 6.0 | 343.3 | 0.115 | 1720 | 3128 |
| Zinc | Zn | 7.19 | 443.2 | 0.093 | 419 | 786.2 |
| Bronze (90 percent Cu 10 percent Sn) | ... | 8.78 | 548.0 | ... | 850–1000 | 1562–1832 |
| Brass (90 percent Cu 10 percent Zn) | ... | 8.60 | 540.0 | ... | 1020–1030 | 1868–1886 |
| Brass (70 percent Cu 30 percent Zn) | ... | 8.44 | 527.0 | ... | 900–940 | 1652–1724 |
| Cast pig iron | ... | 7.1 | 443.2 | ... | 1100–1250 | 2012–2282 |
| Open-hearth steel | ... | 7.8 | 486.9 | ... | 1350–1530 | 2462–2786 |
| Wrought-iron bars | ... | 7.8 | 486.9 | ... | 1530 | 2786 |

*Density compared with air.

Linde Division, Union Carbide Corp.

# DICTIONARY OF WELDING TERMS

## A

ABRASIVE PADS: Commercial cleaning pad used to remove oxide films from weld joint area prior to welding.

ACETONE: Colorless, volatile, water-soluble flammable liquid used to remove grease and oils from weld joint prior to welding.

ACID PICKLE: Chemical bath made with various acids to remove heavy oxide scale and foreign materials from weld area.

AFTERFLOW TIME: Period of time after welding that shielding gas flows through the welding torch to shield the tungsten from contamination.

AGE: Natural hardening sequence to change the mechanical properties of a material, metallurgical term.

AIR CARBON ARC CUTTING (AAC): Arc cutting process in which metals to be cut are melted by heat of carbon arc. Molten metal is removed by a blast of compressed air.

ALKALINE CLEANER: Mixture of sodium hydroxide and water used for removing oil and oxides from metal. Commonly used on aluminum and magnesium for pre-weld cleaning.

ALPHA ALLOY: Metallurgical term identifying a type of grain structure in titanium.

ALPHA BETA ALLOY: Metallurgical term identifying a type of grain structure in titanium.

ALTERNATOR: Word used in reference to an alternating current generator.

ALTERNATING CURRENT: Electrical current is the flow of electrons through a conductor. In alternating current the electrons flow in one direction, stop, and reverse the flow (alternate directions) for each complete cycle. In the United States, alternating current flows at sixty cycles per second. Therefore, the current stops and starts one hundred and twenty times a second.

AMERICAN IRON AND STEEL INSTITUTE (AISI): Industry association of iron and steel producers. It provides statistics on steel production and use. Publishes steel products manuals.

AMERICAN SOCIETY FOR TESTING MATERIALS (ASTM): Organization that makes specifications for iron and steel products for industry.

AMERICAN WELDING SOCIETY (AWS): Nonprofit, technical society organized and founded for the purpose of advancing the art and science of welding. The AWS publishes codes and standards concerning all phases of welding, and a magazine The Welding Journal.

AMMETER: Instrument for measuring electrical current in amperes.

AMPERAGE: Strength of an electrical current measured in amperes.

ANNEAL: Removal of internal stresses in metal by heating and slow cooling.

ANODE: Positive terminal or pole of a circuit.

ARC BLOW: Deflection of intended arc pattern by magnetic fields. In GTAW, this usually happens in DCEN (DCSP).

ARGON: Inert gas used in GTAW to shield tungsten and weld pool. It is heavier than air, colorless, and tasteless.

ARMATURE: Part of an electric machine which includes the main current-carrying windings.

ARTIFICIALLY AGED: Heating operation in the hardening sequence to change mechanical properties of metal, metallurgical term.

ASPHYXIATION: Loss of consciousness due to a lack of oxygen.

ATMOSPHERE: Envelope in which the welding arc is enclosed. May also be called an inert atmosphere.

AUSTENITIC STAINLESS STEEL: Alloy of iron that contains at least eleven percent chromium with varying amounts of nickel. The grain structure is always austenitic (non-magnetic).

AUTOMATIC VOLTAGE CONTROL (AVC): Controller to automatically regulate the electrode-to-work (arc gap) distance while welding.

AUTOMATIC WELDING: Welding that is completely controlled by a series of controllers without the aid of a welding operator.

## B

BACKGROUND CURRENT: Lowest amount of current (amperage) used to maintain the arc during a pulsing sequence.

BACKUP BAR: Tool or fixture attached to the root of weld joint. Tool may or may not control the shape of the penetrating metal.

BALANCED WAVE: Term describing an alternating current wave form that has equal straight and reverse polarity current values.

BERYLLIUM: Hard, light metallic element used in copper alloys for better fatigue endurance.

BETA ALLOY: Grain structure type in titanium, metallurgical term.

BEVEL: Angular type of edge preparation.

BORE: Inside diameter of hole, tube, or hollow object.

BRASS: Various metal alloys consisting mainly of copper and zinc.

BRIGHT ANNEAL: Process of annealing (softening) metal. Usually carried out in a controlled furnace atmosphere so that surface oxidization is reduced to a minimum. Surface remains relatively bright.

BRIGHT METAL: Material preparation where the surface has been ground or machined to a bright surface to remove scale or oxides.

BRITTLE WELD: Hard weld with little or no ductility.

BRONZE: Various metal alloys consisting mainly of copper and tin.

BURN-THRU: Weld which has melted through, resulting in a hole and excessive penetration.

BUTTERING: Form of surfacing where one or more layers of metal are placed on the weld groove face. Material then becomes transition weld deposit for final weld joint.

## C

CADMIUM: White ductile metallic element used for plating material to prevent corrosion.

CALIBRATION: To determine, check, or rectify potentiometer output.

CARBIDES: Compound of carbon with one or more metal elements.

CARBIDE PRECIPITATION: Movement of chromium from grains into grain boundries in the chrome nickel stainless steels.

CATHODE: Negative terminal or pole of circuit.

CERTIFIED CHEMICAL ANALYSIS: Report of chemical analysis on particular heat, lot, or section of material.

CHARPY IMPACT TEST: Test used to determine resistance to failure caused at a notch at low temperatures.

CHEMICAL COMPOSITION: Composition of material in chemical percentages.

CLADDING: Layer of material applied to a surface for the purpose of improved corrosion resistance.

COLD LAPS: Area of weld that has not fused with the base material.

COLD PUDDLE: Molten weld metal which moves very slowly within the weld joint.

COLD WORK: Increasing the tensile properties of material by working the material without heat.

CONCAVE WELD CROWN: Weld crown that is curved inward.

CONDUIT: Metallic sheath or tube through which the filler wire is moved from wire feeder to weld area.

CONTACTOR: Electric switch in power supply. Activating contactor transfers welding power from welder to welding torch.

CONTAMINATION: Indicates a dirty part, impure argon, improper gas shielding, dirty tungstens, etc.

CONTOUR: Shape of the weld bead or pass.

CONVEX WELD CROWN: Weld crown that is curved outward.

COPPER: Malleable, ductile, metallic element.

CORRECTIVE ACTION: Action to be taken to prevent weld discrepancies.

CORROSION: Eating away of material by a corrosive medium.

CRACKING: Operation of blowing dirt out of a high pressure gas cylinder valve.

CRATER: Depression at the end of a weld that has insufficient cross section.

CRATER CRACKS: Cracking that occurs in the crater.

CRYOGENIC TEMPERATURE: Very cold temperatures. Usually considered to be the temperature of liquified gases.

CUBIC FEET PER HOUR (CFH): Measurement of the amount of gas flow used in GTAW operations.

## D

DEIONIZED WATER: Water which has been specially treated for removal of chemicals.

DEMURRAGE: Monetary charge applied to the user of gas cylinders beyond the agreed rental period.

DEOXIDIZED FILLER MATERIAL: Filler material which contains deoxidizers such as aluminum, zirconium, and titanium for welding steels.

DEPARTMENT OF TRANSPORTATION (DOT): Government organization responsible for establishing and maintaining rules and safety precautions for the safe handling of fuels and gases used in welding.

DERATING: Establishment of lower duty cycle rating for power supplies that have not been designed for alternating current GTA welding.

DESTRUCTIVE TESTING (DT): Series of tests by destruction to determine the quality of a weld.

DEWARS: Specially constructed tank similar to a vacuum bottle for the storage of liquified gases.

DIES: Tools that are used to form wire or metal to a specified shape or dimension.

DIGITAL: Use of numerical digits to establish operational limits.

DIRECT CURRENT: Flow of current (electrons) in only one direction, either to the workpiece or to the electrode.

DIRECT CURRENT ELECTRODE NEGATIVE (DCEN): Direct current flowing from tungsten electrode to the work.

DIRECT CURRENT ELECTRODE POSITIVE (DCEP): Direct current flowing from the work to the tungsten electrode.

DIRECT CURRENT REVERSE POLARITY (DCRP): See preferred terminology Direct Current Electrode Positive.

DIRECT CURRENT STRAIGHT POLARITY (DCSP): See preferred terminology Direct Current Electrode Negative.,

DISSIPATE HEAT: Remove heat from weld zone by use of jigs or fixtures.

DOWNSLOPE: See "slopers."

DRIVE ROLLERS: Specially designed rollers for various types and sizes of filler wire to be fed through a mechanized wire feeder.

DROOPERS: Name given to some GTAW power supplies, as the slant of the volt-ampere curve is drooping.

DROSS: Oxidized metal or impurities within the metal.

DRY RUNS: Operations to determine proper tracking of tungsten throughout the length of the weld.

DUCTILITY: Property of material to deform permanently, or to exhibit plasticity without breaking while under tension.

## E

ELECTRODE: Electrode used in GTAW is tungsten, to which may be added various elements to aid in its function.

ELECTRODE EMISSION: Ability of electrode to emit electrons. Thoria is added to the tungsten in some cases to improve electron emission.

ELECTRONIC POTENTIOMETERS: Electrical devices using electronic circuits to control and regulate electrical current flow.

ELECTRONS: Negative charged particles.

ELONGATION: Permanent elastic extension which metal undergoes during tensile testing. Amount of extension is usually indicated by percentages of original gauge length.

EXTRA LOW INTERSTITUAL: Very small amount of carbon, oxygen, hydrogen, and nitrogen.

## F

FEATHERED TACKWELD: Tackweld which has been tapered on both ends to assist in obtaining the proper penetration during a groove joint root pass weld.

FERRITE TEST: Test of austenitic stainless steel weld deposit to determine amount of ferrite.

FERRITIC STAINLESS STEEL: Group of iron-chromium and carbon alloys that are nonhardenable by heat treatment.

FERRO-MAGNETIC: Material that can possess magnetization in absence of an external magnetic field.

FERROUS METALS: Group of metals containing substantial amounts of iron.

FILLET WELD: Weld of approximately triangular cross section joining two surfaces approximately at right angles to each other in a lap joint, T-joint, or corner joint.

FILLET WELD LEG: Leg lengths of largest isosceles right triangle which can be inscribed within fillet weld cross section.

FILLET WELD THROAT: Distance measured from bottom part of weld to crown surface.

FILLER WIRE: Welding material made in the form of wire which is added to the molten pool for chemical and/or dimensional requirements.

FLAG-TAG: Small flag attached to ends of straight length filler material. Type of wire is identified on the flag.

FLOWMETER: Mechanical device used for measuring inert gas rate of flow. Usually measurement is in cubic feet per hour (CFH).

FLUX: Protective material used in GTAW to prevent oxidization of molten metal.

FUSES: Electrical circuit device designed to fail at a predetermined current level to give protection from over-current.

FUSION: Melting together of filler metal and base metal, or of base metal only.

FUSION WELDING: Process which uses fusion to complete the weld.

## G

GAS COOLED TORCH: GTAW torch which uses shielding gas as cooling medium. Gas flows over current cable and then through torch and nozzle.

GAS ENVELOPE: Shape and pattern of shielding gas over the weld area.

GAS LENS: Specially designed screen assembly that attaches to welding torch and gas nozzle to maintain a longer gas envelope.

GAS NOZZLE: Assembly made from glass, metal, or ceramics of various designs. Attaches to torch body to direct inert gas flow (envelope) over weld area.

GRIDS: Welder control station with adjustable potentiometers or switches on multi-operator welding system.

GRIT BLASTING: Process for cleaning or finishing metal by the use of an air blast that blows particles of an abrasive (small pieces of steel, sand, steel balls) against the workpiece.

GROUND TUNGSTEN: Tungsten electrode which has been ground on outside diameter to remove any surface defects after manufacture.

## H

HARD TOOLING: Specially designed tooling or fixturing to hold the parts of weldment during welding operation.

HARD SURFACING: Hard material applied to surface of softer material for protection from abrasion or wear.

HEAT RESISTANT: Materials which are resistant to oxidation at high temperatures.

HEAT SINKS: Tooling applied adjacent to weld zone to absorb heat and to prevent heat flow into parent material.

HELIARC: Original registered tradename of the Linde Company used to identify GTAW.

HELIUM GAS: Second lightest element. Considered chemically inert. Used as shielding gas to increase penetration of weld and to increase welding speed.

HIGH FREQUENCY INTENSITY CONTROL: Rheostat mounted on high frequency generator or power supply to regulate the force of the high frequency spark.

HIGH FREQUENCY POINTS: Device to control the output of the high frequency spark generator.

HIGH FREQUENCY VOLTAGE: Several thousand volts with high frequency (millions of cycles per second) is connected across the arc gap. The high voltage causes the arc gap to become ionized, thus becoming a path for the electrons to flow.

HOT SHORTNESS: Metal which exhibits weakness (has low strength level) when hot.

HOT START: Programmed period of time at the start of the weld sequence. Uses higher welding current than normally used to achieve deeper penetration.

HOT WIRE PROCESS: Process developed by the Linde Company to heat welding wire before insertion into the molten pool. Process is used in automatic welding operations.

HYDRAULIC SYSTEMS: Pressure systems which use fluids, such as oil, to move or hold weld parts or tooling.

HYDROGEN: Lightest known element. Sometimes used in mixtures of argon for the welding of stainless steels to achieve a clean weld surface.

HYDROMOUNT: Name given to a portable welding machine cart with a supply of cooling water for the welding torch.

HYDROSTATIC PRESSURE: Pressure obtained by water.

## I

ICICLES: Intermittent sections of weld drop-through extending below the normal contour of full penetration groove weld.

IMPURITY LEVEL: Level of impurities established in metal and gases to reduce contamination of weld.

INCONEL: Nickel alloys that contain substantial percentages of chromium and iron.

INDUCTOR-CAPACITOR: Inductor controls rate of change of current in a circuit. Capacitor stores energy. Control of these two areas diminishes ripple in a full wave rectifier power supply.

INERT GAS: Gas which does not normally combine chemically with base metal or filler material.

INGOT: Large block of metal usually cast in metal mold. Forms basic material for further processing.

INSULATED GAS NOZZLE: Specially designed torch gas nozzle for use with pilot arc.

INTER-GRANULAR CORROSION: Occurs when welded austenitic stainless steels have been sensitized during welding, then subjected to some type of acid solution to induce corrosion.

INTERPASS TEMPERATURE: In multiple pass weld, minimum or maximum temperature specified for the deposited metal before the next weld pass is started.

IONIZATION POTENTIAL: Energy required to remove an electron from an atom, thereby making it an ion. Shielding gases have different ionization potentials.

IZOD TEST: Test used to determine notch impact values. Often used in conjunction with Charpy impact test.

## K

KINETIC ENERGY: Energy of body or system with respect to the motion of the body or of the particles in the system.

## L

LACK OF FUSION: Fusion which is less than complete.

LACK OF PENETRATION: Joint penetration which is less than specified.

LINEAR POROSITY: Cavity-type discontinuities formed by gas entrapment along a line during solidification of liquid melt.

LIQUIFIED GAS: Gas which has been changed into a liquid for ease in storage and handling.

LITERS PER MINUTE: Metric measurement of the amount of gas flow used in the GTAW operation.

## M

MACHINE DUTY CYCLE: Period of time established by the machine manufacturer for operation within the machine's design specification.

MACRO TEST: Visual test of a weld or parent metal structure cross section with low magnification or with the naked eye.

MAGNE TEST: Inspection tool designed to test stainless steel welds for ferrite content.

MAGNESIUM: Light, ductile, silver-white metal with high strength-to-weight ratio.

MAGNETIC AMPLIFIER: Component in GTAW power supplies used to change the slope of the volt-ampere curve.

MAGNETIC FIELD: Area where the magnetic lines of force have been established. Magnetic fields may be used to deflect the arc for oscillation of the molten pool.

MAGNETIC PARTICLE INSPECTION: Non-destructive inspection test using magnetized part and finely divided iron particles to outline discontinuities in parent metal or weld.

MANIPULATOR: Positioning tool used to locate welding equipment at the weld area for longseam or circumferential welding.

MANUAL WELDING: Welding operation performed manually.

MARTENSITE: Structure obtained when steel is heated and cooled to achieve its maximum hardness.

MECHANICAL PROPERTIES: Includes such areas as tensile strength, ductility, brittleness, elasticity, hardness, toughness, and malleability.

METAL INSERTS: Used in making groove welds in pipe to control the chemistry of the weld and to aid in forming correct penetration contour.

METALLIC VAPORS: Vapors rising from molten pool as various gases evolve from the melt. Often attach to tungsten, causing electrode contamination.

MICRO CRACKS: Very small cracks within metal. Located and seen only with the aid of very high magnification.

MICRO SWITCH: Switch often used in foot or hand controls for initiating welding cycles. Switch operates by depressing lever a few thousandths of an inch.

MISMATCH: Junction point of butt joint where the top or bottom edges are not even.

MODE: Set-up of a welding machine as to manual, remote, or automatic operation.

MODULE: Unit which contains all the required circuits for a specified sequence. Made as controls for various areas such as pulsers, slope, and spot welding.

MOTOR GENERATOR: Welding power supply using an electrical motor to drive a welding generator.

MULTI-OPERATOR SYSTEMS: Master power supply which provides power to a series of welder controlled grids or control panels.

## N

NATIONAL ELECTRICAL MANUFACTURERS ASSOCIATION (NEMA): Industrial association of manufacturers of electrical machinery. Publish standards and industry statistics including welding.

NEOPRENE: Oil resistant synthetic rubber.

NICK-BREAK TEST: Destructive test of the weld to determine internal quality.

NITROGEN GAS: Colorless, odorless, gaseous element.

NON-CONSUMABLE ELECTRODE: Tungsten electrode which is not melted by heat nor becomes part of the molten metal.

NON-DESTRUCTIVE TEST: Test which does not require destruction of the part to determine quality.

NON-FERROUS: Any metal which does not contain iron.

NON-TOXIC: Not harmful or poisonous.

NOTCH SENSITIVITY: Resistance of metal to notch failure when subjected to rapid loading or stress.

NOTCH TOUGHNESS: Ability of metal to resist failure (cracking) at a notch during loading (stress).

NOZZLE: Ceramic or metal tube of various sizes and shapes which attaches to the welding torch to direct the gas flow over the weld area.

NUMERICAL CONTROL: Numerical program established on tape or disk which controls the parameters and sequences of the welding operation.

NYLON: Thermoplastic polyamide which can be molded into welding torch parts of extreme toughness, strength, and elasticity.

## O

OPAQUE: Material that does not allow light to pass.

OPEN CIRCUIT VOLTAGE: Voltage between the output terminals of an operating welding machine when no current is flowing in the circuit.

OSCILLATE: Moving welding torch back and forth across weld joint.

OUTGASSING: Sequence of allowing hydrogen gas to rise through the molten weld metal to the surface. When welding aluminum with a pulser, the pulsing action should be rapid to accomplish this sequence.

OUT-OF-POSITION WELDING: Welding that is performed in a non-standard position such as flat, vertical, and overhead.

OUTPUT RATING: Output limits of power supply with regard to current, open circuit voltage, ranges, power factor, and duty cycle.

OVALITY: Amount of "out-of-round" condition when referring to pipe, tubing, or round objects.

OVERLAP: Protrusion of weld metal beyond the toe, face, or root of the weld.

OVERLAY: Weld placed over the top of another metal for dimensional requirements or to add physical or mechanical properties.

OXIDE FILM: Film formed on base material as a result of exposure to oxidizing agents, atmosphere, chemicals, or heat.

OXIDATION: Process of reaction with an oxidizing agent.

OXYGEN ANALYZER: Instrument used to determine the amount of oxygen within an area. The measurement is generally termed in parts per million (PPM).

## P

PARAMETERS: Information required to complete a weld. These include areas such as amperage, voltage, wire speed, weldment travel speed, type of current, etc.

PART PER MILLION (PPM): Numerical method of identifying the purity or contamination of a gas.

PASSES: Single weld beads are often termed passes.

PENETRAMETERS: Metal shims with specified size holes used during the exposure of the radiograph to verify the sensitivity and the proper exposure of the film.

PENETRANT INSPECTION: Visual surface inspection completed with dye and developer. May also use fluorescent dye and black light for observing the results.

PENETRATION: Minimum depth a groove or flange weld extends from its face into the joint, exclusive of reinforcement.

PERIMETER: Outer edge of circumferential part.

PHOSGENE GAS: Poisonous, colorless, very volatile, suffocating gas. Made when trichloroethylene vapor enters arc and is heated.

PHYSICAL PROPERTIES: Properties of metals relating to electrical, thermal conductivity, and expansion rates.

PILOT ARC: Low current continuous arc between electrode and gas nozzle which aids to ionize the gas and help start the main welding arc.

PLASMA: Gas that has been heated to a partially ionized condition enabling it to conduct an electrical current.

PLASMA ARC CUTTING: Arc cutting process which severs or cuts metal by melting a localized area with a constricted arc. Process to remove the molten metal with a high velocity jet of hot, ionized gas issued from the torch orifice.

POLARITY: Direction of current flow. Current flow from the electrode to the workpiece is (DCEN) or (DCSP). Current flow from the workpiece to the electrode is (DCEP) or (DCRP).

POST FLOW: Period of time at the end of the weld cycle where shielding gas flows around tungsten during cooling to prevent contamination of the electrode.

POSTHEAT: Heat which is applied at the end of the weld cycle to slow down cooling rate to prevent cracking and to relieve stresses.

POSITIONER: Mechanical device for holding workpiece for welding in the desired position. It may rotate the weldment with a controlled speed for circular welds.

PORE: Cavity type discontinuity formed by gas entrapment during solidification of weld metal.

POROUS: Series of pores within a weld.

PRECIPITATE: Movement of elements within metal from grain into the grain boundary.

PREHEAT TEMPERATURE: Specified temperature that base metal must be heated to immediately before welding is started.

PRIMARY VOLTAGE: Incoming voltage of alternating current supplied by utility company.

PROCEDURE: Sequence of events to be accomplished to make a weldment. This may include but is not limited to: assembly sequence, cleaning procedure, tooling, tackwelding, preheating, types of welding current, and levels.

PROGRAMMER: Electronic, or mechanical sequencer to start and finish various portions or all of the sequences required to complete a weld.

POUNDS PER SQUARE INCH (PSI): Measurement of pressure.

PULSER: Electronic device for sequencing and controlling amounts of current and time periods of operation.

PURGING: Operation to remove air (atmosphere) from welding area and replace it with an inert atmosphere.

PURGING DAMS: Used in conjunction with purging of pipe to reduce area to be purged.

## Q

QUALIFIED PROCEDURE: Welding sequence that has met all of the testing requirements of the fabrication specification.

QUALIFIED WELDER: Person who has demonstrated ability to produce a weld within the requirements of the fabrication specification.

QUALIFIED WELDING OPERATOR: Person who has demonstrated the ability to operate a welding machine with a qualified procedure to produce a weld within the requirements of the fabrication specification.

## R

RADIATION: The energy that a welding arc radiates (sends out) in the form of intense ultra-violet light which will burn the human body if not adequately protected.

RADIOGRAPHIC INSPECTION: Non-destructive testing method used to determine interior quality of weld.

RATED LOAD: Load (welding current) which may be obtained from power supply within limits established by the machine duty cycle.

REACTOR CORE: Iron base which has wire wound around it. The main purpose of the reactor core is to control the rate of change of current in a circuit.

RECTIFICATION: As applied to alternating current arc, refers to the device which acts as a one-way valve to convert one half of the AC cycle to current flowing the same as the other half cycle.

RECRYSTALLIZED: New grain structure which has been obtained by heating a cold-worked metal.

RED HEAT TEMPERATURE: Temperature point in some metals where the material becomes brittle.

REFERENCE ARC VOLTAGE: Actual welding arc voltage established on the automatic welding control circuit. By reference to this voltage, the arc voltage sensing control circuit maintains desired arc voltage.

REGULATOR: Mechanical device to reduce and control pressure.

REVERSE POLARITY: Electron flow is from workpiece to electrode.

RHEOSTAT: Adjustable resistor such that its resistance may be changed without opening the circuit in which it is connected.

RIMMED STEEL: Steel not completely deoxidized during manufacture. Outer rim of ingot contains high quality steel used to make steel welding wires.

RIPPLE FACTOR: Percentage of ripple (waviness) in flow of straight polarity direct current.

ROTARY SELECTOR: Mechanical method used by some power supply manufacturers to establish welding current output control.

ROUTING: Removal of metal from a weld using a routing tool. The tool may have various shapes and forms. The cutters may be tool steel or tungsten carbide.

RUST INHIBITOR: Chemical added to water to inhibit (reduce) formation of rust particles.

## S

SATURABLE REACTOR: Electronic device used to regulate the output of amperage flow in the secondary circuit of a transformer power supply.

SEAMER: Machine designed to hold material for the purpose of making longseam welds.

SEAMWELDER: Seamer machine with attached welding equipment for the purpose of making longseam welds.

SEMI-AUTOMATIC WELDING: Welding operation where the welding operator controls the sequences and adjusts various parameters as required.

SEMI-KILLED STEEL: Steel which has been partially deoxidized during solidification in the ingot mold.

SENSITIVITY: Response of the automatic voltage controlled welding torch to varying arc lengths.

SENSITIZATION RANGE: Temperature range in stainless steels where carbide precipitation may take place.

SEVERN GAUGE: Inspection tool designed to test stainless steel welds for ferrite content.

SHIELDING GAS: Inert gas used to shield molten metal from atmosphere contamination.

SHOP-AID-TOOLING: Tooling in the way of clamps, screws, metal/forms, etc., that may be used to hold parts in alignment for welding.

SHORT CIRCUIT: Situation when electrode touches the grounded weldment. Often used where no high frequency spark is available and arc is started by touching electrode to weldment.

SILICON: Non-metallic element used in steel making. If present in molten pool, it will rise to the surface. It may cause weld pool movement control problems and should be removed from the surface of the completed weld before making another pass.

SILICON CONTROLLED RECTIFIER (SCR): Solid state diodes used to change alternating current into direct current.

SINE WAVE: Visual expression of one cycle of alternating current with the rise and fall of both negative and positive portions of the cycle.

SINGLE PHASE CURRENT: Single sine wave of alternating current.

SLOPERS: Used in semi and automatic welding operations to slope welding current upward to the required current level and downward from the required current level.

SODIUM HYDROXIDE: Caustic soda used in a five percent concentration with water for removal of heavy oxide films on aluminum.

SOLID STATE: Use of solid components for generation of power and control of sequential circuits.

SOLUABLE DAM: Paper dam used to plug tubes and pipes for purging. Flushing the tubes after welding dissolves the dam.

SOLUTION HEAT TREATED: Material which has been heated to a pre-determined temperature for a suitable length of time to allow some element in the material to enter into "solid solution." Alloy is then quickly cooled to hold the element in this solution.

SPATTER: Small pieces of metal which have been

ejected from molten pool and attached to base material outside the weld.

SPITTING: Particles of tungsten electrode emitted from the electrode during alternating current welding. Caused by improper firing or rectification of the alternating current during change from straight polarity to reverse polarity cycles.

SPOT WELDS: Controlled weld cycle to produce sheet metal weld with specific characteristics.

SQUARE WAVE: Shape of alternating current cycle. Design is obtained by adding additional switching within the power supply to rapidly change straight to reverse polarity and vice versa.

STABILIZED MATERIAL: Aluminum which has been strain-hardened and then heated to predetermined low temperature to slightly lower the strength and increase ductility.

STABILIZER CONTROL: Electronic circuit used in some alternating current power supplies to assist positive cycle firing of current each time sine wave passes through the zero point. Stabilizer control prevents rectification and tungsten spitting.

STAINLESS STEEL: Alloy of iron containing at least eleven percent chromium and some nickel that resists almost all forms of rusting and corrosion.

STEEL: Alloy of iron and carbon with varying amounts of other alloying elements for specific mechanical properties.

STEP-OVER-DISTANCE: Measurement of weld bead placement when surfacing a material to obtain a relative flat crown bead.

STRAIGHT POLARITY: Arrangement of direct current arc welding leads in which the work is positive and electrode is negative.

STRAIN HARDENED MATERIAL: Material which has been strained by stretching, pulling, or forming to produce a grain structure with higher mechanical properties.

STRENGTH-TO-WEIGHT RATIO: Term used to define material strength in relationship to material weight.

STRINGER BEAD: Weld bead made without oscillation (side to side motion).

SUCKBACK: Concave root surface in a full penetration groove weld. Also termed a concave root surface.

SURFACING: Applying material to the surface of another material for protection from chemicals, heat, wear, rust, etc.

SWAGING: Operation or process of changing the shape of material with mechanical tools such as hammers or dies.

## T

TACHOMETER: Device used in welding to determine speed in "inches per minute." May define travel speed of the torch, part, or filler material.

TACKWELD: Weld made to hold parts of a weldment in alignment until final weld is made.

TAPS: Male and female connection devices designed to complete a welding circuit for various amounts of welding current.

TEFLON*: Insulating material used to make gaskets and insulators for welding torches. Teflon tape is also used on threaded inert gas connections to make an effective seal.

TEMPERING BEADS: Stringer bead welding passes made on crown of completed weld to reduce the grain size of the crown bead. Often ground off of the part later. The tempering reduces the crown bead tensile strength and improves ductility.

TEMPERS: Defines ferrous materials with various mechanical properties that have been made by a series of basic treatments. Usually materials with decreased hardness and increased toughness.

TENSILE TEST: Test to determine mechanical properties of metal. Test is done by placing specially designed piece of metal in a tension machine and applying a load until the part fails.

TERMINAL BLOCKS: Blocks of copper or bronze for attaching welding power and ground cables.

THERMAL-CUT-MATERIAL: Material which has been cut by a melting process. May include oxy-fuel gas, plasma-arc, or carbon arc processes.

THERMAL OVERLOAD PROTECTOR: Heat sensing device which will open welding circuit on welding machine when certain temperature is reached. Used to protect machine from damage due to excessive heat.

THERMAL TREATMENT: Treatment of metal by the use of heat.

THORIATED TUNGSTEN: Tungsten electrode to which thorium has been added to improve service factors.

THORIUM: Grayish-white, lustrous, radioactive metallic element which when added to tungsten improves electrode's operating characteristics.

THREE PHASE CURRENT: Type of alternating current used to operate most welding power supplies.

TIMERS: Mechanical or electronic devices used to sequence various welding machine and other equipment functions.

TIP GUIDES: Metal tube that attaches to cold wire or hot wire drive system. Guide has specific diameter hole to guide welding wire into the molten pool.

TITANIUM: Light, silvery, lustrous, very hard, corrosion-resistant metallic element.

TOLERANCE: Permissable variation of a characteristic, variable, or parameter.

TOLUENE: Colorless, water-soluble, flammable liquid having a benzene-like odor. Used for removing oils, grease, and paint from material.

TORCH CABLE: Stranded copper alloy cable which provides welding current from power supply to welding torch.

TRACTOR: Electrical-mechanical vehicle used to transport welding equipment along a weld joint.

TRANSFORMER: Device used in welding power supplies and equipment to change both voltage and current from one level to another.

TRANSVERSE CRACK: Crack which extends across weld joint or parent metal.

TRICHLOROETHYLENE: Colorless, POISONOUS liquid generally used in vapor state to remove oils, grease, and paint from material.

TUBE SHEETS: End plates of tube bundle assembly.

TUNGSTEN: Rare metallic element with very high melting point, approximately 3410 °C.

TUNGSTEN INERT GAS WELDING (TIG): Original American Welding Society definition of the process now called Gas Tungsten Arc Welding.

## U

ULTRA-SONIC INSPECTION: Non-destructive test method used to determine interior quality of a weld.

UNDER-BEAD CRACKING: Cracking which occurs in the parent metal adjacent to weld. Usually caused by presence of hydrogen being absorbed into weld and parent metal during welding operation.

UNDER-CUT: Groove melted into base metal next to the toe or root of the weld and left unfilled by weld metal.

## V

VARIABLES: Factors which may affect welding sequence and weld quality.

VARIABLE VOLTAGE POWER SUPPLY: Term given to constant current power supply. Varying the welding voltage (arc gap) changes the output current level of the power supply.

VISUAL INSPECTION: Inspection of the material or weld conducted by visual means.

VOIDS: Holes or areas within the completed weld.

VOLT-AMPERE-CURVE: Operating characteristics of welding power supply established by the manufacturer's design.

VOLTS-ALTERNATING-CURRENT (VAC): Letters used on welding power supply data sheets, labels, and operating manuals to identify the amount of primary voltage required to connect the machine to the utility power.

VOLTMETERS: Amount of arc gap or distance between the electrode tip and the workpiece. Low voltage indicates a close gap, and a high voltage indicates a longer gap.

## W

WARRANTY: Period of time a manufacturer guarantees the product.

WASH BEADS: Beads made with an oscillation (side-to-side) technique to widen the weld bead.

WATER COOLED TORCH: Torch which uses water to cool the welding torch head and power cable. These torches operate at a higher duty cycle than gas cooled torches.

WAVE BALANCER: Circuit added to an alternating current power supply to change the amount of the straight or reverse polarity cycle. This aids in obtaining deeper penetration or a better cleaning action when welding aluminum or magnesium.

WELD CHEMISTRY: Final chemical content of the weld.

WELD CYCLE: Complete series of events involved in making of weld.

WELDER: Person who performs a manual or semi-automatic welding operation.

WELDING ARC VOLTAGE: Actual voltage across the arc during welding.

WELDING OPERATOR: Person who operates automatic welding equipment.

WELDING TORCH: Device designed to hold the electrode, provide gas shielding of the weld, and insulate the welder from the welding current.

WELD JOINT: Junction of members or edges of mating parts to be joined.

WELD ROOT: Deepest part of the weld into the parent metal.

WELD SHAVER: Rotary tools designed to remove weldcrowns from groove weld.

WELD PUDDLE: Molten part of the weld during the welding sequence.

WELD ZONE: Area immediately adjacent to the weld joint.

WHISKERS: Defines the shape of the solidified penetration of a groove weld where the weld metal has been contaminated in the atmosphere.

WIRE FEEDER: Electro-mechanical device designed to feed spooled filler wire into a weld puddle at a controlled rate of speed.

WROUGHT MATERIAL: Material made by processes other than casting.

## Z

ZIRCONIUM: Metallic element which when added to tungsten electrodes improves the electrode's operating characteristics.

# INDEX

## T



## U

## X



## Z